Ken Ring's

IRELAND WEATHER ALMANAC

2026

BY THE SAME AUTHOR

New Zealand Weather Almanac (1999-Present)
Australia Weather Almanac (2006-Present)
Ireland Weather Almanac (2010-Present)
How to Get Your Kid to Like Maths
Predicting Weather by The Moon
New Meteorological Techniques
We Used to Call It Summer
The Romany's Apprentice
Moon and WeatherLore
Secrets of the Moon
Notes From the Pit
The Lunar Code
Fingermaths
Supertramp
A Fiery Tale

Paperback ISBN: 978-1-0670445-8-9
Hardback ISBN: 978-1-0670445-9-6
eBook ISBN: 978-1-0671244-0-3

Email: enquiries@predictweather.com
Web: www.predictweather.com

CONTENTS

INTRODUCTION

ABBREVIATIONS....1
THE NATURE OF WEATHER....2
ORIGINS OF A STORM....4
HEAVY RAIN....6
PRECIPITATION CYCLES....6
THE ATMOSPHERE....8
WEATHER GLOSSARY....8
SPEED READING OF CLOUDS....9
FINAL WORD....11

GRAPHS

TEMPERATURES....13
WIND....24
PRECIPITATION (RAIN/SNOW)....25
WHEN WILL THE SUN BE OUT....28
GRAPHS, NOTES & TRENDS....28
SEASONAL OVERVIEWS....30

DAILY FORECASTS

JANUARY....39
FEBRUARY....59
MARCH....77
APRIL....97
MAY....117
JUNE....137
JULY....157
AUGUST....177
SEPTEMBER....197
OCTOBER....217
NOVEMBER....237
DECEMBER....257

APPENDICES....277

RAINFALL EXPECTATIONS – WTR/DRR 10MM VARIATION
WTR/DRR 25% VARIATION TO LONG TERM AVERAGE
ESTIMATED SUNSHINE TRENDS
ESTIMATED MAXIMUM AVERAGE MONTHLY TREND
ESTIMATED AVERAGE TEMPERATURE TREND

GARDENING GUIDE....285
FISHING CALENDAR....296

INTRODUCTION

DISCLAIMER

The forecasts in this work are best-endeavour opinions of the authors and associates. No guarantees of 100% accuracy is claimed. Neither can the author and/or publisher be held accountable for decisions made on the basis of information in these pages.

ABBREVIATIONS

MOON SYMBOLS

1st Q	Moon in first-quarter phase.
3rd Q	Moon in third-quarter phase.
^	Northern declination Moon at furthest point north for month.
A	Apogee Moon furthest from earth for the month.
F	Full moon.
N	New moon.
P	Perigee Moon at closest point to the earth out of the 27 day cycle.
P(n)	Priority closest e.g. P3 means third closest for year.
V	Southern declination Moon at furthest point south for the month.
XhN	Moon crossing the equator heading north.
XhS	Moon crossing the equator heading south.

RAIN SUMMARY TABLES

d	Dry most of the day.
h	Heavy rain.
l	Light, intermittent showers.
n	Fog, mist, dew, or non-recordable showers.
p	Passing Showers
r	Rain.
s	Showery.

SUNSHINE SUMMARY TABLES

F	Fine, sunny (8+ hours sunshine)
pc	Partly cloudy (4-7 hrs sunshine)
c	Cloudy (1-3 hrs sunshine)
o	Overcast (0 hrs sunshine)

TEMPERATURE SUMMARY TABLES

clr	Cooler than average temperatures.
a-c	Average to cooler temperatures.
av	Average temperatures.
a-w	Average to warmer temperatures.
wrm	Warmer than average temperatures.

THE NATURE OF WEATHER

Weather is determined by astronomical and physical reasons which are under relatively fixed control of cycles. It is not determined by any activities by Man. Climate scientists claim that the heavier-than-air "trace" gas, carbon dioxide, which is absent in 99.96% of the atmosphere; and methane, which is absent in 99.9999%, have a controlling effect on the climate. This has not been confirmed scientifically.

Approximately 80-85% of the weather, in this author's opinion, is caused by the moon, changing and repeating its position above the earth. Planning for major droughts or floods is therefore possible, by comparing the position of past lunar cycles, but not by measuring any emissions of a negligible gas.

Here is the weighting system. There is space weather, climatic weather, and everyday weather.

1. **CHANGES TO THE WHOLE PLANET:**
   (Irrespective of geography)
   • Deviation of Earth's rotational axis to the vertical.
   • Variation of Earth's orbit.
   • Magnetosphere
   • Heliosphere

2. **CLIMATE AT LOCATION:**

   (long-term trends in weather patterns)
   • Distance from the equator (angle of the Sun's rays = latitude)
   • Direction of prevailing winds
   • Topographic factors – the shape of the land, proximity to other land masses, elevation, and distance from the nearest coast.

3. **WEATHER AT LOCATION:**

   (trends and daily occurrences, observable and recordable)
   • Influence of Moon on tides (including air tides), cycle of rain from ocean, and back to ocean. (80-85%), also oceanic cycles due to lunar influence, e.g. El Nino.
   • Level and intensity of cloud cover (retains local heat, determines what, where, when, of rain (which in turn is subject to ocean and air tides)
   • Solar winds, cycles of Sunspots (contributing to temperature of seasons), hemispherical and seasonal factors.

4. **NOT IN THE MIX AT ALL, EITHER FOR CLIMATE OR WEATHER**
   • Land use changes, deforestation and greenification through agriculture.
   • Volcanic activity ongoing, which does not alter climate.
   • Ice-reflectivity feedback, constant.
   • Industrial and commercial pollution.
   • other activities of Man e.g. diet.

Immediate weather is modified by cycles of solar activity (position and number of occurring Sunspots), and the cyclic position of the Moon relative to the earth. Solar influences can determine floods and drought periods.

The weather has no brain, so it does not know that there is a small percentage of earth that is not covered in the ocean, called the Land. To the moon, it is all oceans. From the height rain falls from, up to 12km, it is immaterial whether there is land below or ocean. That's why rainforests make no difference to rainfall amounts. Forests grow where it rains often, not vice versa.

About 85% of all the weather is controlled, directly or indirectly by the Moon. If you live by the ocean, then the influence is more obvious and direct. If you don't live on or near the coast, then weather systems take a little longer to reach you. But you can see, in a broad sense, the influence of the moon. For example, on almost every full moon night, you will see the full moon in all its glory. And that's because, on the night of the full moon, the moon clears the sky of clouds close to midnight, by dragging more air underneath it. This pushes away the cold air of space and prevents condensation of water vapor which makes the clouds visible. The moon controls the weather every night, however, it is just not as obvious.

When the moon is in the sky i.e. above the horizon, it is less likely to rain. This I would regard as a rule. The main times of rain (if rain is about), are when the moon is underfoot (IC), overhead (MH), or on the horizon (ASC=rising or DES=setting). The in-between positions i.e. every three hours, also have rain potential. Sir Isaac Newton identified that when the moon reached a 45° angle to the horizon, a gravitational force corridor was established, and gravity was lessened, i.e. the moon effect was enhanced.

Just before it rains, there is a noticeable chill in the air. The animals can sense it, horses with their nostrils. There are also extra wind gusts in the hours of moonrise and moon set. This was noticed and recorded by the Greek philosopher Theophrastus in 300 BC. It still applies today. When the trees suddenly rustle for no reason, a useful exercise is to check out where the moon is.

Jupiter, Saturn, Uranus and Neptune, in that order, affect the Sun, and the sun and moon together affect the oceans. Venus and Mercury affect the tides, but not as much as the sun and moon. Jupiter and Saturn mainly, are responsible for what we call the sunspot cycle. That is when the sun gets overactive and seems to put out visible blotches which some cultures like the Chinese have been watching and tabulating for hundreds of years, for they coincide with extra radiation levels. This in turn translates into increased thunderstorms, and growth, agriculturally.

The sunspot cycle is about 12 years, which is slightly more than the Jupiter cycle of 11.8 years although this can vary between 9 and 14 years. There are variations in the positions of the other planets, and it's the oceans that get affected first. Because the moon is the tiny player in the big team, it gets affected by everything that the big team does, in the same way that a baby is affected by the moods of the whole family. Because the moon is the closest planet to us and so to the oceans, everything in terms of planet aspects involves the moon. There is the timing of the tides, the height of the tides, the speed of the tides, all factors in how the ocean behaves, and all that gets imparted to the weather because all weather begins in the ocean.

That is not a known viewpoint, or if it is, it hasn't reached the ears of meteorologists yet, who still look to the sky for signs of change. That's why, in my opinion, they're wasting their time talking about cloud-seeding, weather modification, and geoengineering, those are arguably scams that people get away with, using research funds made available by western governments, thinking by mistake that the weather originates there. But by the time a weather event reaches the sky, where we on the land see it, it's already half-over, it's obsolete, and just the final dregs of the system.

ORIGINS OF A STORM

All adverse weather comes from *under the boat*. I have never heard a sailor say, aw, a violent storm was going on above me, whilst we were sailing *on a flat, calm sea*. Usually, the flat sea accompanies a high-pressure system. Higher air pressure pushes the sea flat. It is as if your hand was the size of a bath and pushed down on the bath's surface. But the flatter sea arrives first and allows more stable air to gather and consolidate over it, exerting more pressure.

It has been further observed that the deeper a low-pressure system is in the air, the faster the wind blows, imparting more energy, and the longer the wind blows over the same area, the bigger the swells, which are the strong underwater currents. But they have interpreted backwards. It begins in the sea. If currents increase in the deep of the ocean, then waves will increase at the surface and disturb the air over long distances, enough to generate those faster winds.

Strong winds will always accompany a swell. That's because the waves, which arise from the swell, cause the wind, The etymology says it all. Wind, weather, and wing are words that come from the word *wave*, not the other way around. Fishermen have noticed that the direction in which the wind is blowing matters. The word *west* comes from wind as well. So does the word weather.

When there appears to be an onshore wind coming off the sea, what's happening is that underwater waves are being whipped up, which disturbs marine worms out of their burrows, dislodges shellfish such as mussels, limpets, and cockles from their home on rocks, and forces small fish, crabs, and other forms of marine life out of weed beds. The currents change first. How could a surface wind possibly affect a seabed? Surface currents do not transmit downwards to the depths of the ocean.

To prove this theory, put some sand along the bottom of a bath and then splash along the surface, and see if the sand moves. That is like the wind. Now place floating things, like corks, on the surface, and generate underwater currents with your hand, and see if the floating things are affected. Which moves more?

There is no question that the energy transmission in the ocean is always upward. It means that winds are more likely to be created by currents and swells than vice versa. It would be reasonable to conclude that a clear day indicates little wind BECAUSE the sea is calm, indicating deeper currents have lessened. Deep sea fish may have a sense of impending stillness in the tides and may plan to be somewhere when small creatures that they prey on emerge from nooks and crannies.

An experienced fisherman can tell when there is a storm coming. The ocean 'speaks' to him and the sound of the swell reaches him first. Deep ocean currents are a fact, but science does not link them to winds. Seismic activity generates ocean currents, and earthquakes are energy waves up to 20,000 mph through solid rock. It is not likely that ocean currents go downwards to generate earthquakes, and besides, we already know that some earthquake epicenters originate about 400 or 500kms below the seafloor.

It is more likely that the wave regimes that earthquakes are part of, transmit through rock to vibrate the ocean floor enough to generate the ocean currents, and on into the air via oscillating waves to generate unstable airflows. The fact that larger earthquakes happen around full moons and what we call perigees, together with stronger currents and king tides, attest to that, also cyclones form around these times. The moon and larger gravitational forces are responsible for all three. So, when it comes to what happens at sea and in the air above the surface of the sea, it is more feasible that ocean currents are in charge, rather than above-surface winds.

As soon as the temperatures reverse, the winds reverse. Sailors who monitor Sea Surface Temperatures (SSTs) know this to be more reliable than isobaric maps because sea-surface temperatures change before changes in air pressure, which gives warning of storms before they happen. Just by monitoring the temperature of the water, you can tell what the weather will do, in the next few hours or longer.

SUMMARY

To recap, a typical storm begins with the flat sea, the Sun shining above, heating the surface, accompanied by non-mixing of waters at various depths, relatively warm sailing conditions, and SSTs (sea surface temperatures) starting high. The surface of the sea is warm because the layer of water constituting the surface is not being replaced by cooler water. Layers of water closer to the surface are the warmest, being closer to the warmth of the Sun. Then, as mixing begins below the surface, cool water is brought up from below, causing surface cooling. Further mixing brings more cooler waters up from the depths and there is an interchange with the warm waters close to the surface. Due to the drop in temperature at the surface, clouds may form as cooler air mixes with the remainder of warm air. Because warm air always moves to cold, the air begins to move, and wind "picks up". There is a storm brewing. Warmer air continues to supply the area which has now considerably cooled, and the surface is described as windy, and then as a gale. The clouds increase and darken with the load of water that they gather. Condensation accompanies cooling, supplied by the movement of air. The surface of the sea begins to ripple strongly as the swell increases. Each wave has a convex and concave side. The convex has a higher pressure and the concave has a lower pressure. Waves are created, with higher pressure moving up each crest. The waves are joined together, and big areas of high pressure are juxtaposed with big areas of low pressure, and strong winds are generated which go from high pressure to low pressure. The storm keeps up if it is fed by the waves and currents, which are the dynamics of the moon, interacting with gravitational factors that produce underwater earthquakes and faster swells, e.g. new moon, full moon, perigees, and when the moon crosses the equator, called lunar equinox.

The storm dissipates in the reverse order. Winds lessen and the sea surface temperature rises. Mixing below the surface ceases gradually. The Sun comes out as clouds seem to dissipate and the ocean surface flattens. It is all controlled by the size and direction of the swells. Sailors know that to escape a storm, you sail into the incoming wind, which is into the whitecaps. They have known it for thousands of years, and there are ancient accounts in ancient texts, but they may not have realized why.

QUESTIONS FOR SCIENCE

If we ignore the moon then how could a huge celestial body between a third and a quarter the size of Earth, and only ten Earth circumferences away, not have any effect on Earth, sea, and the air? Perhaps astronomers may have let us down over the years. They say that the moon only affects the tides, but nothing else. If it affects something like tides on the earth, then why does it not affect other things that are on the earth too, like humans? And how could all three layers (air, sea, and land) not be interconnected - considering they completely and perfectly interface over most of the surface of the planet? How could it be that only one could get away with having a tide, but not the other two, when all three are joined? Aren't joined things part of a joined system that acts together? Science should answer all questions, for that is the essence of scientific pursuit.

HEAVY RAIN

Cyclones are formed in countries when the sun within 6° latitude of the equator warms the waters that must be 26°C -28°C, over a wide ocean area. And that produces massive evaporation. The heat from a full moon day, especially when the moon is close, in certain areas can create a tropical cyclone close to the equator. When the full moon's heating capacity diminishes, then so will the cyclone. The huge accumulation of water in the sky, ready to fall and cause destruction, can only come from the sea.

That is why they immediately begin to fall apart, the minute that full moon moves away. And that's why cyclones don't form south of Fiji or in Europe. The water is just not warm enough That's why you never get cyclones occurring in Ireland. Cyclones travel from either north or south depending on which hemisphere they are formed in. They will very rarely cross the equator.

Cyclones, hurricanes, and typhoons predominate during equinox months, which are mainly when the atmosphere undergoes a twisting action, through the plane of what we call the ecliptic, due to the increase in speed of the Sun relative to the orbit of the earth. If you can picture the midpoint of a pendulum, that happens in both March and September. Furthermore, with spring and autumn full and new moons, king-tides occur when the moon rises at or near the equator. This represents the midpoint of yet another pendulum, the Sun on the equator being the first, and now the Moon on the equator for the tides. Sun and moon are moving faster relative to earth around equinoxes, so atmospheric disturbance increases.

The end of 2025 is expected to reveal El Nino conditions in the Pacific region, which typically suppresses activity during cyclonic seasons. Currents along the equatorial band, between Tahiti and Darwin stall, allowing for warmth to build up, as it did in 1998. That was an El Nino year, along with 1982/3, 1986/7 and 2016. 1997/8 is still regarded as a powerful El Nino-Southern Oscillation event. Further down the track the Gulf Stream also slowed and stalled, allowing for colder Arctic conditions to affect the northern areas of the globe, for a couple of years.

There are twice as many weak La Nina years as there are weak El Nino years. But there are about equal moderate to strong La Nina and El Nino years. La Nina can be viewed as the planks on a deck, with the El Nino akin to the gaps between them. La Nina years are two times wetter than El Nino years.

It is worth noting that the El Niño/La Nina index has stayed close to the neutral range in 2023 and 2024. In 2025 the index briefly edged up into the El Nino episode range, and then swung over towards La Nina, and will edge over more in 2026, while staying in the neutral range. Fully fledged El Nino is expected around 2030. In between now and then, we may experience La Nina conditions, which means wetter years.

PRECIPITATION CYCLES

We return now to what makes more rainfall and what makes less, over a long period. First, we must go back to the cycles of the sun. With each passage of Jupiter, when Jupiter and Saturn sit on one side of the sun in conjunction, there is usually a major drought in the area of Earth that they call, say, eastern Australia. Those years were 1901, 1920, 1940, 1961, 1981, 2020, and they are about 20 years apart. Halfway into this decadal period, Jupiter and Saturn face each other on opposite sides of the sun, and we call that an opposition, and those become flood years.

Flood years have higher tides than drought years, and that leads to more rainfall which affect the rivers, and everything falls into place accordingly from the oceans to the farmland. We can note that the conjunctions i.e. drought years, were 1862, 1901, and every 12, 24, and about 36 years, so the next worst was 1959, and then around 2020 (but more broadly 2016-2023), we call them the solar minimum years, they were years of world-wide droughts in the drier regions because of low rainfall and low agricultural yields.

Solar cycles determine annual temperatures and amounts of annual rain. Moon cycles determine the timing of weather events. Cycle #23 began in 1998, peaked around 2000/1 (solar maximum), and died off (solar minimum) from 2007-2009. After a 2–3-year period of relative inactivity Cycle #24 began around midway through 2010 and peaked in 2014. It "bottomed out" in 2020-21. 2020/2021 saw the end of Sunspot cycle #24. Cycle #25 began in 2022, is expected to peak in 2026, and then bottom out again in 2032.

By combining lunar and solar influences, temperature (solar) may lead to amounts of rainfall (due to evaporation rates), whereas the timing of severe weather events may follow strict lunar cycles.

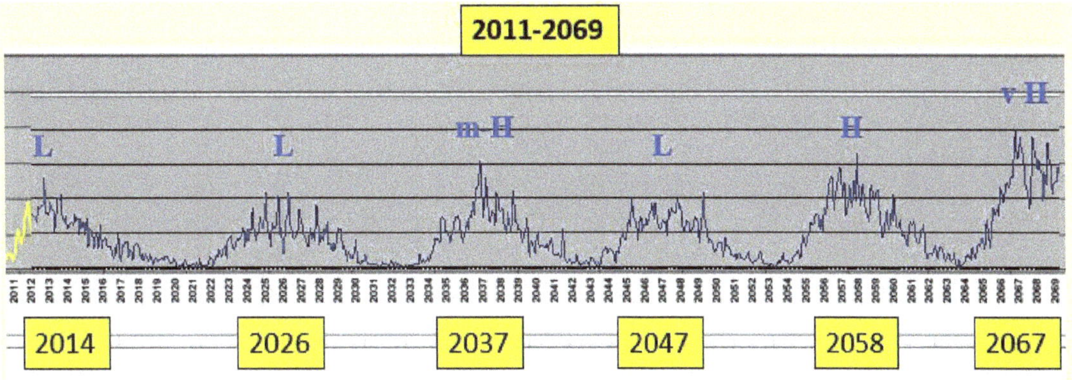

Contrary to what is expected, globally warm periods have fewer and weaker hurricanes while cooler periods have more and stronger hurricanes. It is the Sun's heat that causes evaporation which brings rain. The amount of heat that the Sun puts out determines the amount of evaporation, which must then fall within about 7 days, near the water source. Coastal locations are generally wetter than places that are some distance inland. Without a source of nearby water, a place quickly dries out. The amount of rain falling is due to the amount of solar radiation which causes evaporation. Therefore, rain amounts depend on the sunspot cycle. The amount of rain falling is also dependent on the extent of the air tide, which can be likened to the volume of water in each sea tide. As the solar cycle of radiation is cyclic then so too are rain intensities.

Sea tide timing is predictable, subject to the positions of the Moon and Sun. The volume of water flowing into an estuary depends on wind direction subject to the shape and size of surrounding hills, wind forces increased or confused by surface currents, and less visible factors like the proximity of underwater eruptions, fissures, earthquakes, and time of the day due to temperature gradients. Amounts of rain are influenced by the Sun's warmth for evaporation and then how much cold air comes to the layers of the atmosphere containing clouds.

Sometimes colder air is at a higher or lower altitude than rain-filled clouds, and expected rains must wait for the marrying of the two factors. Until then a cloud may look like rain but heat from the ground is sufficient to buoy up the rain-laden cloud and prevent rain from falling. Analyzing sunspot cycles is at least better than having nothing at all, by looking to see which previous years correspond to the same points of the cycle.

THE ATMOSPHERE

Temperature, volume, and pressure became the focus of climate science a few centuries ago when the Ideal Gas law was discovered. This states that the three factors are dependent, and if given a closed system, and played around with in the laboratory, then by changing one variable you can make others change as well. These principles were applied to the air in the atmosphere regarded as an enclosed gas. It is supposed to be enclosed because it is held here by gravity. This is where science broke down due to a fundamental misunderstanding.

This is not a laboratory, and the earth does not have a lid by which one can change one factor and watch the changes in the other two. In other words, the ideal gas law, which sounds logical, cannot be applied to the atmosphere, which is open to space. Nor can it be shown to work in the laboratory without enclosing the gas in some container. Gravity is still working in the laboratory when you remove the lid of your container. Therefore, we could arguably scrap the ideal gas law, as it applies to our environment, and the idea of temperature (which is a feeling) along with it.

What of volume and pressure? Well, except in isolated and controlled cases, we cannot watch the effects of air pressure on the air, because it is generated too locally close to the ocean surface to be of any use to the atmosphere. Besides, any wind is generated by currents in the sea that work their way to the surface, and create high and low-pressure areas, and wind between high and low pressure, which is measured on the ocean's surface but may only tell half the story.

Which leaves us with volume. Science has not come up with a way of accurately quantifying the amount of air in the world, and therefore telling us that there is a danger of it running out. If every tree was instantly destroyed, we would still have 4 million years' worth of oxygen in the atmosphere, because air contains 21% oxygen produced by photosynthesis in the biomass in the ocean, near the surface. That means 99.9% of our vegetation, which produces oxygen for the survival of every species of life, operates mostly in the sea, and away from the air. That means CO2, or anything else introduced to the air, can have little or no effect.

The sun warms the earth, but the oceans cool it. Even on the hottest summer's day, a swim in the ocean will be refreshing because the water is always cooler than the air above it. What's more, it can never get hotter than the air above. Everything is in balance and will remain so, and what we call the weather is what we perceive as the outcome of planetary cycles that began thousands of years ago.

WEATHER GLOSSARY

To get the most out of the forecasts, it helps to be able to read a weather map and to understand the language used. A high air pressure – a fair-weather sign, is when the isobaric values increase inwards towards the center. A low-pressure zone or depression, indicates worsening conditions, and when isobaric values decrease towards the center. Closer isobars = stronger winds.

TERM	MEANING	WEATHER
Apogee	Moon furthest from Earth for month	Winds generally light. Calm conditions.
Perigee	Moon closest to Earth for month	Spells trouble.Potential Earthquake risk.
Cold front	Cold air advancing, pushing away warm air	Changeable
Depression	Area of low barometric pressure	Unpredictable, bad weather, strong winds
Ridge of high pressure	Edge of anticyclone	Good weather, cooler nights in winter
Stationary front	Slow-moving weather	No change for 2-3 days
Northern declination	Moon at northernmost point for month. Moon rises in the north.	Slow moving weather system, northerlies, rain in north, colder conditions
Southern declination	Moon at southernmost point for month	Slow moving system, southerlies, generally milder, precipitation in south.
Moon crossing equator	Faster moving weather system, moonrise is due east.	Turbulent, changeable, potential electrical storms, gales or winds expected.

SPEED READING OF CLOUDS

METHOD

For many years, there have been people who have looked at the clouds and determined the coming weather. The techniques are relatively simple to master. Generally, the weather in the early morning does not change much during the day. Whatever reading is gained from looking at the clouds, should usually hold for 24 hours. When we look up and take note of the types of cloud shapes that are forming, and then we look again in an hour, what is developing or fading away is noted. Your guru skills have already noted the sky, at least an hour or two before you announce a prediction.

TYPES OF CLOUD

There are essentially 3 types of clouds, which have Latin names to easily describe their shapes –

1. high (cirrus, literally "mares's tails"),
2. middle layer (cumulus, literally "piles')
3. low cloud (stratus, literally "lines").

HIGH
(20,000-40,000 FEET)

Cirrus; the cloud of change

After a run of clear or bad weather, light streaks of very high cirrus clouds are often the first signs of change. Because of their height, they only move with the rotation of the earth. Bad weather is about 36 hours away. Check in about an hour; if cirrus increases across the sky, it indicates an approaching depression, but if curls and wisps dissolve or vanish, fine weather continues. The higher the cloud, the finer the weather. Rapid cirrus, in bars, or at right angles to bars, mean gale or wind. The bands generally precede a storm by 20-48 hours.

When cirrus become lines (cirrostratus), expect a quick change.

If cirrus streaks point upwards, it means rain. If downwards, wind and dry weather.

If mare's tails are followed by altocumulus (literally "high piles") the next day, it means rain within 12 hours. Cirro-stratus, (literally "spread-out hair"), also means rain. When cirrostratus and stratocumulus gather over the sea, it means rain and probably wind on the second day. Cirrocumulus (literally "piles of hair") is a 24-hour warning. It is also called mackerel sky and means fair weather for that day but rains a day or two after.

Before thunder, cirrocumulus can appear as dense and compact masses, in close contact. If their outline is soft and delicate, if fleeces gently merge, it means a continuance of fine weather. Thunder is usually mainly noise but very little rain. But if the clouds are dense, and associated with cirrus, it means lightning in summer or gales in winter. High cirrocumulus commonly appears a few hours before thunderstorms. The harder and more definite the outline, the more unsettled the coming weather. Cirrocumulus in the early morning, which diminishes means a fine and warm day.

MIDDLE
(6,000-20,000 FEET)

Altocumulus (literally "*high* piles") is a sign of coming thunder. It is also called tower clouds, or Cauliflowers. It consists of turret-like upward growths. A thicker cloud generally arrives some hours later. Altocumulus often occurs together with altostratus, (literally "high, spread out"), and is the cloud that carries rain and snow. It can cover the whole sky, and when it increases, rain and snow over a wide area may be the result. It may also bring drizzle.

Cumulonimbus, (literally "*gathered* piles of rain"), is a sign of heavy rains. They are cumulus clouds which are large enough to produce showers. From a distance, they appear as tall piles, but to stand underneath and look up is to see the dark look of rain. They bring thunderstorms, heavy rain, and occasionally hail. They can be anything from 1000 to 60,000 feet high. Towering thunder clouds that look like mountains, early on a summer's day can become intense storm activity during the heat of the afternoon and often become torrents of rain in the evening.

When the cumulus rain clouds look smaller toward sunset than they were at noon, they are diminishing, and you can expect fair weather. The formation of cumulus clouds to leeward during a strong wind indicates the approach of a calm with rain. But cumulonimbus clouds forming rapidly in the mountains on warm afternoons may mean an intense lightning storm is on its way in minutes.

Cumulus (or ice cream clouds"), are the most common clouds and are caused by moderate, generally westerly winds. All have a well-defined flat base, between 1500 and 2500 feet high. If they appear early in the morning they may develop into bigger shower clouds. If they do not form till late afternoon, they may die away to be followed by a clear sky night. If clouds open and close, rain will continue.

If the clouds are like sheep's wool, deep and dense, the edges being very white, the surrounding sky bright and blue; they are of a frosty coldness, and will speedily fall in hail, snow or rain.

Strato-cumulus, (literally "spread-out gatherings"), are the doors to a cold front. 1500-6500 feet. They cover the sky, sometimes in parallel bands but they usually bring little or no rain. When at sea, the strato-cumulus clouds appear on the horizon, it is a sign that the weather will break up. If over the land in the winter, they may produce a light sprinkling of snow.

If long stripes of cloud drive at a slow rate, high in the air, and gradually become larger, the sky having been previously clear, expect rain.

LOW
(500-2,000 FEET)

Stratus, or hill cloud. Nimbostratus (literally "rain-layer") is a cloud of light rain. Sometimes simply called 'nimbus', these are low, uniform, basic rainclouds, 300-2000 feet. Thick enough to blot out the sun, they can disgorge continuous rain or snow. If a little cloud appears in a clear sky, especially if it comes from the west or somewhere in the south; there is a storm brewing. If there are long points, the look of tails or feathers hanging from rain clouds, close to the horizon, with little wind in the summer, thunder may be expected but the storm may be short. Resembling fog and often covering hills, stratus may bring drizzle.

Harbingers of fine weather are when the day breaks out from the morning fog. Stratus at night is followed by a fine day if the barometer stays high. When mountains extend north and south, if fog or mist comes from the west, expect fair weather. If mist comes down from the top of the mountains, expect rain in summer, and snow in winter.

GENERAL SUMMARY

If two layers of clouds move in opposite directions, then bad weather is on the way.

Clouds do not indicate the presence of water vapour, which is everywhere at once. Clouds represent masses of air of a different temperature. The wind is the movement of an air mass and is not necessarily the movement of clouds. Clouds do not move, that is an illusion. They are disintegrating and reforming at the same time, but we view them at a distance. A fog is a close cloud, and if we were to stand in the middle of a fog or mist we experience the air as still around us.

The wind blows rain in, but also blows it away again. It is more likely to rain when the wind drops. The times when the moon is on the horizon, either rising or setting, and when it is directly overhead or underfoot, are times of greatest rainfall. Atmospheric pressure is highest when the tide is lowest, and vice versa. Atmospheric turbulence, and therefore wind and rain, is more likely at mid-tide going out or coming in.

FINAL WORD

Long range forecasting is different from regular meteorology. It is more trends-based, and it relies on archived records. Even then rain statistics from earlier years may not tell the full story. Statistics taken from e.g., Cork Airport, may describe results that are at variance to what falls in the county because the airport is not always representative. A possibility is that whereas rain may arrive, winds around the rain gauge, especially at an airport, can push rain quickly away, which then gives a subsequent low readout and a possible misread for the region, even though the winds have already blown the rain in.

A region can easily get confused with its main city, because of population numbers. For that reason, the reader is advised to consider notes about a whole region when planning, and regarding any weather event within a 50-80-mile range as a potential for their area also, even if it is listed for a neighboring district. Often the amount of rain may be localized, and it is not uncommon for each hill valley to have a different microclimate to the next one over. In a forested and hilly terrain, overall amounts of rain for one small area is practically impossible to determine accurately beyond a general trend. A flat region is much easier as rain is more likely to be widespread due to less quickly changing temperatures.

Quite another problem is that rain gauges often shift locations, to suit the convenience of meteorological staff. It would seem from what has been said that we should give up before we begin. But it is worth the exercise because if this book is used correctly, to spot trends rather than look for literal amounts, then good planning decisions may be made.

A 24-hour error is the most obvious because we cannot know from our archived data files of weather 20-50 years away, incorporating at least two lunar cycles, which side of midnight historical data was gathered. Most gathering stations display a rain reading taken in the 24hrs before 9 am, but did the rain on the day happen just before 9 am, at say, 8.55 am and hence on the day in the listing, or, say, 7 pm some 14 hours before, qualifying as previous day's rain?

The scientist Benjamin Franklin (1706 - 1790), the inventor of the lightning conductor, was also a lunar weather forecaster. His regular *Poor Richard's Almanac* ran for 25 years, and he had 10,000 subscribers. Franklin hedged his bets slightly by stating: 'I should ask the indulgence of the reader for a day of grace on either side of the date for a specific weather prediction.'

Similarly, temperatures are gathered as maximums and minimums, one of each per day per town. Suppose reading is taken at 4 pm (often the practice) but let's say on a particularly hot day a very high maximum is reached at 5 pm. The hotter value will miss the cut, and the 4 pm reading will be recorded as the day's maximum. But if the next day may bring a cold maximum due to an anomalous and temporary southern wind shift, nevertheless the previous day's 4.30 pm reading will be the relevant maximum figure when taken over the previous 24 hours. The error in temperatures can be large because, over a month, this error in average maximums can be magnified. Before long the "normal" is invalid, and the breaking of records is simply fiction media hype to sell newspapers.

Also, as cold air layers have to match with rain-laden clouds for rain to fall, there can sometimes be up to about a 24-hour error in rain arrival. Weather can be generated between 200ft and 8 miles up, which can incur the potential overshoot of a rain system over a radius of up to 50-60 miles (80-100km). So, if rain is forecast for a coastal location but drops 100 km out to sea, it may arguably still qualify as a successful forecast. It is not uncommon for one forecast report heard on radio or TV to cover half the country in one sentence, yet there will be many microclimatic events within even one small area, especially in hilly terrain.

Most historical data is taken at airports, and these are usually well-suited for wind which makes it easier for planes to take off and land safely, and as windier places dry out more rapidly, it means they are typically drier areas than the surrounding hills. Unless your location is at or near the airport the weather may turn out slightly different to the area forecast in this book. This is not a problem in an unchanging landscape but may be one in a varying terrain whereby a farmer living high on a hill or in a valley can't understand why the airport-derived data he finds in this book doesn't tally with what weather occurred on his farm, yet in the same breath he may be aware of his small microclimates around mounds, hills, valleys, and flats.

The almanac process may be the meteorology of the future. There is a respected place now in our society for differing viewpoints. We already have alternative music and alternative medicine. From the old man on the hill who could tell from the spitting of logs in his fire when snow was coming to the way the sailors hauled in sail believing (correctly) that the appearance of dolphins heralded approaching storms, there has always been room for alternative weather. Because weather is so much a part of our lives, we need all the tools, modern and ancient, and all the past observations about cycles that we can acquire.

THANKS

I am hugely grateful to the support team of Glenys and Victoria, for the many hours of work on project design, research and layout, without which the almanac would not be possible. My heartfelt thanks, also to Vicki in dispatch. We also acknowledge Met Eireann for historical archival records. We hope, as always, that the information provided here will assist in your planning.

GRAPHS

MAX/MIN TEMPERATURES FOR COUNTIES

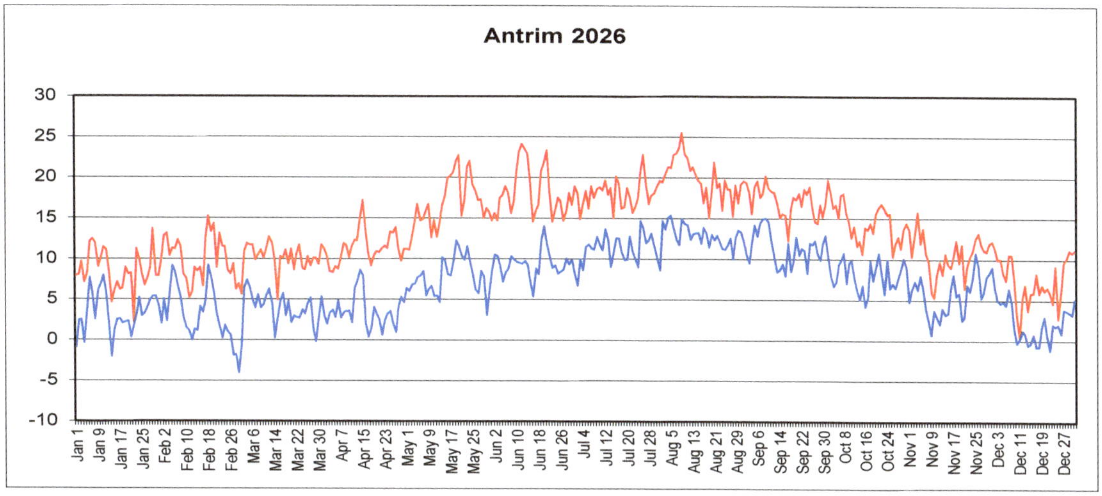

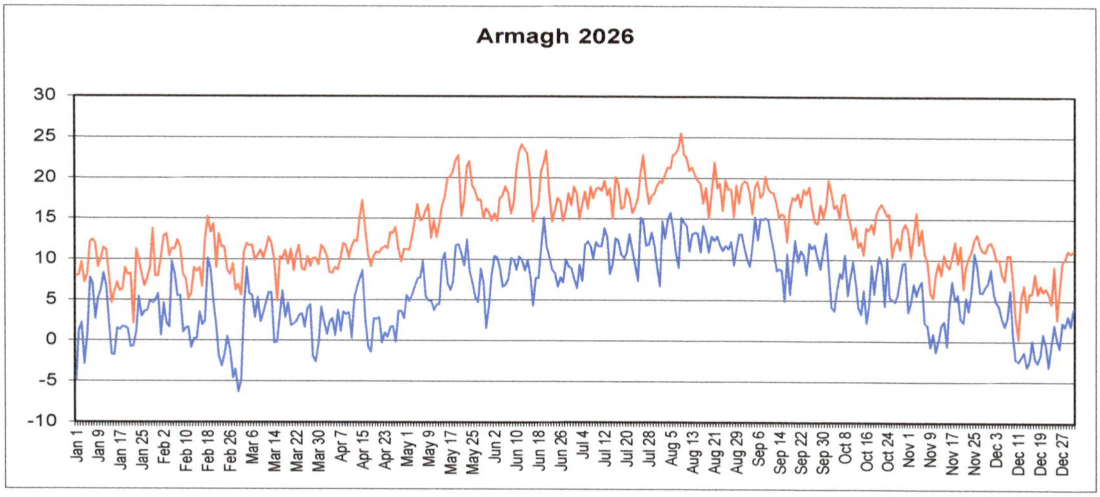

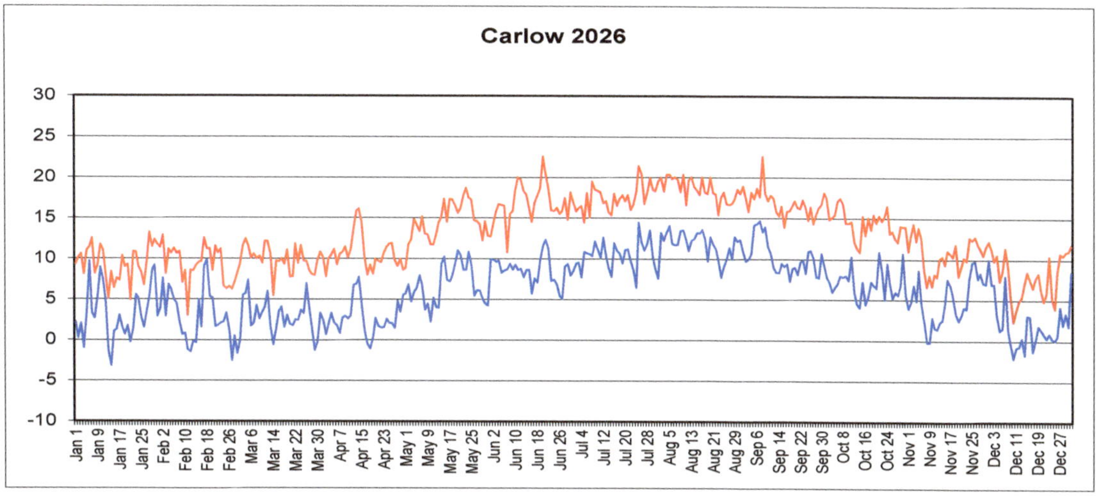

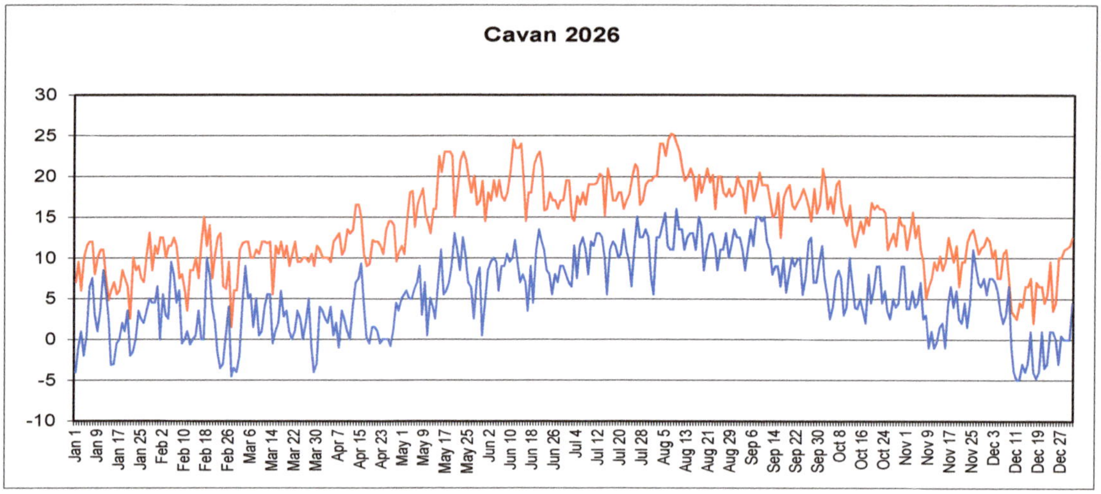
Cavan 2026
30
25
20
15
10
5
0
-5
-10
Jan 1
Jan 9
Jan 17
Jan 25
Feb 2
Feb 10
Feb 18
Feb 26
Mar 6
Mar 14
Mar 22
Mar 30
Apr 7
Apr 15
Apr 23
May 1
May 9
May 17
May 25
Jun 2
Jun 10
Jun 18
Jun 26
Jul 4
Jul 12
Jul 20
Jul 28
Aug 5
Aug 13
Aug 21
Aug 29
Sep 6
Sep 14
Sep 22
Sep 30
Oct 8
Oct 16
Oct 24
Nov 1
Nov 9
Nov 17
Nov 25
Dec 3
Dec 11
Dec 19
Dec 27

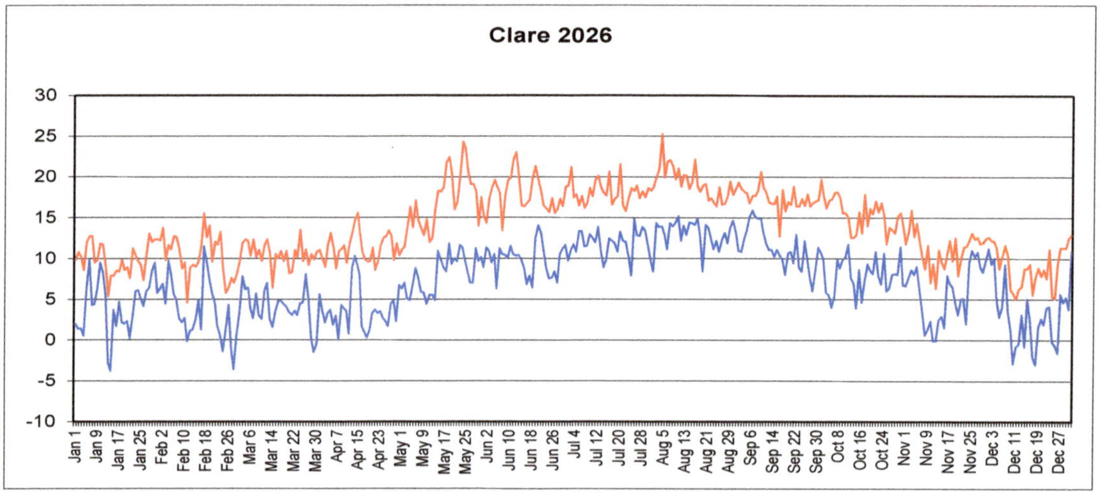
Clare 2026
30
25
20
15
10
5
0
-5
-10
Jan 1
Jan 9
Jan 17
Jan 25
Feb 2
Feb 10
Feb 18
Feb 26
Mar 6
Mar 14
Mar 22
Mar 30
Apr 7
Apr 15
Apr 23
May 1
May 9
May 17
May 25
Jun 2
Jun 10
Jun 18
Jun 26
Jul 4
Jul 12
Jul 20
Jul 28
Aug 5
Aug 13
Aug 21
Aug 29
Sep 6
Sep 14
Sep 22
Sep 30
Oct 8
Oct 16
Oct 24
Nov 1
Nov 9
Nov 17
Nov 25
Dec 3
Dec 11
Dec 19
Dec 27

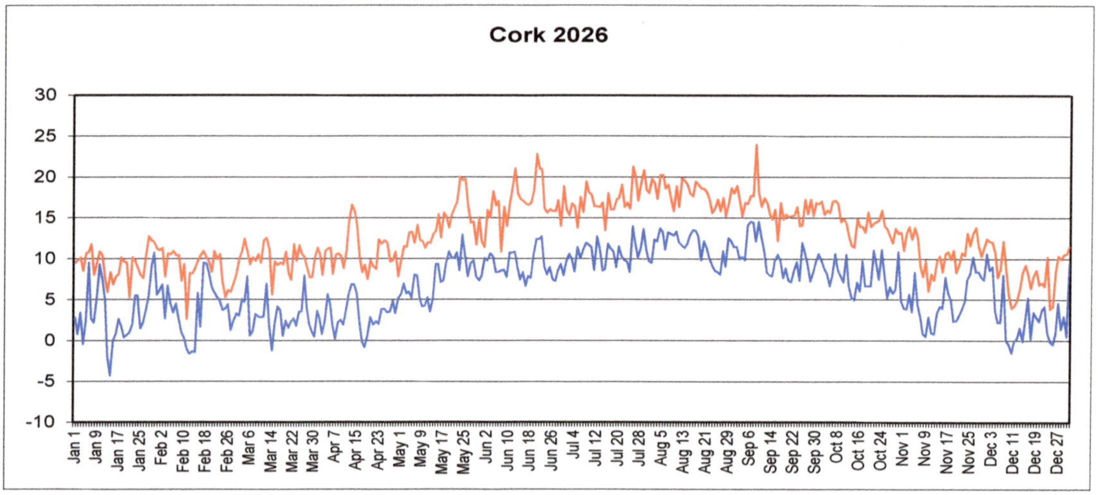
Cork 2026
30
25
20
15
10
5
0
-5
-10
Jan 1
Jan 9
Jan 17
Jan 25
Feb 2
Feb 10
Feb 18
Feb 26
Mar 6
Mar 14
Mar 22
Mar 30
Apr 7
Apr 15
Apr 23
May 1
May 9
May 17
May 25
Jun 2
Jun 10
Jun 18
Jun 26
Jul 4
Jul 12
Jul 20
Jul 28
Aug 5
Aug 13
Aug 21
Aug 29
Sep 6
Sep 14
Sep 22
Sep 30
Oct 8
Oct 16
Oct 24
Nov 1
Nov 9
Nov 17
Nov 25
Dec 3
Dec 11
Dec 19
Dec 27

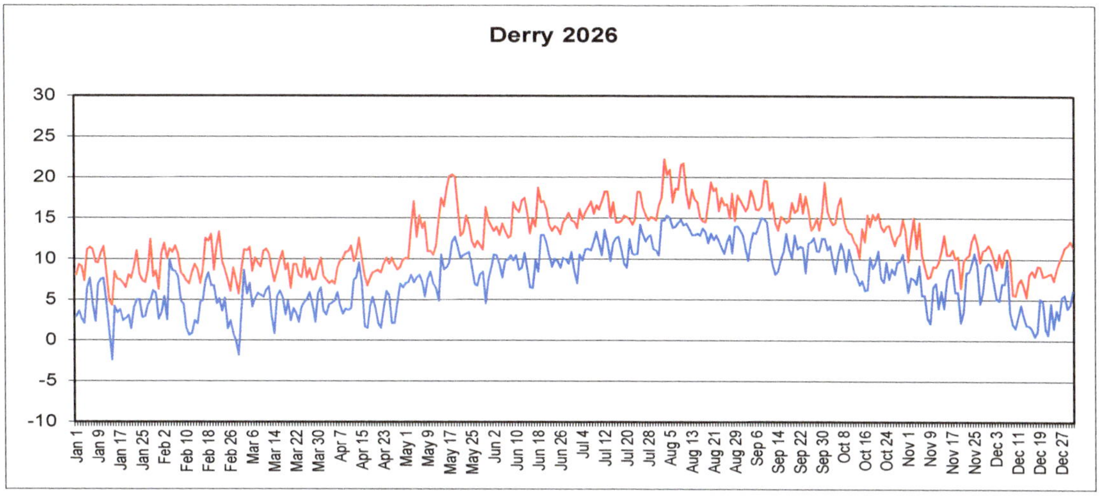

Derry 2026
30
25
20
15
10
5
0
-5
-10
Jan 1 Jan 9 Jan 17 Jan 25 Feb 2 Feb 10 Feb 18 Feb 26 Mar 6 Mar 14 Mar 22 Mar 30 Apr 7 Apr 15 Apr 23 May 1 May 9 May 17 May 25 Jun 2 Jun 10 Jun 18 Jun 26 Jul 4 Jul 12 Jul 20 Jul 28 Aug 5 Aug 13 Aug 21 Aug 29 Sep 6 Sep 14 Sep 22 Sep 30 Oct 8 Oct 16 Oct 24 Nov 1 Nov 9 Nov 17 Nov 25 Dec 3 Dec 11 Dec 19 Dec 27

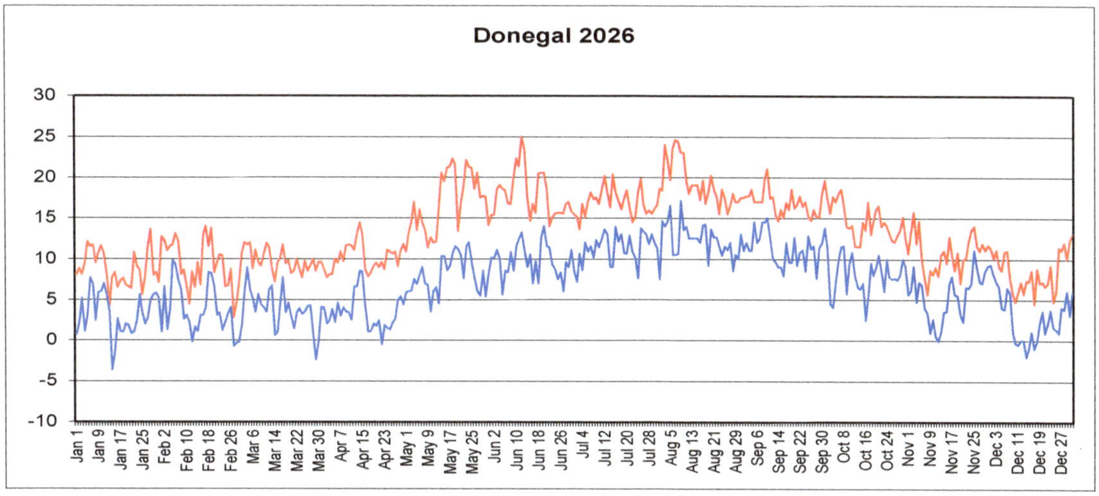

Donegal 2026
30
25
20
15
10
5
0
-5
-10
Jan 1 Jan 9 Jan 17 Jan 25 Feb 2 Feb 10 Feb 18 Feb 26 Mar 6 Mar 14 Mar 22 Mar 30 Apr 7 Apr 15 Apr 23 May 1 May 9 May 17 May 25 Jun 2 Jun 10 Jun 18 Jun 26 Jul 4 Jul 12 Jul 20 Jul 28 Aug 5 Aug 13 Aug 21 Aug 29 Sep 6 Sep 14 Sep 22 Sep 30 Oct 8 Oct 16 Oct 24 Nov 1 Nov 9 Nov 17 Nov 25 Dec 3 Dec 11 Dec 19 Dec 27

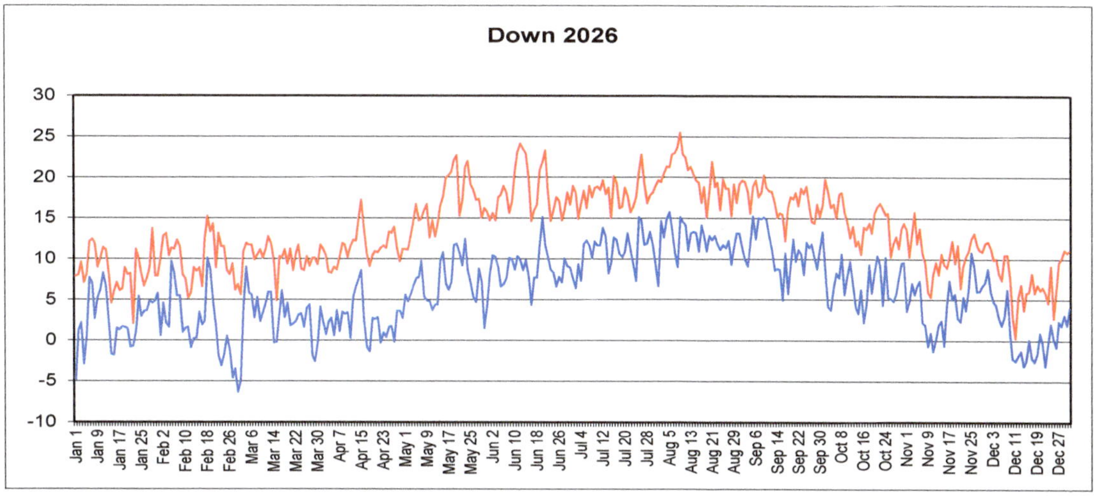

Down 2026
30
25
20
15
10
5
0
-5
-10
Jan 1 Jan 9 Jan 17 Jan 25 Feb 2 Feb 10 Feb 18 Feb 26 Mar 6 Mar 14 Mar 22 Mar 30 Apr 7 Apr 15 Apr 23 May 1 May 9 May 17 May 25 Jun 2 Jun 10 Jun 18 Jun 26 Jul 4 Jul 12 Jul 20 Jul 28 Aug 5 Aug 13 Aug 21 Aug 29 Sep 6 Sep 14 Sep 22 Sep 30 Oct 8 Oct 16 Oct 24 Nov 1 Nov 9 Nov 17 Nov 25 Dec 3 Dec 11 Dec 19 Dec 27

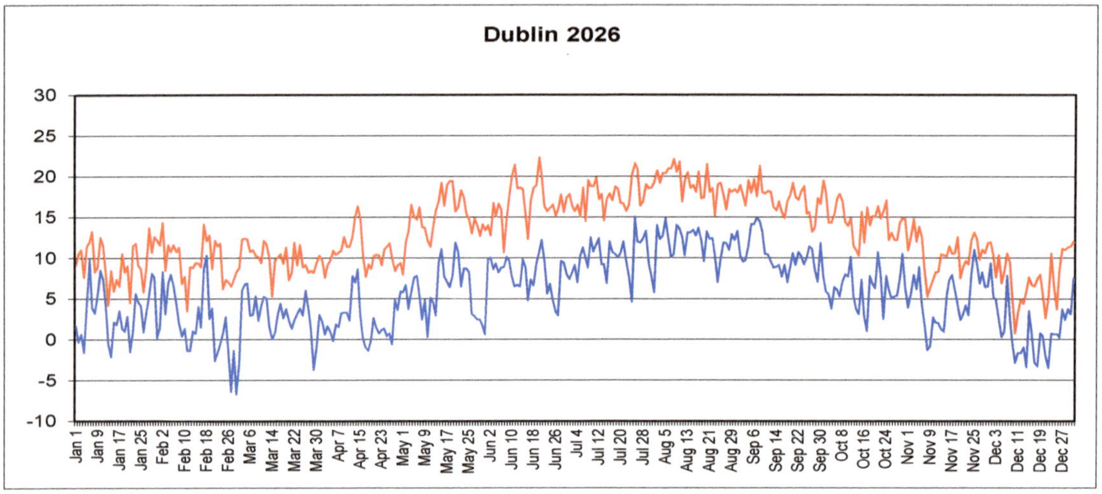
Dublin 2026
30
25
20
15
10
5
0
-5
-10
Jan 1
Jan 9
Jan 17
Jan 25
Feb 2
Feb 10
Feb 18
Feb 26
Mar 6
Mar 14
Mar 22
Mar 30
Apr 7
Apr 15
Apr 23
May 1
May 9
May 17
May 25
Jun 2
Jun 10
Jun 18
Jun 26
Jul 4
Jul 12
Jul 20
Jul 28
Aug 5
Aug 13
Aug 21
Aug 29
Sep 6
Sep 14
Sep 22
Sep 30
Oct 8
Oct 16
Oct 24
Nov 1
Nov 9
Nov 17
Nov 25
Dec 3
Dec 11
Dec 19
Dec 27

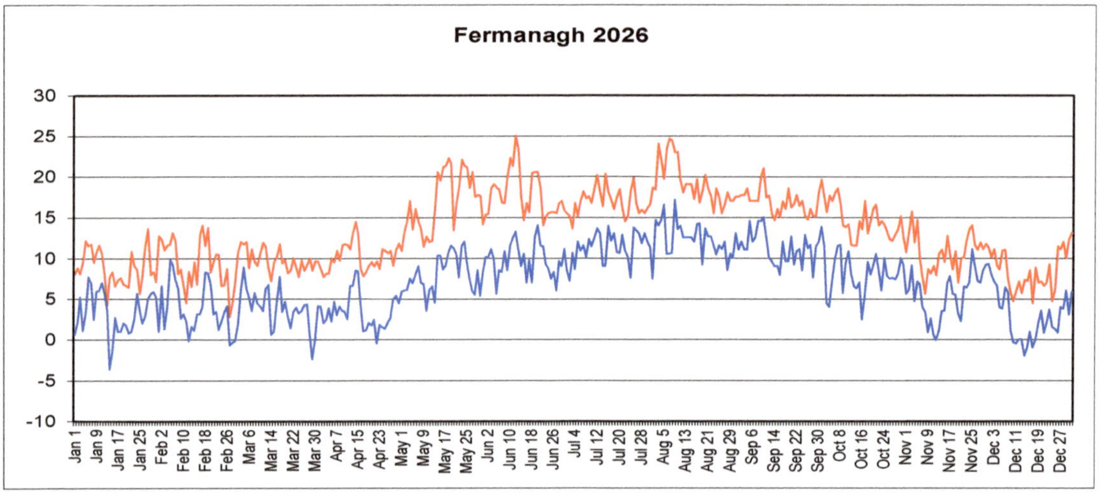
Fermanagh 2026
30
25
20
15
10
5
0
-5
-10
Jan 1
Jan 9
Jan 17
Jan 25
Feb 2
Feb 10
Feb 18
Feb 26
Mar 6
Mar 14
Mar 22
Mar 30
Apr 7
Apr 15
Apr 23
May 1
May 9
May 17
May 25
Jun 2
Jun 10
Jun 18
Jun 26
Jul 4
Jul 12
Jul 20
Jul 28
Aug 5
Aug 13
Aug 21
Aug 29
Sep 6
Sep 14
Sep 22
Sep 30
Oct 8
Oct 16
Oct 24
Nov 1
Nov 9
Nov 17
Nov 25
Dec 3
Dec 11
Dec 19
Dec 27

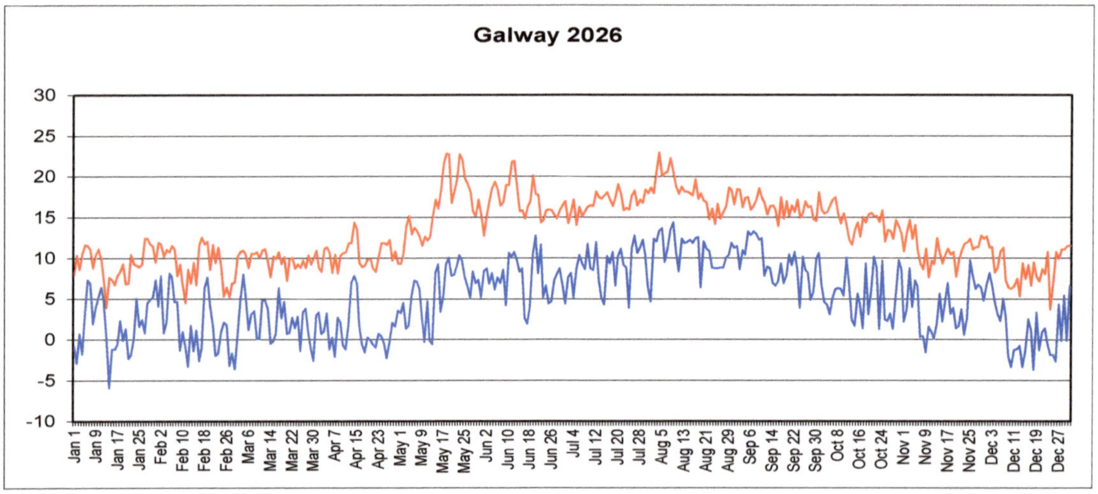
Galway 2026
30
25
20
15
10
5
0
-5
-10
Jan 1
Jan 9
Jan 17
Jan 25
Feb 2
Feb 10
Feb 18
Feb 26
Mar 6
Mar 14
Mar 22
Mar 30
Apr 7
Apr 15
Apr 23
May 1
May 9
May 17
May 25
Jun 2
Jun 10
Jun 18
Jun 26
Jul 4
Jul 12
Jul 20
Jul 28
Aug 5
Aug 13
Aug 21
Aug 29
Sep 6
Sep 14
Sep 22
Sep 30
Oct 8
Oct 16
Oct 24
Nov 1
Nov 9
Nov 17
Nov 25
Dec 3
Dec 11
Dec 19
Dec 27

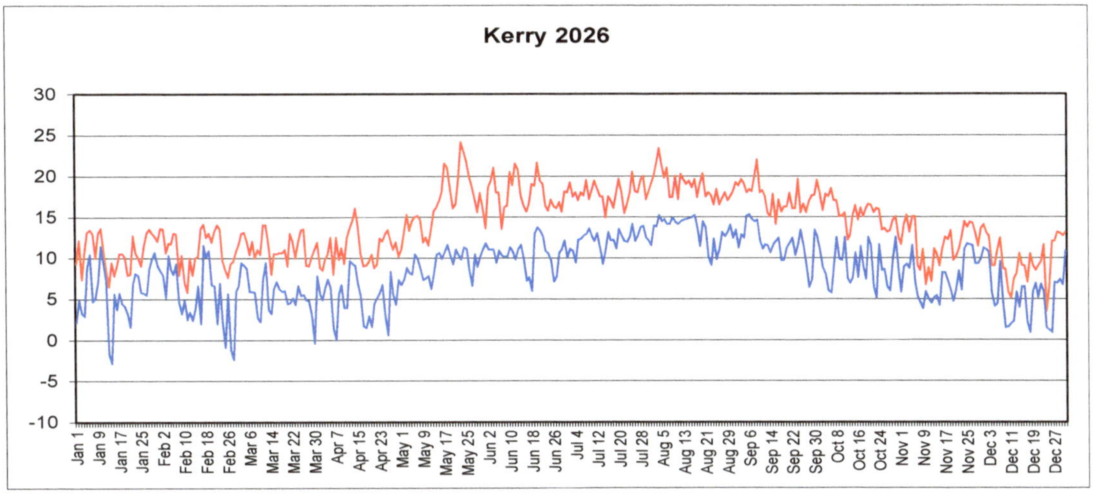
Kerry 2026
30
25
20
15
10
5
0
-5
-10
Jan 1
Jan 9
Jan 17
Jan 25
Feb 2
Feb 10
Feb 18
Feb 26
Mar 6
Mar 14
Mar 22
Mar 30
Apr 7
Apr 15
Apr 23
May 1
May 9
May 17
May 25
Jun 2
Jun 10
Jun 18
Jun 26
Jul 4
Jul 12
Jul 20
Jul 28
Aug 5
Aug 13
Aug 21
Aug 29
Sep 6
Sep 14
Sep 22
Sep 30
Oct 8
Oct 16
Oct 24
Nov 1
Nov 9
Nov 17
Nov 25
Dec 3
Dec 11
Dec 19
Dec 27

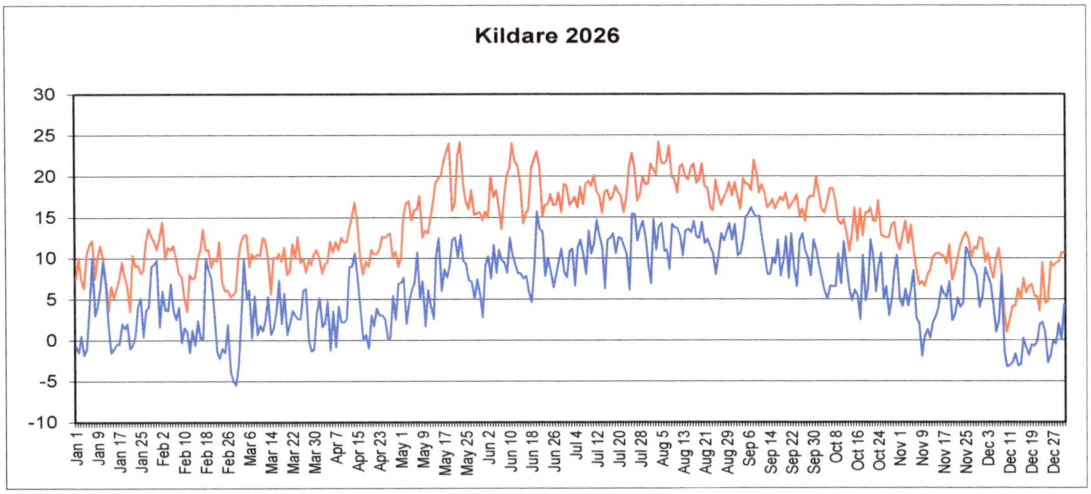
Kildare 2026
30
25
20
15
10
5
0
-5
-10
Jan 1
Jan 9
Jan 17
Jan 25
Feb 2
Feb 10
Feb 18
Feb 26
Mar 6
Mar 14
Mar 22
Mar 30
Apr 7
Apr 15
Apr 23
May 1
May 9
May 17
May 25
Jun 2
Jun 10
Jun 18
Jun 26
Jul 4
Jul 12
Jul 20
Jul 28
Aug 5
Aug 13
Aug 21
Aug 29
Sep 6
Sep 14
Sep 22
Sep 30
Oct 8
Oct 16
Oct 24
Nov 1
Nov 9
Nov 17
Nov 25
Dec 3
Dec 11
Dec 19
Dec 27

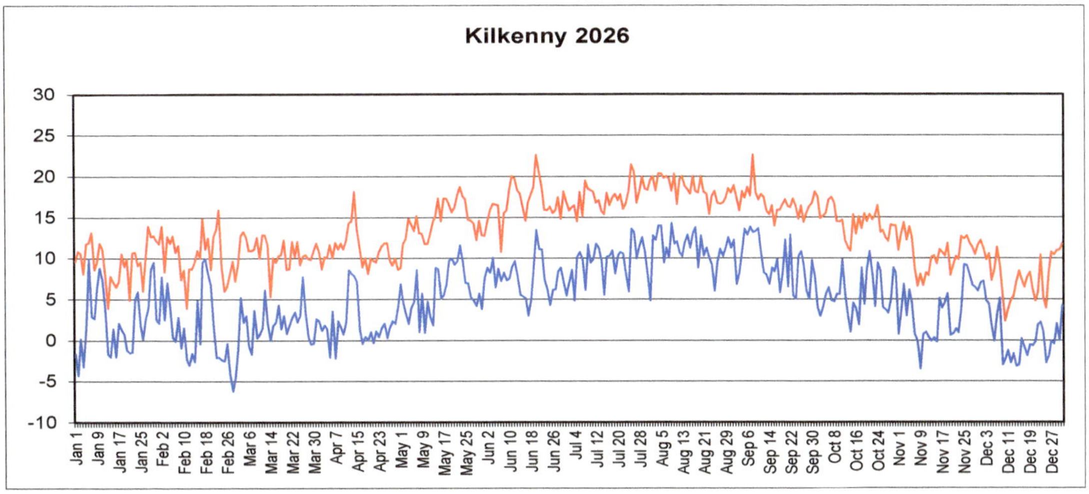
Kilkenny 2026
30
25
20
15
10
5
0
-5
-10
Jan 1
Jan 9
Jan 17
Jan 25
Feb 2
Feb 10
Feb 18
Feb 26
Mar 6
Mar 14
Mar 22
Mar 30
Apr 7
Apr 15
Apr 23
May 1
May 9
May 17
May 25
Jun 2
Jun 10
Jun 18
Jun 26
Jul 4
Jul 12
Jul 20
Jul 28
Aug 5
Aug 13
Aug 21
Aug 29
Sep 6
Sep 14
Sep 22
Sep 30
Oct 8
Oct 16
Oct 24
Nov 1
Nov 9
Nov 17
Nov 25
Dec 3
Dec 11
Dec 19
Dec 27

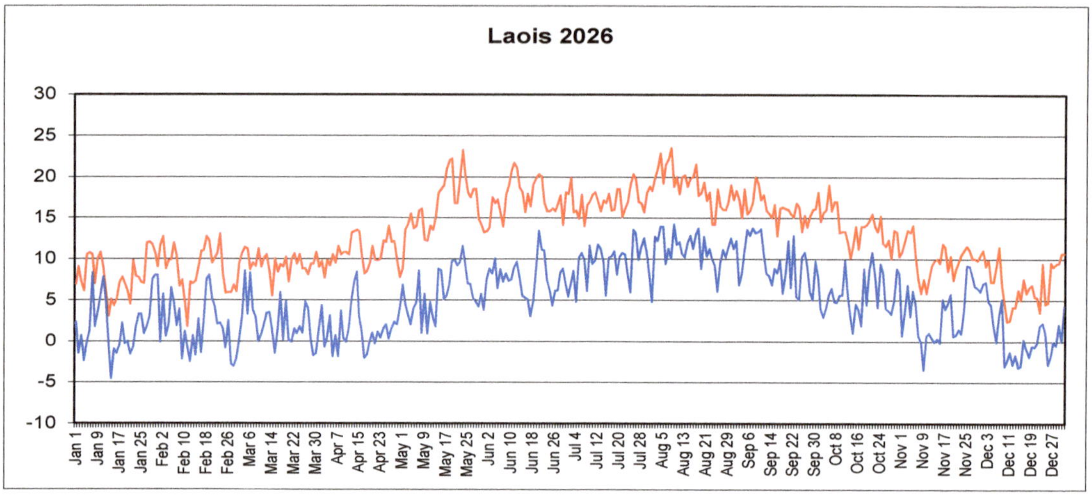
Laois 2026
30
25
20
15
10
5
0
-5
-10
Jan 1
Jan 9
Jan 17
Jan 25
Feb 2
Feb 10
Feb 18
Feb 26
Mar 6
Mar 14
Mar 22
Mar 30
Apr 7
Apr 15
Apr 23
May 1
May 9
May 17
May 25
Jun 2
Jun 10
Jun 18
Jun 26
Jul 4
Jul 12
Jul 20
Jul 28
Aug 5
Aug 13
Aug 21
Aug 29
Sep 6
Sep 14
Sep 22
Sep 30
Oct 8
Oct 16
Oct 24
Nov 1
Nov 9
Nov 17
Nov 25
Dec 3
Dec 11
Dec 19
Dec 27

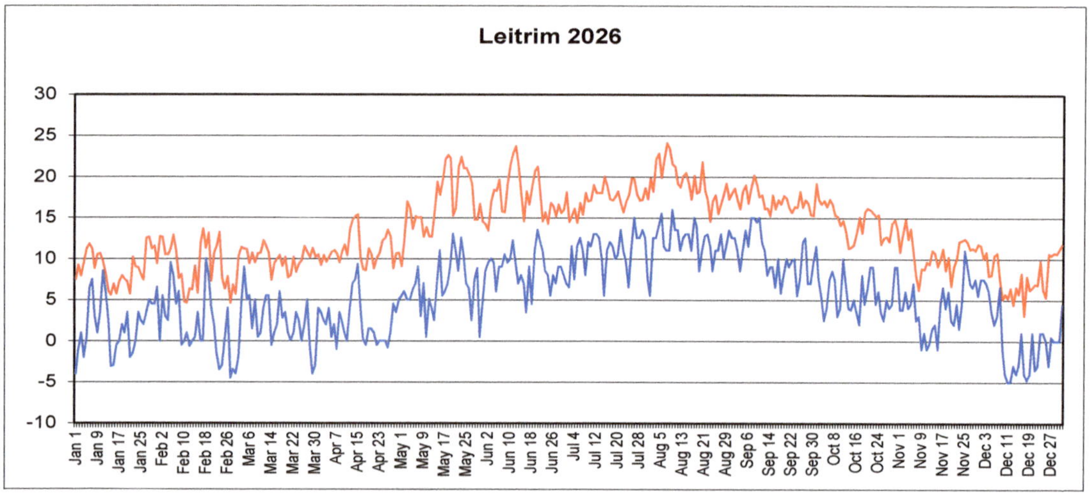
Leitrim 2026
30
25
20
15
10
5
0
-5
-10
Jan 1
Jan 9
Jan 17
Jan 25
Feb 2
Feb 10
Feb 18
Feb 26
Mar 6
Mar 14
Mar 22
Mar 30
Apr 7
Apr 15
Apr 23
May 1
May 9
May 17
May 25
Jun 2
Jun 10
Jun 18
Jun 26
Jul 4
Jul 12
Jul 20
Jul 28
Aug 5
Aug 13
Aug 21
Aug 29
Sep 6
Sep 14
Sep 22
Sep 30
Oct 8
Oct 16
Oct 24
Nov 1
Nov 9
Nov 17
Nov 25
Dec 3
Dec 11
Dec 19
Dec 27

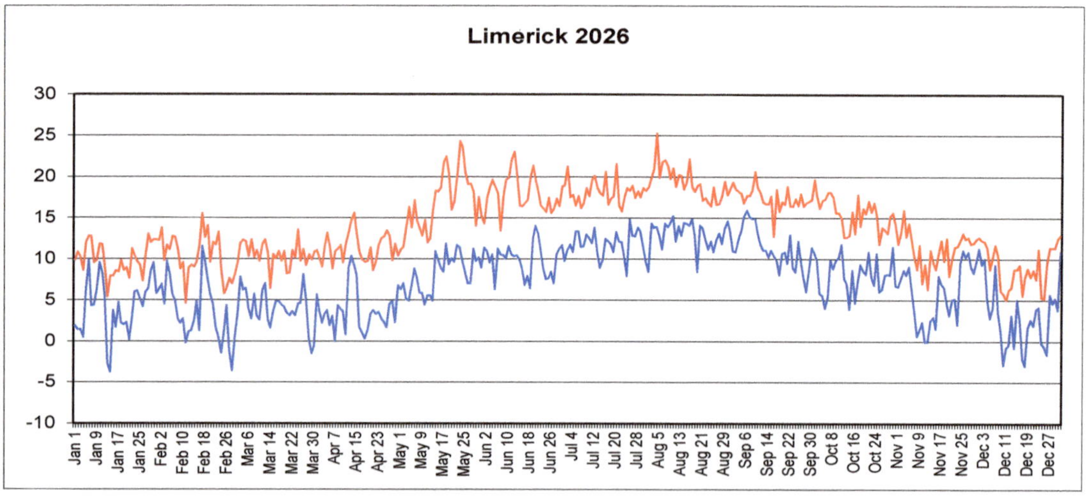
Limerick 2026
30
25
20
15
10
5
0
-5
-10
Jan 1
Jan 9
Jan 17
Jan 25
Feb 2
Feb 10
Feb 18
Feb 26
Mar 6
Mar 14
Mar 22
Mar 30
Apr 7
Apr 15
Apr 23
May 1
May 9
May 17
May 25
Jun 2
Jun 10
Jun 18
Jun 26
Jul 4
Jul 12
Jul 20
Jul 28
Aug 5
Aug 13
Aug 21
Aug 29
Sep 6
Sep 14
Sep 22
Sep 30
Oct 8
Oct 16
Oct 24
Nov 1
Nov 9
Nov 17
Nov 25
Dec 3
Dec 11
Dec 19
Dec 27

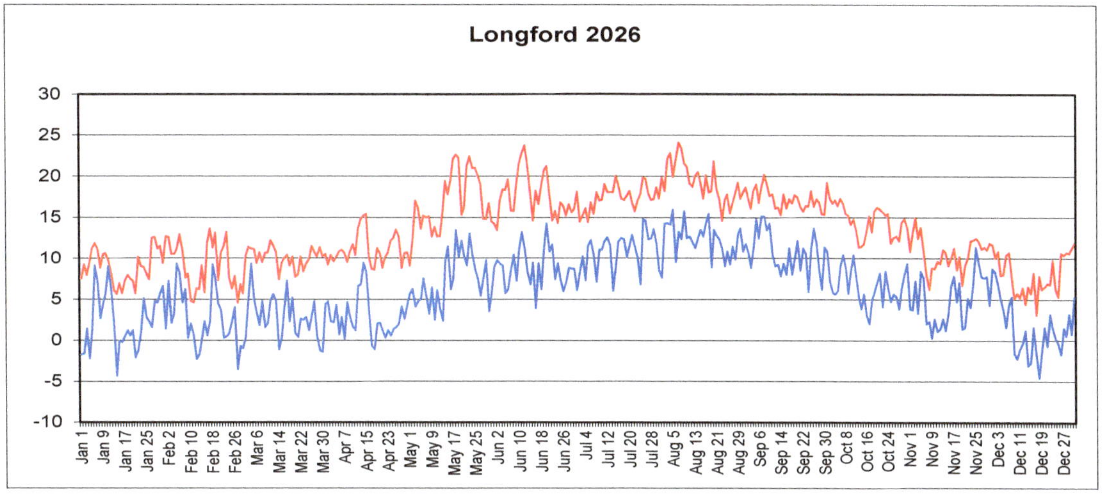

Longford 2026
30
25
20
15
10
5
0
-5
-10
Jan 1
Jan 9
Jan 17
Jan 25
Feb 2
Feb 10
Feb 18
Feb 26
Mar 6
Mar 14
Mar 22
Mar 30
Apr 7
Apr 15
Apr 23
May 1
May 9
May 17
May 25
Jun 2
Jun 10
Jun 18
Jun 26
Jul 4
Jul 12
Jul 20
Jul 28
Aug 5
Aug 13
Aug 21
Aug 29
Sep 6
Sep 14
Sep 22
Sep 30
Oct 8
Oct 16
Oct 24
Nov 1
Nov 9
Nov 17
Nov 25
Dec 3
Dec 11
Dec 19
Dec 27

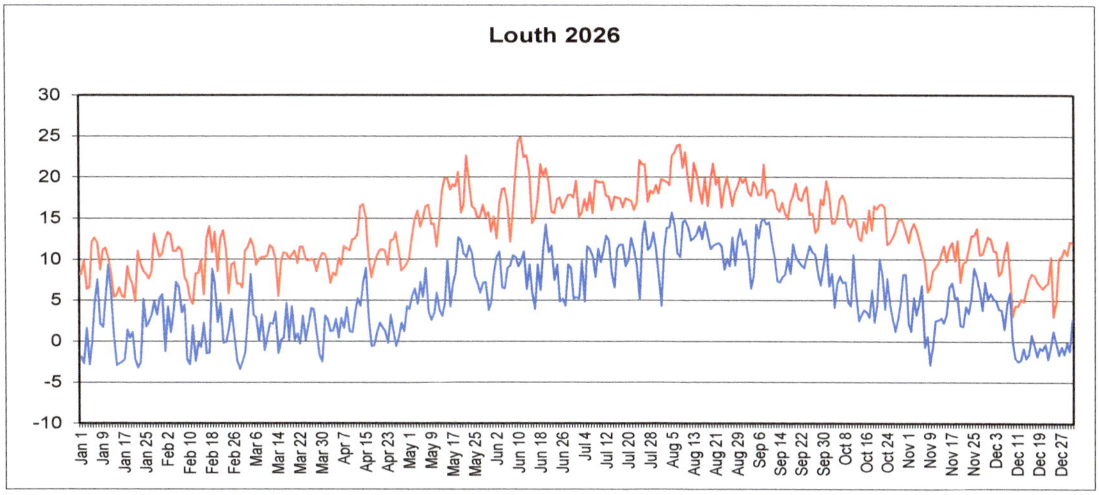

Louth 2026
30
25
20
15
10
5
0
-5
-10
Jan 1
Jan 9
Jan 17
Jan 25
Feb 2
Feb 10
Feb 18
Feb 26
Mar 6
Mar 14
Mar 22
Mar 30
Apr 7
Apr 15
Apr 23
May 1
May 9
May 17
May 25
Jun 2
Jun 10
Jun 18
Jun 26
Jul 4
Jul 12
Jul 20
Jul 28
Aug 5
Aug 13
Aug 21
Aug 29
Sep 6
Sep 14
Sep 22
Sep 30
Oct 8
Oct 16
Oct 24
Nov 1
Nov 9
Nov 17
Nov 25
Dec 3
Dec 11
Dec 19
Dec 27

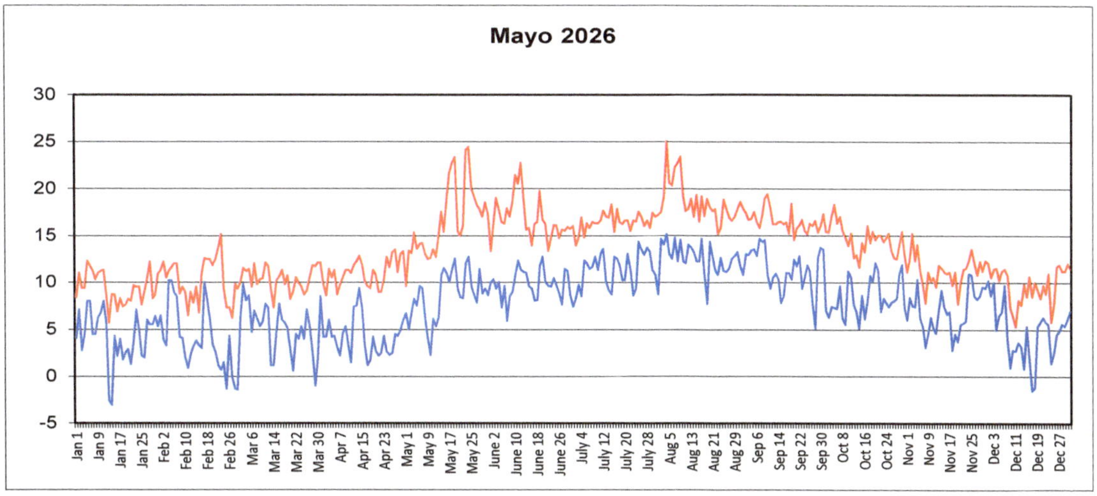

Mayo 2026
30
25
20
15
10
5
0
-5
Jan 1
Jan 9
Jan 17
Jan 25
Feb 2
Feb 10
Feb 18
Feb 26
Mar 6
Mar 14
Mar 22
Mar 30
Apr 7
Apr 15
Apr 23
May 1
May 9
May 17
May 25
June 2
June 10
June 18
June 26
July 4
July 12
July 20
July 28
Aug 5
Aug 13
Aug 21
Aug 29
Sep 6
Sep 14
Sep 22
Sep 30
Oct 8
Oct 16
Oct 24
Nov 1
Nov 9
Nov 17
Nov 25
Dec 3
Dec 11
Dec 19
Dec 27

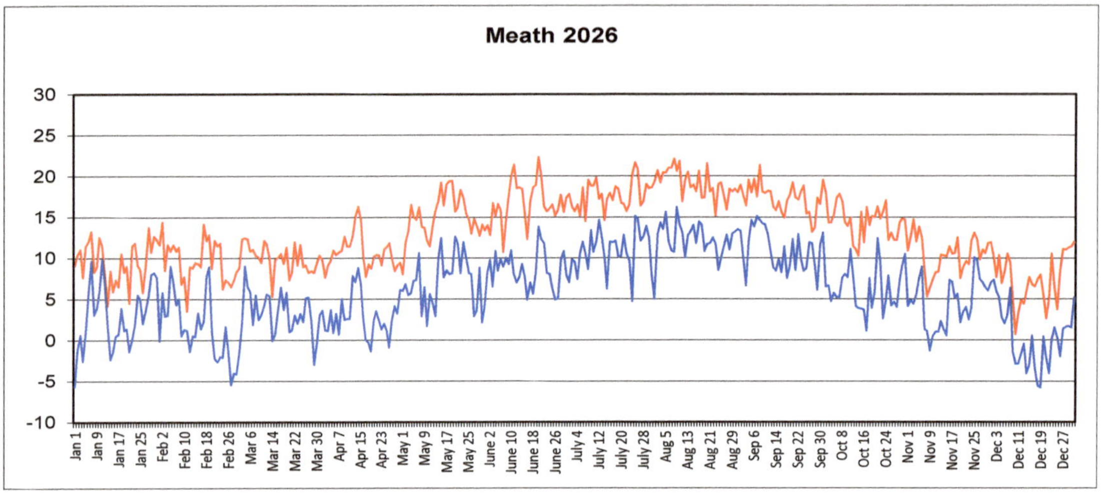
Meath 2026
30
25
20
15
10
5
0
-5
-10
Jan 1
Jan 9
Jan 17
Jan 25
Feb 2
Feb 10
Feb 18
Feb 26
Mar 6
Mar 14
Mar 22
Mar 30
Apr 7
Apr 15
Apr 23
May 1
May 9
May 17
May 25
June 2
June 10
June 18
June 26
July 4
July 12
July 20
July 28
Aug 5
Aug 13
Aug 21
Aug 29
Sep 6
Sep 14
Sep 22
Sep 30
Oct 8
Oct 16
Oct 24
Nov 1
Nov 9
Nov 17
Nov 25
Dec 3
Dec 11
Dec 19
Dec 27

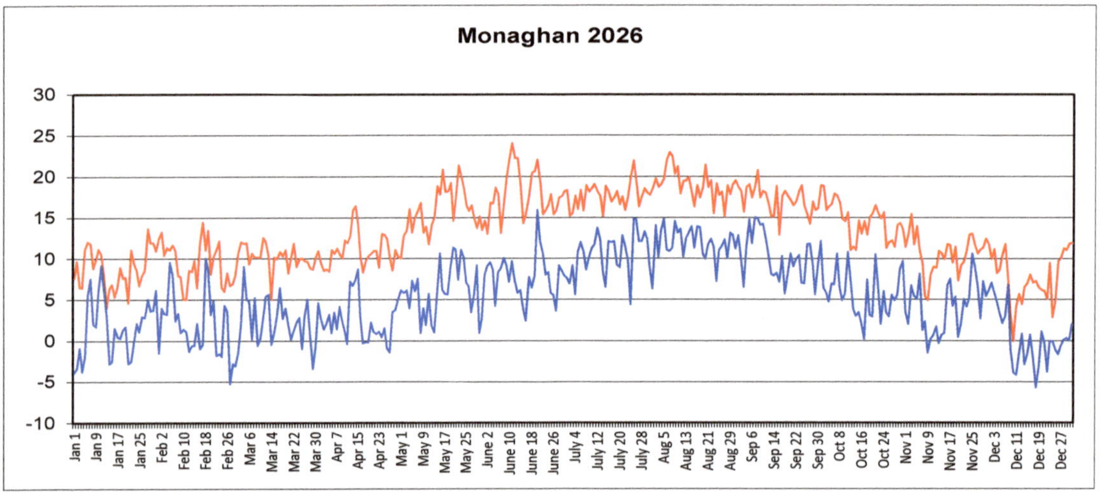
Monaghan 2026
30
25
20
15
10
5
0
-5
-10
Jan 1
Jan 9
Jan 17
Jan 25
Feb 2
Feb 10
Feb 18
Feb 26
Mar 6
Mar 14
Mar 22
Mar 30
Apr 7
Apr 15
Apr 23
May 1
May 9
May 17
May 25
June 2
June 10
June 18
June 26
July 4
July 12
July 20
July 28
Aug 5
Aug 13
Aug 21
Aug 29
Sep 6
Sep 14
Sep 22
Sep 30
Oct 8
Oct 16
Oct 24
Nov 1
Nov 9
Nov 17
Nov 25
Dec 3
Dec 11
Dec 19
Dec 27

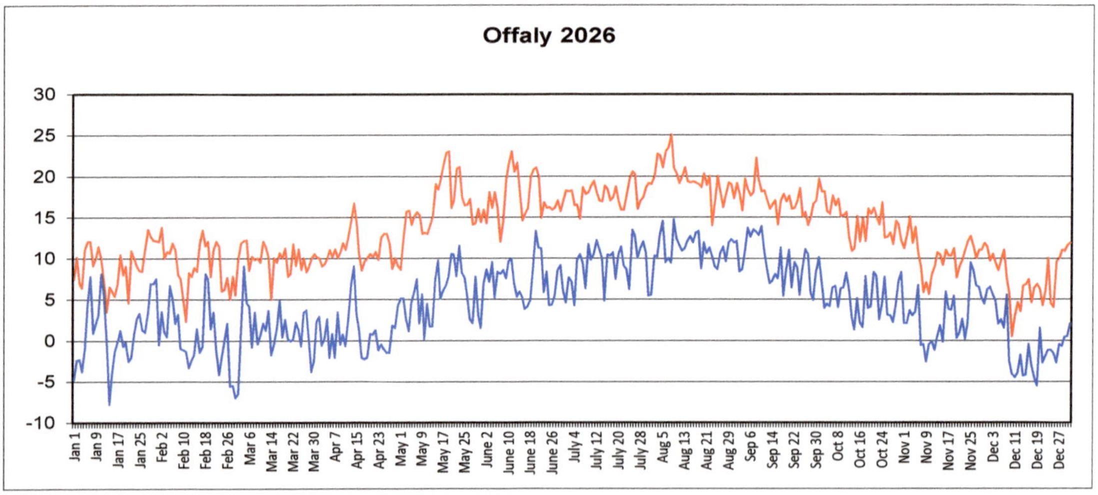
Offaly 2026
30
25
20
15
10
5
0
-5
-10
Jan 1
Jan 9
Jan 17
Jan 25
Feb 2
Feb 10
Feb 18
Feb 26
Mar 6
Mar 14
Mar 22
Mar 30
Apr 7
Apr 15
Apr 23
May 1
May 9
May 17
May 25
June 2
June 10
June 18
June 26
July 4
July 12
July 20
July 28
Aug 5
Aug 13
Aug 21
Aug 29
Sep 6
Sep 14
Sep 22
Sep 30
Oct 8
Oct 16
Oct 24
Nov 1
Nov 9
Nov 17
Nov 25
Dec 3
Dec 11
Dec 19
Dec 27

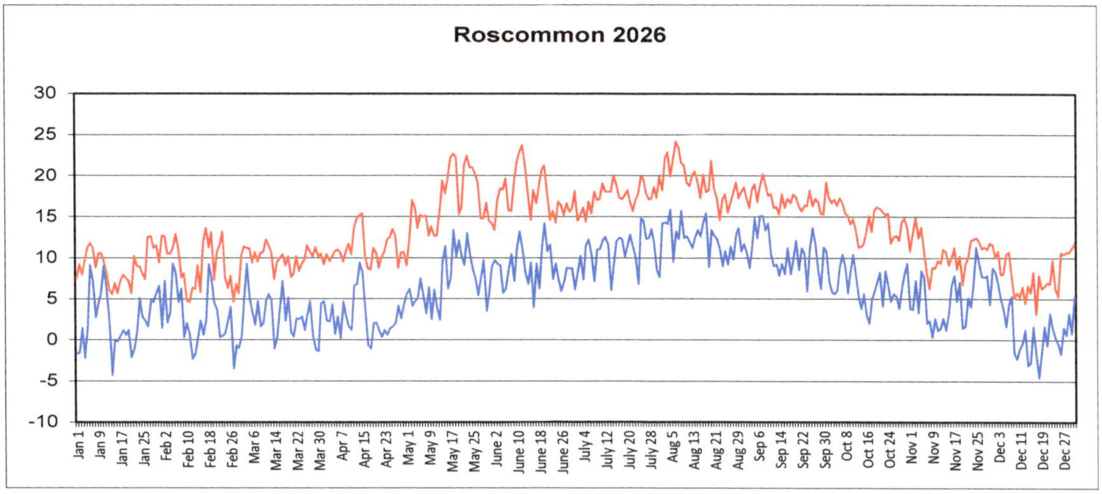
Roscommon 2026
30
25
20
15
10
5
0
-5
-10
Jan 1
Jan 9
Jan 17
Jan 25
Feb 2
Feb 10
Feb 18
Feb 26
Mar 6
Mar 14
Mar 22
Mar 30
Apr 7
Apr 15
Apr 23
May 1
May 9
May 17
May 25
June 2
June 10
June 18
June 26
July 4
July 12
July 20
July 28
Aug 5
Aug 13
Aug 21
Aug 29
Sep 6
Sep 14
Sep 22
Sep 30
Oct 8
Oct 16
Oct 24
Nov 1
Nov 9
Nov 17
Nov 25
Dec 3
Dec 11
Dec 19
Dec 27

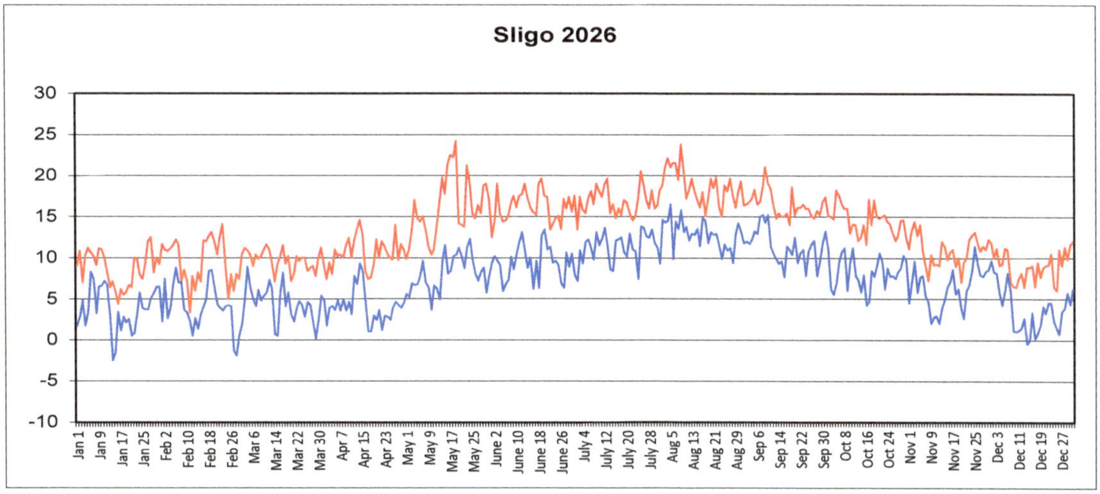
Sligo 2026
30
25
20
15
10
5
0
-5
-10

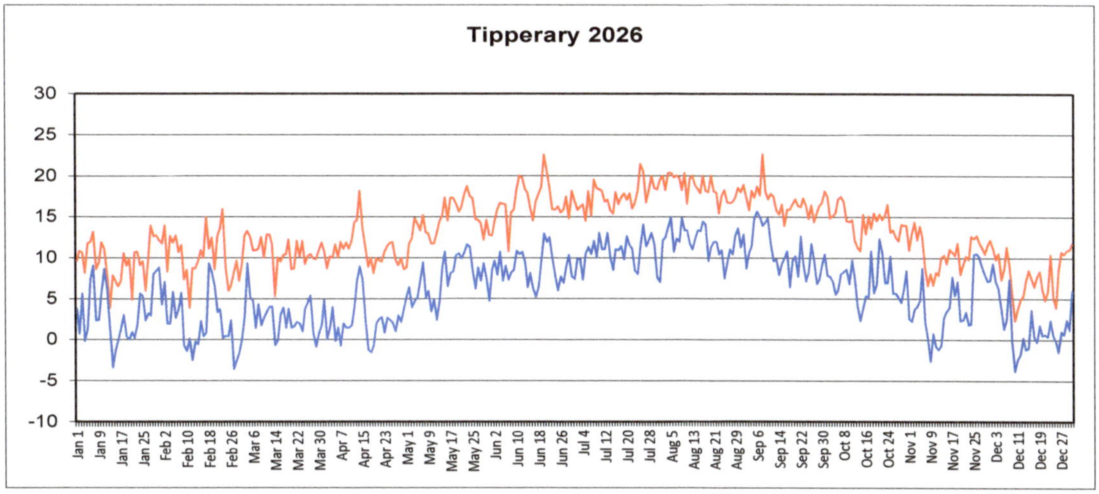
Tipperary 2026
30
25
20
15
10
5
0
-5
-10

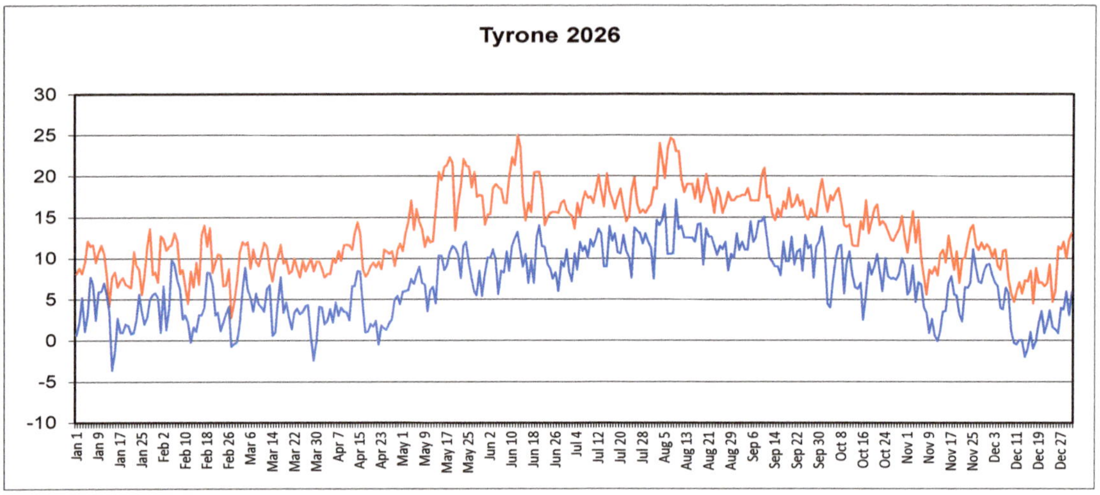
Tyrone 2026
30
25
20
15
10
5
0
-5
-10
Jan 1
Jan 9
Jan 17
Jan 25
Feb 2
Feb 10
Feb 18
Feb 26
Mar 6
Mar 14
Mar 22
Mar 30
Apr 7
Apr 15
Apr 23
May 1
May 9
May 17
May 25
Jun 2
Jun 10
Jun 18
Jun 26
Jul 4
Jul 12
Jul 20
Jul 28
Aug 5
Aug 13
Aug 21
Aug 29
Sep 6
Sep 14
Sep 22
Sep 30
Oct 8
Oct 16
Oct 24
Nov 1
Nov 9
Nov 17
Nov 25
Dec 3
Dec 11
Dec 19
Dec 27

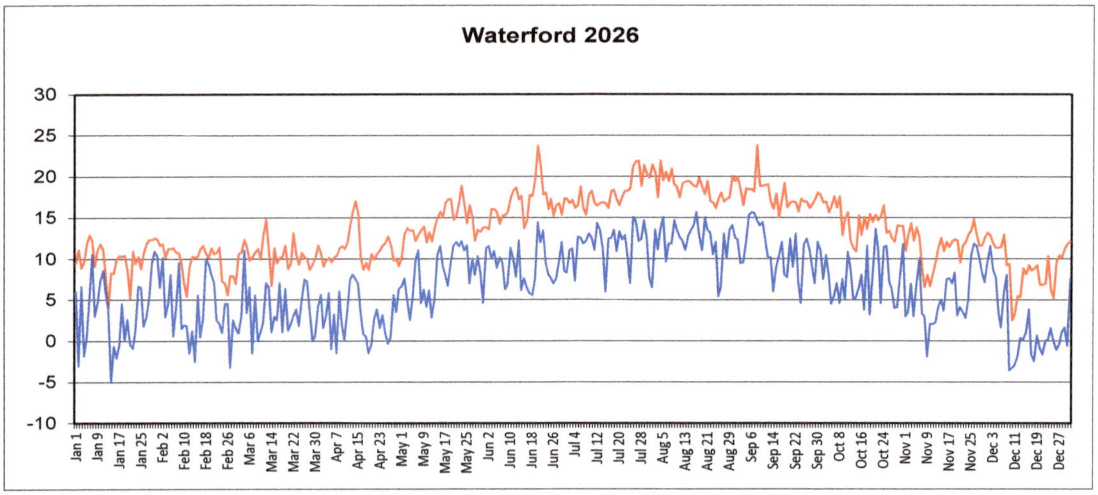
Waterford 2026
30
25
20
15
10
5
0
-5
-10
Jan 1
Jan 9
Jan 17
Jan 25
Feb 2
Feb 10
Feb 18
Feb 26
Mar 6
Mar 14
Mar 22
Mar 30
Apr 7
Apr 15
Apr 23
May 1
May 9
May 17
May 25
Jun 2
Jun 10
Jun 18
Jun 26
Jul 4
Jul 12
Jul 20
Jul 28
Aug 5
Aug 13
Aug 21
Aug 29
Sep 6
Sep 14
Sep 22
Sep 30
Oct 8
Oct 16
Oct 24
Nov 1
Nov 9
Nov 17
Nov 25
Dec 3
Dec 11
Dec 19
Dec 27

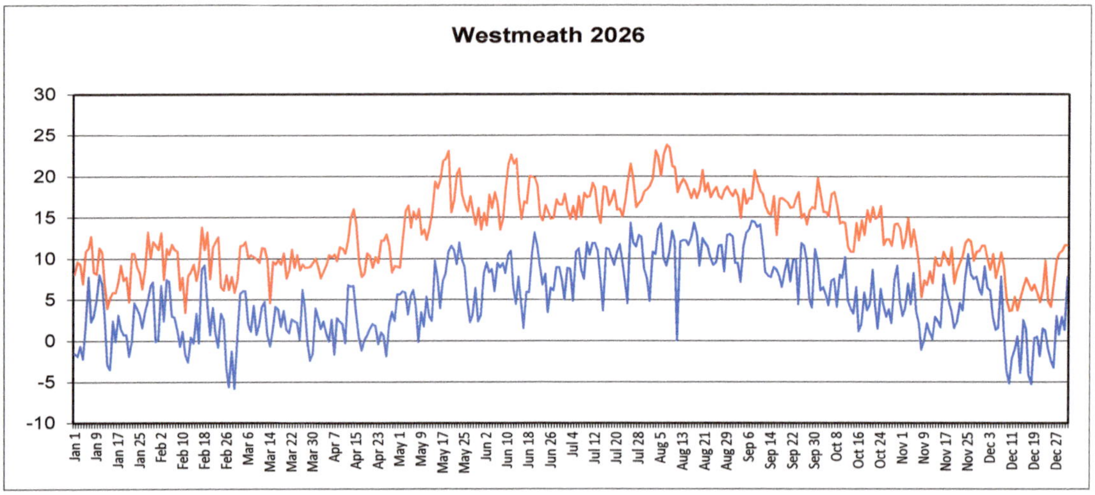
Westmeath 2026
30
25
20
15
10
5
0
-5
-10
Jan 1
Jan 9
Jan 17
Jan 25
Feb 2
Feb 10
Feb 18
Feb 26
Mar 6
Mar 14
Mar 22
Mar 30
Apr 7
Apr 15
Apr 23
May 1
May 9
May 17
May 25
Jun 2
Jun 10
Jun 18
Jun 26
Jul 4
Jul 12
Jul 20
Jul 28
Aug 5
Aug 13
Aug 21
Aug 29
Sep 6
Sep 14
Sep 22
Sep 30
Oct 8
Oct 16
Oct 24
Nov 1
Nov 9
Nov 17
Nov 25
Dec 3
Dec 11
Dec 19
Dec 27

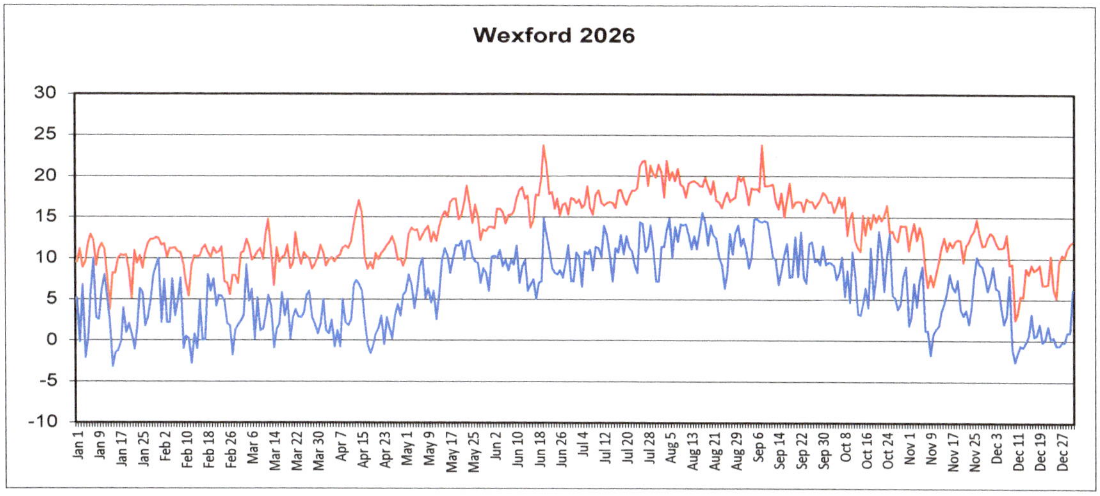

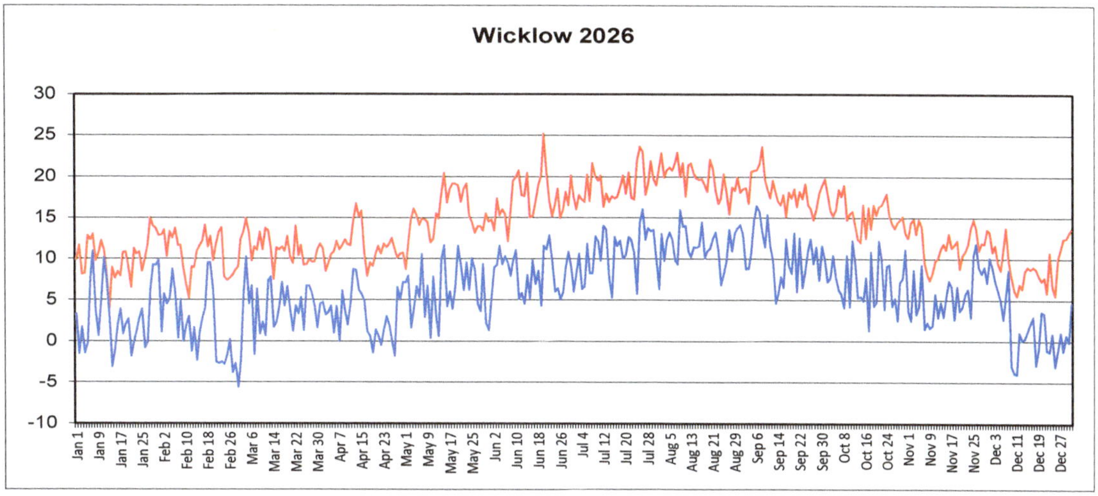

TEMPERATURE TRENDS

AVERAGED ACROSS THE WHOLE COUNTRY

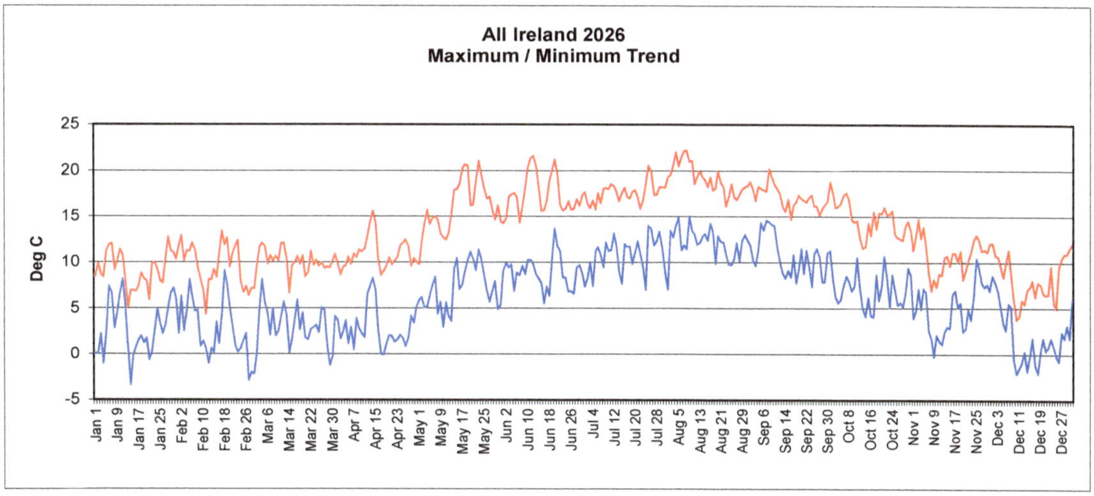

WIND

The graph below shows estimated daily wind force trends, for 2026, as calculated for Knock Airport, County Mayo. This western county has been chosen because of the availability of continuous historical data, and because most wind systems in Ireland come from the west. Other areas may show corresponding trends even if dissimilar force values.

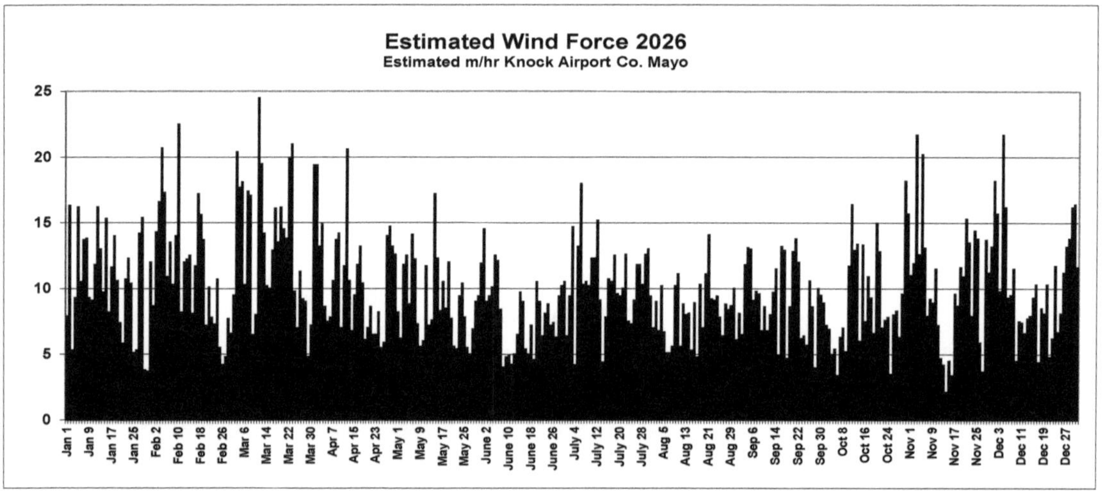

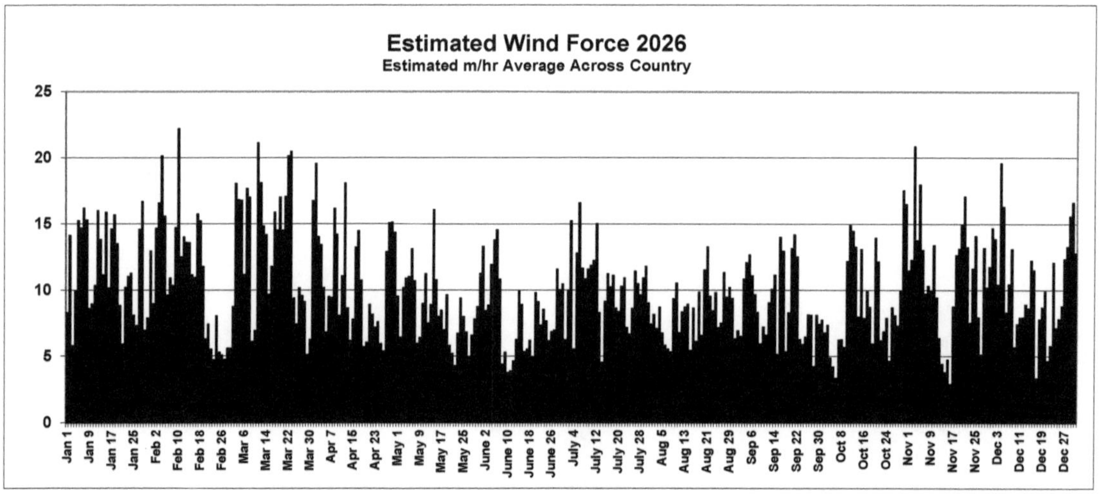

Averaging wind force values across the months reveals that March may be the windiest followed by January and February, with May, August and June the calmest. The order of overall average windiest months is estimated to be March, January, February, April, November, December, July, October, September, May, August, then June.

This makes the average windiest season, in order of windiest, winter, spring, autumn, then summer.

PRECIPITATION (RAIN/SNOW)

WHEN WILL RAIN AND LOW TEMPERATURES PRODUCE SNOW?

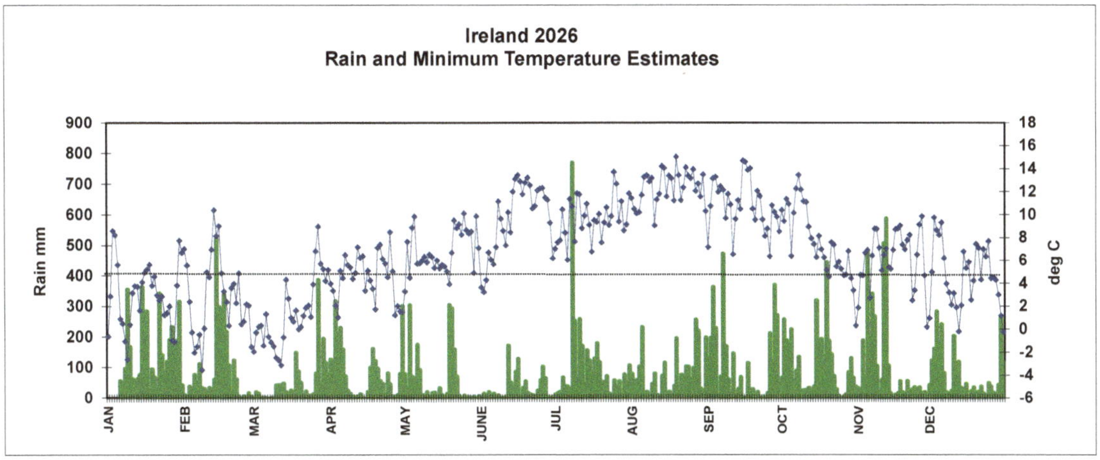

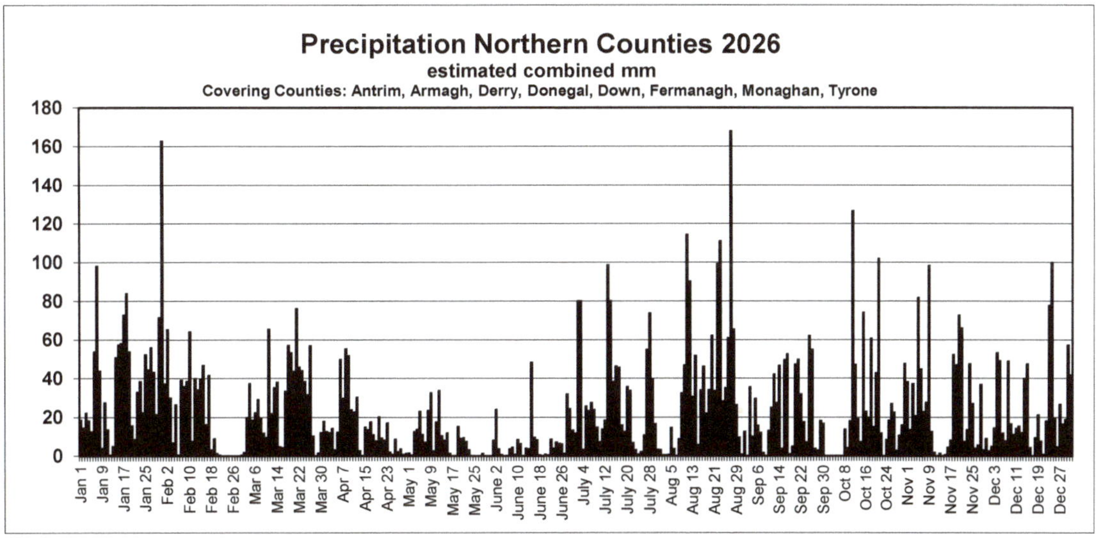

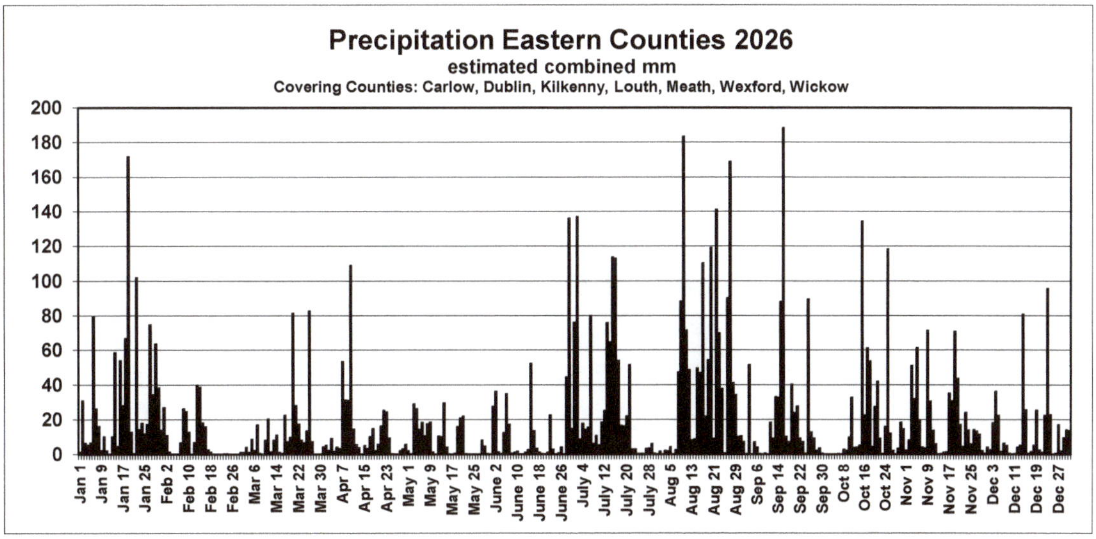

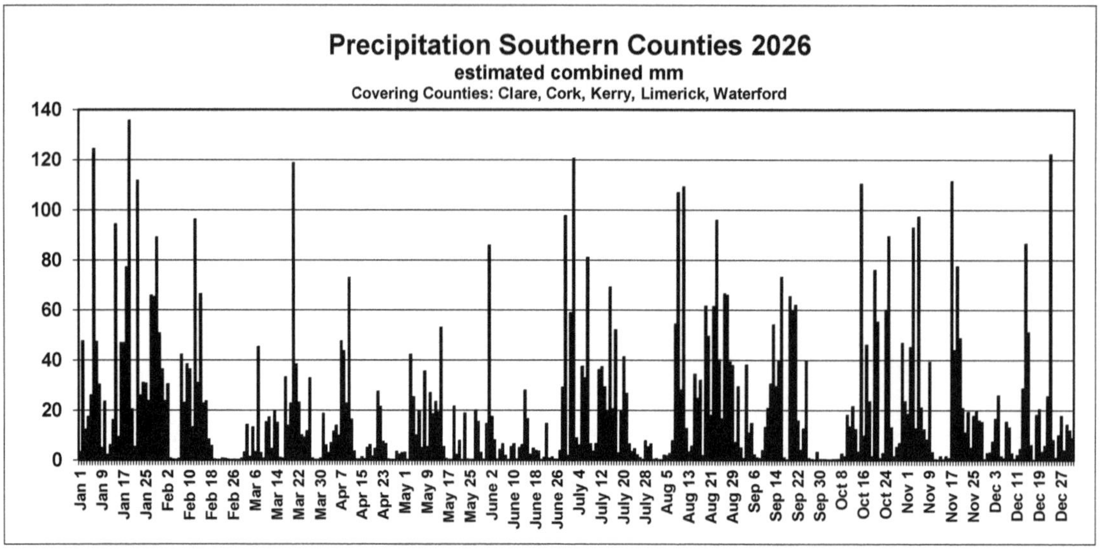
Precipitation Southern Counties 2026
estimated combined mm
Covering Counties: Clare, Cork, Kerry, Limerick, Waterford
0 20 40 60 80 100 120 140
Jan 1 Jan 9 Jan 17 Jan 25 Feb 2 Feb 10 Feb 18 Feb 26 Mar 6 Mar 14 Mar 22 Mar 30 Apr 7 Apr 15 Apr 23 May 1 May 9 May 17 May 25 June 2 June 10 June 18 June 26 July 4 July 12 July 20 July 28 Aug 5 Aug 13 Aug 21 Aug 29 Sep 6 Sep 14 Sep 22 Sep 30 Oct 8 Oct 16 Oct 24 Nov 1 Nov 9 Nov 17 Nov 25 Dec 3 Dec 11 Dec 19 Dec 27

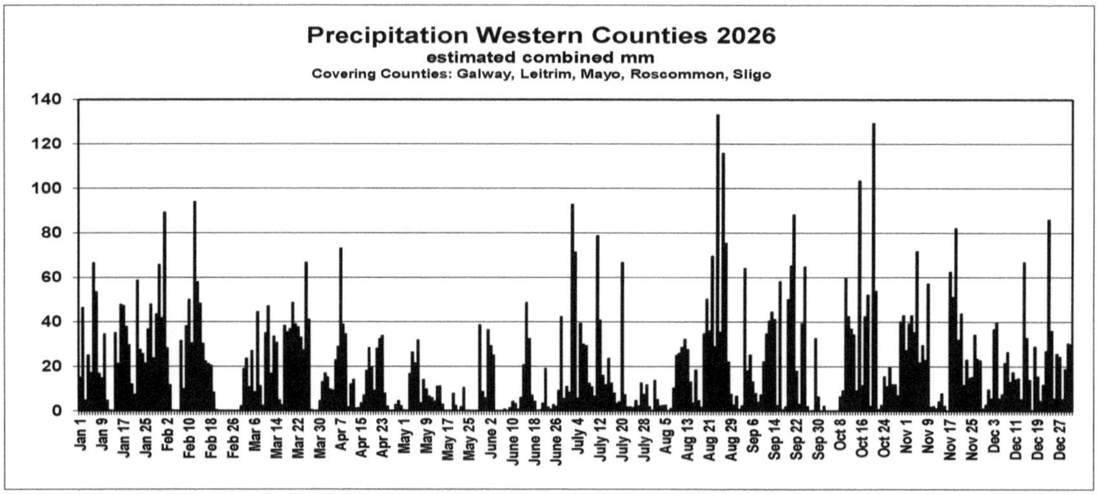
Precipitation Western Counties 2026
estimated combined mm
Covering Counties: Galway, Leitrim, Mayo, Roscommon, Sligo
0 20 40 60 80 100 120 140
Jan 1 Jan 9 Jan 17 Jan 25 Feb 2 Feb 10 Feb 18 Feb 26 Mar 6 Mar 14 Mar 22 Mar 30 Apr 7 Apr 15 Apr 23 May 1 May 9 May 17 May 25 June 2 June 10 June 18 June 26 July 4 July 12 July 20 July 28 Aug 5 Aug 13 Aug 21 Aug 29 Sep 6 Sep 14 Sep 22 Sep 30 Oct 8 Oct 16 Oct 24 Nov 1 Nov 9 Nov 17 Nov 25 Dec 3 Dec 11 Dec 19 Dec 27

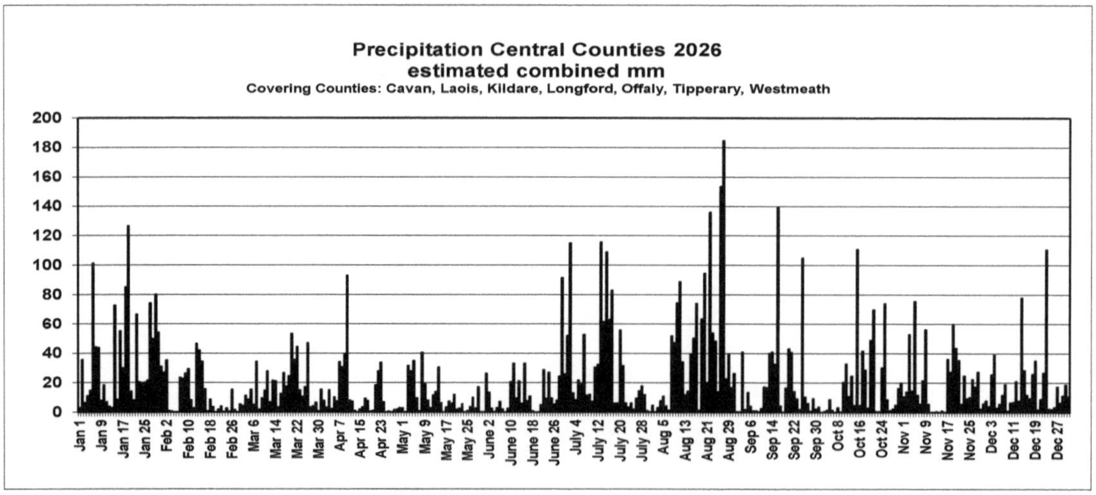
Precipitation Central Counties 2026
estimated combined mm
Covering Counties: Cavan, Laois, Kildare, Longford, Offaly, Tipperary, Westmeath
0 20 40 60 80 100 120 140 160 180 200
Jan 1 Jan 9 Jan 17 Jan 25 Feb 2 Feb 10 Feb 18 Feb 26 Mar 6 Mar 14 Mar 22 Mar 30 Apr 7 Apr 15 Apr 23 May 1 May 9 May 17 May 25 June 2 June 10 June 18 June 26 July 4 July 12 July 20 July 28 Aug 5 Aug 13 Aug 21 Aug 29 Sep 6 Sep 14 Sep 22 Sep 30 Oct 8 Oct 16 Oct 24 Nov 1 Nov 9 Nov 17 Nov 25 Dec 3 Dec 11 Dec 19 Dec 27

RAIN TRENDS 2026 -2027

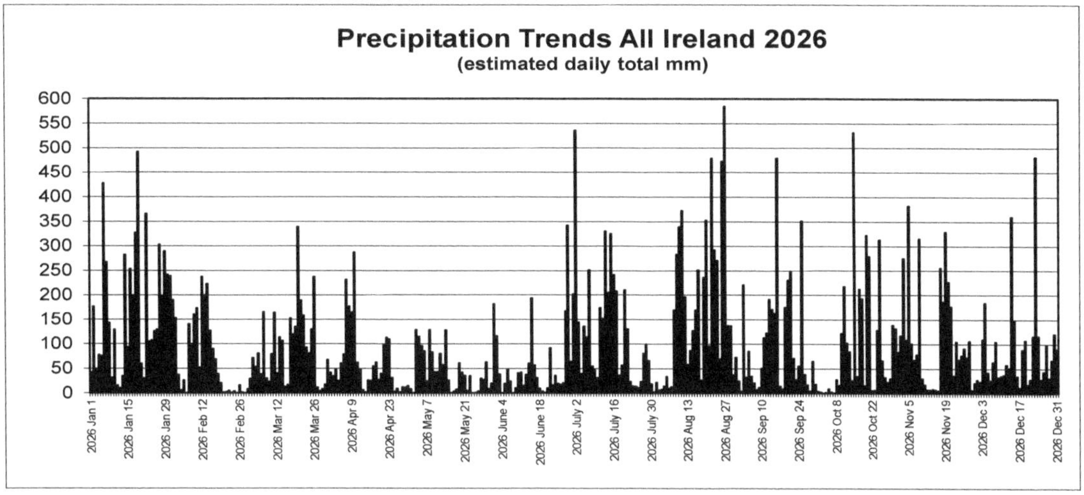

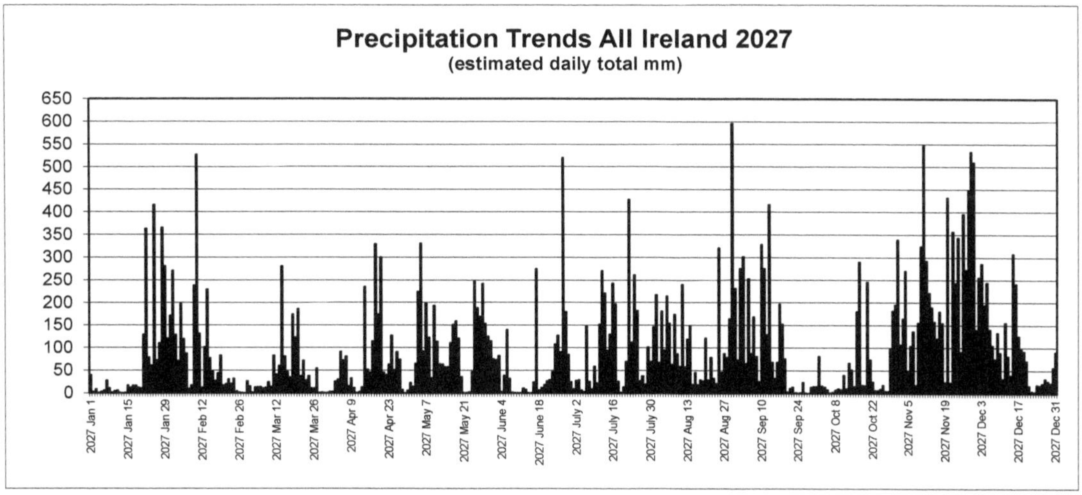

WHEN WILL THE SUN BE OUT?
AVERAGED ACROSS THE WHOLE COUNTRY.

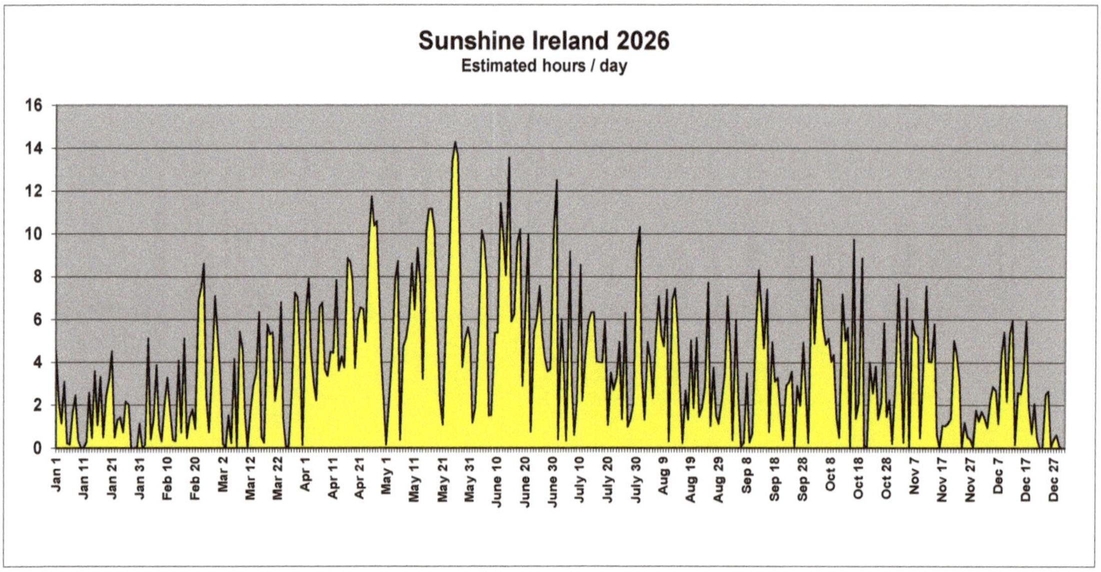

GRAPHS, NOTES, AND TRENDS

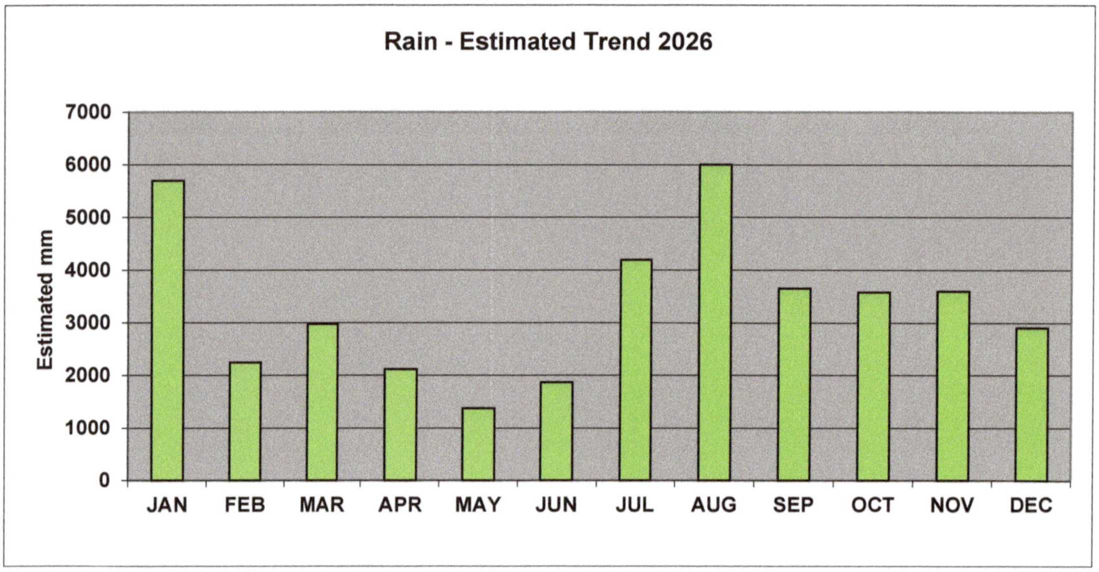

The **most precipitation** for the whole country in 2026 may be **August followed by January and then July, with May, June and February** likely to be the driest. The largest amount of precipitation in one region/month may be in January and in August, particularly in the northern and western counties. The driest months for the year may be in May and June almost everywhere, but particularly in the northern and western counties.

.

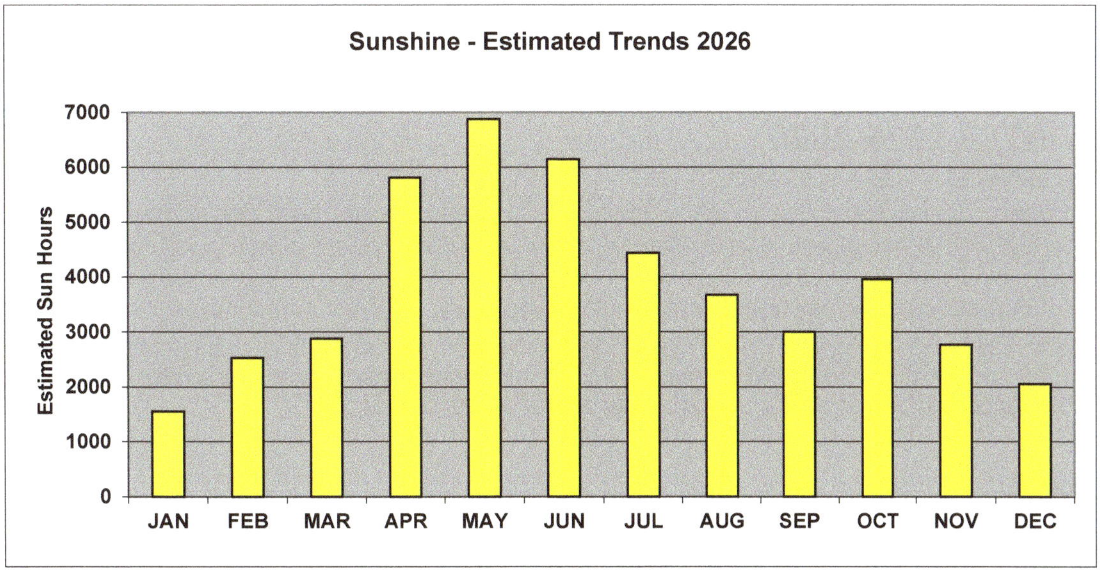
Sunshine - Estimated Trends 2026
Estimated Sun Hours
7000
6000
5000
4000
3000
2000
1000
0
JAN
FEB
MAR
APR
MAY
JUN
JUL
AUG
SEP
OCT
NOV
DEC

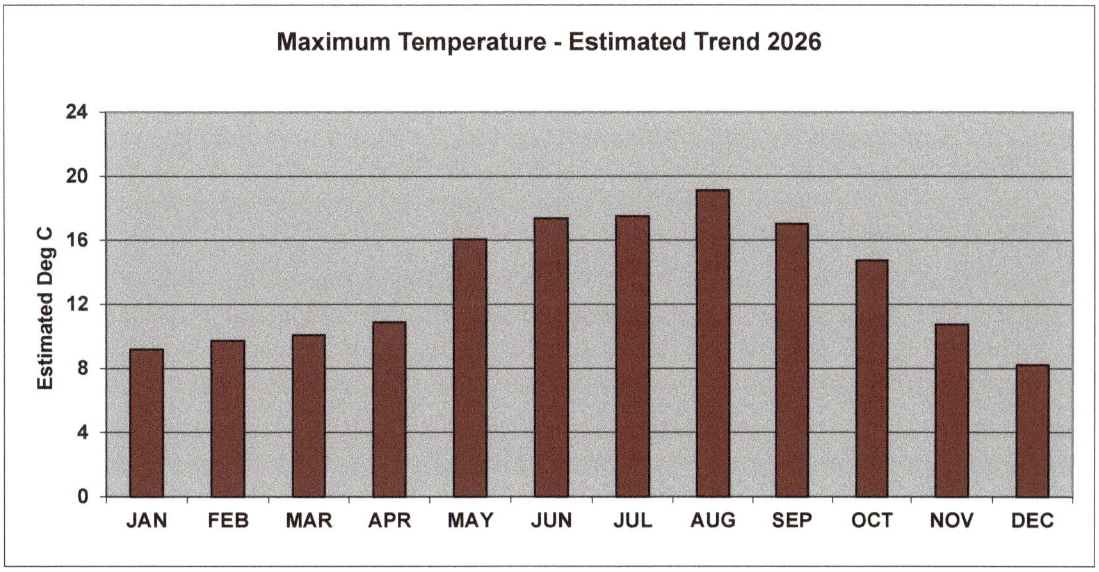
Maximum Temperature - Estimated Trend 2026
Estimated Deg C
24
20
16
12
8
4
0
JAN
FEB
MAR
APR
MAY
JUN
JUL
AUG
SEP
OCT
NOV
DEC

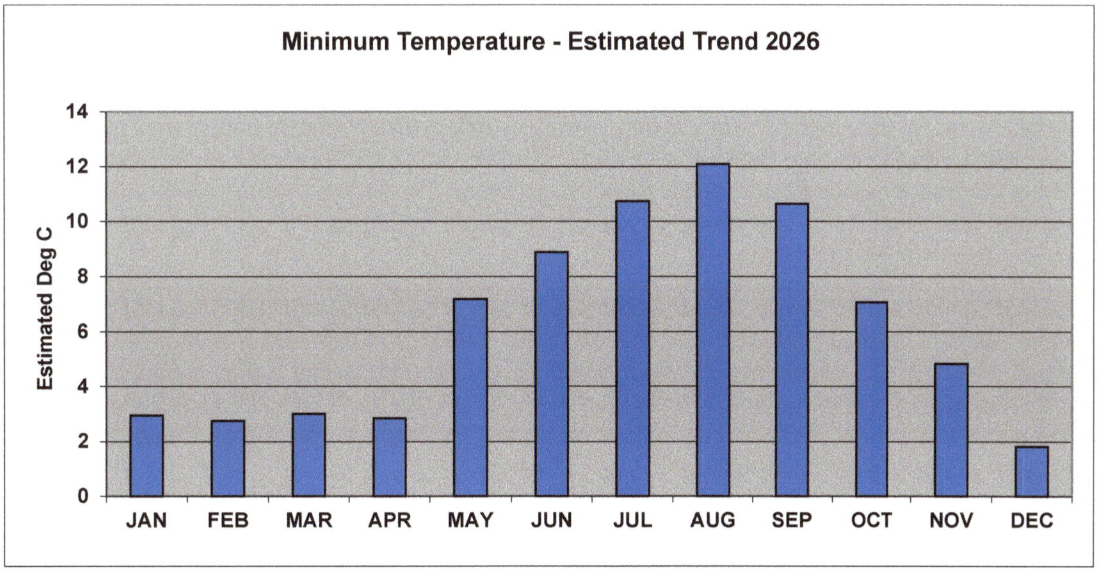
Minimum Temperature - Estimated Trend 2026
Estimated Deg C
14
12
10
8
6
4
2
0
JAN
FEB
MAR
APR
MAY
JUN
JUL
AUG
SEP
OCT
NOV
DEC

SEASONAL OVERVIEWS

WINTER 2025-2026

(December 2025 – February 2026)

Many districts may experience an unusually wetter than normal winter, with many districts receiving around 20-30% more than normal rainfall than an average winter except in eastern parts around Dublin which may receive closer to their usual winter average rainfall amounts. Despite this, winter sunshine hours may be above average for the season almost everywhere, with many areas possibly up to 15-20% sunnier than what is average for winter and accompanying the extra sunshine hours, temperatures may also be warmer than normal. Winds may also be above normal almost everywhere with winter quite stormy particularly during mid-January and late February, into early March.

RAINFALL:

The number of rain days (as measured at 0.2mm or more recorded each day) may be above average with the country as a whole averaging between 65-75 rain days compared to the norm for a typical winter of between 50-65 rain days. Overall, the winter season may be up to 25-30% wetter than average across the country as a whole, with the far southwest of Munster up to 120% wetter than average, while the southeast only around 10% wetter. Overall, the Munster Province may average around 44% wetter, the Connaught Province around 28% wetter, Ulster averaging around 18% wetter, and the Leinster Province closer to the winter average but still slightly wetter by around 12% more than normal. Despite most parts of the country likely to receive above average rainfall this season, some parts of the Leinster Province, particularly in the Dublin area, may go against the trend and actually receive almost average amounts of rainfall typical for the winter season.

The winter season may begin relatively dry at first with little rainfall over the last week of December 2025. However, January may see rain almost daily in most parts of the country, with very heavy rain at the end of the first week of January and during a particularly stormy spell during the second half of the month, daily rain may cause some localised flooding in the west and south of the country at times.

In contrast, most of February's precipitation may fall over the first half of the month, with the last ten days being mostly dry everywhere. Finally, winter may end on a wet note with some rain almost daily during the first three weeks of March up to and on the New Moon around the 20th, when heavy rain may prevail in many parts of the country.

SUNSHINE:

By the end of winter, expected sunshine hours will be around 15-20% above average for this season for the country as a whole, with only the western coastal fringes of Connaught cloudier than average but only by around 5% less than the seasonal norm. All areas except around the Bay of Donegal may expect above average sunshine, with most areas sunnier by around 15-20% and only the Sligo area slightly cloudier but only by around 5-10% below the seasonal norm. Overall, the Ulster Province may average around 20-25% sunnier, the Leinster and Connaught Provinces around 10-15% sunnier, and the Munster Province areas around 15-20% sunnier.

The winter breakdown by month may see December at around 30% sunnier than normal, January slightly cloudier but still within 5% of the norm, and February sunnier once more by around 20% more than average. Most days in January may be very dull and gloomy, with the sunniest days likely at the end of the third week, midway between the New Moon and the First Quarter. The sunniest spells in February may be in the last week of the month, either side of the First Quarter. The dullest spell of the winter season may be over the second and fourth weeks of January.

TEMPERATURE:
Mean air temperatures will be generally around 1.5–2°C warmer than normal for the season overall, with maximum both maximum and minimum temperatures averaging around 1.5-2°C more than average. However, some parts in the south of the country may be up to 2.5°C warmer for average minimum temperatures, and eastern parts only around 1°C warmer for minimum temperatures. There may be a very short spell of particularly cold weather around mid-January and at the end of February, when frosts may be severe in many areas.

***Maximum Temperatures*:** The warmest spell in December 2025 may be over the first half of the month, between the Full Moon and the New Moon, while the coolest period may be in the last few days of December, from the First Quarter to New Year's Eve. January 2026 may be the warmest towards the end of the first week for a couple of days, and again in the last 5 days of the month, while coolest mid-month between Apogee and the New Moon, from the end of the second week to early in the third week. February may be the mildest between the New Moon and the First Quarter over the third week and early into the fourth week, while coolest midway through the second week, between Apogee and Southern Declination, and again in the last few days of the month after Northern Declination.

***Minimum Temperatures*:** The mildest spell in December 2025 may be in the second half of the first week of December just after the Full Moon, and again briefly at the start of the third week, and again a couple of days before Christmas Day. The coldest spells may be midway through the first week just before and on the Full Moon, and again at the New Moon at the end of the third week and finally, in the last 5 days of the month from the First Quarter to the end of the month. The mildest spell in January may be over the second week, in the days up to and over the Third Quarter.

The coldest spells in January may be in the first half of the first week either side of the Full Moon, again between Apogee and Southern Declination at the end of the second week into the start of the third week, and finally just before Perigee at the end of the third week into the start of the fourth week. February may be the coldest over the second half of the second week between Apogee and the New Moon, and again in the last week of the month for several days either side of the First Quarter. In contrast, the mildest minimum temperature spells may be towards the end of the first week of February for a couple of days, and again immediately after the New Moon midway through the third week.

Frosts: There may be up to 30-35 days with frosts observed during the winter season, as well as up to 3 days with air frosts around the coasts.

SPRING 2026

(March – May)
Spring may see near average rainfall amounts overall with the west of the country generally wetter than average, while the east and north drier than average. The spring season may also be around 10-15% sunnier than average while temperatures overall nearer to average. The windiest period may be in the first few weeks of spring, in the second half of March. Spring may be the windiest around mid-March near the start of the third week, particularly in the northwest.

RAINFALL:
The number of rain days (as measured at 0.2mm or more recorded each day) may be around 75-80 days for the country as a whole, compared to the average range for spring of between 35-40 wet days. Despite the higher than average number of rain days, the rainfall quantities may be less than normal resulting in a drier than average spring in the east and north, while slightly wetter but still within 10% of the seasonal norm in the west and south. Overall, the spring season may average at around 2% drier than average across the country as a whole, with the Ulster Province around 7% drier and the Leinster Province up to 20% drier than normal. In contrast, the Connaught and Munster Provinces may average around 5-6% wetter, with a mix of above and below average rainfall areas throughout the regions.

The far northwest and southwest areas of Ireland may be up to 30-50% wetter than average, while central areas may be up to 30% drier than average. Looking at the spring months individually, April and May are expected to be generally drier than average except in the west, while March may be very wet in many areas. The wettest period of spring may be in the second half of March between the New Moon and the First Quarter, while the driest spell may be in the second half of May after the New Moon.

SUNSHINE:

Overall, sunshine hours may follow a similar trend to winter and be above normal for the season, averaging around 10-15% sunnier than normal. Nearly all areas may have a sunnier season overall, with the Ulster, Connaught, and Leinster Provinces all averaging around 10-15% sunnier, while the Munster Province may average around 5-10% sunnier. Only the area in the Cork district may see below average sunshine this season overall, but still only cloudier by around 5% below the norm. March may be cloudier by an average of 10-15% less than normal sunshine, while April and May both may be sunnier by around 20% more than average.

The sunniest period in March may be from midway through the third week to midway through the fourth week, as well as in the last few days of the month, with the cloudiest spell in March expected over the first week and either side of the First Quarter in the last week. April may be very sunny on most days, particularly over the second half of the month, with the cloudiest spell expected midway through the first and second weeks, and in the east and south at the very end of the month. May should be very dull in the first couple of days, at the end of the first week, and at the end of the third week, with good sunny days dominating most of the remaining month.

TEMPERATURE:

Overall, spring temperatures may be close to average for this season, with maximum temperatures averaging between 0.5-1°C warmer in the Connaught Province, while the Ulster slightly warmer and Leinster and Munster slightly cooler but all still within half a degree of average. Similarly, minimum temperatures may also be close to average overall for the season, with the Munster Province possibly up to 0.5-1°C warmer than normal, Leinster Province cooler by a similar variation, and Ulster and Connaught provinces both slightly cooler but still within half a degree of their seasonal norm. Generally, March may see near average temperatures for the month, April may be much colder for both maximum and minimum temperatures throughout, and May around 0.5-1.5°C warmer overall. Temperatures may reach up to 20°C in some inland and western areas during the second half of May, while the coldest period of spring may be in the last week of March and into early April, when frosts may be widespread and down to -10°C in some parts.

***Maximum Temperatures*:** March may bring average maximum temperatures across the country with most districts within half a degree of the monthly norm, while April may be up to 1-1.5°C colder than average and May up to 1.5°C warmer than normal. The warmest spell during March may be midway through the first week around the Full Moon, and again towards the end of the second week, from the Third Quarter for a few days.

The coolest spell during March may be right at the start of the month brief, and again at the end of the second week, and finally towards the end of the third week, just after the New Moon. April may be warmest over the second week and again in the last week for a few days after the First Quarter, while coolest in the first couple of days at the Full Moon and then again at the New Moon for a few days midway through the third week. May maximum temperatures may be the warmest after the New Moon over the third week and into the first half of the fourth week at the First Quarter. In contrast, May's coolest spells are likely over the first couple of days at the Full Moon again and midway through the second week from the Third Quarter and during the next few days.

Minimum Temperatures: March may be around 0.5°C cooler than average in the Leinster Province, while Munster up to 1°C warmer, and elsewhere slightly warmer but still within half a degree of average. April may be much cooler throughout the country, averaging around 1-1.5°C below normal across the country. May warms once more to averaging around 0.5°C warmer than average overall, with some parts in the Connaught and Munster Provinces up to 1-1.5°C milder.

Overall, the coldest minimum temperatures may be expected in March over the first couple of days and then again at the end of the second week, and finally in the last few days of the month, while mildest midway through the first week around the Full Moon, and again midway through the second week, just after the Third Quarter. April may be the mildest in the second half of the second week between the Third Quarter and the New Moon, while coldest for a couple of days at the New Moon and in at the end of the first week around Apogee. During the month of May, minimum temperatures may be at their lowest over the second half of the second week for a few days after the Third Quarter, while at their mildest over the third week and into the first half of the fourth week, between the New Moon and the First Quarter.

Frosts: March may see up to 12-15 days with frosts, April 10 days and May 4-5 days with frosts. Overall, there may be up to 45-50 days with ground frosts during the season, which is more than the average of 32 days.

SUMMER 2026

(June – August)

This summer will be unusually wetter than normal, with most districts receiving above normal rain, averaging around 45-50% wetter across the country overall, with some parts in the east possibly receiving up to twice as much rainfall than average, while only around 20-30% wetter in the north. Sunshine hours however, may be close to average for the summer season, averaging around 5-10% cloudier than normal, except in the west, which may be sunnier by a similar variation. Similarly, while temperatures may also be overall near to average overall, most areas may see temperatures still within 10% of what is typical for this time of year. Summer may be the windiest in mid-July towards the end of the second week, particularly in the northwest.

RAINFALL:

The number of rain days (as measured at 0.2mm or more recorded each day) may be around 80 days across the country as a whole, compared to the summer average of between 30-40 days. Overall, the Ulster Province may receive up to 25-30% more rainfall than average and the Connaught Province averaging around 30% wetter. The Munster Province may be up to 60% wetter this summer, and the Leinster Province up to 80-85% wetter overall. Some districts in the Leinster Province may see up to twice as much rainfall than average for the season, particularly in the Dublin and Edenderry areas. Similarly, the far southwest of Munster may also receive up to twice as much rainfall than usual for the summer season.

Only a small area around the Donegal Bay region may see near average rainfall for the season as a whole, but may still be wetter but within 10% of the norm. Rainfall amounts may progress as summer continues, with June seeing the least rain, July up to twice as much rain as June, but still mostly dry over the last week of the month. August however, may be the wettest month of the season, particularly over the second half of August, which may see heavy rain in some parts of the country on several days, with the wettest spell of summer likely in the last fortnight of August. During this final wet spell of summer in the latter half of August, heavy rain may again cause the risk of some local flooding, particularly between the First Quarter and the Full Moon.

SUNSHINE:

Overall, summer may see slightly cloudier than normal conditions by the end of the season, but still averaging with 5% of the norm. It may be cloudier almost everywhere, particularly in the east and south. By Provinces, Ulster, Leinster, and Munster may all average around 5% cloudier than normal, while the

Connaught Province may be slightly sunnier by a similar variant. Looking at the season by monthly breakdown, June may be up to 20% sunnier than average, July around 5% cloudier than normal, and August up to 20-25% cloudier. June may be cloudiest in the first couple of days of the month and over the last couple of days of the first week, briefly early in the fourth week, and finally in the last few days of the month. The sunniest spells in June may be between the Third quarter and New Moon during the second and third weeks. July may be cloudiest over the first and fourth weeks, while sunniest over the third week and in the last couple of days of the month. August may be the sunniest over the first half of the month, cloudiest over the second half of the month, and particularly gloomy early in the third week and start of the fourth week.

TEMPERATURE:

This summer temperatures may be close to average for the season overall. Maximum temperatures may be warmer than average by around 1°C in August, below average by a similar variant in July. June however, may be slightly warmer but within 1°C of the norm in the north and west, while around 0.5-1°C cooler in the east, and the southern parts generally average overall. Minimum temperatures however, may be slightly cooler but within half a degree of the norm in June and July, while August may be warmer than average by around 1-1.5°C above the average, resulting in the summer season minimum temperatures closer to the average overall. The warmest spell of summer may be mostly over the first half of August, when maximum temperatures may rise above 20°C in most districts, but overall only around 10 days throughout the summer season may rise higher than 20°C overall.

***Maximum Temperatures*:** The warmest spell in June may be over the second half of the second week between the Third Quarter and the New Moon, and again just before the First Quarter, in the second half of the third week. The coolest spell of June may be in the first couple of days of the month. July may be the warmest in the fourth week, over Apogee and Southern Declination while coolest in over the first week and early in the fourth week just after the First Quarter. The warmest spell of August (as well as the year overall), may be over the first ten days of August, either side of the Third Quarter and before the New Moon. The coolest spell of August may be in the second half of the third week just before the First Quarter, and again in the last few days of the month, either side of the Full Moon.

***Minimum Temperatures*:** The mildest spells in June may be either side of the First Quarter, from the end of the third week to early in the fourth week, as well as briefly midway through the second week. The coolest spells in June may be for a few days after the New Moon and Perigee midway through the third week, and again just before Southern Declination towards the end of the fourth week. July may be coolest over the first half of the first week, again for a couple of days at the New Moon at the turn into the third week, and just before Apogee early in the fourth week. July may be the warmest over most of the second week, and again from Apogee to the Full Moon in the last week of the month. August may be the warmest midway through the first week just before the Third Quarter, and midway through the second week just before Perigee, while coolest briefly right at the start of the month and then again between Apogee and the Full Moon over the fourth week of the month.

Frosts: No frosts may occur over this summer season, with only the coldest spell in mid-June around the New Moon having the highest potential for any frosts. After this, no further risks of frosts are anticipated until the first week of October.

AUTUMN 2026

(September – November)

Autumn may be wetter overall by an average of 10% more than normal, with the east and north around 5% wetter than average, and the south and west wetter by up to 20% more. Despite this, sunshine hours may also be above average everywhere, averaging around 15-20% sunnier than normal and temperatures may also be slightly warmer than the seasonal average. Autumn may be the windiest in the second half of November.

RAINFALL:
The number of rain days (as measured at 0.2mm or more recorded each day), may be around 80 days across the country as a whole, compared to the autumn average of between 35-55 wet days. However, like in spring, despite the number of rain days being more than average, the overall rainfall amounts may be less than average overall, particularly in the north and central parts. Overall, the Ulster Province may see average rainfall for the season, with below average rain in the east, and above average rain in the west, but most districts still within 10% of their seasonal norms. Further south, a similar pattern of below average and above average rainfall amounts may prevail through the Leinster Province but also averaging within 5% of the norm overall for the region.

In contrast, the Connaught and Munster Provinces may average at around 15-20% wetter for the season, with some parts around the Donegal Bay region to the far southwest of Munster slightly drier but still within 10% of the seasonal average. All other parts in the two provinces may expect to be wetter than average by around 10-20% more than average, but particularly wetter in the Galway and Killarney areas, which may be up to 50% wetter than average. Each month of spring may see a similar amount of rainfall recorded across the country as a whole with September and October receiving high rainfall in the periods between the New and Full Moons, with November wettest during the first week and between the First Quarter and the Full Moon over the second half of the month. The driest period during spring may be over the last few days of September and throughout the first week of October, from the Full Moon in September to the New Moon in October.

SUNSHINE:
The above average seasonal conditions return this autumn, with the country as a whole expected to see around 15-20% more sunshine than is typical for the season. All provinces may expect above average sunshine hours, with the Leinster Province around 10-15% sunnier, Ulster and Connaught Provinces around 15-20% sunnier, and the Munster Province up to 20-25% sunnier; and it may be up to 30% sunnier in some central areas, while only around 10% sunnier in the far northwest and Dublin areas. Looking at the sunshine hour trends by monthly breakdown, September may be around 15% cloudier than normal, while October and November may both be around 40-50% sunnier than average.

The sunniest spell in September may be mid-month for a few days during the New Moon period, while the cloudiest spell expected between the Third Quarter and the New Moon from midway through the first week to midway through the second week. It may also be particularly dull either side of the Full Moon in the last week of the month. October however, may be particularly sunny over the first week, and towards the end of the second week, while regularly dull over the last ten days of the month, particularly just after the First Quarter at the end of the third week. November may follow a similar pattern to October, with the sunniest spells over the first half of the month and the second half very dull. November may be very dull from the Full Moon in the fourth week to the end of the month. In contrast, November may be the sunniest in the second half of the first week.

TEMPERATURE:
Autumn temperatures may return to above average throughout the country by the end of the season, with most parts of the country seeing both maximum and minimum temperatures averaging around 0.5-1°C above the seasonal norm, with some parts in the west up to 1-1.5°C warmer than average. Only some parts of the west and north may be cooler than average. Overall, September and October may be warmer than average by around 0.5-1°C more than their monthly averages, while although November may also be warmer than average, it may also be the coldest month of the season averaging around 5°C for minimum temperatures and only around 10-11°C for maximum temperatures. As expected, September should be the warmest month of the season. Maximum temperatures should be above average for all the three months, with some areas in the west almost 2°C warmer in October. In contrast, minimum temperatures may be slightly cooler but still within half a degree of their norms in the west and east during October.

Maximum Temperatures**:** September may be the warmest early in the second week of the month in the build up to the New Moon, while coolest just after the New Moon over second half of the second week and early into the third week. October may be warmest over the first week, while coolest at the end of the second week. November may be warmest over the first week during the Third Quarter phase, as well as at the end of the month midway through the fourth week at the Full Moon to Northern Declination period. It may be the coolest either side of the New Moon early in the second week, when daily maximum temperatures may not reach double digits in most districts for several days. Overall, the warmest spell of autumn may be in the first half of September, while the coldest spell may be in the second week of November.

Minimum Temperatures**:** The mildest spell in September may be from midway through the first week to midway through the second week between the Third Quarter and New Moon. The coolest minimum temperatures over September may be over the second half of the month, from after the New Moon to the First Quarter, over most of the third week, and again for a couple of days shortly after the Full Moon at the end of the month, with minimum temperatures mostly in the single digits. It may be particularly cold overnight around the 24th, but not cold enough for frosts. October may see a fluctuation of warmer and cooler days with a warm day follows swiftly by several cooler days, then back again.

The coolest minimum temperature spells may be midway through the first week for a few days after the Third Quarter, at the end of the second week midway through the New Moon and First Quarter, and in the second half of the fourth week for a few days after the Full Moon. Some frost patches may appear briefly early in the month around the Third Quarter, mostly in the northwest. November may be the coldest over the second week around the New Moon to Southern Declination, in a spell that may see the return of widespread patches of frost, particularly in northern and inland parts. The mildest minimum temperatures for November may occur over the last week of the month, from just after the Full Moon for several days, when minimum temperatures into the low double digits may prevail briefly in many areas.

Frosts: There may be around 10 days with ground frosts this season, compared to the average of 16-20.

2026 OVERALL WEATHER SUMMARY

Winds: Overall, winds this year may be above normal almost everywhere, with the strongest winds possibly during the second half of the second week in January.

Rain: Wetter overall. The wettest spells of the year are likely to occur over the second half of January and August.

Dry Spells: The longest dry spell of the year may be from the second half of May to the last week of June.

Sun: Sunnier overall. The sunniest month may be May and the sunniest period may be during the fourth week of May and mid-June. The dullest month may be January, with the dullest periods over the second and fourth weeks of January as well as in the first half of February.

Maximum Temperatures**:** Average overall. The hottest spells in the year may be over the middle of June between the Third Quarter and the First Quarter, spanning much of the second and third weeks, and again over the first ten days of August, either side of the Third Quarter and in the build up to the New Moon. The coolest spell of the year with the lowest overall maximum temperatures may be in the middle of January between Apogee and the New Moon over much of the third week and again midway through the second week of February. It may also be unusually cold midway through the second week of December 2026, between New Moon and Southern Declination and Apogee period.

Minimum Temperatures**:** Average overall. The coldest period of the year may be in the middle of January around Apogee, the second week of February either side of Southern Declination and in the last week of February, either side of the First Quarter. The mildest spell of the year may be the last week of July in the build up to the Full Moon, in the middle of the second week of August just before the Third Quarter, and over the first 10 days of September, particularly between the Third Quarter and the New Moon.

Frosts: Frosts are likely on at least 85-90 days during the year, including December 2026.

RELATIVELY DRIER SPELLS FOR 2026

January	Despite some rain in most districts every day, the driest spell may be only for a couple of days near the end of the second week just after the Third Quarter.
February	For a couple of days midway through the first week, and then again over the last 10 days of the month from after the New Moon to the First Quarter to the end of the month.
March	Most days may see some rain activity this month, with only the first couple of days of March possibly mostly dry and again in the last 3-4 days at the end of the month. It may also be moderately dry around the start of the third week for a couple of days.
April	Mostly around the end of the second week for a couple of days and then again over the last week of the month, from the First Quarter onwards.
May	Over the second half of the month, from the New Moon to the Full Moon covering a relatively dry spell over the midway through the third week to the end of the month. It may be particularly dry for a few days at the New Moon to Northern Declination, and again immediately after the First Quarter for the next few days.
June	During the second half of the month, from midway through the third week to almost the end of the fourth week, with just a brief showery spell around the First Quarter.
July	Despite some light shower activity still possible, the last 10 days of the month, from the First Quarter to the Full Moon, may be the driest overall.
August	The first week may be the driest overall, in the build up to the Third Quarter.
September	The last few days of September for a couple of days after the Full Moon.
October	Much of the first week of October may be mostly dry everywhere, with being one of the longest dry spells of the year, with up to 7 consecutive rain free days, similar to the drier spells of February and May.
November	From midway through the second week into the first couple of days of the third week between the New Moon and First Quarter.
December	Some rain activity is likely almost daily during December with the least precipitation briefly mid-month at the First Quarter, half way through the third week.

NEXT THREE YEARS

Comparing estimated rain over the next three years averaged for the whole country; 2026 should have most rain and 2028 the least, with 2027 in the middle position, but very similar to 2026.

REGIONAL EXPECTATIONS BY REGIONS:

West: Wettest 2026, driest 2028, and middle 2027

South: Wettest 2027, driest 2028, and middle 2026

East: Wettest 2027, driest 2028, and middle 2026

North: Wettest 2026, driest 2028, and middle 2027

Central: Wettest 2027, driest 2028, and middle 2026

2028 may be the best year for sunshine amounts, with 2026 the cloudiest, and 2027 in the middle position.

For max temps, 2026 and 2027 may be very similar and the warmest overall, with 2028 slightly cooler.

For min temps, 2028 may be the coldest overall, with 2027 the mildest, and 2026 in the middle position.

JANUARY

PHASES OF THE MOON

1st	Perigee #6
2nd	Northern Declination
3rd	Full Moon
9th	Crossing Equator
10th	Third Quarter
13th	Apogee
16th	Southern Declination
18th	New Moon
24th	Crossing Equator
26th	First Quarter
29th	Perigee #10
29th	Northern Declination

1	2	3	4	5	6	7	8	9	10	11	12	13	14	15	16	17	18	19	20	21	22	23	24	25	26	27	28	29	30	31
Thur	Fri	Sat	Sun	Mon	Tue	Wed	Thur	Fri	Sat	Sun	Mon	Tue	Wed	Thur	Fri	Sat	Sun	Mon	Tue	Wed	Thur	Fri	Sat	Sun	Mon	Tue	Wed	Thur	Fri	Sat
P6	^	F						XhS	3Q			A			V		N						XhN		1Q			P10 ^		

MONTHLY SUMMARY

1st-10th Atlantic depressions near the western coasts of Ireland bringing widespread wet and windy conditions throughout. Widespread frosts at first, becoming milder towards the end of this spell.

1st Chance of widespread thunderstorm activity over the western half of the country.

1st-4th Widespread frosts.

2nd-7th Period of high rainfall mostly in the western half of the country, occasionally accompanied by thunderstorms.

3rd-7th Scattered patches of fog possible during this period.

4th Wintry showers of hail likely.

5th-8th Very windy spell.

6th-7th Heavy rain and scattered thunderstorms possible, particularly in the western half of the country.

7th Some snowfalls likely, mostly in the west.

8th-9th Chance of further thunderstorms and hail showers.

Over the next four weeks, the climate of Ireland may be dominated by deep Atlantic depressions near or over the country, bringing spells of heavy rain, very dull conditions and strong winds, often in a south-westerly direction, resulting in above normal temperatures for this time of year on many days. Generally, the first couple of weeks ahead and the last week of this outlook period may see well below normal air pressure, while the third week may bring the only significant high-pressure spell. Overall, rainfall totals in the next four weeks may be above average almost everywhere, particularly in western and southern areas of Connacht and Munster, with up to 20-25 wet days possible overall for the country as a whole. Only a short spell towards the end of this outlook period, early in February, may be relatively dry. The above average rainfall may accompany much cloudier than normal conditions almost everywhere, except in some southern parts, which may be closer to average to slightly sunnier than normal. Temperatures during this outlook period may be up to 1-2°C warmer than normal for this time of year, with only the first upcoming week likely to be colder than average. As such, there may only be around half the number of days when air frosts are observed compared to normal. During this outlook period, thunderstorm activity may occur on 10 days, mostly in the south and west, hail on 13 days, snow on 9 days, ground frosts on 14-16 days, and fog on 4 days.

11th-13th A mild southerly airflow may at first bring a mostly dry and cloudy spell across the country, with a easterly change later, bringing a colder turn with wintry showers in the north at first, moving later into the east.

13th-18th Cold spell with frequent frosts. Chance of hail showers and snow.

13th-15th Chance of thunderstorm and hail activity mostly in the south and west.

14th Chance of widespread fog. This may possibly be one of the coldest periods of winter.

14th-2nd Feb A prolonged spell of unsettled weather likely with the passage of frontal systems bringing frequent bands of rain during this period that may cause some localised flooding, particularly in parts of the south and west. Wintry showers of sleet or snow also likely in the first half of this period, before milder conditions return near the end of January.

13th-18th Colder spell with wintry showers and frequent frosts.

14th-17th Spells of frequent thunderstorms with hail likely, more widespread in southern parts.

14th-19th The prevailing southwest to westerly winds may reach up to gale force during this period.

17th-19th Chance of thunderstorm activity and hail mostly in the south and west.

18th A depression to the north of Ireland around this time may extend a cold front over the country, with bands of heavy rain and thunderstorms at times. Temperatures may drop significantly in a short space of time with the passage of the front, and showers may be very wintry and heavy behind the front, particularly in the western half of the country.

19th Very strong winds over the northern half of Ireland and particularly blustery in the northwest. Chance of snow. Very heavy rain likely around this time.

20th Chance of hail showers.

21st-22nd Widespread fog patches likely. Fairly sunny after the fog lifts.

22nd-31st Milder temperatures and occasional sunny spells likely to return for the remainder of this month.

25th A deep depression may near the northwest coast. Chance of fog in numerous areas.

27th Thunderstorms likely, particularly in the south and west.

28th Milder day likely.

28th-31st Very dull and overcast spell.

30th Chance of very heavy rain in the country, particularly in southern districts.

IRELAND RAINFALL ESTIMATES
January 2026

d=mainly dry, n= nonrecordable, l=light shrs, s=significant shrs, r=rain, h=hvy falls

		ULSTER PROVINCE					CONNAUGHT PROVINCE						LEINSTER PROVINCE				MUNSTER PROVINCE						
		Alderg	Hillsbo	Armag	Loughl	Malin I	Sligo	Ardtarı	Belmul	Drums	Galway	Roscoı	Markre	Edende	Kilkenı	Dublin	Shannı	Newpo	Tralee	Killarne	Cork	All	Moon
1st	Thur	l	d	l	l	l	n	l	s	d	d	d	l	d	d	l	l	l	s	d	l	30	P6
2nd	Fri	l	s	d	s	l	s	s	s	s	r	s	s	s	s	d	l	s	r	h	d	124	^
3rd	Sat	l	d	s	d	s	d	d	n	d	l	d	d	d	l	l	r	l	s	d	r	49	F
4th	Sun	n	l	d	s	s	l	s	s	l	s	s	l	l	l	l	s	s	s	s	l	65	
5th	Mon	l	l	n	l	l	s	l	l	s	r	s	s	l	n	d	n	s	l	r	n	70	
6th	Tues	r	h	l	s	l	s	s	s	h	h	h	r	r	r	r	h	r	h	h	s	283	
7th	Wed	r	r	r	s	r	s	s	s	r	h	r	s	l	s	s	s	s	r	r	h	193	
8th	Thur	s	d	s	l	r	l	l	s	l	s	l	l	d	s	l	r	l	l	l	l	55	
9th	Fri	l	d	l	l	l	s	l	l	l	l	l	s	d	d	d	d	s	n	l	d	38	XhS
10th	Sat	l	l	d	l	l	l	l	r	s	s	s	d	l	l	d	l	s	s	s	s	68	3Q
11th	Sun	s	d	s	d	d	d	d	l	n	d	d	l	d	n	l	l	d	l	n	l	22	
12th	Mon	d	d	d	d	d	d	d	d	d	d	d	d	d	d	d	d	d	l	l	l	9	
13th	Tues	d	d	d	d	l	d	d	d	d	d	d	d	d	d	l	d	d	d	l	l	8	A
14th	Wed	s	s	s	s	s	r	s	s	s	r	r	s	r	s	d	r	r	h	h	h	195	
15th	Thur	l	d	l	l	r	l	l	s	l	l	l	l	n	l	d	s	l	l	l	l	49	
16th	Fri	s	r	n	s	s	r	s	s	h	r	h	r	s	s	s	s	h	s	r	s	173	V
17th	Sat	s	l	r	s	r	r	s	l	s	r	r	r	l	s	l	s	r	r	r	s	161	
18th	Sun	r	h	l	s	r	s	s	s	r	r	r	s	r	r	r	h	s	h	h	h	212	N
19th	Mon	r	r	r	s	s	s	s	s	s	s	s	s	r	h	r	r	s	h	h	h	264	
20th	Tues	l	d	s	d	d	l	d	s	d	l	n	d	n	l	r	r	l	h	l	r	79	
21st	Wed	d	d	d	l	l	l	l	l	d	n	n	n	d	d	d	l	l	d	l	l	16	
22nd	Thur	s	r	d	s	l	r	s	r	r	r	h	r	r	h	s	s	r	h	h	r	223	
23rd	Fri	s	s	s	s	l	s	s	r	l	l	l	l	l	l	l	s	s	r	s	s	89	
24th	Sat	l	l	l	l	l	s	l	l	l	r	l	l	l	l	s	s	s	s	s	s	71	XhN
25th	Sun	l	s	l	l	r	l	l	l	r	r	s	r	l	l	l	s	l	s	s	l	99	
26th	Mon	s	s	s	s	s	s	s	s	l	s	s	s	s	l	l	s	s	s	r	l	94	1Q
27th	Tues	s	r	s	r	s	h	r	s	r	r	r	s	r	r	s	s	h	s	h	s	215	
28th	Wed	s	l	s	l	s	l	l	s	l	s	l	s	s	s	s	h	n	r	h	s	135	
29th	Thur	s	s	l	r	d	r	r	s	s	r	r	r	r	r	r	h	s	h	h	h	216	P1 ^
30th	Fri	h	h	r	h	l	h	h	s	h	r	h	h	r	l	l	s	h	r	h	r	284	
31st	Sat	r	l	h	l	h	s	l	h	l	l	l	s	l	l	s	r	l	s	s	s	153	
	Estimate:	150	163	137	132	169	167	132	166	171	242	187	157	134	164	106	238	216	299	389	225	3741	
	Average:	80	91	75	125	120	131	133	134	105	117	130	131	93	80	63	102	167	175	127	131	2309	
	Trend:	w tr	w tr	w tr	av	w tr	w tr	av	w tr	w tr	w tr	w tr	w tr	w tr	w tr	w tr	w tr	w tr	w tr	w tr	w tr	w tr	

IRELAND SUNSHINE ESTIMATES
January 2026

F=fine (8-12 hours of sunshine), pc= partly cloudy (4-7 hours), c=cloudy (1-3 hours), o=overcast (0 hours)

		ULSTER PROVINCE					CONNAUGHT PROVINCE						LEINSTER PROVINCE				MUNSTER PROVINCE						
		Alderg	Hillsbo	Armag	Loughl	Malin I	Sligo	Ardtar	Belmul	Drums	Galway	Roscoı	Markre	Edende	Kilkenı	Dublin	Shannc	Newpo	Tralee	Killarne	Cork	All	Moon
1st	Thur	pc	pc	pc	pc	pc	pc	pc	c	pc	c	c	pc	pc	pc	pc	c	pc	c	c	pc	91	P6
2nd	Fri	c	c	c	c	c	o	c	o	c	c	o	c	c	c	pc	c	c	o	o	o	36	^
3rd	Sat	c	c	c	c	c	c	c	c	o	c	c	c	o	o	o	o	o	c	c	o	28	F
4th	Sun	c	c	c	c	c	o	c	o	c	c	c	c	c	pc	pc	c	pc	o	o	c	53	
5th	Mon	o	o	o	o	o	o	o	o	o	o	o	o	o	o	o	o	o	o	o	o	4	
6th	Tues	o	o	o	o	o	o	o	o	o	o	o	o	o	o	o	o	o	o	o	o	4	
7th	Wed	c	c	c	c	c	o	c	o	c	o	o	c	c	c	c	c	c	o	o	o	32	
8th	Thur	c	c	c	c	c	c	c	pc	c	c	c	c	c	c	c	o	c	c	c	c	50	
9th	Fri	o	o	o	o	o	o	o	o	o	o	o	o	o	o	o	o	o	o	o	o	7	XhS
10th	Sat	o	o	o	o	o	o	o	o	o	o	o	o	o	o	o	o	o	o	o	o	1	3Q
11th	Sun	o	o	o	o	o	o	o	o	o	o	o	o	o	o	o	o	o	o	o	o	1	
12th	Mon	o	o	o	o	o	c	o	o	o	o	o	o	o	o	o	o	o	o	o	o	7	
13th	Tues	c	c	c	c	c	pc	c	pc	c	pc	pc	c	c	o	o	c	c	c	c	o	59	A
14th	Wed	o	o	o	o	o	c	o	o	o	o	o	o	o	o	o	o	o	o	o	o	11	
15th	Thur	c	c	c	c	c	c	c	c	pc	c	c	c	pc	pc	pc	c	c	c	c	pc	69	
16th	Fri	c	c	c	c	c	c	c	o	c	o	o	c	c	c	c	o	o	o	o	o	21	V
17th	Sat	c	c	c	c	c	c	c	c	c	c	c	c	c	pc	pc	c	c	c	c	pc	63	
18th	Sun	o	o	o	o	o	o	o	o	o	o	o	o	o	o	c	o	o	o	o	o	11	N
19th	Mon	c	c	c	c	c	c	c	c	c	c	c	c	c	c	pc	c	c	o	o	c	43	
20th	Tues	c	c	c	c	c	c	c	c	c	c	c	c	c	c	c	c	c	c	c	c	64	
21st	Wed	pc	pc	pc	pc	pc	pc	pc	c	pc	pc	pc	pc	pc	pc	c	pc	pc	pc	pc	pc	94	
22nd	Thur	o	o	o	o	o	o	o	o	o	o	o	o	o	o	c	o	o	o	o	o	9	
23rd	Fri	o	o	o	o	o	o	o	c	o	c	c	o	o	c	o	c	c	pc	pc	c	29	
24th	Sat	o	o	o	o	o	o	o	c	c	c	c	o	c	c	c	c	c	c	c	c	26	XhN
25th	Sun	o	o	o	o	o	o	o	o	o	c	o	o	o	c	o	c	c	c	c	c	14	
26th	Mon	c	c	c	c	c	c	c	c	c	c	c	c	c	c	o	o	c	c	c	c	48	1Q
27th	Tues	c	c	c	c	c	o	c	c	c	c	c	c	c	c	c	c	c	o	o	pc	36	
28th	Wed	o	o	o	o	o	o	o	o	o	o	o	o	o	o	o	o	o	o	o	o	0	
29th	Thur	o	o	o	o	o	o	o	o	o	o	o	o	o	o	o	o	o	o	o	o	0	P1 ^
30th	Fri	o	o	o	o	o	o	o	o	o	o	o	o	o	o	o	o	o	o	o	o	1	
31st	Sat	c	c	c	c	c	c	c	o	c	o	o	c	c	c	c	o	c	o	o	c	22	
	Estimate:	51	51	51	51	51	46	51	39	52	39	40	51	52	53	54	38	45	36	36	50	932	
	Average:	46	46	44	41	37	47	47	46	43	51	53	54	51	53	56	49	43	44	44	53	946	
	Trend:	av	av	av	av	more	av	av	av	av	less	less	av	av	av	av	less	av	av	av	av	average	

JANUARY **1** THURSDAY

A high pressure system to the southwest of Ireland and another high to the east of the country brings a light to moderate southeasterly flow over the country and a showery trough over eastern and southern regions, while the north and west are likely to be mostly dry and cold. Chance of fog in north and east, and frosts in the west.

Belfast: Cold, dry, severe frosts, misty patches possible, some sunny spells
Dublin: Partly cloudy, brief showers, cold and misty air
Cork: Cold, showery, some sunny spells, gentle breezes to mostly calm air
Galway: Partly cloudy, dry, cold, frosts, breezy at times

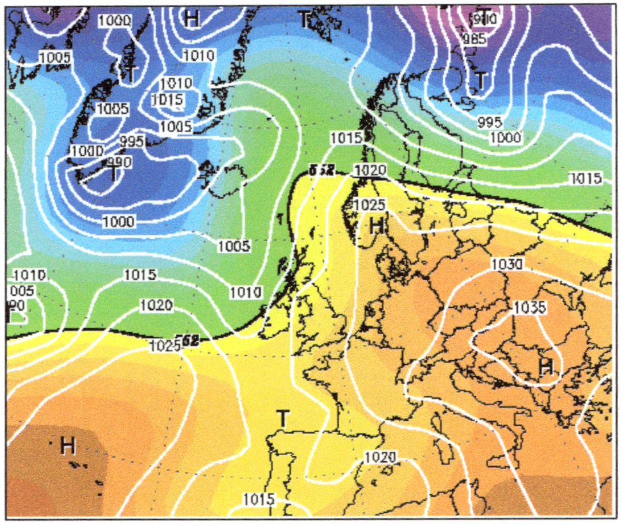

RAIN POTENTIAL

FROST/SNOW

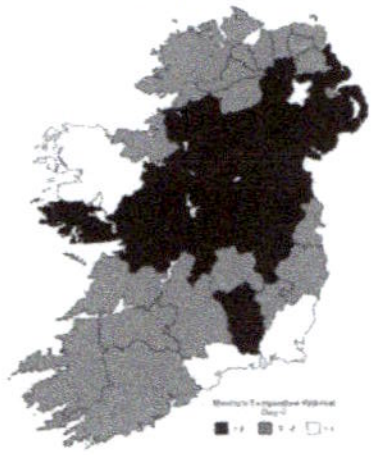

JANUARY **2** FRIDAY

High pressure moves up over the country bringing a light southerly flow over central and southern regions, while low pressure to the northwest of Ireland brings stronger and more gustier winds over the north and west. Mostly sunny with patches of fog in the east and south, while chance of gale gusts and wintry drizzle and showers in the north and west, where some light snow may be observed.

Belfast: Mostly cloudy, milder, blustery at times, occasional showers
Dublin: Partly cloudy, dry, some windy spells, frosty overnight, chance of fog patches
Cork: Overcast, cold, dry with occasional threats of rain, calm air, possible mists
Galway: Cloudy, rain, frosty overnight with chance of snow, windy at times

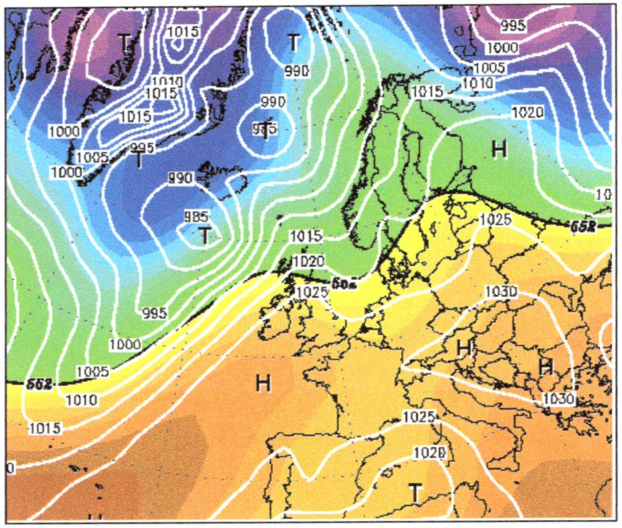

RAIN POTENTIAL

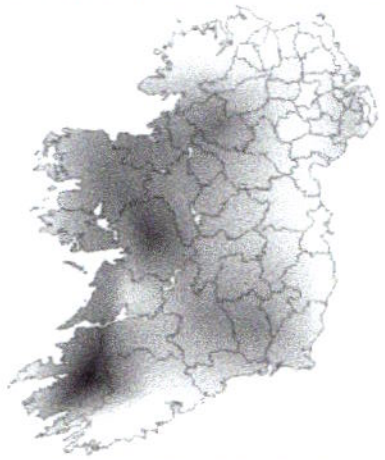

FROST/SNOW

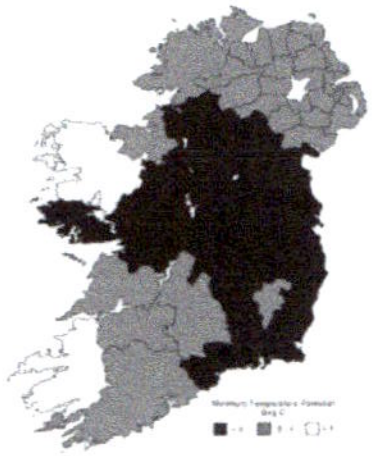

JANUARY **3** SATURDAY

A depression to the north of Ireland moves southward across the country bringing stormy conditions with widespread cloud and outbreaks of rain almost everywhere, starting in the west at first and spreading eastward as the day progresses, with heavy falls in some places.

Belfast: Cloudy, cool, mostly dry with passing threats of rain, strong winds at times
Dublin: Overcast and blustery, squally showers, chance of snow flurries
Cork: Overcast, milder temperatures, rain, windy spells developing
Galway: Strong winds, cloudy, cold, brief squally showers

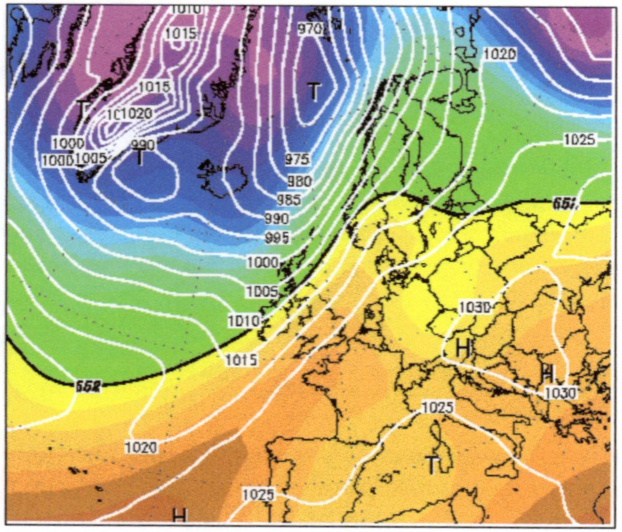

RAIN POTENTIAL

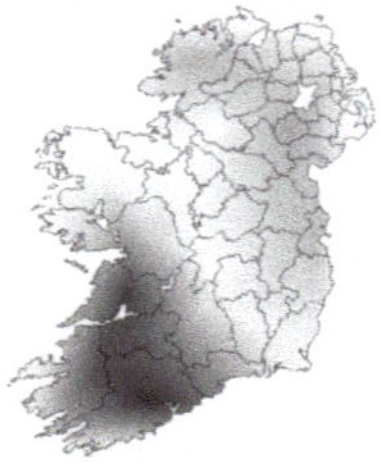

FROST/SNOW

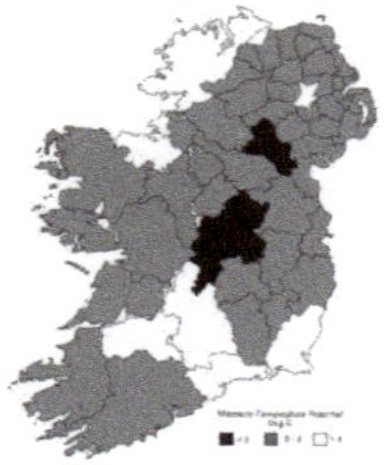

JANUARY **4** SUNDAY

Low pressure maintains unstable conditions across the country, despite winds easing everywhere. A line of heavy showers may cross the west at first, and the southern parts, possibly falling as snow in some places, while remaining cloudy but mostly dry in the north and east. Cold with patches of frosts and fog in many places, but becoming milder and breezier with a southwest change.

Belfast: Cloudy, cold, frosts, winds easing to light, misty drizzle patches likely
Dublin: Cloudy spells, isolated showers, cold, chance of freezing fog, frosts
Cork: Changeable, mix of showers and sunny breakthroughs, cold, frosts patches
Galway: Cloudy, some showers, light breezes, chance of freezing mist, sleet or snow

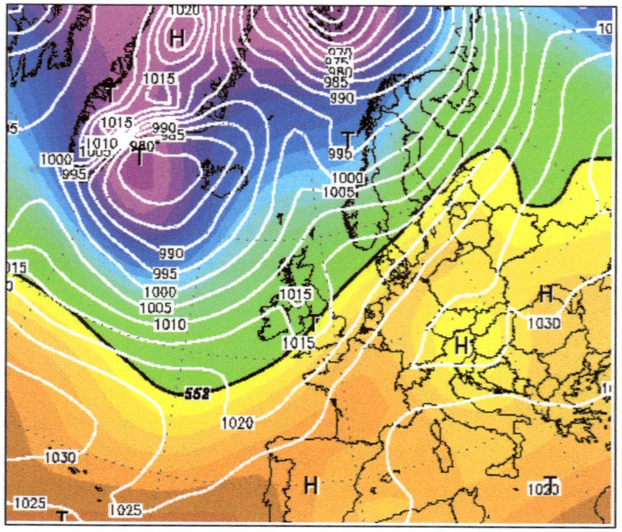

RAIN POTENTIAL

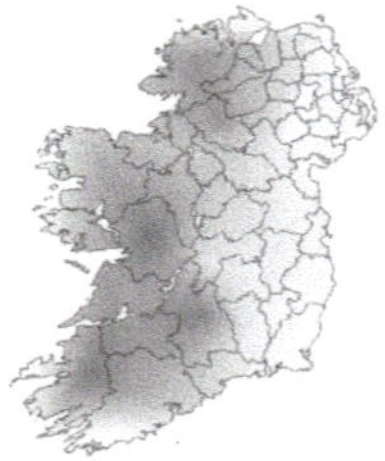

FROST/SNOW

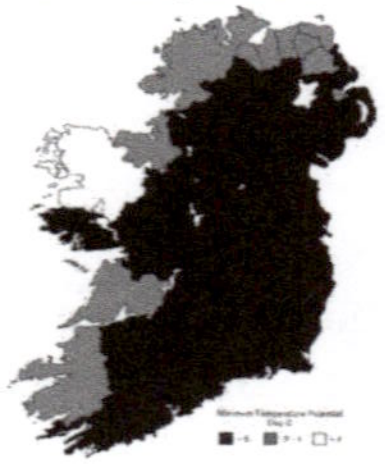

JANUARY **5** MONDAY

Low pressure continues to maintains a southwesterly flow across the country, strengthening to blustery and very unsettled in many areas. Mostly gloomy and dull air throughout, but some brief sunny spells likely for a time mostly in the east and south. Patches of rain may be observed in most districts as the day progresses, with heavier spells in the northwest for a time.

Belfast: Dull and gloomy, stormy air, cold, drizzly showers
Dublin: Overcast, milder, mostly dry with passing threats of rain, windy spells
Cork: Light drizzle patches, widespread cloud, slightly warmer, breezy at times
Galway: Overcast, breezy, rain, milder temperatures

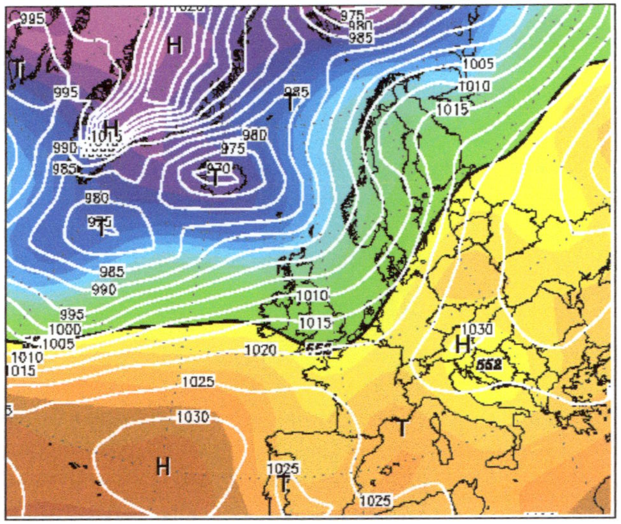

RAIN POTENTIAL

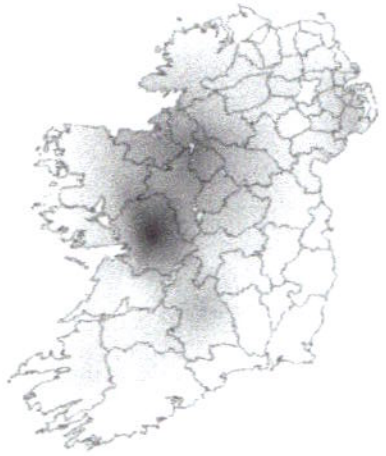

FROST/SNOW

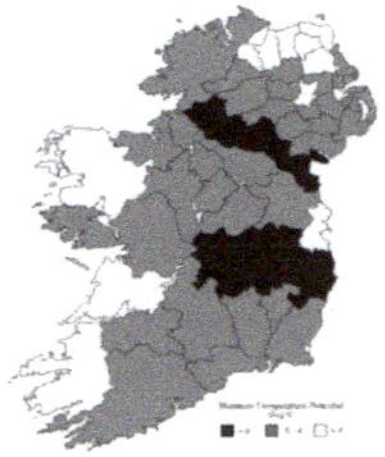

JANUARY **6** TUESDAY

A depression to the north of Ireland continues to project unstable and stormy conditions throughout the country. Rain likely to be very heavy for a time, particularly in the west while winds may be strong and gusty for much of the day before easing later.

Belfast: Storm, overcast and dull, heavy rain, milder temperatures
Dublin: Stormy air, overcast, rain, milder temperatures
Cork: Stormy air, overcast, milder, rain
Galway: Stormy air, overcast and dull, heavy rain

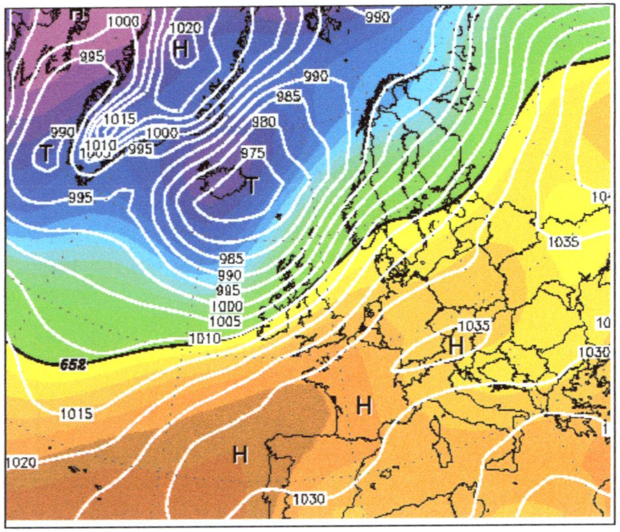

RAIN POTENTIAL

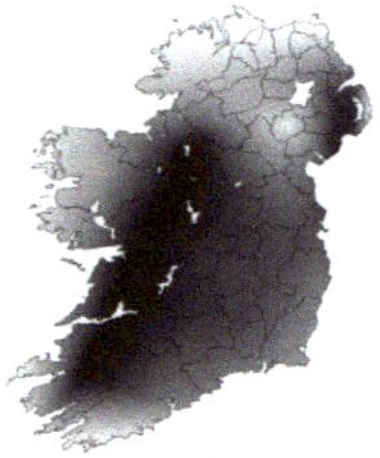

FROST/SNOW

JANUARY **7** WEDNESDAY

A low pressure system to the northwest of Ireland continues to bring a fresh to strong southwesterly flow throughout, with a mix of sunny spells and scattered showers in most parts of the country, with more heavier rain likely in southern parts, and occasionally accompanied by thunderstorm activity. Some precipitation may fall as snow, particularly in the north. Winds may be up to gale force for a time on exposed coasts.

Belfast: Mild temperatures, cloudy, rain, fluctuating winds with strong gusts at times
Dublin: Changeable, mix of sunny spells and showers, thunderstorms possible
Cork: Unsettled and blustery air, overcast, heavy rain
Galway: Overcast, heavy rain, hail showers likely, chance of snow, gust winds

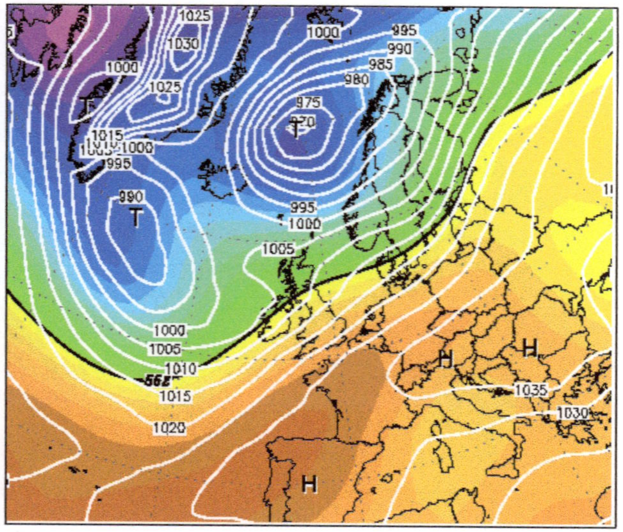

RAIN POTENTIAL

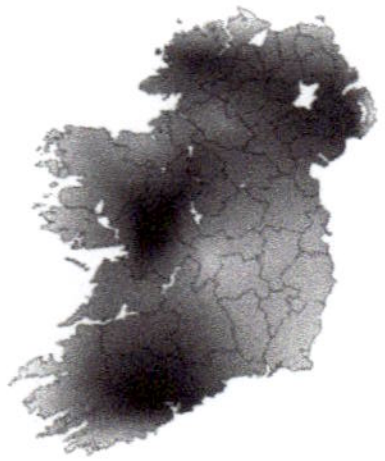

FROST/SNOW

JANUARY **8** THURSDAY

A depression to the north of Ireland continues to extend stormy conditions throughout, with forceful winds to gale force at times, and with winds particularly strong and gusty along the west and northern coasts. A band of thunderstorms cross the country during the day, accompanying outbursts of rain in most districts, as well as the chance of hail at times and some snow.

Belfast: Cloudy, mostly dry, passing threats of rain, colder, unsettled strong winds
Dublin: Stormy air, cloudy, colder, showers, chance of thunderstorms
Cork: Strong blustery winds, cloudy, isolated showers, cooler
Galway: Blustery winds, cloudy, colder, scattered showers, snow and hail possible

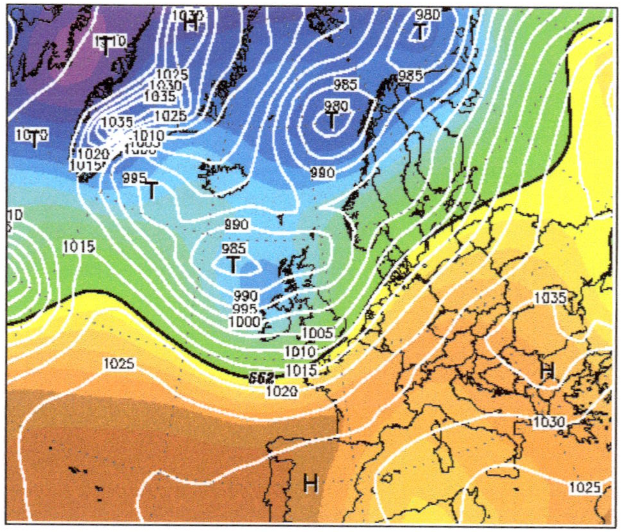

RAIN POTENTIAL

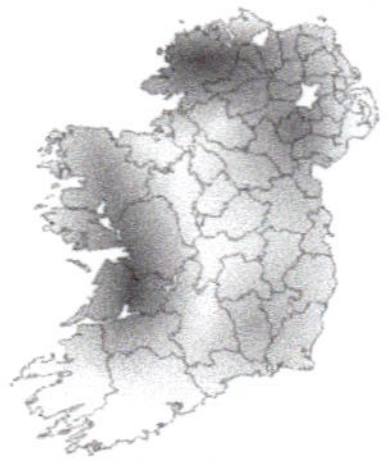

FROST/SNOW

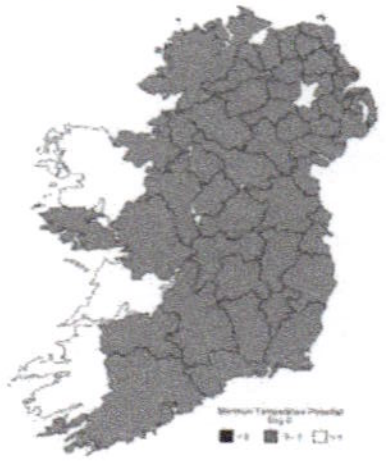

JANUARY **9** FRIDAY

The depression that brought storms to Ireland over the previous couple of days drifts away to the west, while another depression moves closer to the northeast of the country, with a trough of high pressure between the two systems and a high pressure system to the south prevailing. Overall, milder temperatures everywhere, with the occasional sunny spell and the odd patches of light showers or drizzle. Winds generally lighter for most of the day, in a mostly westerly flow.

Belfast: Dull and gloomy, winds easing to mostly calm air, dry, chance of fog patches
Dublin: Overcast and dull, dry, cold, light breezes, chance of drizzle patches
Cork: Overcast, mostly dry with chance of drizzle patches, milder, lighter winds
Galway: Winds easing, overcast, occasional showers

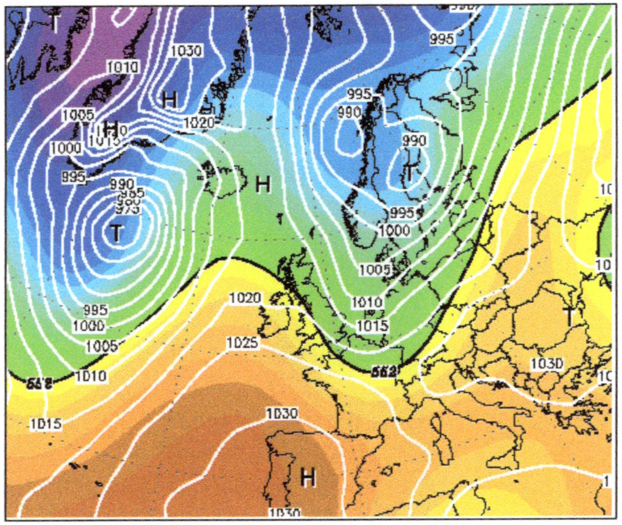

RAIN POTENTIAL

FROST/SNOW

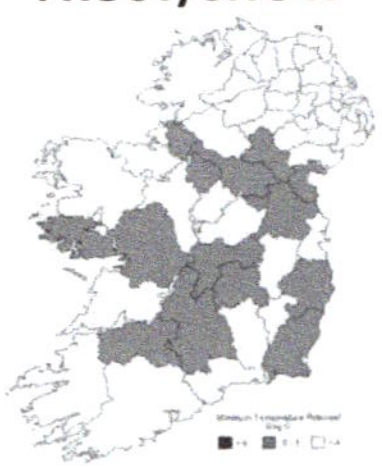

JANUARY **10** SATURDAY

High pressure moves to the southeast of Ireland while low pressure prevails to the west. Dull and overcast almost everywhere, with only a small chance of a few sunny spells mostly in the northwest later in the day. Patches of rain and drizzle are likely in many districts. Winds generally light in a southerly flow at first with the chance of mist or fog patches overnight, while more blustery later in the day, particularly in the northwest. Chance of the passage of a band of thunderstorms during the day, with possible hail showers.

Belfast: Overcast, drizzly showers at times, mild temperatures, fog patches likely
Dublin: Overcast, slightly warmer, dry with passing threats of rain, light breezes
Cork: Overcast, showers, milder temperatures, chance of misty patches
Galway: Overcast, rain and drizzles, slightly warmer, chance of evening fog patches

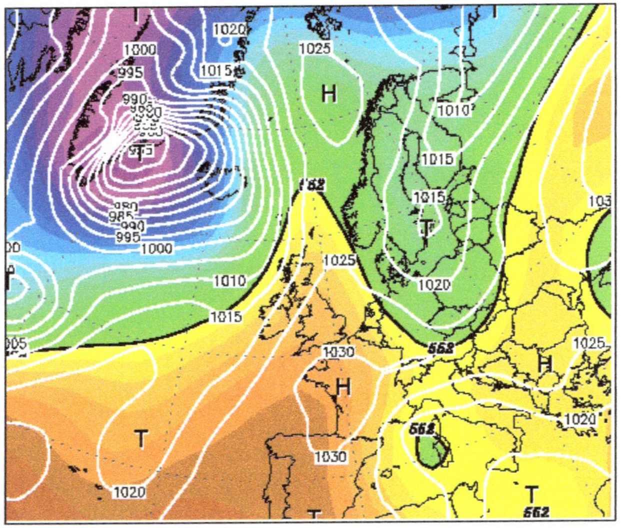

RAIN POTENTIAL

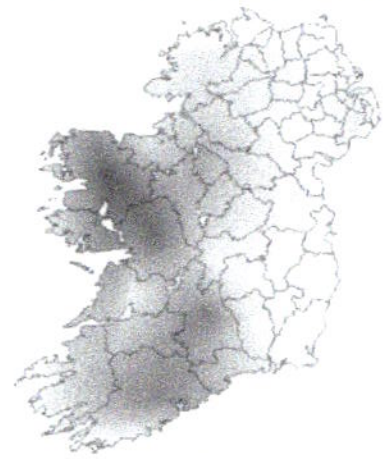

FROST/SNOW

JANUARY **11** SUNDAY

Another depression prevails to the west while a high pressure ridge flows up over the country. Widespread cloud throughout. Mostly dry in the northwest and southeast and through central parts, while light rain and drizzle may prevail elsewhere. Some precipitation may fall as snow overnight, particulary in the west. Light southerly breezes prevail, but later strengthen with a southeasterly change.

Belfast: Overcast, slightly warmer, dry with odd threat of rain, light breezes
Dublin: Light breezes, dull and gloomy air, drizzly showers, mild temperatures
Cork: Light showers, dull and gloomy, windy at times, chance of overnight fog
Galway: Overcast, dry with passing threats of rain, moderate winds

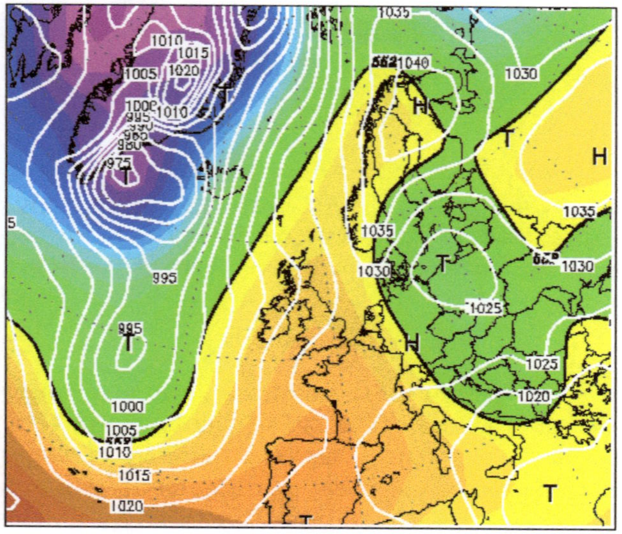

RAIN POTENTIAL

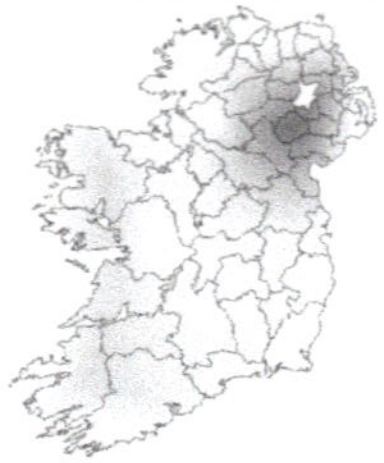

FROST/SNOW

JANUARY **12** MONDAY

A weak high pressure ridge brings a mostly dry day through the north and east of the country, while a weak front associated with a depression to the west of Ireland brings some showers to the south and west. Widespread cloud prevails throughout the country, with mild temperatures overall. Light to moderate southeasterly breezes prevail, later building to gale force for a time, before easing overnight.

Belfast: Dry, overcast, moderate breezes freshening later in the day
Dublin: Colder, overcast, mostly dry, passing threats of rain, strengthening breezes
Cork: Overcast, cooler, occasional showers, moderate breezes
Galway: Dull and gloomy air, colder, dry, threats of rain, windy at times

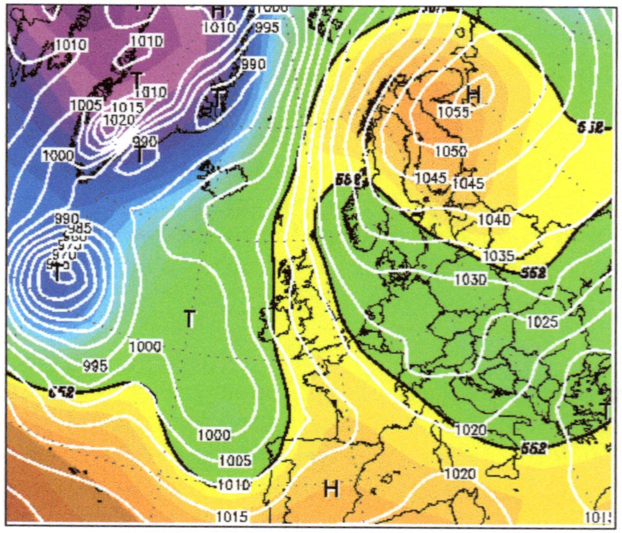

RAIN POTENTIAL

FROST/SNOW

JANUARY 13 TUESDAY

A weak depression spreads over the country in a strong to gusty southeasterly flow for most of the day before easing into the evening. Mostly dry at first but later spotty showers may develop, particularly in the southern half of the country. Some showers may be very wintry, as well as fall as snow. Very cold overnight with ground frosts in many districts.

Belfast: Colder, blustery, some sunny spells, mostly dry, chance of snow flurries
Dublin: Overcast, cold, drizzly showers, chance of snow, frosts, blustery
Cork: Overcast, colder, freshening winds, some showers, chance of snow
Galway: Partly cloudy, cold, dry, increasing threats of rain, chance of snow, blustery

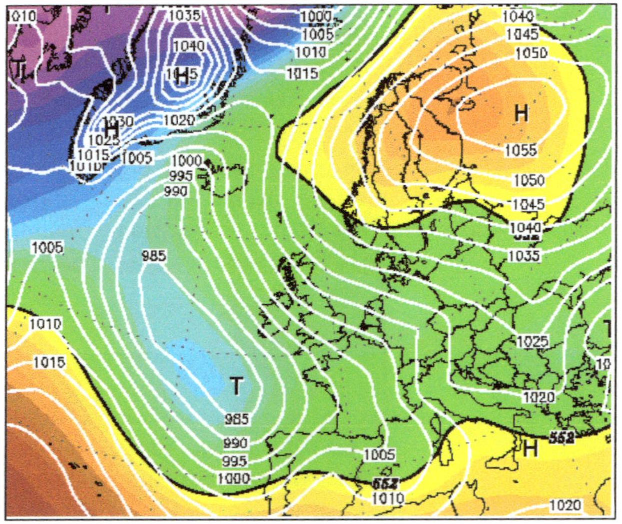

RAIN POTENTIAL

FROST/SNOW

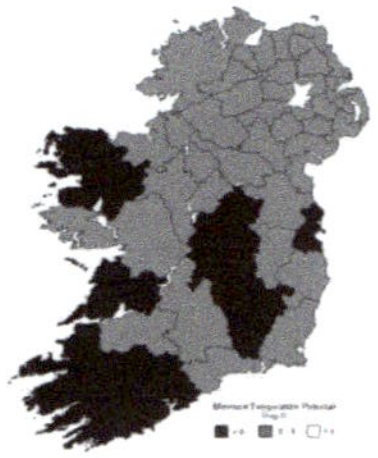

JANUARY 14 WEDNESDAY

A low pressure trough moves up over the country bringing a very cold, dull and gloomy day throughout. Winds generally very light and variable in direction, with scattered patches of fog overnight and severe frosts in many districts. Later in the day widespread rain develops and crosses the country from the west at first, moving into the east by the end of the day. Chance of thunderstorms and hail with the passage of wet fronts, as well as some snowfalls likely, particularly in the west.

Belfast: Very cold, dull and gloomy, frosts and snow likely, chance of fog
Dublin: Overcast, winds easing, mostly dry with frosts, chance of snow flurries
Cork: Overcast, severe frosts overnight, heavy rain, possible snow, mostly calm air
Galway: Overcast, rain, cold, severe frosts, snow likely

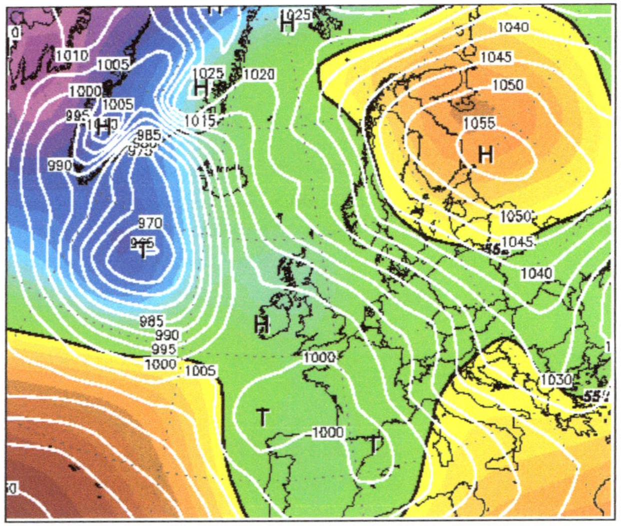

RAIN POTENTIAL

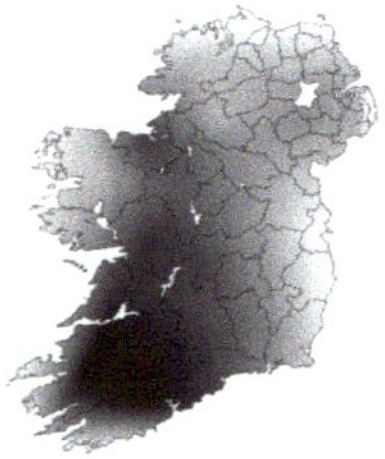

FROST/SNOW

JANUARY 15 THURSDAY

An intense depression moves onto the country with a fresh to strong southwest flow, possibly to gale force for a time, particularly in the northwest. Despite a stormy air throughout, the east and northeast may remain mostly dry except for the odd patch of freezing drizles, while showers may be expected elsewhere. Chance of heavier showers, some of them wintry, in the west, as well as possible snow at times.

Belfast: Fluctuating breezes, mostly dry, occasional sunny spells, cold, frosts
Dublin: Blustery air, mostly dry, cold, some sunny spells, chance of freezing drizzles
Cork: Storm winds for a time, isolated showers, cold, some sunny spells
Galway: Cloudy, thunderstorms, wintry showers, chance of hail, snow and gales

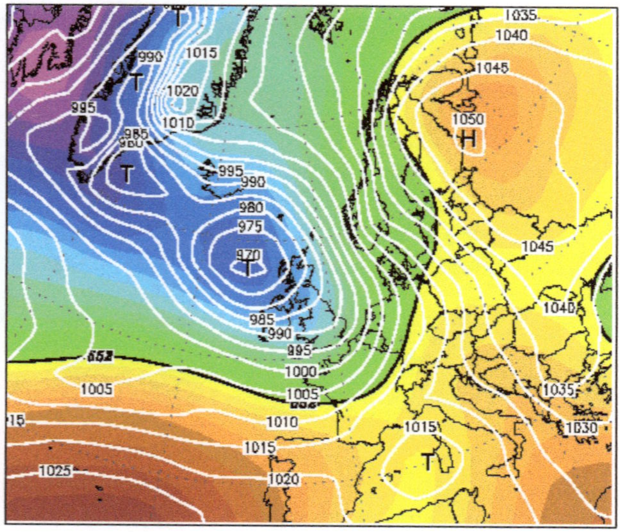

RAIN POTENTIAL

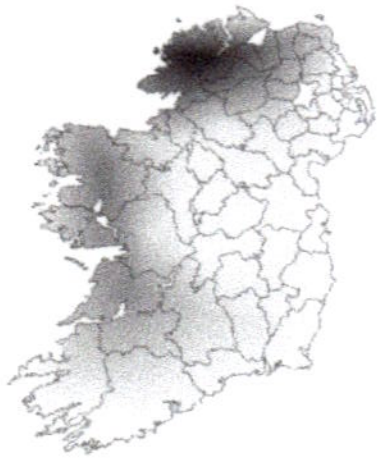

FROST/SNOW

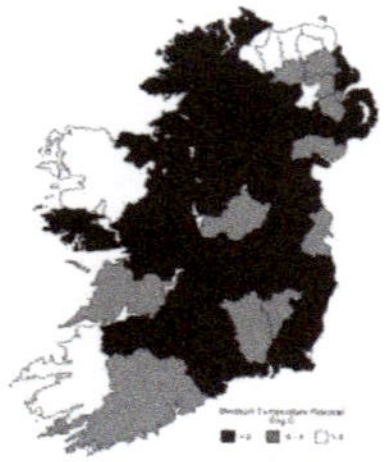

JANUARY 16 FRIDAY

A lingering depression continues to spread widespread a stormy air throughout the country with widespread rain prevailing, particularly wintry and accompanied by thunderstorms, hail and snow, in the west. Winds ease to light and drizzly in southern parts, while remaining strong and gusty, sometimes possibly up to gale force, elsewhere.

Belfast: Unsettled winds, cloudy, cold, rain
Dublin: Cloudy and cold, scattered showers, windy spells
Cork: Winds easing, overcast, cold, showery, chance of overnight misty patches
Galway: Unsettled air, overcast, rain, frosty, chance of thunderstorms, hail and snow

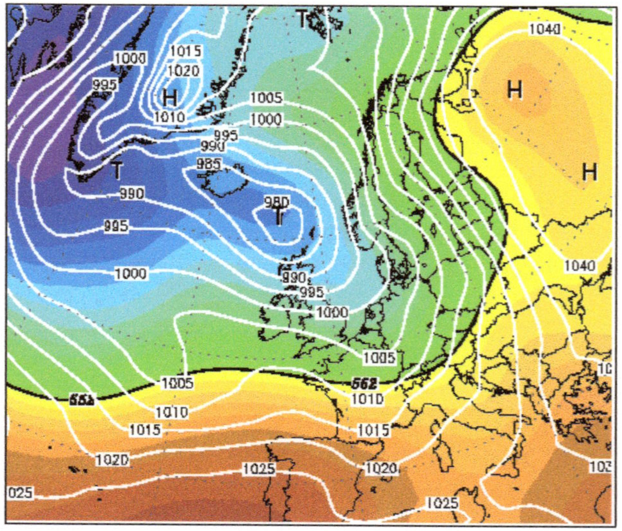

RAIN POTENTIAL

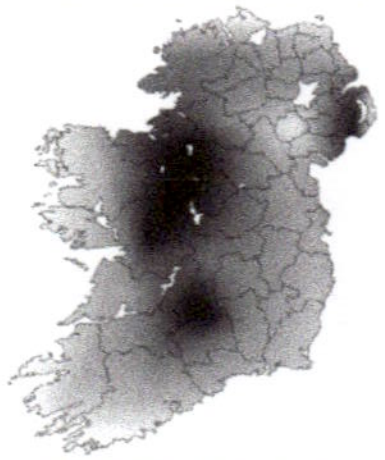

FROST/SNOW

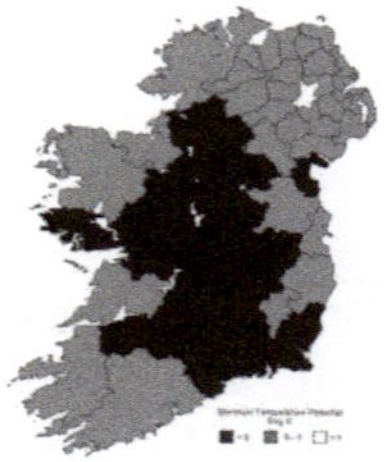

JANUARY 17 SATURDAY

Low pressure continues to produce stormy conditions throughout the country, with outbreaks of heavy rain and wintry showers and possible hail in a number of areas, particularly in the west and possible pockets of flooding in central parts. Snow showers are also possible, particularly on the higher reaches. Some sunny spells may break through by the end of the day, with very sharp frosts likely overnight. Generally southwest to westerly winds prevail, possibly reaching to storm force level gusts at times.

Belfast: Mostly cloudy, some showers, cold, stormy air at times
Dublin: Changeable, partly cloudy, light showers, cold, fluctuating breezes
Cork: Changeable, mix of showers and sunny breakthroughs, cold, unsettled winds
Galway: Stormy air, cloudy, cold, rain, chance of snow

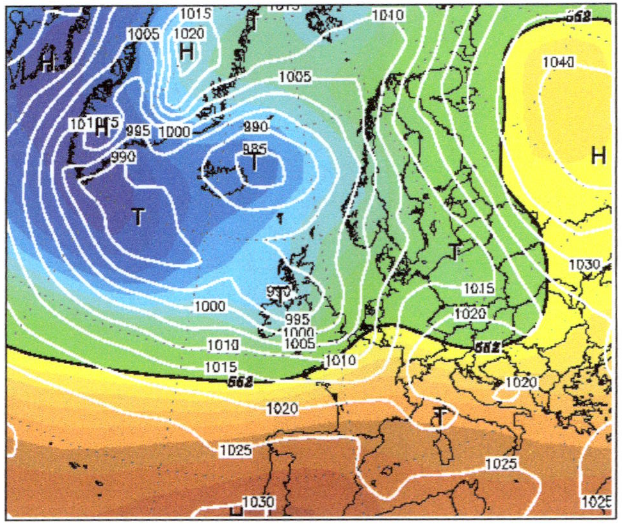

RAIN POTENTIAL

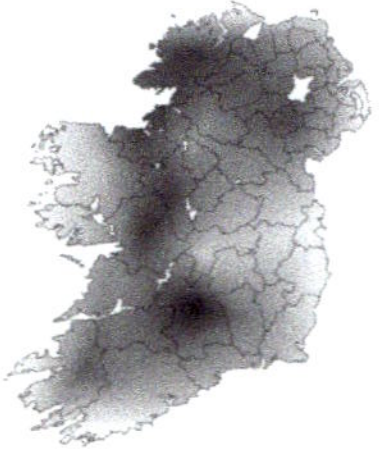

FROST/SNOW

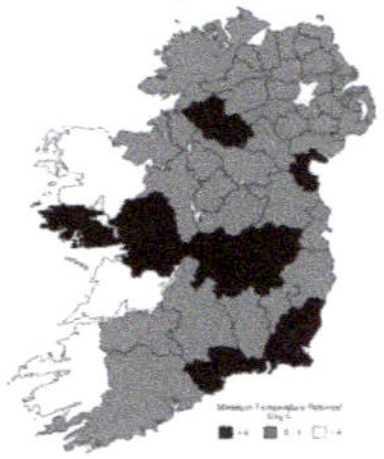

JANUARY 18 SUNDAY

Widespread storm conditions associated with an North Atlantic depression persist throughout, with wintry precipitation, including hail or sleet, and with particularly heavy rain in the south and southwest. Winds however, may ease to moderate in a southeastery flow, before rebuilding to storm force for a time in the west, with a southwest change.

Belfast: Overcast, heavy rain, cold, chance of hail or sleet
Dublin: Fresh breezes, cloudy, rain, milder day cold night
Cork: Heavy precipitation, chance of snow, overcast, milder, breezy
Galway: Dull and gloomy, breezy, rain, thunderstorms possible

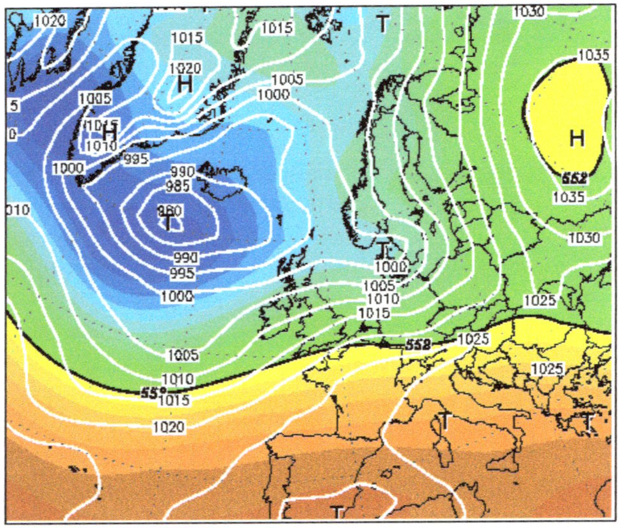

RAIN POTENTIAL

FROST/SNOW

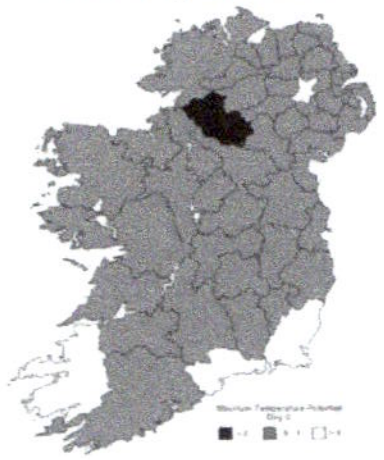

JANUARY 19 MONDAY

The depression reinvigorates over the north of England intensifying storm activity throughout Ireland. Generally, strong west to southwest winds prevail, reaching up to storm force for a time in northern parts. Widespread and frequently heavy rain, as well as showers of hail, sleet, or snow prevails everywhere, with particularly heavy falls in the southern half of Ireland, and bands of thunderstorm likely mostly in the west.

Belfast: Stormy air, cloudy, rain, cold, chance of sleet or hail
Dublin: Stormy, rain, some sunny breakthroughs, cold
Cork: Storm, cloudy, heavy precipitation, snow possible, chance of gale gusts
Galway: Cloudy, rain, strong blustery winds, thunderstorms and snow possible

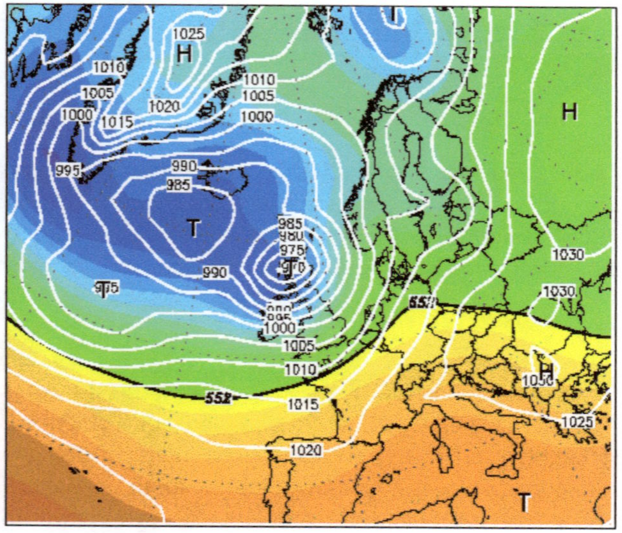

RAIN POTENTIAL

FROST/SNOW

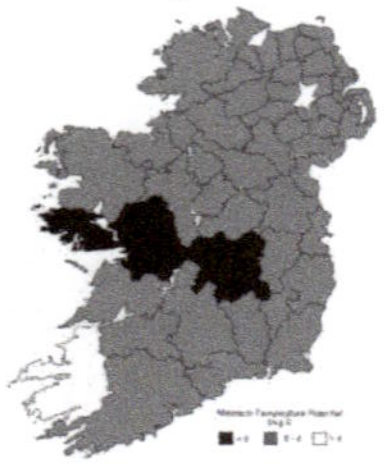

JANUARY 20 TUESDAY

Low pressure continues to dominate the country bringing widespread cloud and continued rain activity, sometimes wintry, and including the chance of further thunderstorms and overnight outbreaks of snow and heavy rain for a time in the south. However, storm conditions begin to ease as the day progresses, with some breakthroughs of sun now and then, mostly through central parts and in the far northern areas and later in the east.

Belfast: Cloudy and cold, mostly dry, winds easing to light
Dublin: Cold, overnight rain, brief sunny spells, blustery winds easing later in the day
Cork: Storm easing, brief sunny spells, outbreaks of rain
Galway: Cloudy, cold, showers, chance of overnight snow and thunderstorms

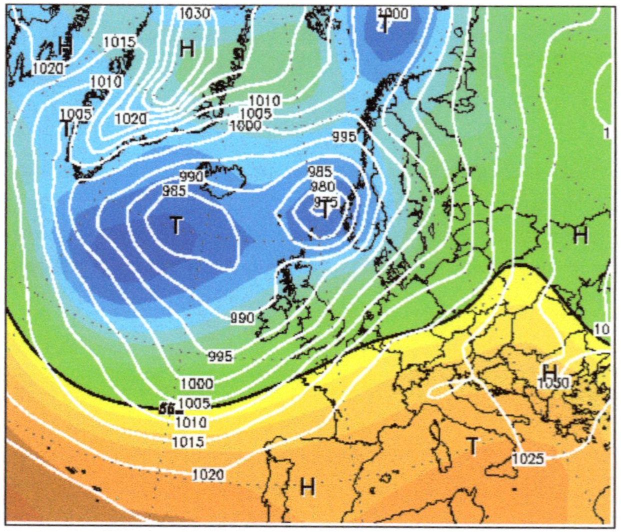

RAIN POTENTIAL

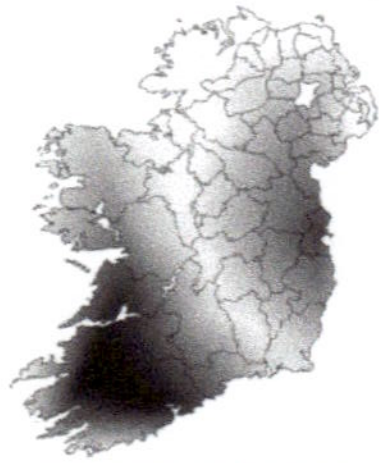

FROST/SNOW

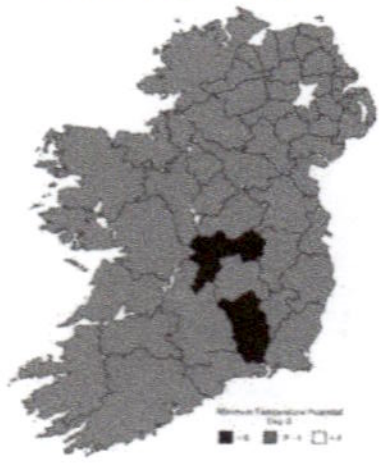

JANUARY **21** WEDNESDAY

Low pressure weakens with winds easing to light and variable at first, backing westerly later in the day. Very cold with widespread patches of frost, scattered fogs, and some occasional sunny spells. Chance of some lingering showers for a time, mostly in the western half of the country.

Belfast: Cold, frosts, dry, occasional sunny spells, chance of fog, possible snow
Dublin: Partly cloudy, cold, frosts, light breezes to mostly calm air
Cork: Mostly calm air, some sunny spells, colder, brief showers, chance of fog
Galway: Cold with frosts, some sunny spells, isolated showers, light breezes

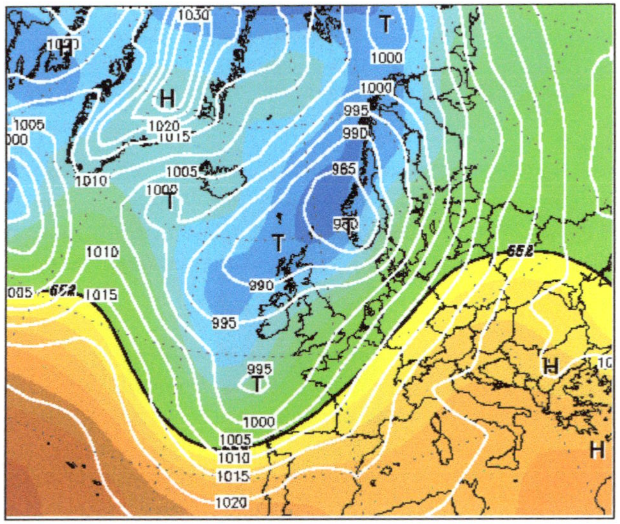

RAIN POTENTIAL

FROST/SNOW

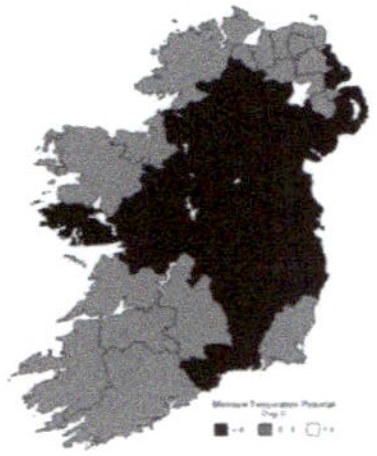

JANUARY **22** THURSDAY

A rain-embedded trough between two depressions crosses the country spreading widespread rain throughout as the day progresses starting in the southwest and spreading northwards, after a brief mostly dry, foggy and frosty start to the day. Wintry precipitation of sleet and snow likely in many parts of the country. Chance of heavy rain in the south and west of the country may result in some pockets of brief flooding, while winds strengthen to blustery later in the day.

Belfast: Overcast, severely cold, sleet and snow likely, freshening winds later
Dublin: Milder, cloudy, occasional showers, light breezes
Cork: Overcast, rain, milder day, cold night, light breezes, chance of fog overnight
Galway: Overcast, rain, milder day, overnight frosts, strengthening breezes later

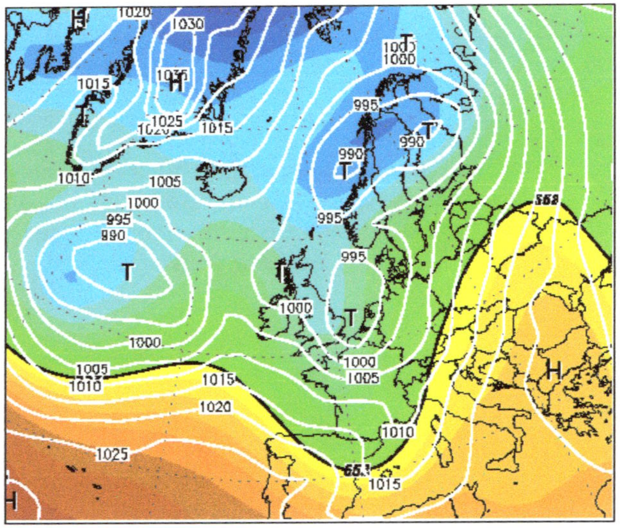

RAIN POTENTIAL

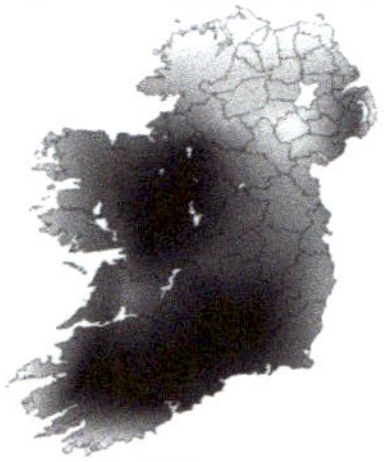

FROST/SNOW

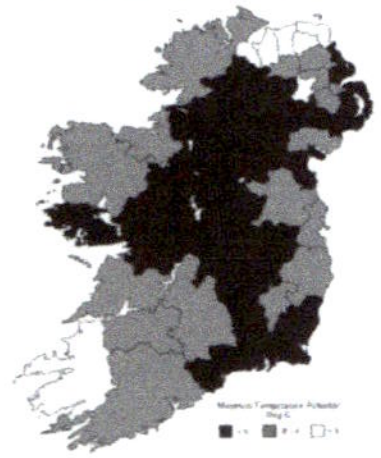

JANUARY **23** FRIDAY

A depression continues to linger immediately to the west of Ireland intensifying winds in a southerly flow throughout the country, generally moderate to fresh but up to gale force at times before easing later. Dull and gloomy with showers everywhere, some rain heavy at times and generally persistent in other areas. Chance of wintry showers including snow for a time in northern parts, while pockets of localised flooding may occur in central parts.

Belfast: Overcast, warmer, scattered showers, stormy winds
Dublin: Overcast, milder, isolated showers, windy at times
Cork: Cloudy, rain, stormy air
Galway: Cloudy, brief showers, strong blustery winds at times

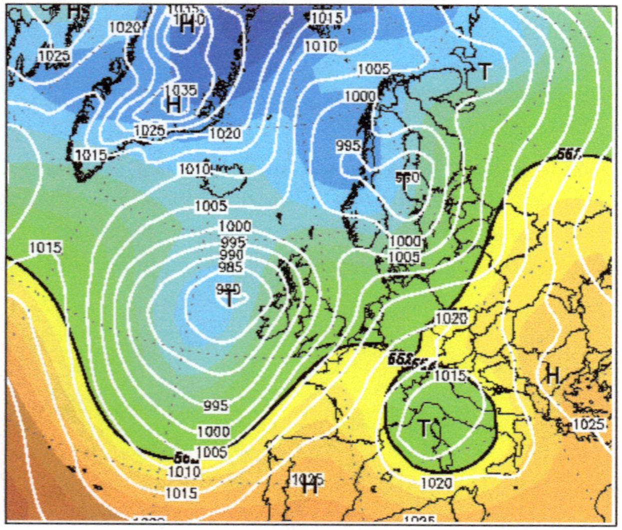

RAIN POTENTIAL

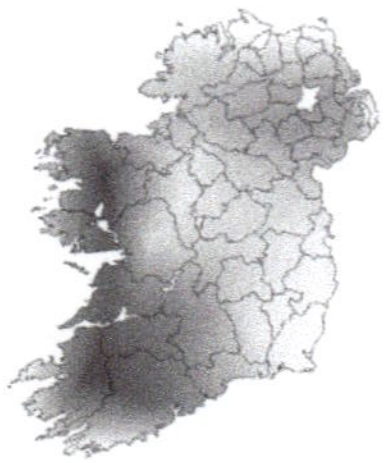

FROST/SNOW

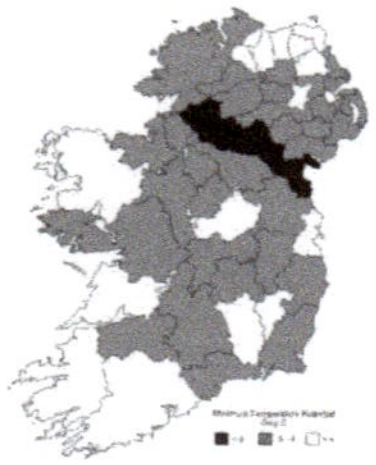

JANUARY **24** SATURDAY

A depression drifts eastward to be centred over Ireland continuing to bring widespread cloud and showers throughout with more sunny spells in southern parts, lesser to the north. Winds are likely to be variable and weaken as the day progresses.

Belfast: Unsettled winds, dull and gloomy air, light showers, chance of misty patches
Dublin: Cloudy, cold, occasional showers, fluctuating winds
Cork: Cloudy, showers, shifting winds
Galway: Cloudy, milder temperatures, rain, fluctuating breezes

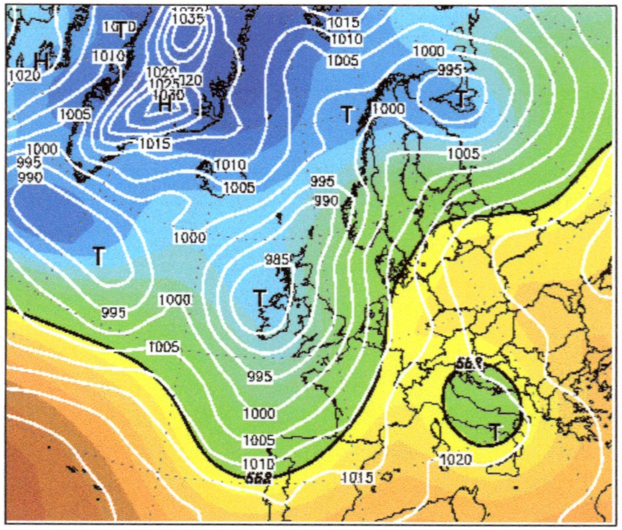

RAIN POTENTIAL

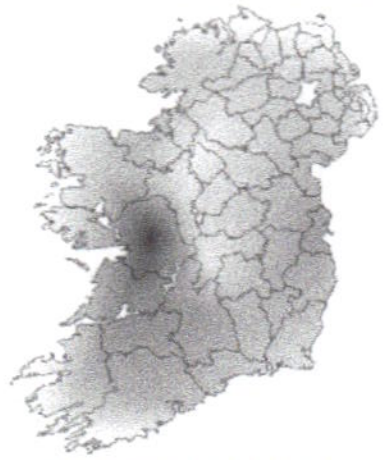

FROST/SNOW

JANUARY 25 SUNDAY

The depression over Ireland moves off to the west of the country while another to the north of Scotland nears, allowing for a low pressure trough to bring moderate to variable winds, sometimes of a south to southwest course to prevail. Dull and gloomy everywhere, with drizzle showers in most districs. Chance of patches of fog in northern parts while winds may be stronger around the southern coasts.

Belfast: Overcast, cold, drizzly showers, chance of fog patches
Dublin: Overcast and cold, shifting windy spells, light showers
Cork: Cold and blustery air, mostly cloudy, brief showers
Galway: Cloudy, rain, fresh breezes

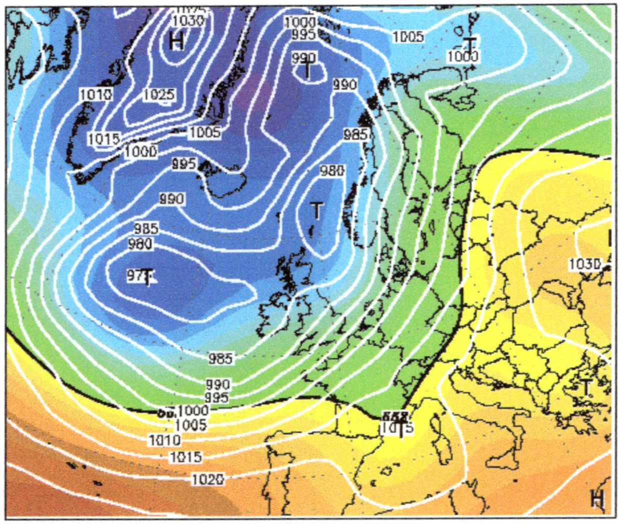

RAIN POTENTIAL

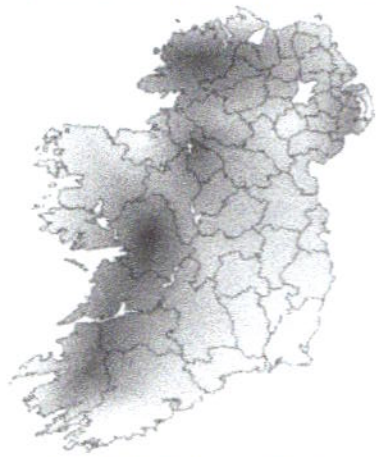

FROST/SNOW

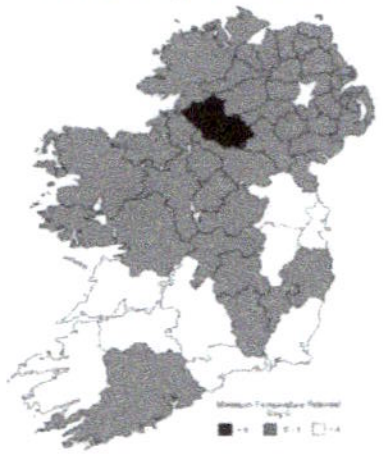

JANUARY 26 MONDAY

The two depressions near Ireland attempt to merge over the country, bringing unsettled conditions throughout. A mix of cloud everywhere, with some sunny spells in the south and west, while cloudier to mostly overcast in the north and east. Winds generally light at first, allowing for the chance of some fog mostly in northern and central parts, but freshening later in the day becoming more variable and moderate, and gustier in the north. Scattered outbreaks of rain almost everywhere, sometimes accompanied by thunderstorm activity.

Belfast: Cloudy and cold, occasional showers, light shifting breezes, chance of fog
Dublin: Overcast, colder, isolated showers, fluctuating and shifting breezes
Cork: Cold, some sunny breakthroughs, brief showers, unsettled breezes
Galway: Cloudy, scattered showers, fluctuating and shifting breezes

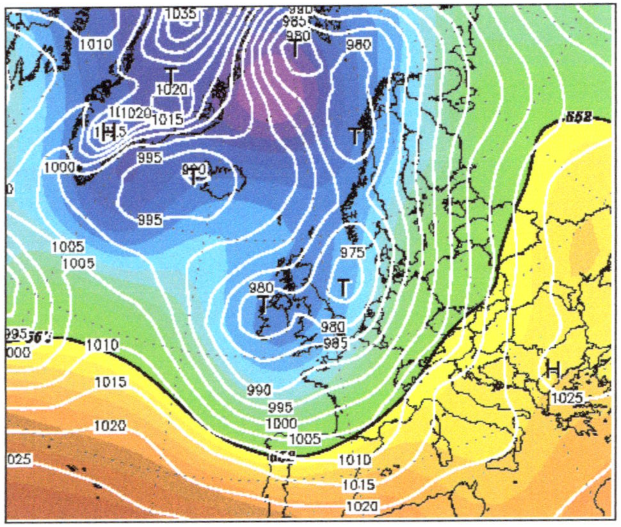

RAIN POTENTIAL

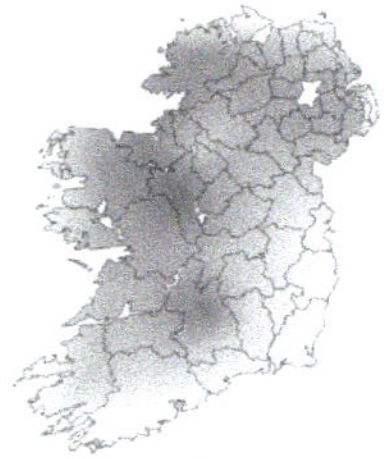

FROST/SNOW

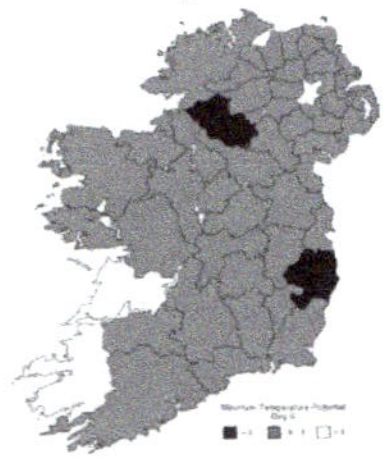

JANUARY **27** TUESDAY

A depression continues to bring unsettled conditions throughout and including the chance of extending a band of thunderstorms crossing the country in a strong and blustery southwest flow. Some thunderstorms may bring heavy showers at times accompanied by hail. Chance of further snow in the north and west, while hail more predominant in the east and south.

Belfast: Cold, cloudy, rain, moderate winds
Dublin: Cloudy, cold, scattered showers, chance of hail and thunderstorms
Cork: Changeable, mix of sunny spells and showers, chance of thunderstorms
Galway: Fresh breezes, cloudy, rain, chance of thunderstorms

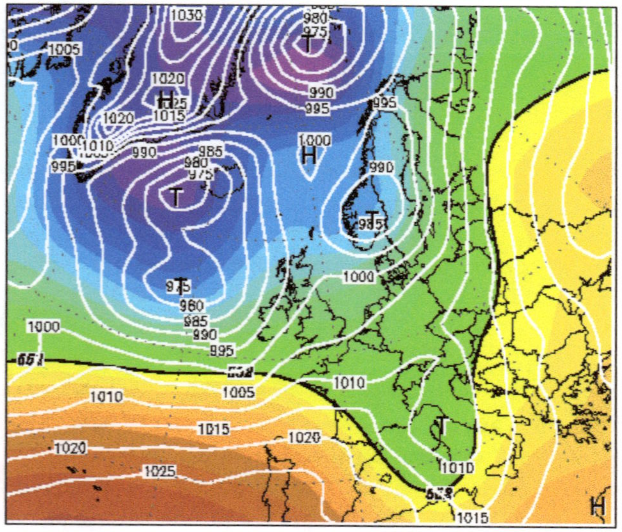

RAIN POTENTIAL

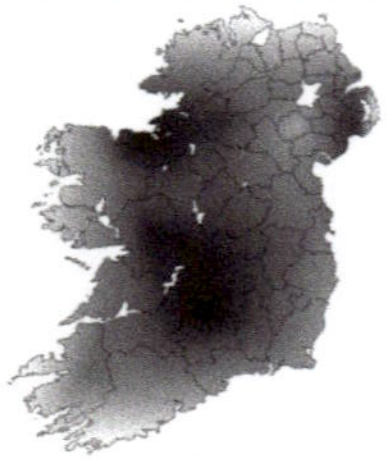

FROST/SNOW

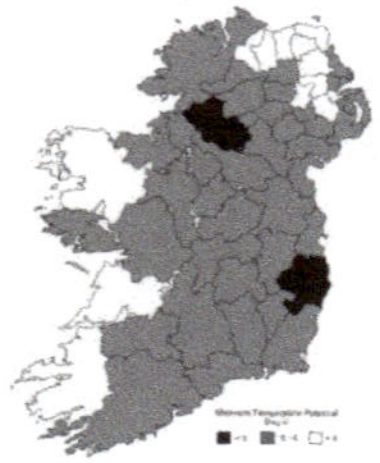

JANUARY **28** WEDNESDAY

A large anticyclone well south of Ireland pushes the depression to be centred to the far north of the country. Despite this, widespread cloud continues, with outbreaks of heavy rain in the west and southwest at first, later moving eastward across the country, while turning patchy later in the day. Winds generally moderate to blustery in a southwest to westerly flow.

Belfast: Overcast, light showers, milder, gentle breezes, fog patches possible
Dublin: Overcast, milder temperatures, rain, fluctuating breezes
Cork: Overcast, rain, windy, milder temperatures, chance of thunderstorms
Galway: Overcast, milder, showers, blustery at times, possible thunderstorms

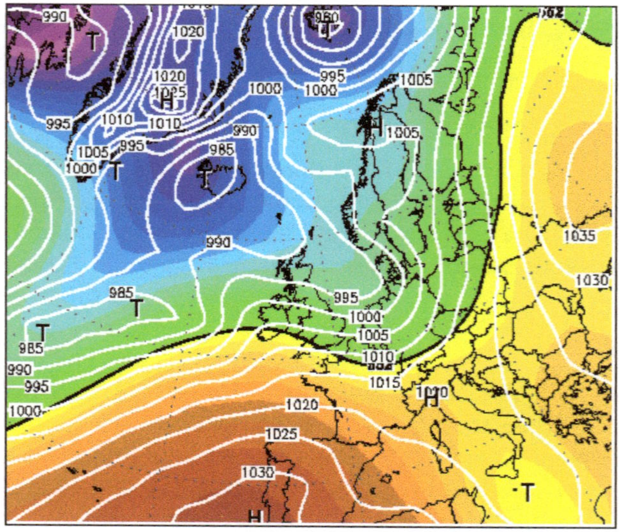

RAIN POTENTIAL

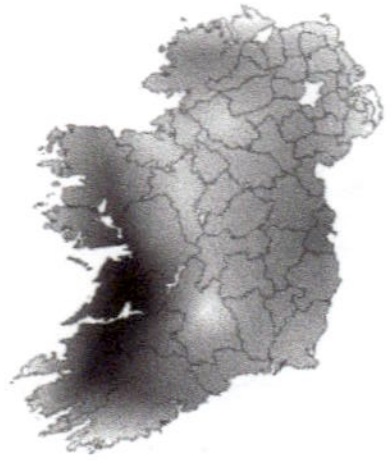

FROST/SNOW

JANUARY **29** THURSDAY

A high pressure ridge still remains to the south of the country directing a light southwest flow over southern parts, while a depression in the Norwegian Sea intensifies, and brings strengthening northwest winds to the north as the day progresses. Light breezes at first with fog possible in the north, and milder temperatures throughout. Dull and gloomy everywhere, with heavy rain in the south and more scattered outbreaks of rain elsewhere, while the far northern coastal fringes may be mostly dry.

Belfast: Overcast, slightly warmer, rain and drizzles, overnight fog possible
Dublin: Dull and gloomy air, rain, blustery winds at times
Cork: Overcast, heavy rain, mild temperatures, fluctuating winds
Galway: Moderate breezes, overcast, rain

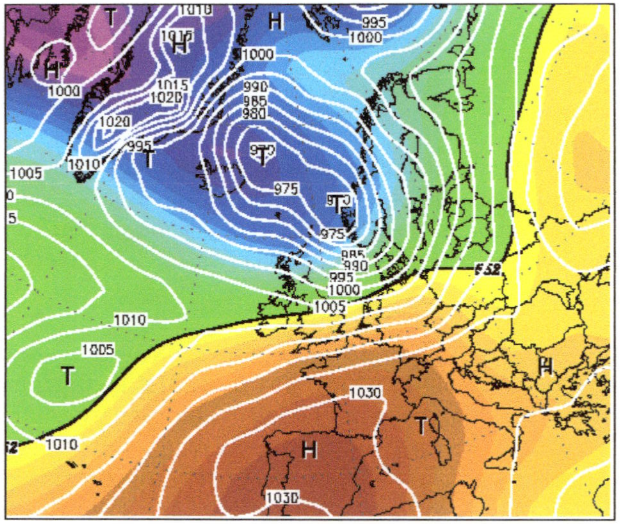

RAIN POTENTIAL

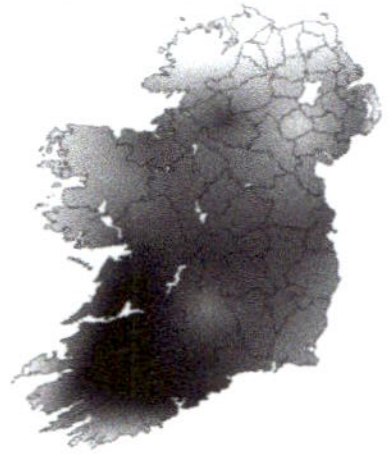

FROST/SNOW

JANUARY **30** FRIDAY

A low pressure trough covers the country bringing a dull and gloomy day with outbreaks of rain and drizzle into most districts, except in the far north and northwest of Ulster, which may remain mostly dry and dull. Chance of thunderstorms, hail and snow at times in central, eastern, and northern districts south of the northern coasts. Winds generally light to moderate in a southwest flow.

Belfast: Gloomy and dark, heavy rain, colder, light breezes, chance of fog
Dublin: Milder day cooler night, overcast, light showers, winds easing to light
Cork: Wet and gloomy air, rain, mild temperatures, light to moderate breezes
Galway: Light breezes to mostly calm air, overcast, rain, mild temperatures

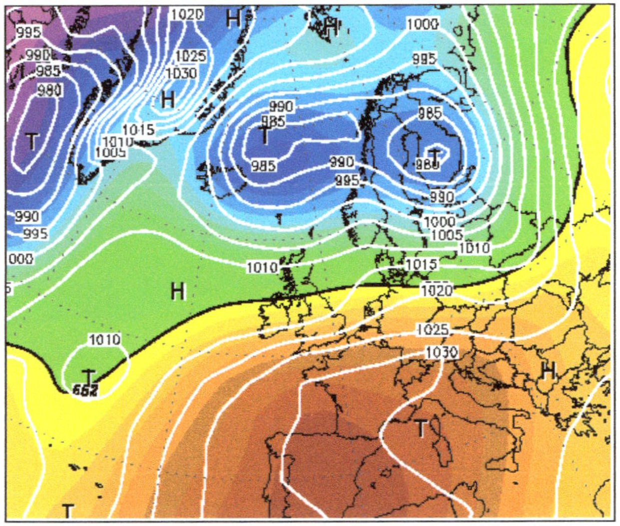

RAIN POTENTIAL

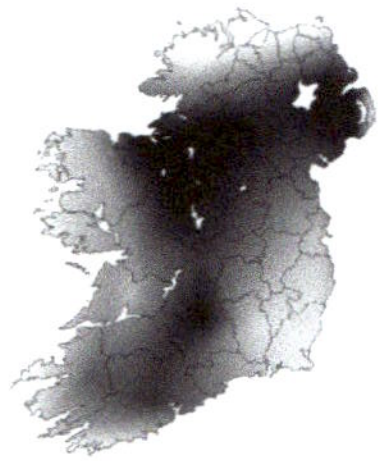

FROST/SNOW

JANUARY 31 SATURDAY

A high pressure ridge still lingers to the south of Ireland while a low pressure trough continues to prevail over over the country. Widespread cloud throughout, more overcast in the west, while some brief sunshine may breakthrough in the south and east. Showers in most districts, with heavier falls, including possible hail in the north and west, particularly around the coasts, and snow in central parts. Winds likely to be variable, light to moderate, before briefly strengthening for a time and easing again later in the day.

Belfast: Cloudy and cold, light showers, chance of overnight fog patches
Dublin: Cloudy, occasional showers, mild day cold night, light breezes
Cork: Cloudy, scattered showers, mild day cold night, moderate breezes
Galway: Overcast, light showers, cooler, fresh breezes at times

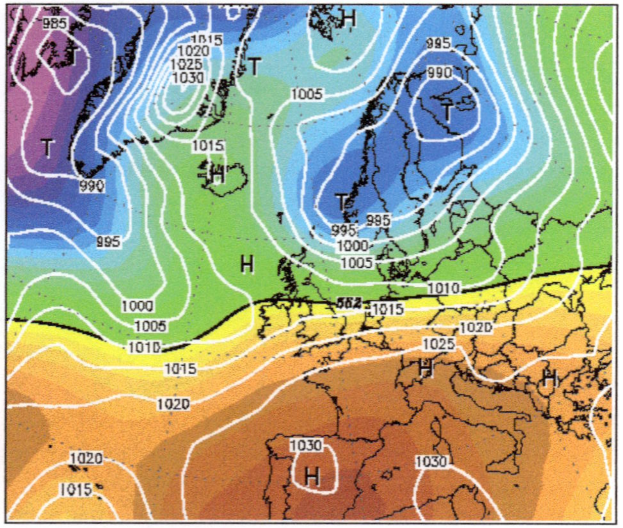

RAIN POTENTIAL

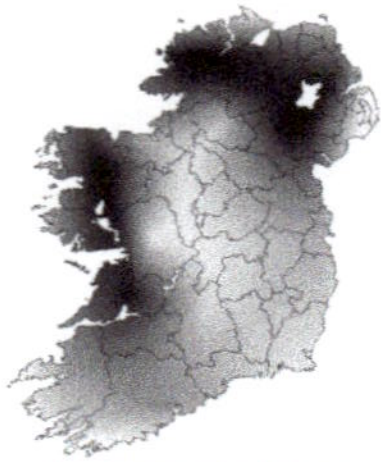

FROST/SNOW

FEBRUARY

PHASES OF THE MOON

1st	Full Moon
6th	Crossing Equator
9th	Third Quarter
10th	Apogee
12th	Southern Declination
17th	New Moon
20th	Crossing Equator
24th	First Quarter
24th	Perigee #14
25th	Northern Declination

1 Sun	2 Mon	3 Tue	4 Wed	5 Thur	6 Fri	7 Sat	8 Sun	9 Mon	10 Tue	11 Wed	12 Thur	13 Fri	14 Sat	15 Sun	16 Mon	17 Tue	18 Wed	19 Thur	20 Fri	21 Sat	22 Sun	23 Mon	24 Tue	25 Wed	26 Thur	27 Fri	28 Sat
F					XhS			3Q	A		V					N			XhN				1Q P14	^			

MONTHLY SUMMARY

1st-10th A large anticyclone to the west of Ireland slowly crosses the country. Colder with severe frosts in numerous areas.

1st-2nd Very dull and overcast conditions likely.

2nd Mild day likely.

2nd-3rd Chance of thunderstorm activity and hail mostly in the south and west. Some snowfalls possible.

3rd Mostly sunny day across the country.

3rd-7th Anticyclonic system to the southeast of the country brings a spell of quieter, sunnier, warmer and mostly dry weather, with only brief showers along the north and northwest coasts. Breezes generally light in a south to southwest flow.

6th High pressure crosses the country. Good sunny spells likely throughout.

8th-10th A cold front crosses the country in a fresher west to northwesterly airstream. Some heavy rain at times and wintry showers, including hail and snow, particularly in the west, and more widespread around the 10th when accompanied by thunderstorm activity.

The next four weeks may be milder and sunnier than of recent, while still being unusually wet in the western half of the country, while much drier than normal in the east. The first and last weeks of this upcoming outlook period may see heavy rain particularly in the west, due to deep Atlantic depressions centred to the northwest of Ireland extending unsettled frontal systems across the country, including the likelihood of gale winds at times. In contrast, the middle two upcoming weeks may be more stable due to high pressure centred to the east of Ireland bringing generally drier and sunnier weather with lighter winds. Overall, rainfall amounts may be above normal during this outlook period in the western half of the country, while below normal in the eastern half, particularly around the coasts. Overall, there may be between 15-20 rain days in the west, but only around 8-12 in the east. Sunshine hours may be above normal almost everywhere, except in some parts of the south and southeast, with the sunniest spell likely to be in the middle of this outlook period, during the spell when high pressure systems are established as the dominant weather pattern. Temperatures may be above normal on most days during this outlook period, except for a short colder than normal spell in the first week and in the third week, when frosts may be widespread and severe at times. Overall, during this outlook period, there may be 5 days on which thunderstorms are observed, 6 days with hail showers, 8 days with snow, 7 days with the development of widespread dense fogs, and 10-12 days with frosts.

11th Widespread snowfall possible.

11th-15th Very wet spell with widespread wintry showers, including hail and snow likely, particularly in the southwest, west, and northern areas. Very strong northwest winds may prevail, drawing down the Artic air with very cold temperatures becoming widespread.

12th-15th Very wet spell. Chance of thunderstorms, particularly in the south of the country.

13th Chance of further widespread snow across the country.

16th-19th Very unsettled spell as deep depressions centred to the north of Ireland continues.

14th A southwest change may occur around this time with winds remaining strong, but temperatures becoming milder.

20th-28th Anti-cyclonic spell. Mostly dry and very sunny spell apart for some light showers around the coasts now and then. Generally milder at first with a mostly light southeasterly flow, later cooler with an easterly change. Spells of frost and fog overnight, more severe in midlands to eastern areas.

21st-24th Widespread fogs overnight, some very dense at times, particularly in inland and eastern parts. Generally a mostly sunny period throughout.

21st-28th Sunnier than normal spell across the country.

22nd-28th Colder spell with widespread frosts likely, some severe in places.

23rd Chance of thunderstorms, particularly in the southwest. Unusually milder day possible.

26th A deep depression near the northwest coast may produce very unsettled conditions throughout with the chance of thunderstorms, particularly in the southwest.

26th-28th Sunnier spell.

27th-28th Colder turn. Widespread and dense overnight fog patches likely.

IRELAND RAINFALL ESTIMATES

February 2026

d=mainly dry, n= nonrecordable, l=light shrs, s=significant shrs, r=rain, h=hvy falls

		ULSTER PROVINCE					CONNAUGHT PROVINCE						LEINSTER PROVINCE				MUNSTER PROVINCE						
		Alderg	Hillsbo	Armag	Loughl	Malin I	Sligo	Ardtari	Belmul	Drums	Galway	Roscoi	Markre	Edende	Kilkenı	Dublin	Shannc	Newpo	Tralee	Killarnı	Cork	All	Moon
1st	Sun	s	s	l	h	s	h	h	h	r	r	r	h	s	l	s	r	h	s	l	l	201	F
2nd	Mon	s	l	s	s	r	s	s	s	s	s	l	s	s	l	l	s	s	s	r	l	104	
3rd	Tues	l	d	l	l	s	l	l	s	l	l	l	d	d	d	l	l	l	l	d	d	26	
4th	Wed	l	d	l	d	l	d	d	d	d	d	d	d	d	d	d	d	l	d	d	d	5	
5th	Thur	d	d	d	d	s	d	d	d	d	d	d	d	d	d	d	d	d	d	d	d	7	
6th	Fri	d	d	d	d	d	d	d	d	d	d	d	d	d	d	d	d	d	d	d	n	1	XhS
7th	Sat	l	s	d	s	s	s	s	s	s	s	s	s	l	n	d	l	s	n	h	l	101	
8th	Sun	s	l	s	l	s	l	l	l	l	d	l	l	l	l	l	h	s	h	s	s	98	
9th	Mon	l	s	l	s	s	s	s	s	s	r	r	s	l	l	d	l	r	s	h	n	121	3Q
10th	Tues	s	l	s	s	r	s	s	r	s	r	s	h	s	l	s	s	r	s	s	s	147	A
11th	Wed	l	d	s	s	d	r	s	s	s	l	s	s	l	d	d	s	s	s	s	s	79	
12th	Thur	d	n	d	h	r	h	h	h	h	h	r	h	d	l	d	l	h	s	h	l	235	V
13th	Fri	l	s	n	s	s	s	s	h	s	s	s	s	s	r	l	h	r	h	l	s	166	
14th	Sat	s	s	s	r	s	r	r	s	r	r	r	r	s	l	l	r	r	h	h	r	216	
15th	Sun	s	s	s	l	s	s	l	s	l	r	l	l	l	l	l	s	s	s	s	s	97	
16th	Mon	l	l	l	s	l	s	s	l	r	l	s	s	l	l	l	l	s	s	s	s	76	
17th	Tues	l	d	l	s	r	s	s	s	l	l	l	l	d	d	d	d	l	d	s	l	53	N
18th	Wed	d	d	d	l	l	s	l	l	l	s	l	l	d	d	d	l	s	s	l	s	40	
19th	Thur	d	d	d	d	l	l	d	s	d	d	n	l	d	d	d	d	l	d	d	d	10	
20th	Fri	d	d	d	d	n	d	d	d	d	d	d	d	d	d	d	d	d	d	d	d	1	XhN
21st	Sat	d	d	d	d	d	d	d	d	d	d	d	d	d	d	d	d	d	d	d	d	0	
22nd	Sun	d	d	d	d	d	d	d	d	d	d	d	d	d	d	d	d	d	d	d	l	1	
23rd	Mon	d	d	d	d	d	d	d	d	d	d	d	d	d	d	d	d	d	d	d	n	0	
24th	Tues	d	d	d	d	d	d	d	d	d	d	d	d	d	d	d	d	d	d	d	d	0	1Q P14
25th	Wed	d	d	d	d	d	d	d	d	d	d	d	d	d	d	d	d	d	d	d	d	0	^
26th	Thur	d	d	d	d	d	d	d	d	d	d	d	d	d	d	d	d	d	d	d	d	0	
27th	Fri	d	d	d	d	d	d	d	d	d	d	d	d	d	d	d	d	d	d	d	d	0	
28th	Sat	d	d	d	d	d	d	d	d	d	d	d	n	d	d	d	d	d	d	d	d	0	
	Estimate:	45	42	48	102	104	134	102	126	104	118	83	111	43	30	22	91	135	134	154	58	1784	
	Average:	58	68	54	91	87	92	94	97	78	88	98	92	70	57	49	76	127	141	103	98	1717	
	Trend:	drr	drr	av	av	w tr	w tr	av	w tr	w tr	w tr	drr	w tr	drr	drr	drr	w tr	av	av	w tr	drr	av	

IRELAND SUNSHINE ESTIMATES

February 2026

F=fine (8-12 hours of sunshine), pc= partly cloudy (4-7 hours), c=cloudy (1-3 hours), o=overcast (0 hours)

		ULSTER PROVINCE					CONNAUGHT PROVINCE						LEINSTER PROVINCE				MUNSTER PROVINCE						
		Alderg	Hillsbo	Armag	Loughl	Malin I	Sligo	Ardtari	Belmul	Drums	Galway	Roscoi	Markre	Edende	Kilkeni	Dublin	Shannc	Newpo	Tralee	Killarne	Cork	All	Moon
1st	Sun	o	o	o	o	o	o	o	o	o	o	o	o	o	o	o	o	o	o	o	o	0	F
2nd	Mon	o	o	o	o	o	o	o	o	o	o	o	o	o	o	o	o	o	o	o	o	3	
3rd	Tues	pc	pc	pc	pc	pc	c	pc	c	pc	pc	c	pc	pc	pc	pc	pc	pc	c	c	pc	96	
4th	Wed	o	o	o	o	o	o	o	o	o	o	o	o	o	o	o	o	o	o	o	o	7	
5th	Thur	o	o	o	o	o	o	o	c	c	o	c	o	c	c	o	o	c	c	c	pc	23	
6th	Fri	c	c	c	c	c	c	c	o	pc	c	c	c	pc	pc	pc	pc	pc	pc	pc	pc	75	XhS
7th	Sat	o	o	o	o	o	o	o	o	o	c	o	o	o	o	c	c	c	o	o	o	16	
8th	Sun	o	o	o	o	o	o	o	o	o	o	o	o	o	o	o	o	o	o	o	o	8	
9th	Mon	c	c	c	c	c	o	c	o	c	c	c	c	c	c	c	c	c	c	c	c	40	3Q
10th	Tues	c	c	c	c	c	c	c	c	c	c	c	c	c	c	pc	c	c	c	c	c	65	A
11th	Wed	o	o	o	o	o	o	o	c	c	c	c	o	c	c	c	c	c	c	c	c	33	
12th	Thur	o	o	o	o	o	o	o	o	o	o	o	o	o	o	o	o	o	o	o	o	6	V
13th	Fri	o	o	o	o	o	o	o	o	o	o	o	o	o	o	o	o	o	c	c	c	8	
14th	Sat	pc	pc	pc	pc	pc	c	pc	c	pc	c	c	pc	pc	pc	pc	c	pc	c	c	c	79	
15th	Sun	o	o	o	o	o	o	o	o	o	o	o	o	o	c	o	c	c	c	c	c	16	
16th	Mon	pc	pc	pc	pc	pc	pc	pc	pc	pc	c	pc	pc	pc	pc	pc	c	pc	c	c	c	98	
17th	Tues	o	o	o	o	o	o	o	o	o	o	o	o	o	o	o	c	c	o	o	o	9	N
18th	Wed	c	c	c	c	c	c	c	pc	c	c	c	c	c	o	o	o	o	o	o	o	31	
19th	Thur	c	c	c	c	c	o	c	o	c	c	c	c	c	c	c	pc	c	o	o	o	33	
20th	Fri	o	o	o	o	o	o	o	pc	o	c	pc	o	o	o	o	o	o	pc	pc	o	24	XhN
21st	Sat	pc	pc	pc	pc	pc	pc	pc	F	pc	F	F	pc	pc	pc	F	F	pc	F	F	c	140	
22nd	Sun	F	F	F	F	F	pc	F	F	pc	pc	pc	F	pc	pc	F	pc	pc	F	F	pc	153	
23rd	Mon	F	F	F	F	F	F	F	F	F	F	F	F	F	F	F	F	F	F	F	F	173	
24th	Tues	o	o	o	o	o	c	o	F	o	pc	F	o	o	o	o	pc	c	pc	pc	o	55	1Q P14
25th	Wed	o	o	o	o	o	o	o	c	o	c	c	o	o	o	o	o	o	pc	pc	o	17	^
26th	Thur	pc	pc	pc	pc	pc	c	pc	o	pc	o	o	pc	pc	pc	F	o	c	o	o	o	68	
27th	Fri	pc	pc	pc	pc	pc	c	pc	F	pc	F	F	pc	pc	pc	F	F	F	pc	pc	pc	138	
28th	Sat	pc	pc	pc	pc	pc	pc	pc	F	pc	pc	pc	pc	pc	c	pc	c	c	F	F	o	111	
	Estimate:	74	74	74	74	74	55	74	84	75	82	84	74	76	78	91	79	78	82	82	63	1526	
	Average:	64	64	62	63	62	62	62	67	60	64	64	65	62	64	70	66	61	62	60	64	1268	
	Trend:	av	av	more	more	more	av	more	more	more	more	more	av	more	more	more	more	more	more	more	av	more	

FEBRUARY **1** SUNDAY

Low pressure prevails over the country with a depression to the northwest and another to the northeast of Ireland and high pressure ridges to the south. Widespread cloud brings a dull and gloomy day throughout, with light breezes at first possibly bringing some patches of freezing fog here and there, with winds strengthening later in the day. General showers or persistent drizzle almost everywhere, and including the chance of snow in many places.

Belfast: Mild day cold night, some showers, dull and gloomy, chance of fog patches
Dublin: Overcast, mild day cold night, scattered showers, light breezes
Cork: Overcast, light showers, mild temperatures, chance of fog patches
Galway: Overcast, rain, moderate breezes, slightly warmer

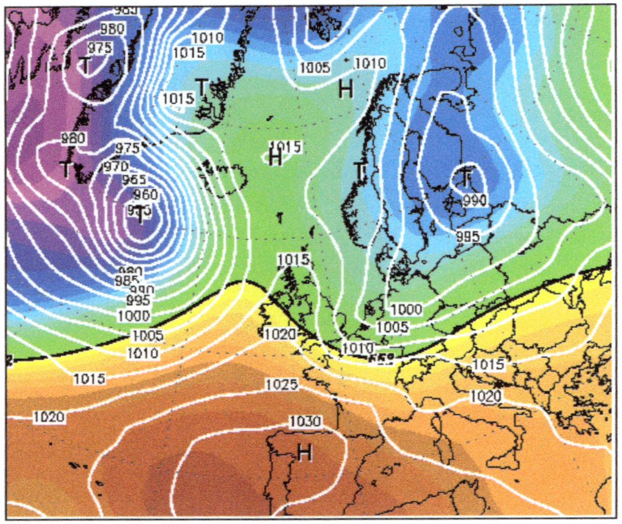

RAIN POTENTIAL

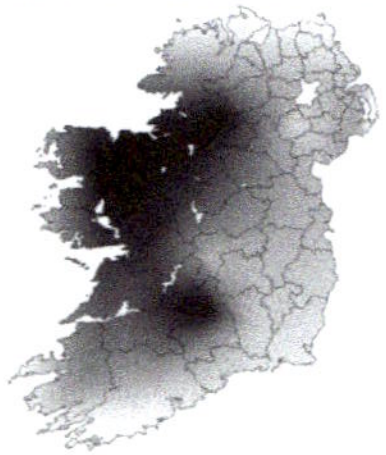

FROST/SNOW

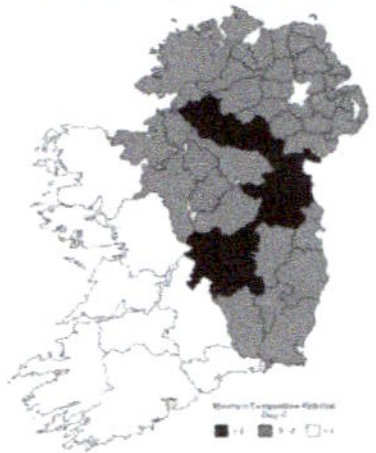

FEBRUARY **2** MONDAY

Low pressure over Greenland extends a fresh to strong southwesterly flow over the country. Widespread cloud, rain and drizzle prevails, as well as the chance of snow in most regions except along the southern coastal fringes. Winds may reach up to gale gusts in the west towards evening.

Belfast: Overcast, milder, fog possible, freshening breezes later in the day, showers
Dublin: Light showers, warmer, mostly cloudy, moderate breezes
Cork: Brief showers, dull and gloomy, light breezes, chance of fog overnight
Galway: Overcast, scattered showers, windy spells

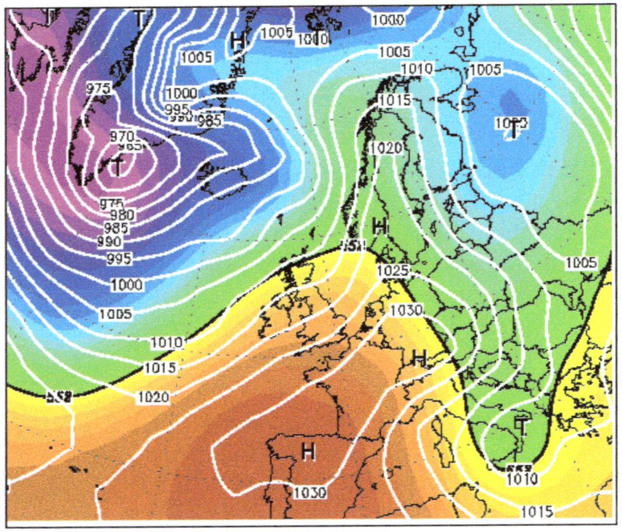

RAIN POTENTIAL

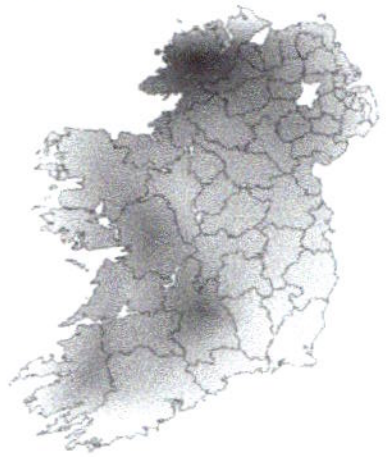

FROST/SNOW

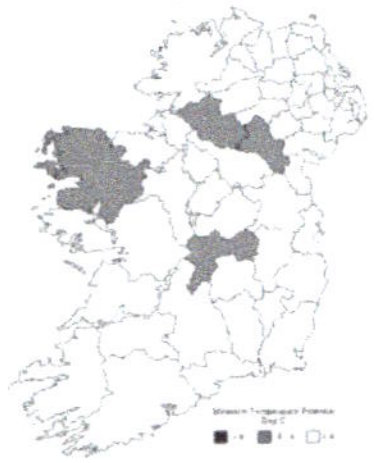

FEBRUARY **3** TUESDAY

A strong and blustery west to northwest flow prevails throughout the country, directed from a series of depressions to the west and north of Ireland. A changeable day with some sunny outbreaks. Mostly dry along the eastern and southern coasts, with brief showers elsewhere, more frequent in the north and west.

Belfast: Occasional sunny spells, dry, mild day cold night, unsettled winds at times
Dublin: Changeable, partly cloudy, isolated showers, colder, freshening winds
Cork: Moderate breezes, dry, mostly sunny, colder
Galway: Unsettled winds, sunny spells, isolated showers, chance of snowflurries

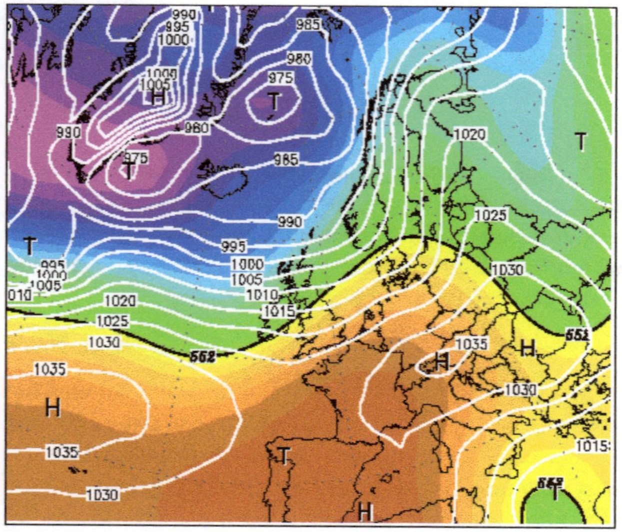

RAIN POTENTIAL

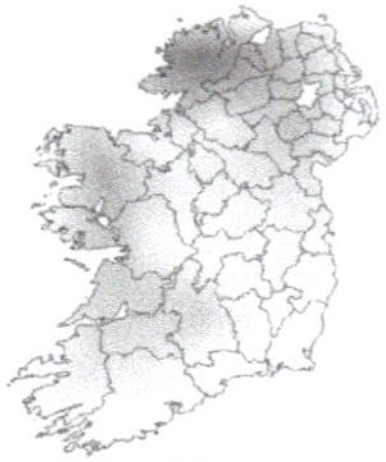

FROST/SNOW

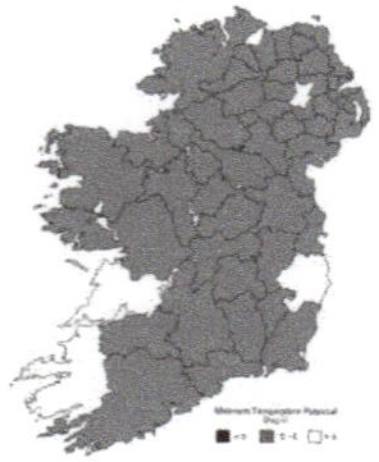

FEBRUARY **4** WEDNESDAY

West to southwest winds strengthen as a depression moves southwards nearer to Scotland, intensifying to gale force at times, particularly along the Ulster and Connaught coasts. However, high pressure ridges to the south of the country. Widespread cloud everywhere, with isolated showers mostly in the north, and chance of some hail showers in central parts. Mostly dry elsewhere.

Belfast: Overcast, cooler, mostly dry, odd threats of rain, stormy air, chance of gales
Dublin: Stormy air, strong winds, overcast, mostly dry with occasional threats of rain
Cork: Dry, mostly overcast, slightly warmer, strengthening breezes
Galway: Stormy air, overcast, mostly dry, threats of rain, gusty and strong winds

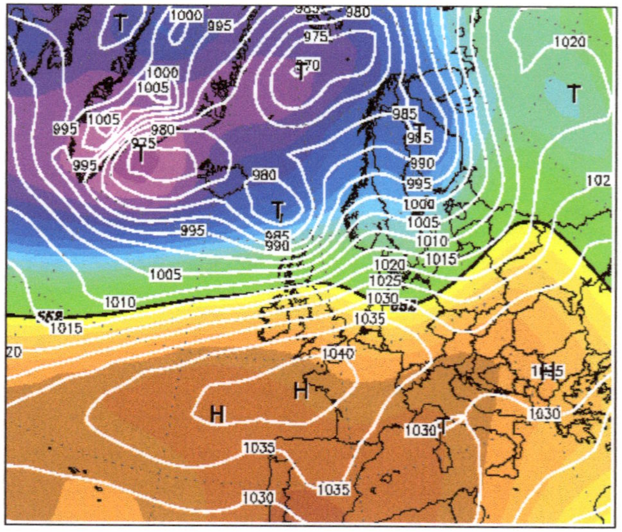

RAIN POTENTIAL

FROST/SNOW

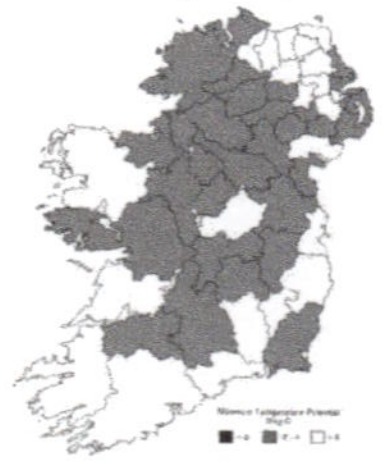

FEBRUARY 5 THURSDAY

A high pressure system to the south of Ireland ridges over the country clearing rain activity in most areas south of Ulster, as well as bringing some brief sunny respites in Munster. Some showers may still fall as snow in the higher reaches of central parts overnight. Winds continue to be fresh to strong and gusty in a southwest flow.

Belfast: Overcast, milder, dry, blustery, passing threats of rain
Dublin: Overcast, dry with passing threats of rain, blustery at times
Cork: Partly cloudy, dry, moderate breezes
Galway: Overcast, windy, dry

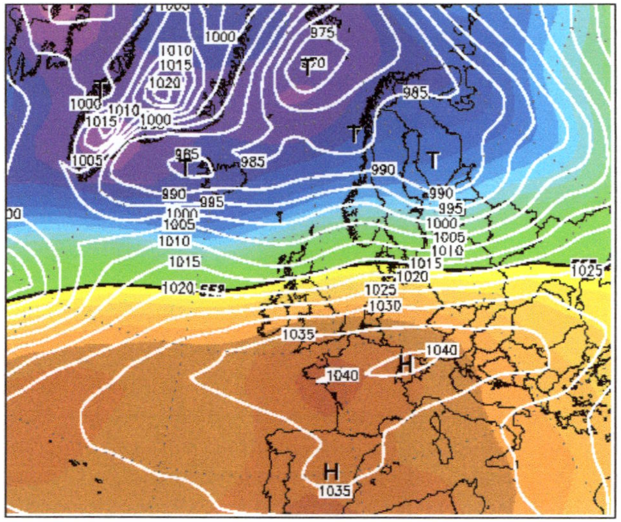

RAIN POTENTIAL

FROST/SNOW

FEBRUARY 6 FRIDAY

A high pressure ridge continues to bring a mostly dry day throughout the country, with increasing outbreaks of sunny spells. Winds fresh to strong at first in a southwest flow, more gusty along the northwest coasts, before easing to light later in the day. Chance of some patches of fog overnight, mostyl in the south. Possible snow flurries and freezing mists in central parts.

Belfast: Cloudy, dry, mild temperatures, blustery winds at times
Dublin: Partly cloudy, dry, mild temperatures, unsettled winds with gusts
Cork: Occasional sunny spells, brief misty patches, mostly calm air, chance of fog
Galway: Fresh winds, chance of gales at times, dry, some sunny breakthroughs

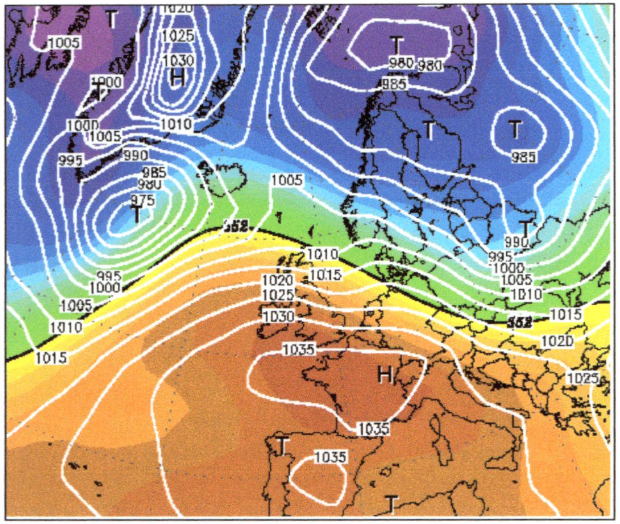

RAIN POTENTIAL

FROST/SNOW

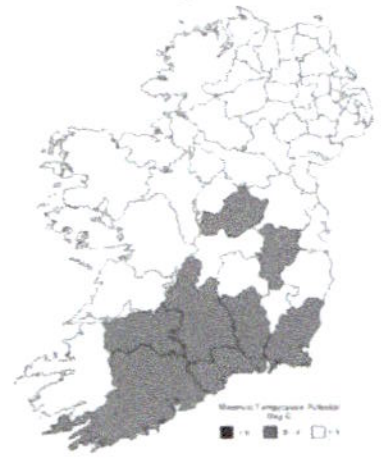

FEBRUARY 7 SATURDAY

High pressure continues to ridge over the southern half of Ireland with a light to moderate southerly flow prevailing. In contrast, a depression to the north of Scotland nears the northern coasts, strengthening winds in northern parts of Ireland, possible reaching gale gusts for a time in the northwest. Mostly dry with some sunny spells at first, followed by increasing cloud and occasional showers in the west at first, moving eastward later in the day.

Belfast: Mild day cool night, dull and gloomy air, winds easing, occasional showers
Dublin: Dry, cloudy, winds moderating
Cork: Overcast, isolated showers, gentle breezes to mostly calm air
Galway: Cloudy, scattered showers, winds easing

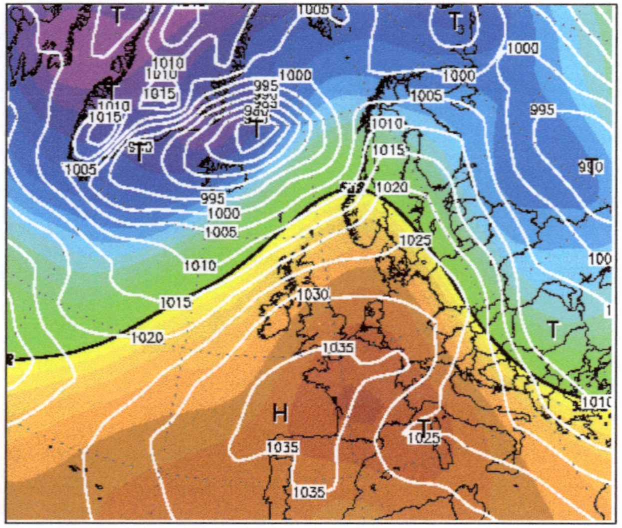

RAIN POTENTIAL

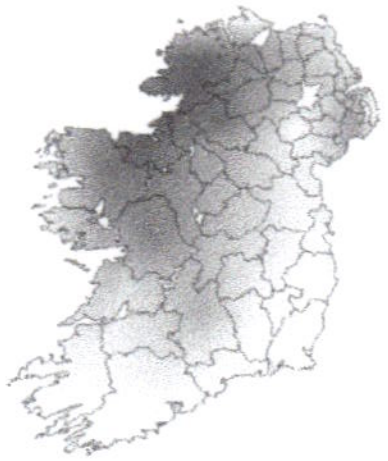

FROST/SNOW

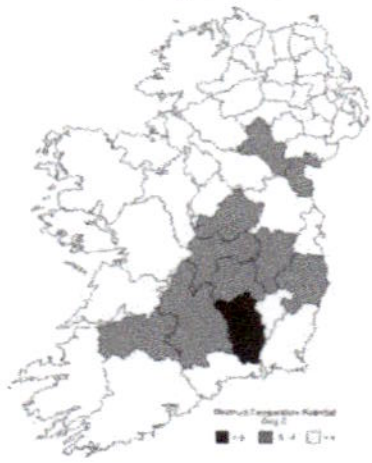

FEBRUARY 8 SUNDAY

Despite the high pressure ridge continuing to linger over Ireland, particularly through southern parts, widespread cloud prevails everywhere, with a very dull and gloomy day likely, after some patches of overnight fog clear. Brief showers return almost everywhere, with some heavier falls possible in the southwest. Winds generally light to moderate but gustier in the north and northwest.

Belfast: Overcast, brief showers, light breezes, chance of overnight patches of fog
Dublin: Overcast, drizzly showers, mild day cold night, moderate breezes
Cork: Moderate breezes, overcast and dull, scattered showers
Galway: Overcast, colder, mostly dry, increasing threats of rain, moderate breezes

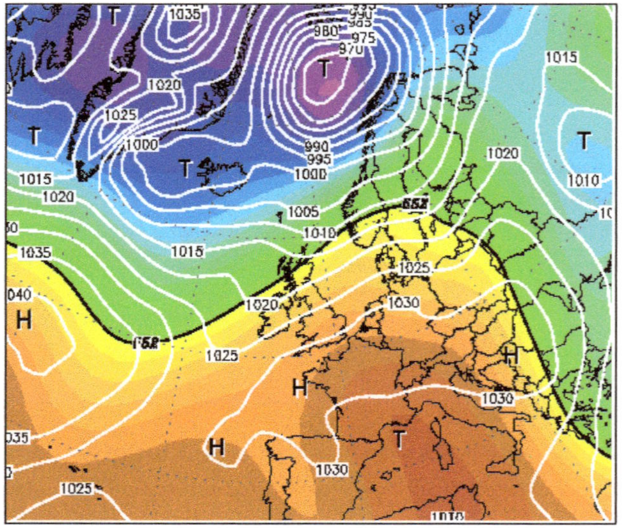

RAIN POTENTIAL

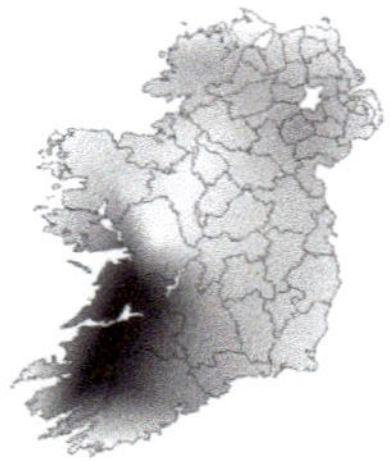

FROST/SNOW

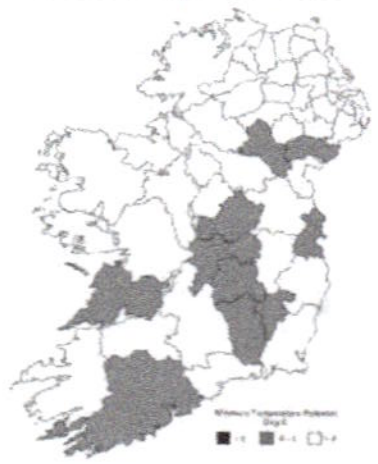

FEBRUARY 9 MONDAY

High pressure continues to prevail over Ireland with a moderate west to northwest flow prevailing, but gustier at times along the western coasts. A mixed day of sunny outbreaks and scattered wintry showers, including the chance of some light snow or flurries.

Belfast: Cloudy, colder, scattered showers, moderate breezes
Dublin: Cloudy, colder, dry, moderate breezes, chance of frost or snow flurries
Cork: Occasional sunny spells, isolated showers, light shifting breezes, colder
Galway: Cloudy, rain, moderate winds, overnight frost possible, chance of snow

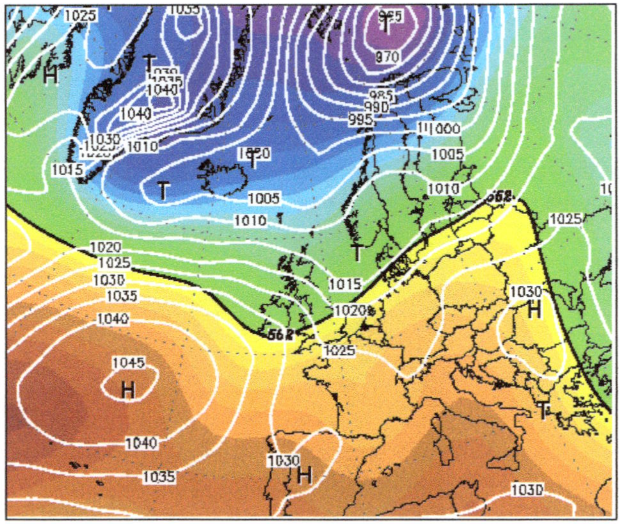

RAIN POTENTIAL

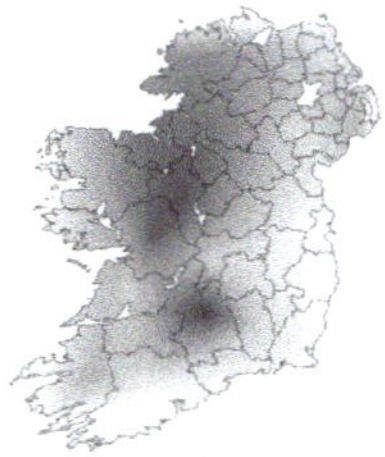

FROST/SNOW

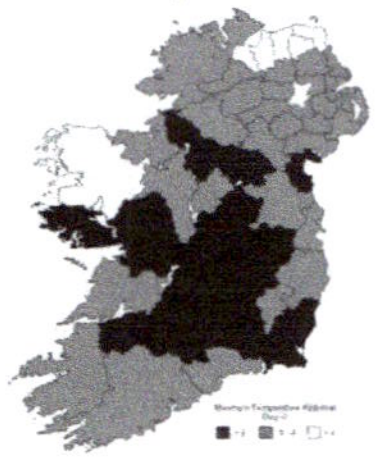

FEBRUARY 10 TUESDAY

An intense North Atlantic depression may swiftly bear down on Ireland around this time bringing the arrival of widespread stormy conditions everywhere. Strong and gusty west to southwest winds prevail, up to gale force and with storm force gusts possible at times. Some sunny spells at first before bands of rain, some accompanied by thunderstorms and hail activity, cross the country. Chance of wintry showers manifesting as snow and sleet in many districts.

Belfast: Stormy, cold, partly cloudy, some showers, strong winds, chance of snow
Dublin: Stormy winds, mix of showers and sunny spells, cold, chance of snow
Cork: Stormy air, cloudy, squally showers, chance of snow
Galway: Storm, cloudy, cold, rain, chance of thunderstorms, hail and snow

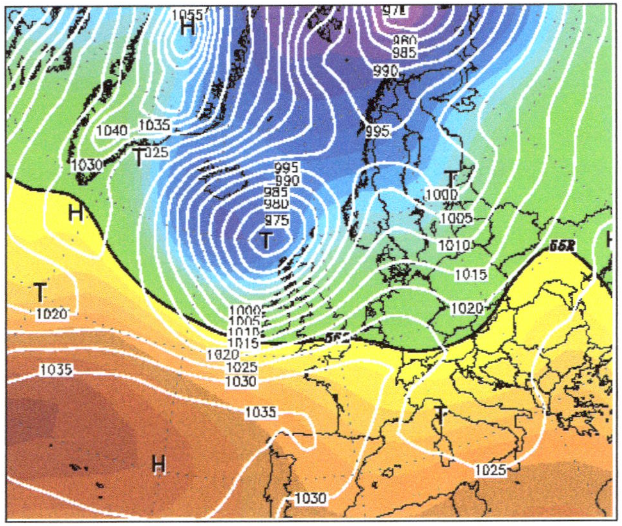

RAIN POTENTIAL

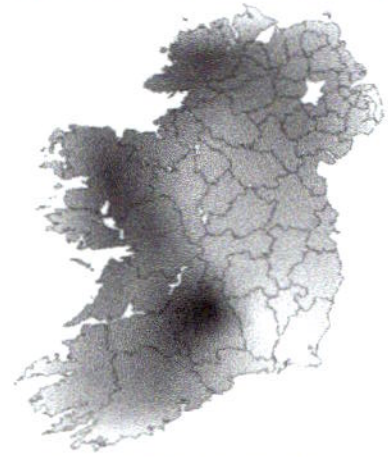

FROST/SNOW

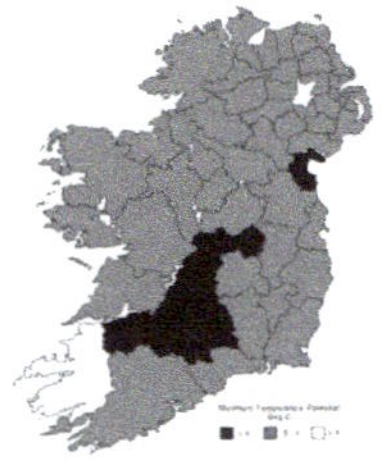

FEBRUARY 11 WEDNESDAY

Stormy conditions may continue throughout the day, with very strong and forceful winds prevailing in a west to northwest flow. Widespread cloud and very cold and frosty everywhere. Mostly dry however, along the northern coasts and in the coastal counties of Leinster, with outbreaks of showers elsewhere, some of them wintry and sometimes possibly as snow.

Belfast: Overcast, colder, stormy winds, mostly dry, chance of brief evening snow
Dublin: Stormy, cloudy, very cold, mostly dry, severe frosts, chance of snowflurries
Cork: Mix of showers and sunny outbreaks, colder, snow likely, windy at times
Galway: Stormy air, cloudy, colder, brief showers, snow and gales possible

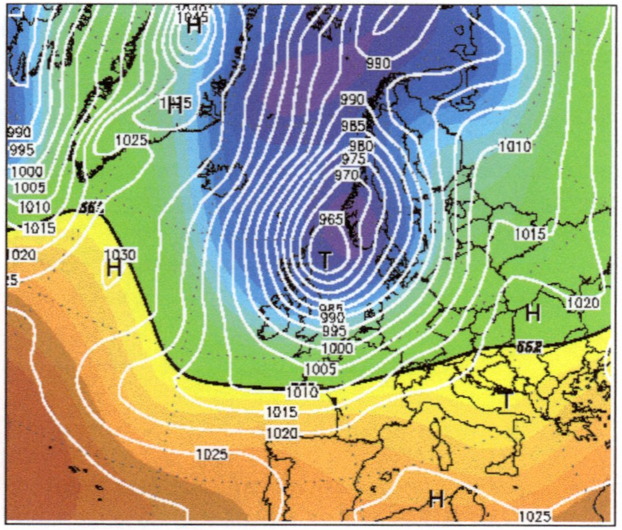

RAIN POTENTIAL

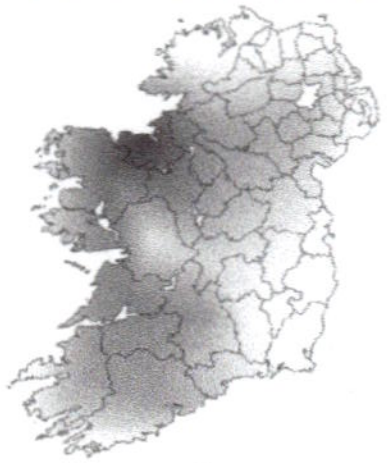

FROST/SNOW

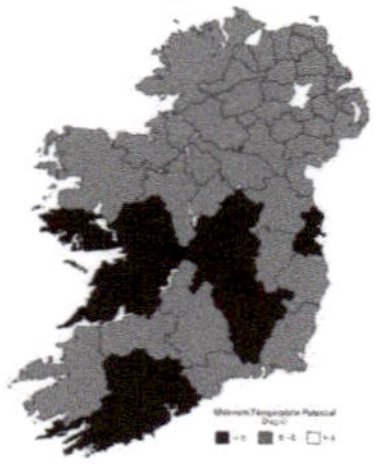

FEBRUARY 12 THURSDAY

Low pressure continues to bring wintry showers throughout most of the western and central parts of the country, while the eastern coasts may be mostly dry. A very dull and gloomy day, with patches of persistent drizzle, and chances of snowflurries amongst the wintry rain outbreaks. Winds ease to mostly light in a west to southwest flow, later backing more southerly and becoming stronger.

Belfast: Overcast, cold, frosts, blustery winds easing later, chance of snowflurries
Dublin: Overcast, dry, milder day, overnight frosts, winds easing to light for a time
Cork: Overcast, cold, scattered showers, frosts and possible patches of fog at times
Galway: Winds easing, overcast, severely cold, heavy precipitation, snow likely

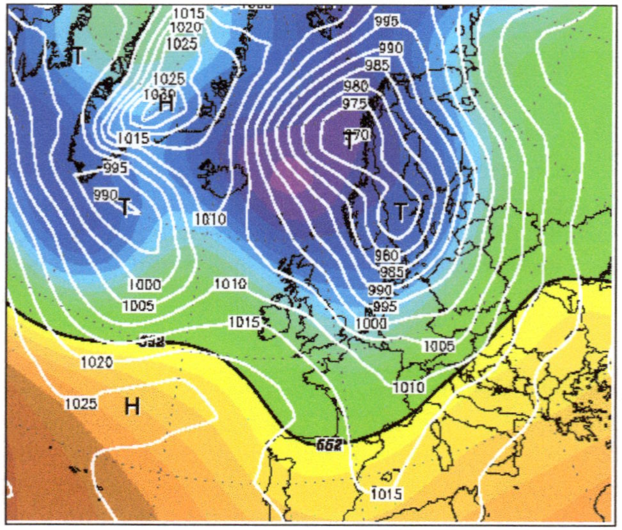

RAIN POTENTIAL

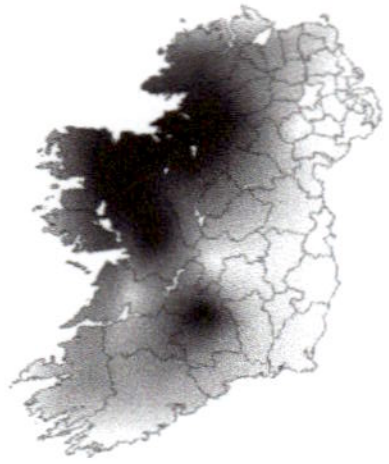

FROST/SNOW

FEBRUARY **13** FRIDAY

A depression centred to the northwest of Ireland maintains very strong south to southwest winds over the country, possibly up to gale force around the northern and western coasts. Storm conditions continue throughout with widespread cloud and wintry showers everywhere, some of them as sleet or hail and possible snow overnight, as well as the chance of bands of thunderstorms crossing the country for a time.

Belfast: Stormy air, overcast, squally showers, cold, chance of sleet or hail
Dublin: Overcast, cold and dull, brief showers, unsettled and blustery winds at times
Cork: Cloudy, scattered showers, cold, frosts, strong winds, chance of snow
Galway: Strong and forceful winds, overcast, cold, squalls, chance of overnight snow

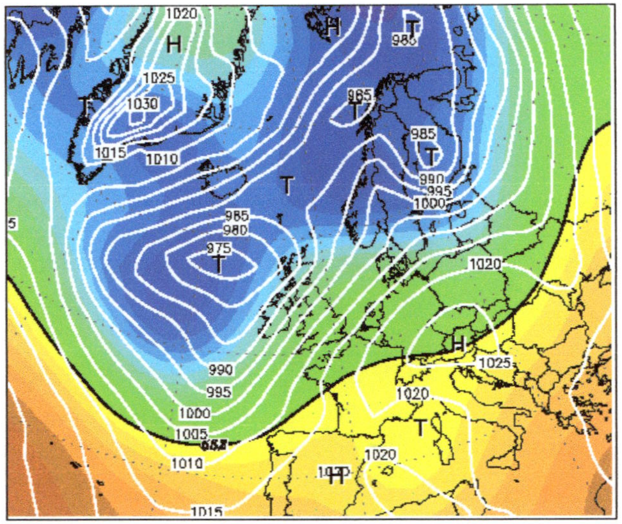

RAIN POTENTIAL

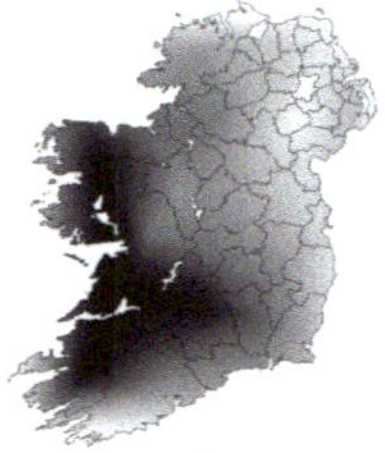

FROST/SNOW

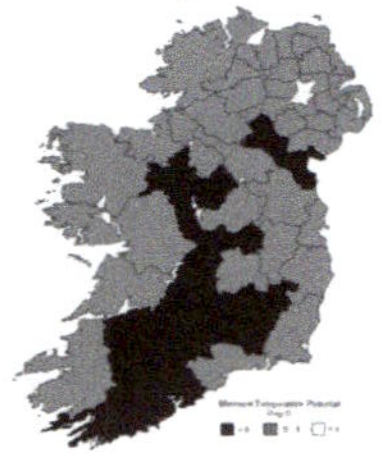

FEBRUARY **14** SATURDAY

A depression maintains its dominance over the country with a strong southwest flow over the northern half of Ireland, while more westerly and light in the south. A changeable day, with a mix of sunny and dry spells, later followed by increased cloud and wintry showers. Chance of snow overnight especially in central parts. Rain may be heavy for a time, particularly in the southwest.

Belfast: Changeable, mix of sunny spells and showers, cold, fresh breezes at times
Dublin: Changeable, mix of sunny spells and light showers, cold, unsettled breezes
Cork: Unsettled and cloudy, cold, frosts, rain, chance of overnight snow
Galway: Winds easing, some sunny spells, rain, chance of thunderstorms and snow

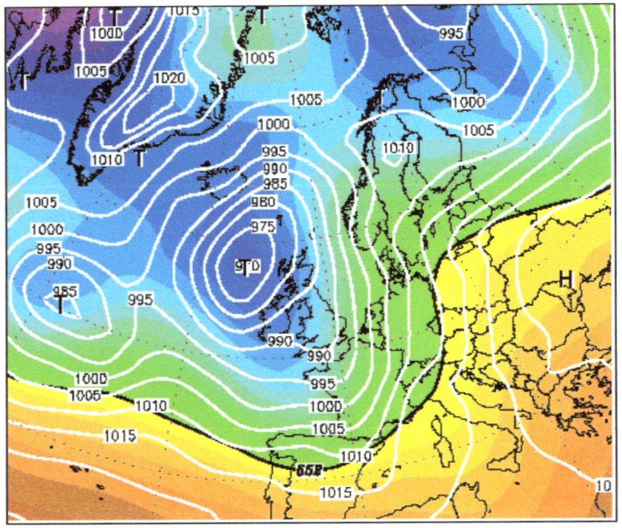

RAIN POTENTIAL

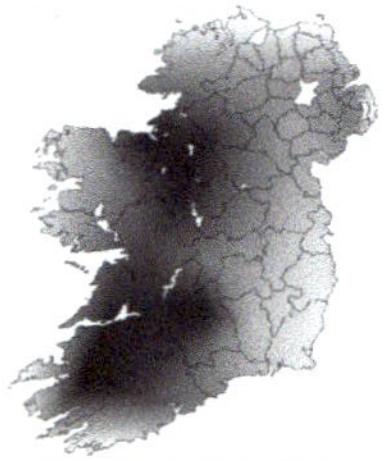

FROST/SNOW

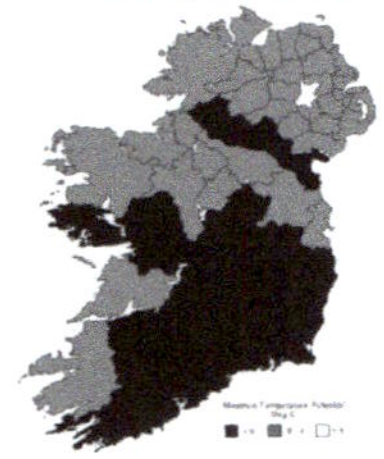

FEBRUARY **15** SUNDAY

Low pressure continues to bring widespread cloud and scattered outbreaks of rain throughout the country, with occasional heavy falls in some western districts, and possible snow overnight in the north. Winds generally light and variable in the north, with chance of fog patches here and there, while fresher and more gusty over the southern half of the country.

Belfast: Dull and gloomy air, passing showers, mostly calm air, fog possible
Dublin: Fluctuating breezes, dull and gloomy air, light showers
Cork: Fluctuating breezes, cloudy, scattered showers, milder temperatures
Galway: Overcast, rain, cold, shifting breezy spells

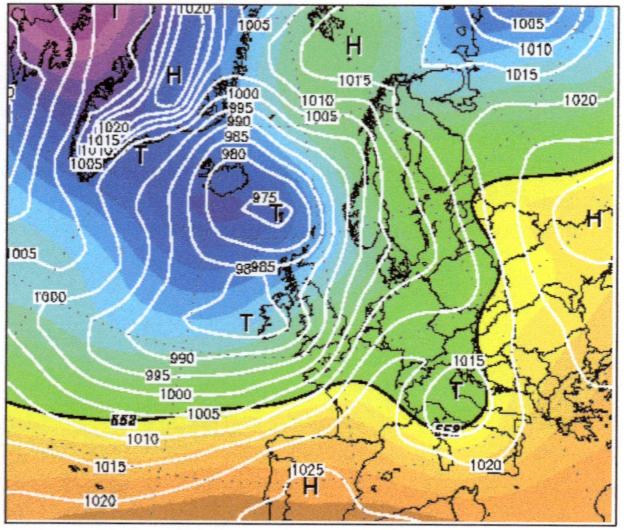

RAIN POTENTIAL

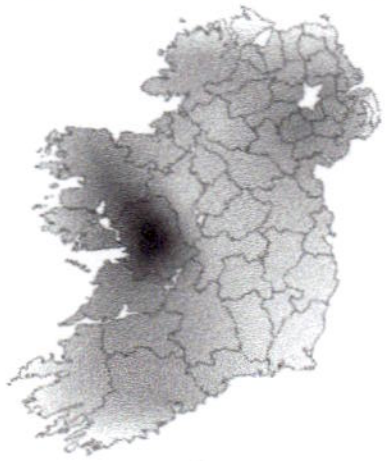

FROST/SNOW

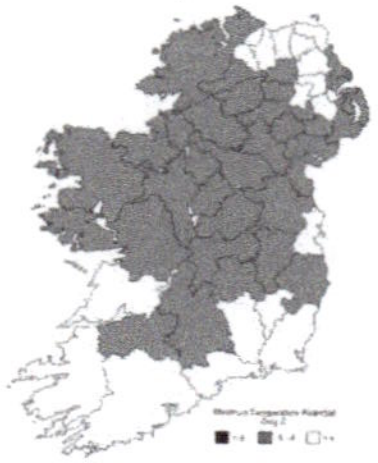

FEBRUARY **16** MONDAY

A fresh and strong northwesterly flow accompanies a low pressure system across the country, easing with a southerly change later in the day. Very cold overnight with frosts and the odd patch of fog, particularly in central and eastern parts. A mixed day with some sunny spells and the occasional showery spell, including the chance of further light snow mostly in northern parts.

Belfast: Cold, overnight showers, occasional sunny spells, light breezes
Dublin: Shifting breezy spells, mostly sunny, isolated overnight showers likely
Cork: Changeable, mix of showers and sunny spells, chance of misty patches
Galway: Mix of showers and sunny spells, milder day, overnight frosts, fresh breezes

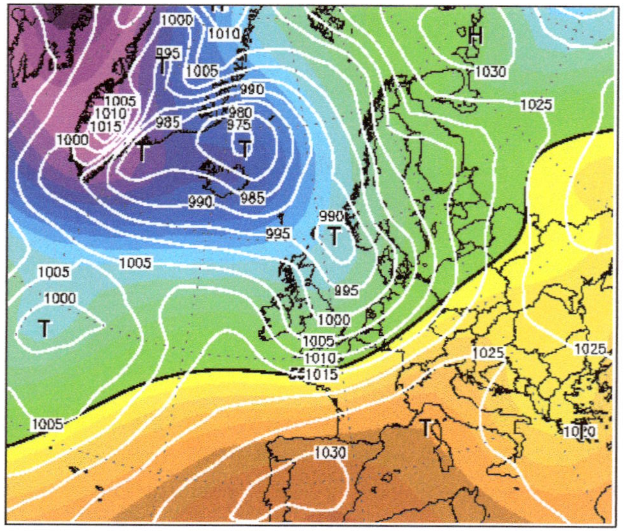

RAIN POTENTIAL

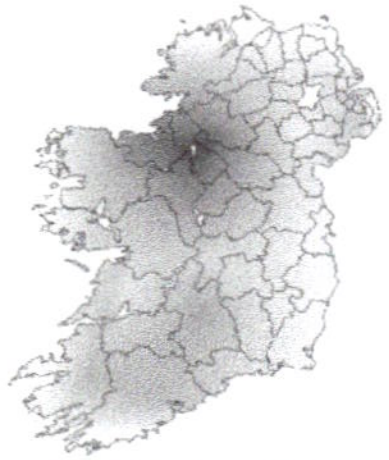

FROST/SNOW

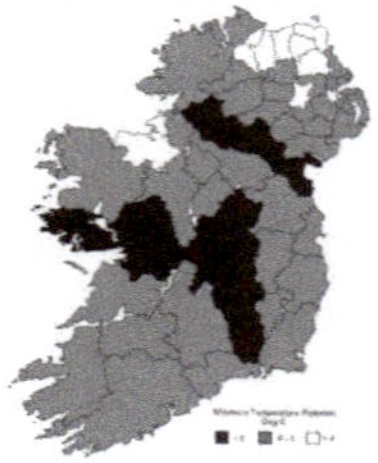

FEBRUARY 17 TUESDAY

A depression centred to the northwest of Ireland extends very strong and gusty southwest winds over the north and west of the country, while a high pressure ridge covers the east and south, allowing for lighter southwesterly winds and the odd patch of fog overnight. Showers likely in most districts, more frequent in the west and north, while isolated in the east and south and chance of snow flurries overnight, mosty in the central, northern, and eastern parts. Milder temperatures.

Belfast: Overcast, milder, mostly dry, windy at times, chance of overnight fog
Dublin: Overcast, mostly dry, threats of rain, milder, windy with occasional gusts
Cork: Overcast, brief showers, fresh breezes
Galway: Overcast, light showers, mild day overnight frosts, blustery at times

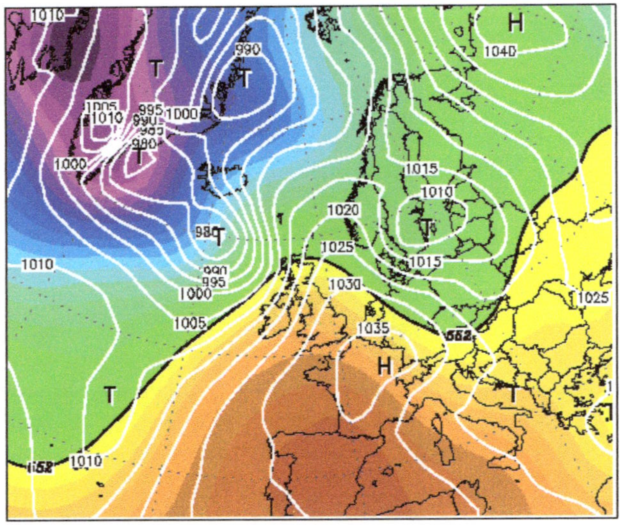

RAIN POTENTIAL

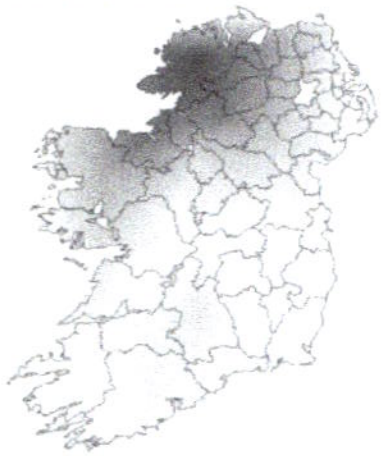

FROST/SNOW

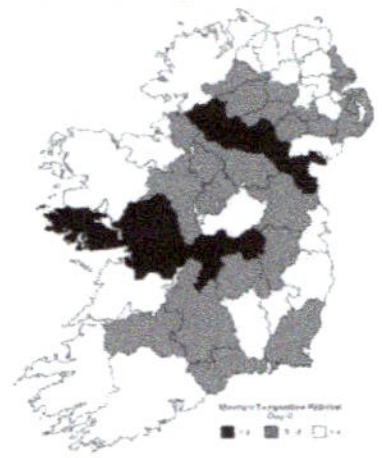

FEBRUARY 18 WEDNESDAY

A high pressure system to the southeast of Ireland extends its ridge over the eastern half of the country, bringing milder temperatures and mostly dry conditions in the Leinster and northeast of Ulster regions.In contrast, a weak depression centres very near the northwest of Ireland bringing further dull and gloomy conditions and showers into the western half of the country. Winds generally moderate in a south to southwest flow, but gustier at at times along the western coasts.

Belfast: Cloudy, warmer, mostly dry, windy spells, chance of overnight misty patches
Dublin: Overcast, milder, dry with passing threats of rain, moderate breezes
Cork: Overcast, scattered showers, windy spells
Galway: Cloudy, occasional showers, milder, windy spells

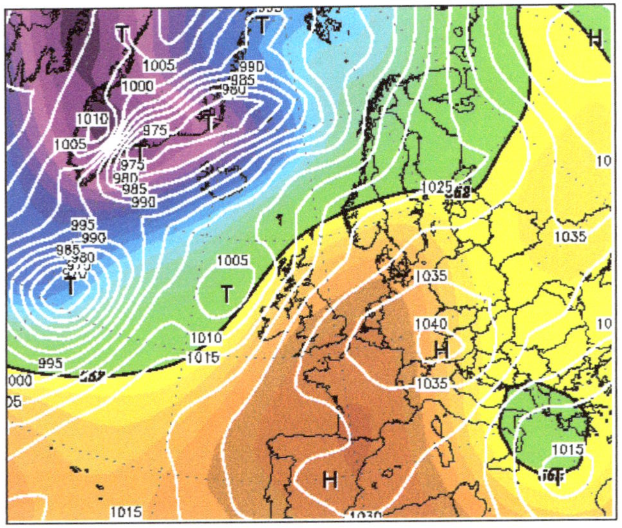

RAIN POTENTIAL

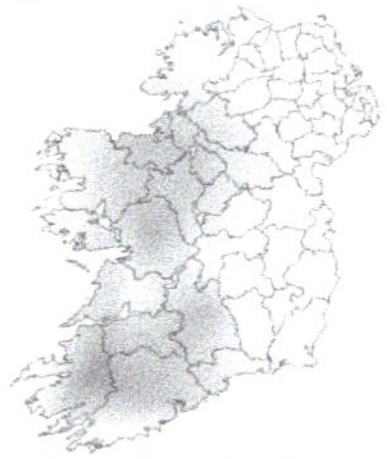

FROST/SNOW

FEBRUARY **19** THURSDAY

A high pressure ridge covers much of the country bringing a moderate to fresh south to southwest flow throughout, fresher and more gusty along the western and northern coasts. Mostly dry day with passing sunny spells everywhere, except for more cloud and some showers in the northwest counties from Galway to Donegal. Two depression centres northwest of Ireland the past couple of day merge into a single intense depression situated west of Iceland around this time, but are not expected to affect Ireland over the next few days.

Belfast: Cloudy, windy, dry, slightly cooler
Dublin: Cloudy, breezy, dry, mild day cold night
Cork: Dull and gloomy air, mostly dry with passing threats of rain, light breezes
Galway: Dry, mild temperatures, occasional sunny spells, moderate breezes

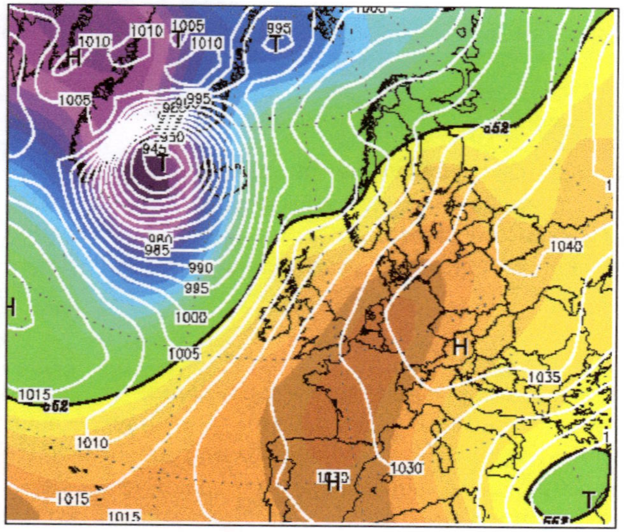

RAIN POTENTIAL

FROST/SNOW

FEBRUARY **20** FRIDAY

A large anticyclonic system covers Ireland directing a gentle east to southeast flow over the country. Overnight drizzle patches, mostly in the north and northwest clear early, with some patches of morning fog likely over the northern half of the country. Mostly dry and cloudy day throughout. Milder temperatures likely in the south and west, while cooler in the midlands, east and north. Longer sunny spells in the west and southwest likely.

Belfast: Dry, milder day colder night, mostly calm air, overcast, chance of fog
Dublin: Colder, overcast, mostly dry, gentle breezes to mostly calm air
Cork: Overcast, colder, dry, threats of rain, mostly calm air, chance of fog
Galway: Cloudy, cooler, mostly calm air, dry

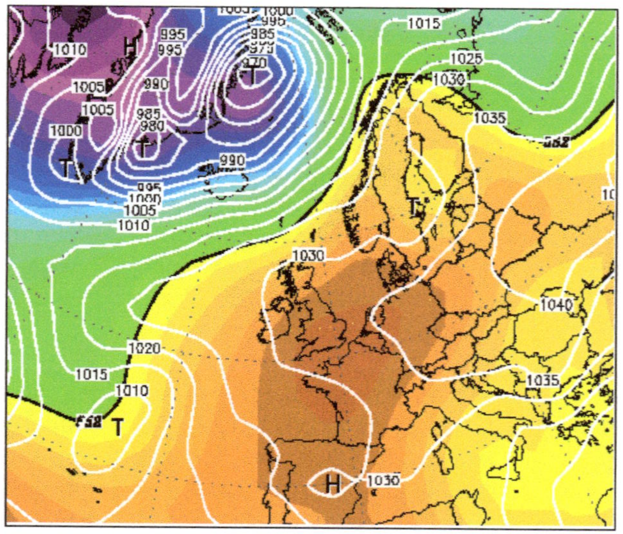

RAIN POTENTIAL

FROST/SNOW

FEBRUARY 21 SATURDAY

A moderate southeasterly flow accompanies high pressure throughout the country, with patches of overnight fog in many areas. Mostly dry and sunny day almost everywhere, although some longer cloudy spells likely in the far south. Chance of frost overnight mostly in the east.

Belfast: Mostly calm air, overnight fog patches, colder, occasional sunny spells
Dublin: Sunny, dry, mostly calm air, mild day overnight frosts, chance of fog patches
Cork: Partly cloudy, dry, mostly calm air, fog patches likely
Galway: Sunny, dry, mild day, cold night, light breezes

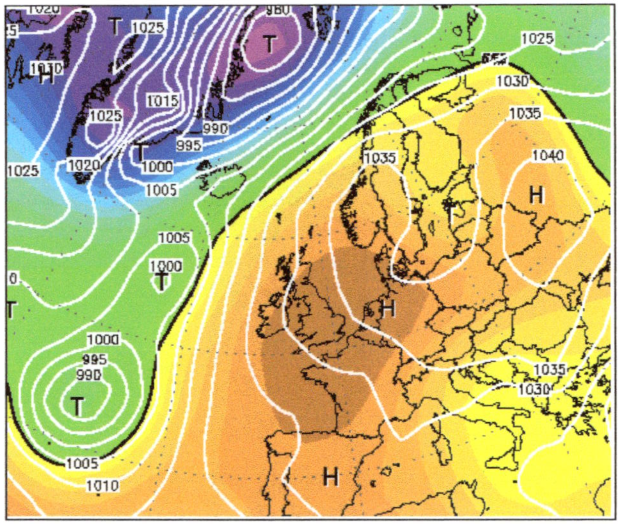

RAIN POTENTIAL

FROST/SNOW

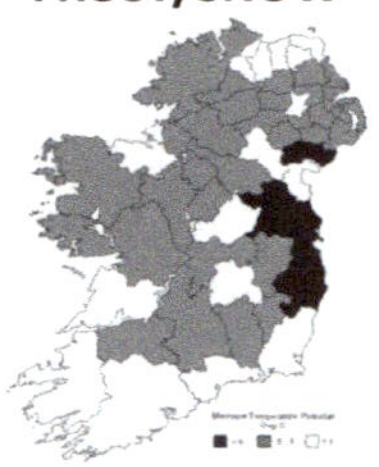

FEBRUARY 22 SUNDAY

High pressure prevails with a light southeasterly flow throughout. Sunny and dry everywhere, with patches of frost and fog likely overnight in many districts, although like to be milder along the southern coasts.

Belfast: Sunny, dry, warmer day, overnight frosts, mostly calm air, chance of fog
Dublin: Sunny, dry, frosts, calm air, chance of fog patches
Cork: Increasing sunny spells, isolated showers, gentle breezes, possible fog
Galway: Mostly sunny and dry, cooler, overnight patches of frost likely, light breezes

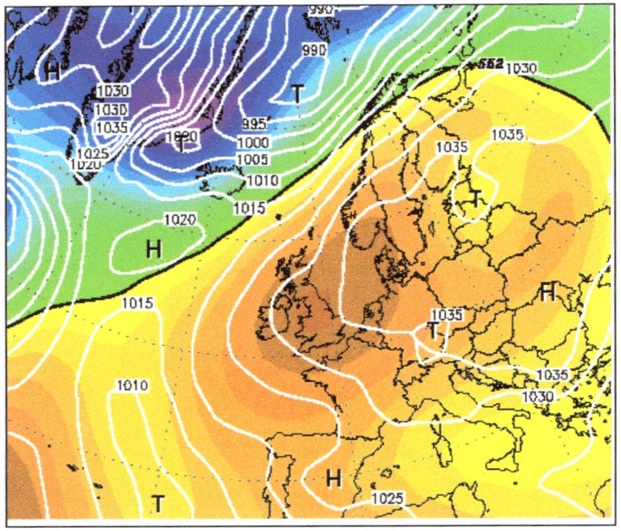

RAIN POTENTIAL

FROST/SNOW

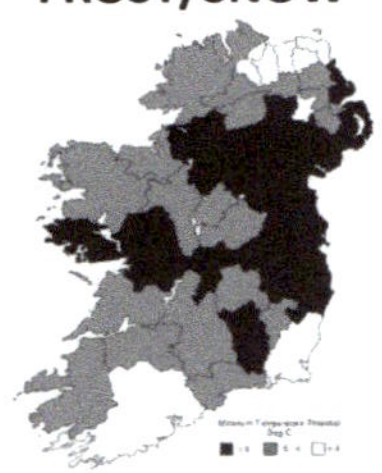

FEBRUARY **23** MONDAY

High pressure continues to dominate the country. A light mostly easterly air prevails, with widespread patches of frost and fog in most areas. Dry and mostly sunny day, except in areas where fog may be slow to clear in some districts, possibly not lifting until the afternoon.

Belfast: Sunny, dry, calm atmosphere, frost and fog likely
Dublin: Dry and sunny, mild day, overnight frosts, calm air, fog possible
Cork: Sunny, mostly dry with chance of fog, possible brief drizzle patches overnight
Galway: Sunny, dry, frosts, light breezes

RAIN POTENTIAL

FROST/SNOW

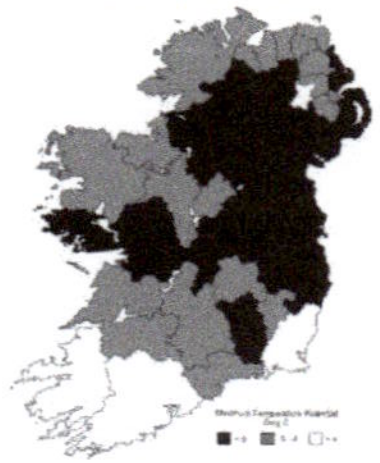

FEBRUARY **24** TUESDAY

Cold, dry, and a very light east to southeast flow continues to prevail throughout the country as Ireland sits under a strong high pressure influence. Frost and fog likely in many districts once more, but cloudier in the northern half of the country, and sunnier in the southern half.

Belfast: Overcast and dull, dry, frosts, mostly calm air, chance of fog
Dublin: Calm and serene air, dry, overcast, colder, fog likely, patches of frost possible
Cork: Colder, overcast, mostly dry, odd threat of rain, mostly calm air
Galway: Partly cloudy and cooler, dry, gentle breezes

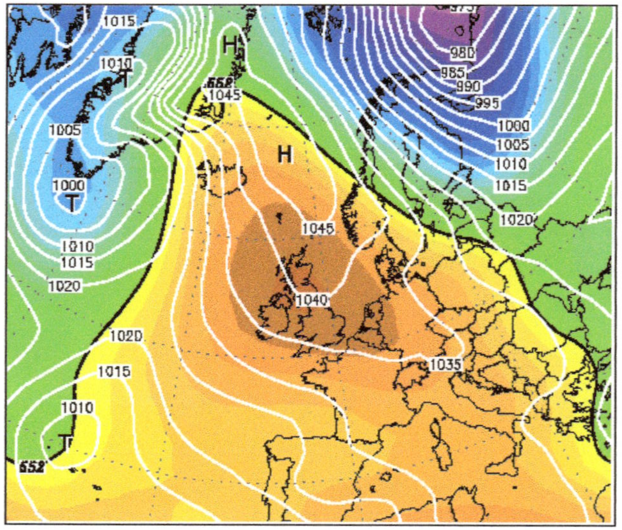

RAIN POTENTIAL

FROST/SNOW

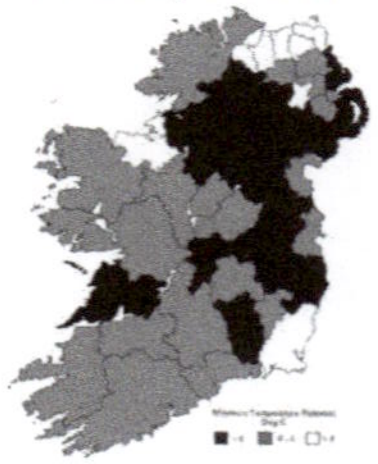

FEBRUARY **25** WEDNESDAY

High pressure maintains a mostly dry and calm influence throughout the country, with severe frosts likely overnight in central parts, and fog patches once more mostly in the east and north. Winds generally slight, in an east to southeast flow. Slightly cloudier than the past few days, except along the western coasts, where long spells of sunshine may be observed.

Belfast: Colder, overcast, dry, passing threats of rain, chance of frost and fog patches
Dublin: Overcast and cold, mostly calm air, dry, fog patches possible
Cork: Calm and gloomy air, dry, cold
Galway: Cloudy, dry, colder, mostly calm air

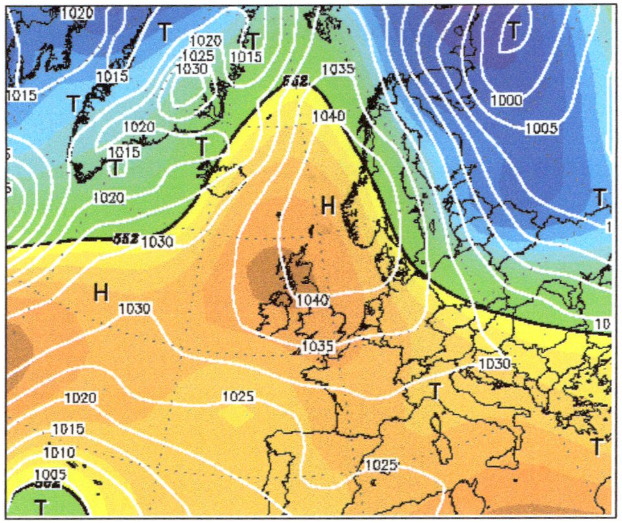

RAIN POTENTIAL

FROST/SNOW

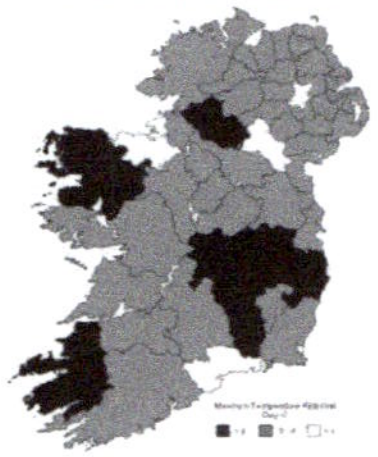

FEBRUARY **26** THURSDAY

The large high pressure system that has dominated the country over the past few days continues to slowly drift over towards Europe, with the high pressure ridge still lingering over the country, directing a light southeasterly to variable flow throughout. Another dry day with good sunshine in the north and east, while cloudier in the west and south. Very cold overnight with particularly sharp frosts in some inland places away from the Atlantic coasts. Misty patches likely overnight.

Belfast: Light breezes to mostly calm air, occasional sunny spells, dry and frosty
Dublin: Sunny, dry, cold, frost and fog likely
Cork: Overcast, mostly dry, odd threat of rain, gentle breezes to mostly calm air
Galway: Overcast, cold, dry, mostly calm and serene air

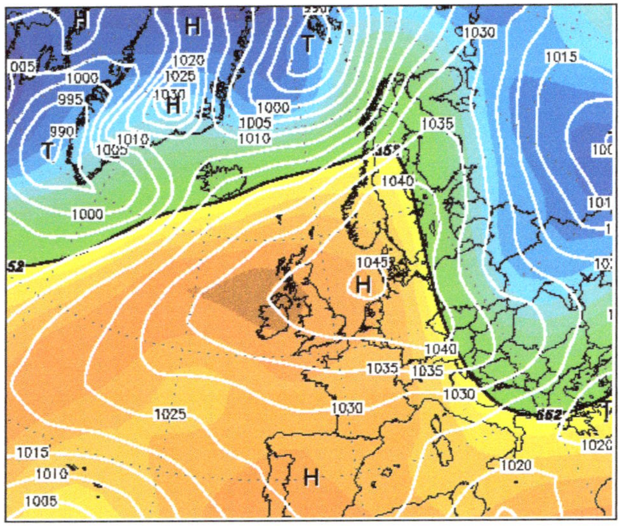

RAIN POTENTIAL

FROST/SNOW

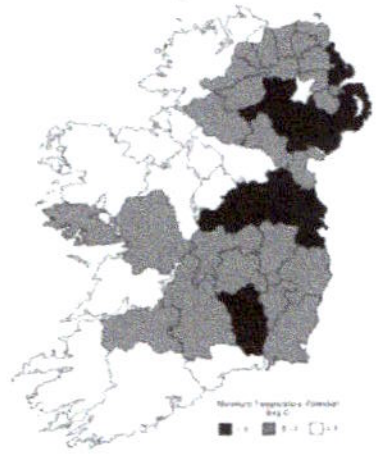

FEBRUARY 27 FRIDAY

The high pressure system influencing Ireland remains almost stalled over the country, bringing severe frosts overnight in most areas except in the far southeast, as well as the likelihood of widespread fog, particularly dense in many inland parts. Dry in all districts, with sunny spells breaking through in the afternoon allowing for warmer temperatures for a time. Winds very light and variable throughout, and frosts and fog returning after dark.

Belfast: Partly cloudy, dry, fog and severe frosts likely, chance of snowflurries
Dublin: Light shifting breezes to mostly calm air, sunny, dry, cold, severe frosts likely
Cork: Light breezes to mostly calm air, dry, cold, pleasant sunny outbreaks
Galway: Sunny, dry, cold, serene atmosphere, overnight frost patches, chance of fog

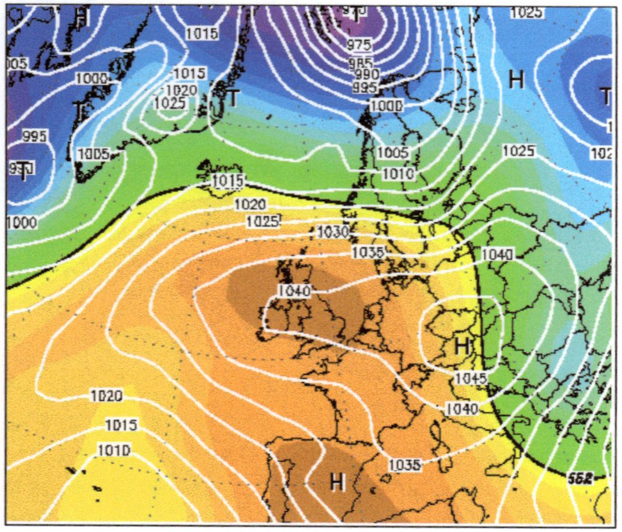

RAIN POTENTIAL

FROST/SNOW

FEBRUARY 28 SATURDAY

Anticyclonic conditions persist. Very light shifting breezes prevail throughout. Dry and cold everywhere with occasional outbreaks of sunny spells, longer sunny spells in the afternoon. Very cold overnight with frost and fog in most districts, more severe frosts in inland parts, and fog likely to be dense in many places. Fog likely to be slow to clear, and accompanied by frosts, return quickly after dark.

Belfast: Dry, colder, occasional sunny spells, mostly calm air, fog and frost likely
Dublin: Cold and frosty, mostly sunny, mostly calm air, chance of fog overnight
Cork: Light breezes to mostly calm air, overcast, cold, dry with odd threat of rain
Galway: Gentle breezes, cold, frosts, possible fog overnight, partly cloudy and dry

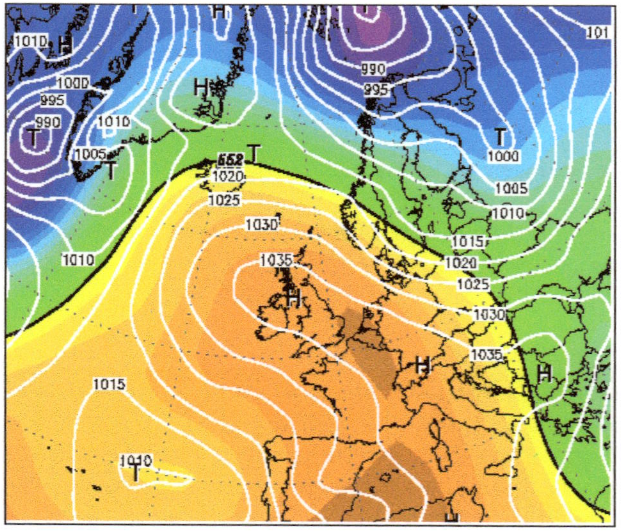

RAIN POTENTIAL

FROST/SNOW

MARCH

PHASES OF THE MOON

3rd	Full Moon
5th	Crossing Equator
10th	Apogee
11th	Third Quarter
11th	Southern Declination
19th	New Moon
19th	Crossing Equator
22nd	Perigee #11
25th	First Quarter
25th	Northern Declination

1	2	3	4	5	6	7	8	9	10	11	12	13	14	15	16	17	18	19	20	21	22	23	24	25	26	27	28	29	30	31
Sun	Mon	Tue	Wed	Thur	Fri	Sat	Sun	Mon	Tue	Wed	Thur	Fri	Sat	Sun	Mon	Tue	Wed	Thur	Fri	Sat	Sun	Mon	Tue	Wed	Thur	Fri	Sat	Sun	Mon	Tue
		F		XhS					A	3Q V								N XhN			P11			1Q ^						

MONTHLY SUMMARY

1st Colder with scattered fogs and widespread frosts likely, some severe in places, while good sunshine hours likely.
1st-3rd An area of low pressure crossing the country brings widespread showers.
2nd-11th Depressions near the northwest of Ireland prevail, returning unsettled conditions throughout.
4th-7th Heavy rain possible in the west and north of the country. Some showers may be wintry and of hail.
7th Very strong and blustery winds possible around the northwest.
7th-11th A southwesterly airflow may bring stormy conditions throughout, with gale winds at times, and above normal temperatures.
11th Very wet across most of the country. Chance of heavy rain in northern and western parts.

Over the next four weeks, despite possibly being wetter and windier than what is typical for this time of year, it may also be sunnier, while temperatures may be closer to the norm. Most days during the next four weeks may see some shower or rain activity, as well as frequent thunderstorm activity, particularly in the western half of the country. Overall, there may be up to 18-20 days when some rain may occur in most parts of Ireland but up to 25 days in the north and northwest, with many areas likely to see above normal rainfall amounts by the end of this outlook period. Despite possibly being wetter overall during the next four weeks, it may also be sunnier than normal by up to 25% more sunshine hours than what is usual for this time of year in some areas. Generally, temperatures during this outlook period may be close to normal but may also be cool enough at times for sleet or snow to occur on a few days, particularly in western parts, while up to twice as many ground frosts observed in some eastern and southern parts compared to what is normal for this time of year. During this outlook period there may be 12 days with thunderstorm activity, 18-20 days with hail showers, 2 days with widespread fogs, 16-20 days with frosts, and snow on 13 days.

12th-27th A series of depressions near the west of Ireland bring frequent passages of rain or showers and blustery winds across the country. During this period, the unsettled conditions may dominate the western half of the country, while eastern parts may see much less rainfall.
13th-15th Chance of hail showers and snow.
14th Chance of widespread frosts.
15th-16th High pressure may prevail.
18th-24th Spell of frequent thunderstorms and hail showers likely.

20th-22nd Chance of heavy rain which may bring some localised flooding, and snowfall in higher reaches.
21st-23rd An intense depression near Ireland may bring very strong winds throughout and high seas around the coasts, particularly in the west. Winds may reach storm force for a time. During this period, a band of storm activity with heavy rain, thunderstorms, hail, and squalls may cross the country. Localised flooding may develop in some eastern areas.
25th-26th Chance of widespread fog patches overnight while heavy rain possible during the day.
26th Unusually warmer than normal day possible.
28th-30th An anticyclonic system near the northwest of the country may bring a short spell of light winds, cold temperatures, mostly dry and sunny weather, with widespread frosts likely.
31st Low pressure may re-establish itself to the northeast of Ireland around this time, with further unsettled conditions developing for the next couple of weeks. A cold northerly flow may become established, and some light rain and occasional thunderstorms and hail activity on most days over the next ten days. A westerly turn during the first week of April may see an increase in daytime temperatures.

IRELAND RAINFALL ESTIMATES

March 2026

d=mainly dry, n= nonrecordable, l=light shrs, s=significant shrs, r=rain, h=hvy falls

		ULSTER PROVINCE					CONNAUGHT PROVINCE						LEINSTER PROVINCE				MUNSTER PROVINCE						
		Alderg	Hillsbo	Armag	Loughl	Malin H	Sligo	Ardtara	Belmul	Drums	Galway	Roscom	Markre	Edende	Kilkenn	Dublin	Shanno	Newpo	Tralee	Killarne	Cork	All	Moon
1st	Sun	d	d	d	l	d	l	l	d	d	l	n	l	d	d	d	d	l	d	d	d	8	
2nd	Mon	d	d	d	l	d	l	l	l	l	s	l	l	d	d	d	d	s	n	d	d	32	
3rd	Tues	l	l	d	s	s	s	s	l	s	r	s	l	l	d	d	l	r	n	l	d	61	F
4th	Wed	l	d	l	d	r	l	d	s	d	l	d	n	d	d	l	l	d	l	n	l	26	
5th	Thur	l	l	d	s	s	s	s	s	s	s	s	s	l	l	d	d	r	s	l	l	73	XhS
6th	Fri	l	d	l	l	s	l	l	l	d	l	n	n	d	d	l	l	l	l	d	l	25	
7th	Sat	l	s	d	s	s	r	s	r	s	s	s	s	l	l	n	l	s	r	h	s	109	
8th	Sun	l	d	s	s	s	l	s	l	n	l	l	l	d	d	d	d	s	l	l	n	34	
9th	Mon	n	d	l	d	l	l	d	l	d	n	d	d.l	d	d	d	d	l	d	d	d	8	
10th	Tues	l	l	l	s	n	r	s	l	s	r	s	s	l	d	d	d	h	l	s	d	90	A
11th	Wed	s	s	s	s	r	s	s	h	s	r	l	s	s	l	s	s	s	s	l	l	117	3Q V
12th	Thur	l	l	s	l	l	l	l	s	l	l	l	s	l	n	d	l	r	l	l	d	52	
13th	Fri	l	l	n	r	s	r	r	s	s	s	s	r	l	l	d	l	r	l	s	n	106	
14th	Sat	l	d	s	s	s	r	s	s	r	l	s	s	l	s	l	l	s	r	s	l	106	
15th	Sun	l	l	n	l	l	l	l	l	l	d	l	l	n	l	d	d	l	d	d	d	10	
16th	Mon	l	d	l	d	d	l	d	d	d	n	d	n	d	d	d	d	l	d	d	d	6	
17th	Tues	l	l	d	s	s	s	s	s	s	r	r	s	l	s	l	s	r	s	r	l	120	
18th	Wed	l	l	s	s	r	s	s	s	s	r	s	s	l	l	l	l	h	l	s	n	102	
19th	Thur	s	s	s	s	r	s	s	s	l	s	s	s	l	n	s	s	r	s	s	l	113	N XhN
20th	Fri	s	s	l	r	s	r	r	r	s	r	s	s	s	h	l	r	r	h	h	h	241	
21st	Sat	s	l	r	s	r	s	s	s	s	s	s	s	l	s	s	r	r	r	r	r	164	
22nd	Sun	s	r	n	s	s	r	s	s	s	s	h	s	s	l	l	s	h	r	s	s	152	P11
23rd	Mon	s	l	s	l	s	s	l	r	l	s	s	s	l	l	l	s	h	n	l	l	92	
24th	Tues	l	l	l	s	s	s	s	s	s	l	s	s	l	n	l	l	s	l	s	l	78	
25th	Wed	s	l	s	h	s	r	h	h	s	s	s	r	l	l	d	l	r	s	d	s	143	1Q ^
26th	Thur	s	s	s	l	s	s	l	s	r	r	s	l	r	s	h	s	r	s	l	s	151	
27th	Fri	l	d	s	d	d	d	d	d	d	d	n	d	n	d	s	l	d	d	n	d	18	
28th	Sat	d	d	d	d	d	d	d	d	d	d	d	d	d	d	d	d	d	d	d	d	0	
29th	Sun	d	d	d	d	n	d	d	d	d	d	d	d	d	d	d	d	d	d	d	d	1	
30th	Mon	d	d	d	d	l	d	d	l	d	d	d	l	d	d	d	d	l	d	d	d	7	
31st	Tues	l	l	d	l	s	s	l	l	s	l	s	l	l	d	l	l	s	d	r	d	50	
Estimate:		73	69	76	121	152	165	121	151	121	151	129	124	53	51	61	79	244	127	143	86	2294	
Average:		67	77	66	96	88	109	111	99	88	95	109	109	77	63	53	79	141	116	84	98	1824	
Trend:		av	av	w tr	w tr	w tr	w tr	av	w tr	w tr	w tr	w tr	w tr	drr	drr	av	av	w tr	w tr	w tr	drr	w tr	

IRELAND SUNSHINE ESTIMATES

March 2026

F=fine (8-12 hours of sunshine), pc= partly cloudy (4-7 hours), c=cloudy (1-3 hours), o=overcast (0 hours)

		ULSTER PROVINCE					CONNAUGHT PROVINCE						LEINSTER PROVINCE				MUNSTER PROVINCE						
		Alderg	Hillsbo	Armag	Loughl	Malin I	Sligo	Ardtari	Belmul	Drums	Galway	Rosco	Markre	Edend	Kilkeni	Dublin	Shann	Newpo	Tralee	Killarn	Cork	All	Moon
1st	Sun	c	c	c	c	c	o	c	pc	c	c	c	c	c	pc	pc	o	c	o	o	o	61	
2nd	Mon	o	o	o	o	o	o	o	o	o	o	o	o	o	o	o	o	o	o	o	o	4	
3rd	Tues	o	o	o	o	o	o	o	o	o	o	o	o	o	o	o	o	o	o	o	o	0	F
4th	Wed	c	c	c	c	c	o	c	o	c	c	c	c	c	c	c	c	c	c	c	o	28	
5th	Thur	o	o	o	o	o	o	o	o	o	o	o	o	o	o	o	o	o	o	o	o	5	XhS
6th	Fri	pc	pc	pc	pc	pc	pc	pc	pc	c	pc	pc	pc	c	c	c	pc	pc	pc	pc	pc	91	
7th	Sat	o	o	o	o	o	o	o	o	o	o	o	o	o	o	o	o	o	o	o	o	2	
8th	Sun	pc	pc	pc	pc	pc	c	pc	c	pc	c	c	pc	pc	pc	pc	c	pc	pc	pc	pc	102	
9th	Mon	pc	pc	pc	pc	pc	c	pc	c	pc	c	c	pc	pc	pc	pc	pc	pc	pc	pc	pc	88	
10th	Tues	c	c	c	c	c	c	c	c	c	c	c	c	c	c	o	o	o	o	o	c	33	A
11th	Wed	o	o	o	o	o	o	o	o	o	o	o	o	o	o	o	o	o	o	o	o	0	3Q V
12th	Thur	c	c	c	c	c	o	c	0	c	o	o	c	c	c	c	o	c	o	o	c	31	
13th	Fri	c	c	c	c	c	c	c	c	c	c	c	c	c	c	pc	c	c	c	c	c	60	
14th	Sat	c	c	c	c	c	o	c	c	pc	c	c	c	pc	pc	pc	c	c	c	c	c	62	
15th	Sun	pc	pc	pc	pc	pc	pc	pc	pc	pc	pc	pc	pc	pc	pc	pc	pc	pc	pc	pc	pc	126	
16th	Mon	c	c	c	c	c	c	c	o	o	o	o	c	o	o	o	o	o	o	o	o	12	
17th	Tues	o	o	o	o	o	o	o	c	o	o	o	o	o	o	o	o	o	o	o	o	6	
18th	Wed	pc	pc	pc	pc	pc	c	pc	c	pc	pc	pc	pc	pc	pc	F	pc	pc	pc	pc	pc	110	
19th	Thur	pc	pc	pc	pc	pc	pc	pc	pc	pc	pc	pc	pc	pc	pc	pc	pc	pc	pc	pc	pc	106	N XhN
20th	Fri	pc	pc	pc	pc	pc	pc	pc	pc	pc	pc	pc	pc	pc	pc	pc	pc	pc	pc	pc	pc	107	
21st	Sat	c	c	c	c	c	c	c	c	c	c	c	c	c	c	c	c	c	c	c	c	42	
22nd	Sun	c	c	c	c	c	c	c	c	c	c	c	c	c	pc	c	c	pc	c	c	pc	63	P11
23rd	Mon	pc	pc	pc	pc	pc	pc	pc	pc	pc	pc	pc	pc	pc	F	pc	pc	pc	pc	pc	F	132	
24th	Tues	c	c	c	c	c	c	c	c	c	c	c	c	c	o	c	o	o	o	o	o	29	
25th	Wed	o	o	o	o	o	o	o	o	o	o	o	o	o	o	o	o	o	o	o	o	0	1Q ^
26th	Thur	o	o	o	o	o	o	o	o	o	o	o	o	o	o	o	o	o	c	c	o	3	
27th	Fri	c	c	c	c	c	pc	c	pc	c	c	pc	c	c	c	c	o	o	o	o	o	48	
28th	Sat	F	F	F	F	F	F	F	F	pc	F	F	F	pc	pc	pc	pc	pc	pc	pc	pc	149	
29th	Sun	pc	pc	pc	pc	pc	F	pc	F	pc	F	F	pc	pc	pc	pc	F	pc	pc	pc	pc	148	
30th	Mon	c	c	c	c	c	pc	c	pc	c	pc	pc	c	c	pc	c	pc	pc	F	F	F	95	
31st	Tues	o	o	o	o	o	o	o	o	o	o	o	o	o	o	o	o	o	o	o	o	2	
Estimate:		90	90	90	90	90	81	90	84	92	82	84	90	92	94	98	80	87	77	77	90	1745	
Average:		93	93	89	95	96	96	97	102	93	105	109	102	106	103	112	104	92	97	95	109	1987	
Trend:		av	av	av	av	av	less	av	less	av	less	less	less	less	av	less	less	av	less	less	less	less	

MARCH **1** SUNDAY

A large high pressure system lingers over the country with a light east to southeast flow prevailing. Mostly dry and cloudy throughout, although increased sunny outbreaks may be more frequent in the west and east. Very cold overnight with patches of severe frosts in inland parts and fog most likely in the north and east. Chance of some brief wintry showers for a time in some central parts, possibly falling as snow flurries overnight.

Belfast: Calm, mostly dry, cold, severe frosts, chance of snowflurries and fog
Dublin: Mostly sunny, dry, cold, severe frosts, calm atmosphere, fog likely
Cork: Overcast, cold, dry with occasional threats of rain, light breezes
Galway: Cloudy, cold with severe frosts overnight, some showers, chance of snow

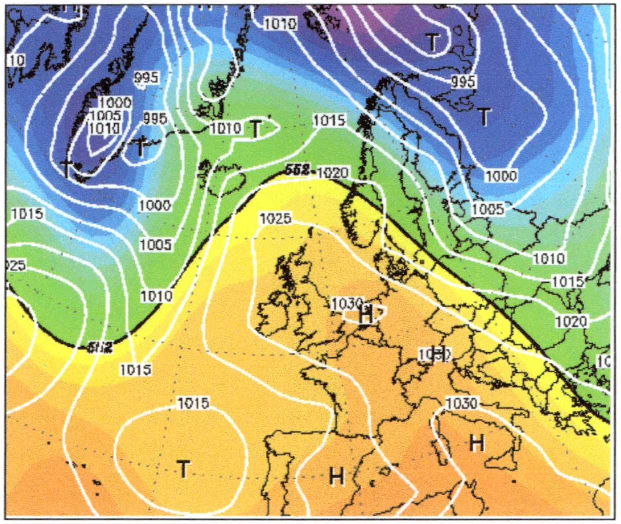

RAIN POTENTIAL

FROST/SNOW

MARCH **2** MONDAY

A weak depression moves up between two highs to the south of Ireland, while low pressure north of Ireland begins to move southward towards the country. Light winds persist in most parts, but becoming fresher with a southwest change. Thicker cloud everywhere, bringing a very dull and gloomy day. Some patches of mist or fog turn may bring outbreaks of drizzle mostly in central to western parts of the country. Further frosts overnight.

Belfast: Overcast, cold, dry, severe frosts, chance of misty patches and fog
Dublin: Overcast, dry and cold, frosts, calm atmosphere, chance of fog
Cork: Dull and gloomy, mostly dry, threats of rain, chance of misty patches
Galway: Overcast, milder, some showers, light breezes strengthening later in the day

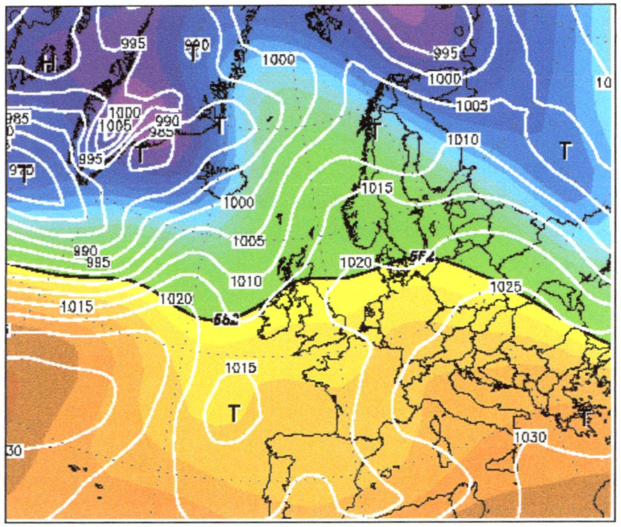

RAIN POTENTIAL

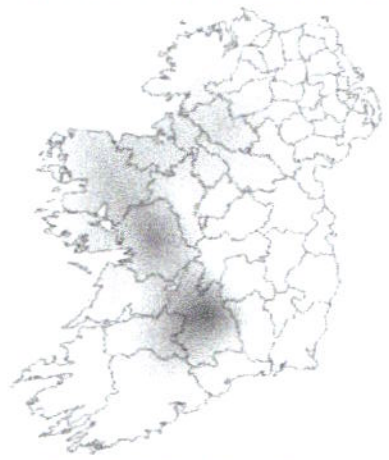

FROST/SNOW

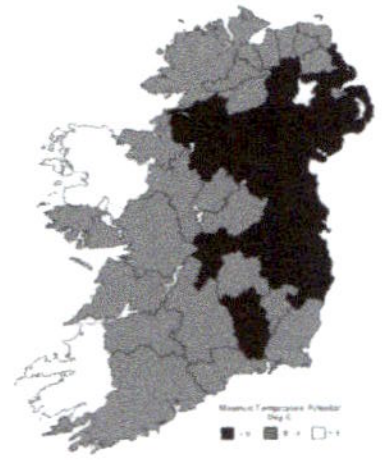

MARCH 3 TUESDAY

A complex of low pressure systems near Greenland extends very strong and blustery westerly winds across Ireland, sometimes reaching gale force with severe gusts at times. Very dull and gloomy everywhere, with patches of showers mostly in the western half of the country, with southern and eastern coastal regions remaining mostly dry. Milder temperatures overnight with little chance of frosts, although some patches of fog overnight likely, particularly in the east and south before winds build later in the day.

Belfast: Overcast, , brief showers, freshening breezes with occasional gusts likely
Dublin: Milder temperatures, overcast and dull, dry, freshening breezes
Cork: Overcast, dry, increasing threats of rain, milder, strengthening winds
Galway: Overcast, rain, blustery

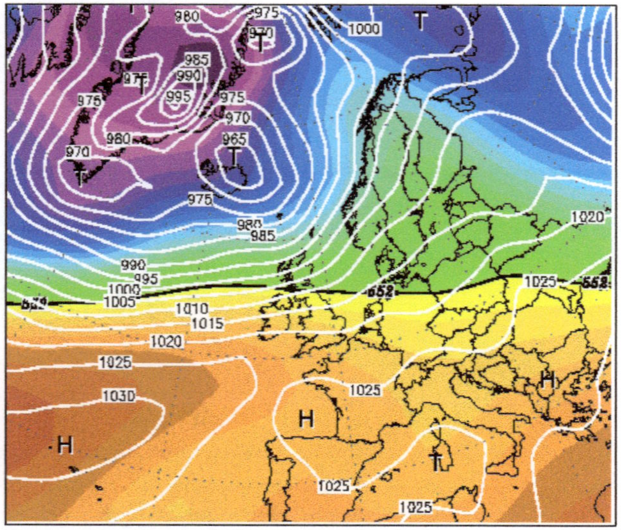

RAIN POTENTIAL

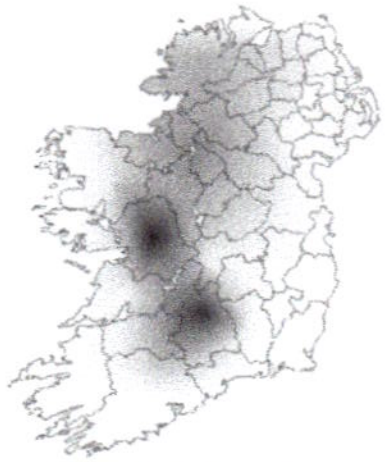

FROST/SNOW

MARCH 4 WEDNESDAY

Intense depressions north of Ireland continue to bring strong and gusty winds across the country in a west to southwest flow. Mostly dry around the Leinster coastal fringes, with showers elsewhere, more frequent and slightly heavier in the northwest. Some brief breakthroughs of sunshine later in the day.

Belfast: Unsettled and blustery winds, cloudy, mostly dry, odd threat of rain
Dublin: Unsettled winds, cloudy, brief squally showers
Cork: Overcast, brief showers, milder, strong and unsettled winds
Galway: Cloudy, isolated squally showers, strong gusty winds

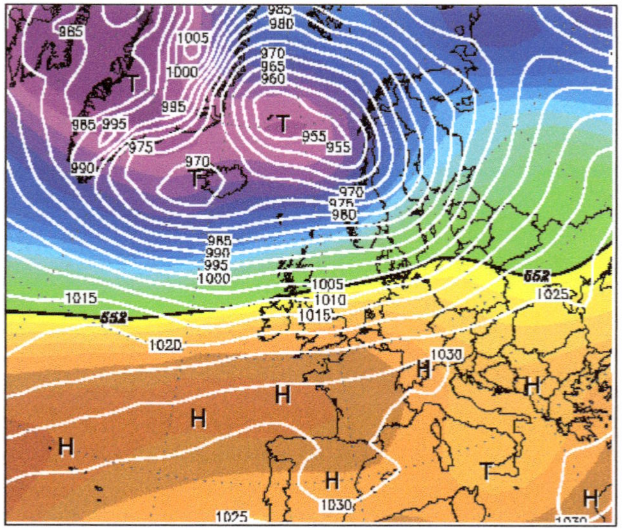

RAIN POTENTIAL

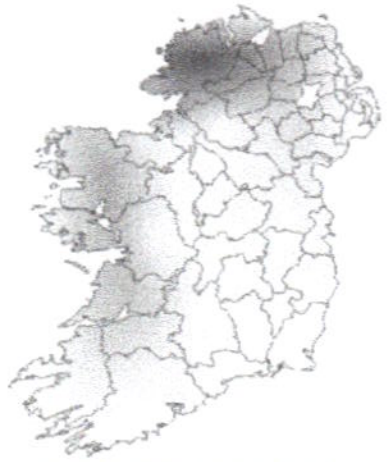

FROST/SNOW

MARCH **5** THURSDAY

Intense depressions to the west of Ireland extend a southwest flow over the country, strong and blustery over the northern half of Ireland while more moderate to the south. Widespread cloud everywhere, with showers and drizzle patches in the west at first, spreading into the east as the day progresses.

Belfast: Overcast, blustery at times, light showers
Dublin: Overcast, dry with passing threats of rain, unsettled winds at times
Cork: Overcast, light showers, cooler day milder night, moderate breezes
Galway: Unsettled and blustery, overcast, squally showers

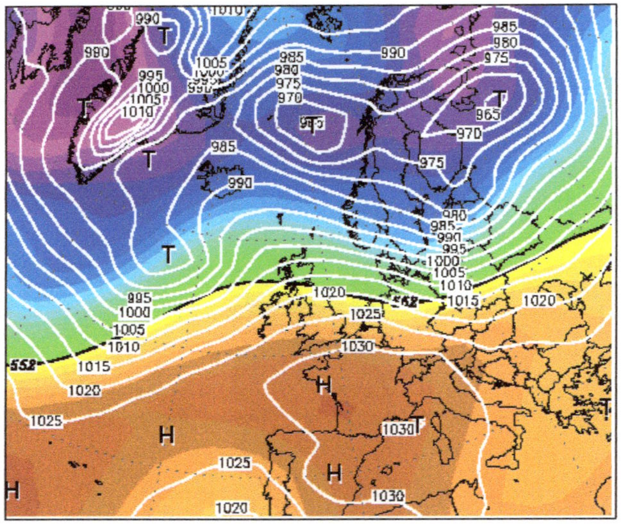

RAIN POTENTIAL

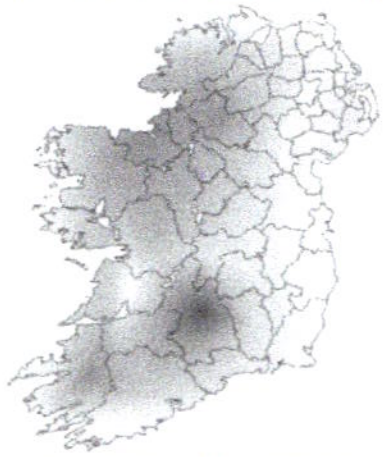

FROST/SNOW

MARCH **6** FRIDAY

Low pressure and southwest winds prevail across the country with showers in the west later moving into the east and then clearing for a time, with some sunny spells developing behind the showers. Some showers may be wintry overnight in some western parts, possibly also falling as light snow or flurries.

Belfast: Sunny, dry, breezy spells easing later, chance of overnight fog patches
Dublin: Cloudy, cooler, light showers, fresh breezes
Cork: Partly cloudy, cool, light showers, moderate breezes
Galway: Unsettled and windy, brief showers, some sunny outbreaks, chance of snow

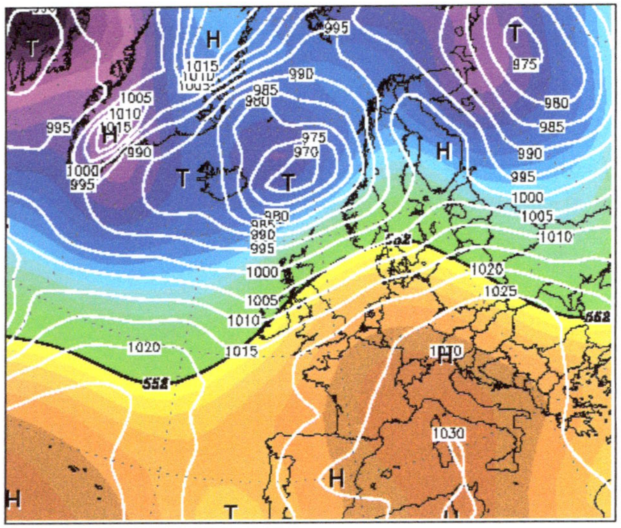

RAIN POTENTIAL

FROST/SNOW

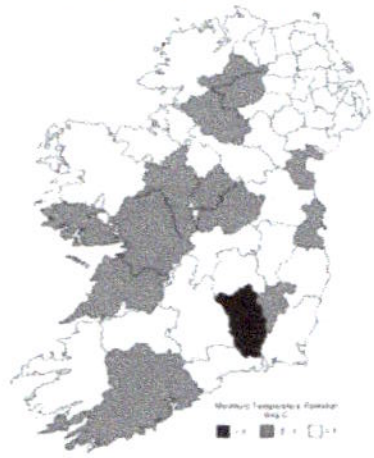

MARCH **7** SATURDAY

An Atlantic Ocean depression slips southwards to be centred to the immediate west of Ireland directing a southwest flow over the country, light at first, but strengthening to gusty for a time in the northwest later in the day. Widespread cloud throughout, with rain entering western districts at first, before spreading eastward as the day progresses, then clearing by evening to mostly dry. Chance of brief snow flurries in the far north for a time.

Belfast: Overcast, scattered showers, breezy at times
Dublin: Overcast, isolated showers, mild day cold night, occasional breezy spells
Cork: Overcast, scattered showers, cold overnight, shifting breezes
Galway: Fluctuating windy spells, overcast, showers

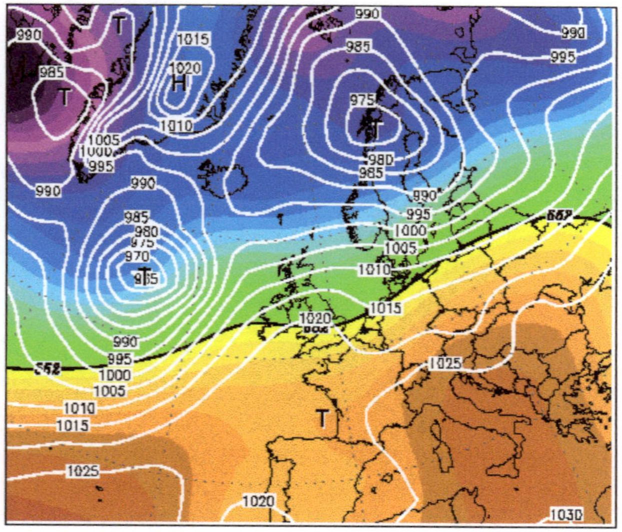

RAIN POTENTIAL

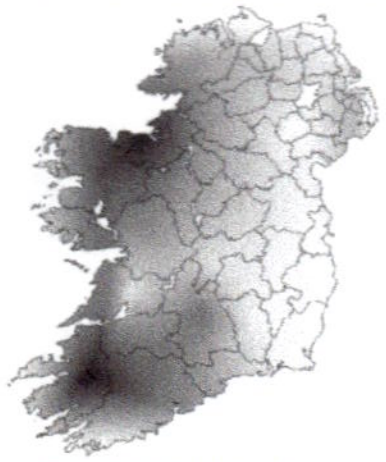

FROST/SNOW

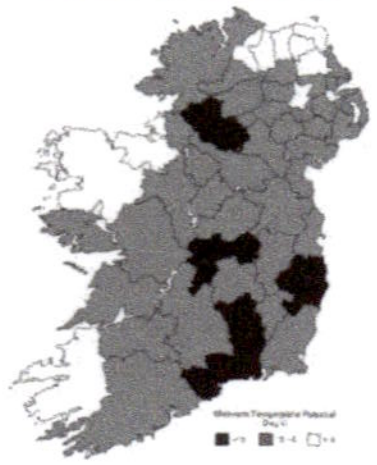

MARCH **8** SUNDAY

A depression continues to bear down on Ireland directing very strong westerly winds over the northern half of the country, with lighter westerly breezes over the southern half. A mixed day with sunny spells and occasional scattered showers; sunniest in the east.

Belfast: Partly cloudy, dry, cold, strong winds with gusts
Dublin: Mostly sunny and dry, cool, blustery at times
Cork: Partly cloudy, isolated showers, blustery at times
Galway: Partly cloudy, light showers, strong and unsettled winds

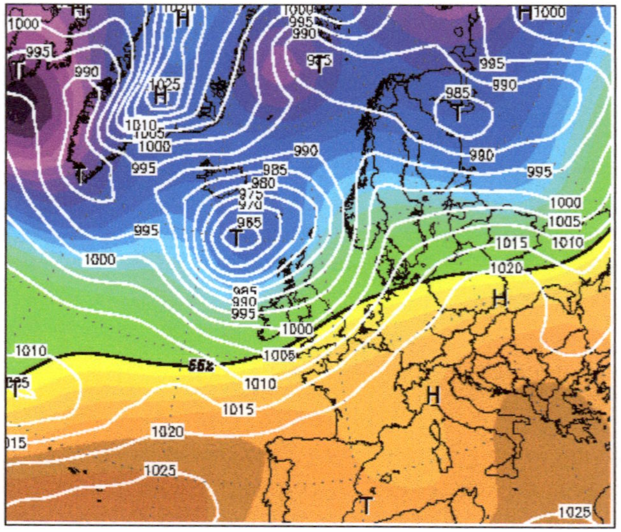

RAIN POTENTIAL

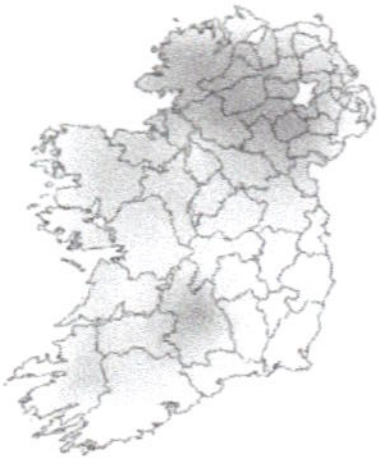

FROST/SNOW

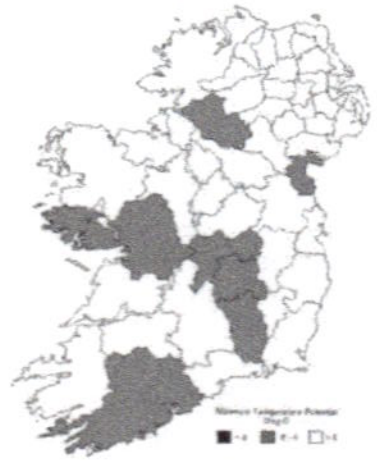

MARCH 9 MONDAY

The depression to the north of Ireland continues to move further east away from the country, while a high pressure ridge moves up over the south. Some lingering showers mostly in the north and northwest, while mostly dry with longer sunnier spells elsewhere, particularly in the south. Overnight frosts likely to return, particularly in the eastern and inland parts. Winds generally light in a west to northwest flow, more unsettled and blustery around the northwestern coasts.

Belfast: Dry, scattered cloudy spells, mild day cold night, moderate breezes
Dublin: Partly cloudy, dry, cold overnight, fresh breezes at times
Cork: Partly cloudy, dry, windy spells
Galway: Partly cloudy, isolated showers, fresh breezes

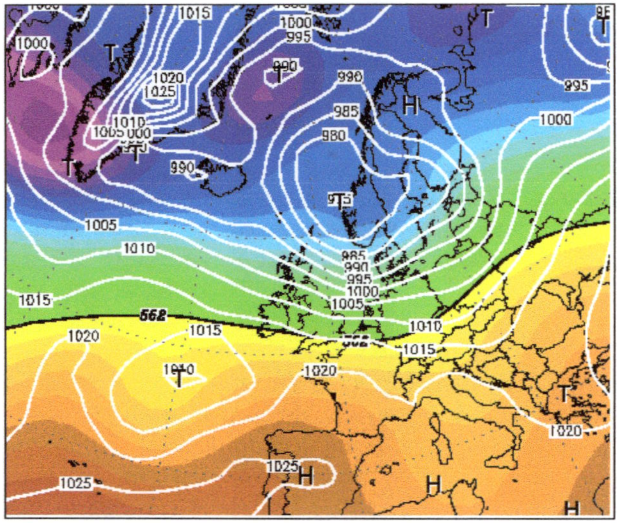

RAIN POTENTIAL

FROST/SNOW

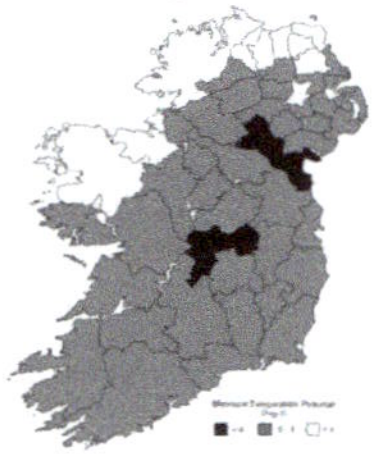

MARCH 10 TUESDAY

A high pressure ridge sits near the southern coasts of Ireland while low pressure prevails to the north. Generally cloudy throughout, with just a few sunny spells here and there. A weak front carried on a light to moderate westerly flow crosses the country bringing rain through central and western parts of the country with the southern, eastern, and far northern coastal fringes remaining mostly dry. Very cold overnight with chance of severe frosts.

Belfast: Light breezes to mostly calm air, cloudy, occasional showers
Dublin: Mostly calm and serene air, cold, overcast, dry with increasing threats of rain
Cork: Cloudy, dry, light breezes, chance of overnight patches of fog
Galway: Fluctuating winds, mostly cloudy, rain, cold overnight

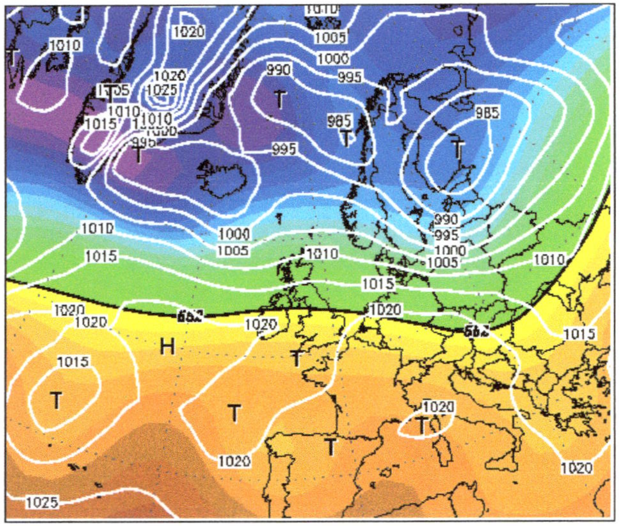

RAIN POTENTIAL

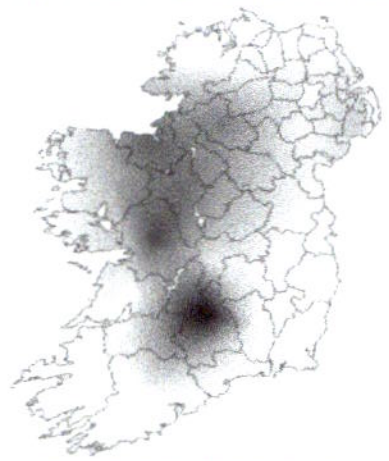

FROST/SNOW

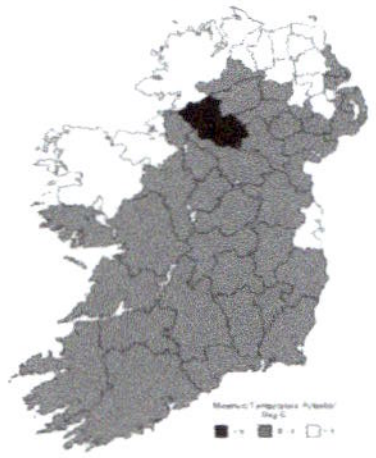

MARCH **11** WEDNESDAY

Low pressure to the northwest of Ireland and a high pressure system sits to the south, with a light southwesterly flow prevailing at first, freshening later to strong and blustery with a southwest change. Dull and gloomy everywhere, with widespread rain activity, heavy in some places for a time, before becoming more patchy later in the day. Chance of thunderstorms and the odd shower of hail, otherwise temperature generally milder than what is normal for this time of year. A few frosts may still be observed, mostly in central parts.

Belfast: Overcast, scattered showers, breezy at times
Dublin: Overcast, occasional showers, slightly warmer, freshening breezes
Cork: Overcast, light showers, milder day, cold night, moderate breezes
Galway: Fresh breezes, overcast, rain

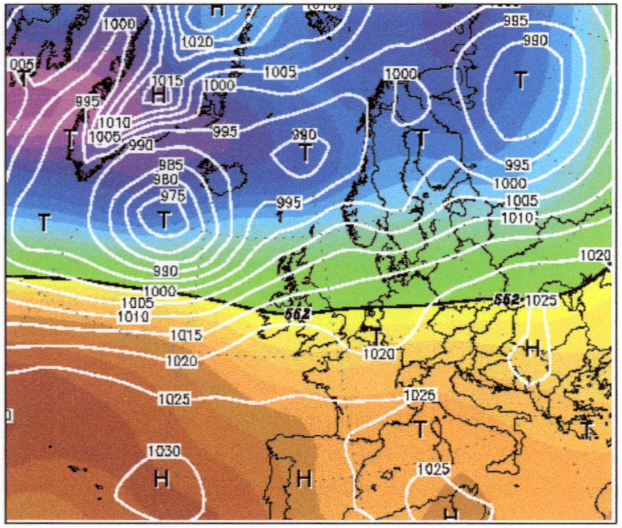

RAIN POTENTIAL

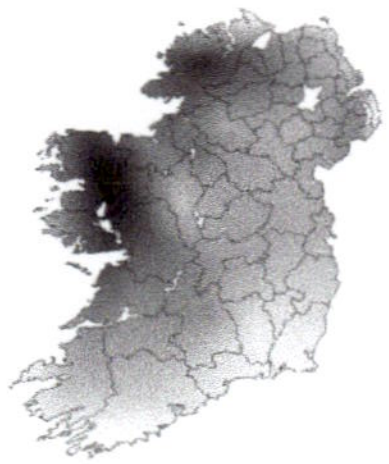

FROST/SNOW

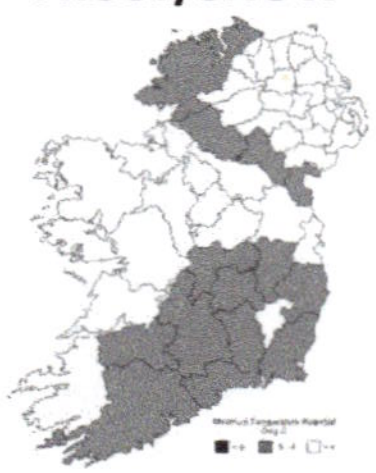

MARCH **12** THURSDAY

The depression to the north of Ireland inensifies and pushes the high pressure ridge away further south of the country, while bringing very strong west to northwest winds over Ireland, possibly up to gale force for a time, particularly in northern and western parts. Very unsettled day throughout. Mostly dry along the southeastern coasts, with a mix of drizzle patches and scattered showers elsewhere.

Belfast: Strong and unsettled winds, cloudy, slightly warmer, isolated showers
Dublin: Mostly cloudy and dry, strong and unsettled winds at times
Cork: Unsettled and blustery air, occasional sunny breakthroughs, dry, milder
Galway: Overcast, showers, strong winds, possible gale gusts at times

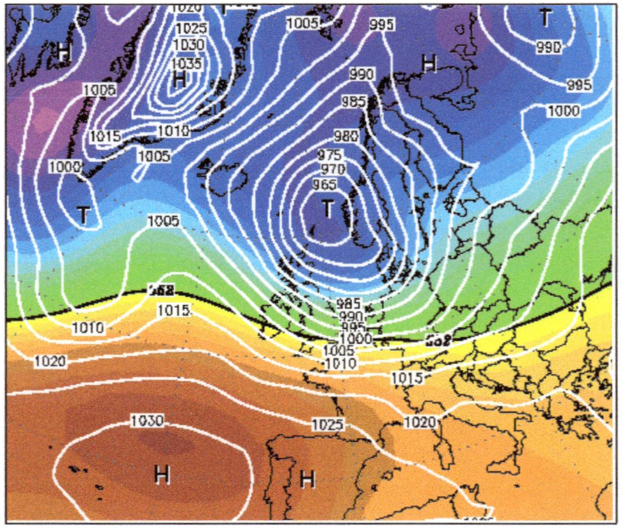

RAIN POTENTIAL

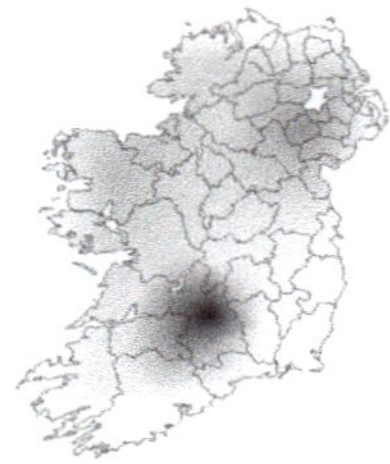

FROST/SNOW

MARCH **13** FRIDAY

Low pressure continues to dominate the country under a strong and unsettled northwest flow. Very unsettled day everywhere, with a mix of some brief sunny spells and bands of showers, some more wintry in the evening particularly in the north and west. Chance of thunderstorms and hail at times, as well as possible overnight snow on the higher reaches.

Belfast: Partly cloudy, light showers, strong and blustery winds at times
Dublin: Unsettled winds, partly cloudy, mostly dry with odd threat of rain
Cork: Cloudy, isolated showers, mild day cold night, unsettled blustery winds
Galway: Unsettled winds, cloudy, rain, chance of thunderstorms, hail and snow

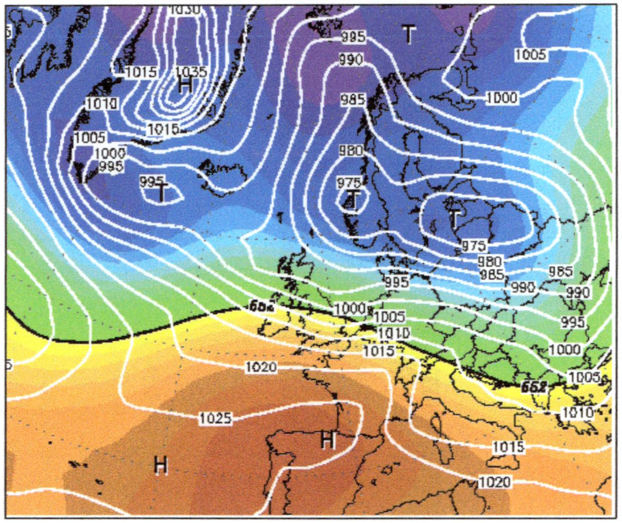

RAIN POTENTIAL

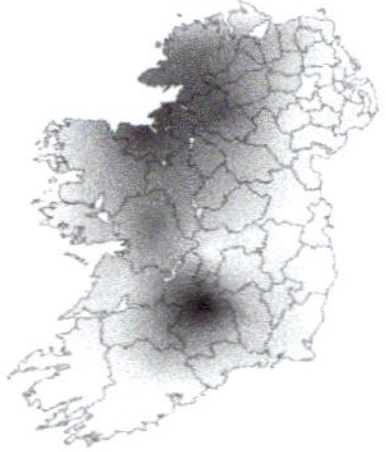

FROST/SNOW

MARCH **14** SATURDAY

A colder northwesterly air with strong and blustery winds at times continues to affect Ireland as low pressure centres to the north of the country prevail. A mix of snow and hail showers may fall overnight, particularly in the midlands, western and northern parts and wintry showers also persist throughout all other parts of the country during the day, as well as the chance of further thunderstorms.

Belfast: Dry, partly cloudy, colder, frosts likely, chance of snow flurries overnight
Dublin: Changeable, cold, mix of showers and sunny spells, chance of light snow
Cork: Mix of showers and some sunny spells, overnight frosts, possible snow
Galway: Cloudy, light showers, cold, frosts, thunderstorms and snow possible

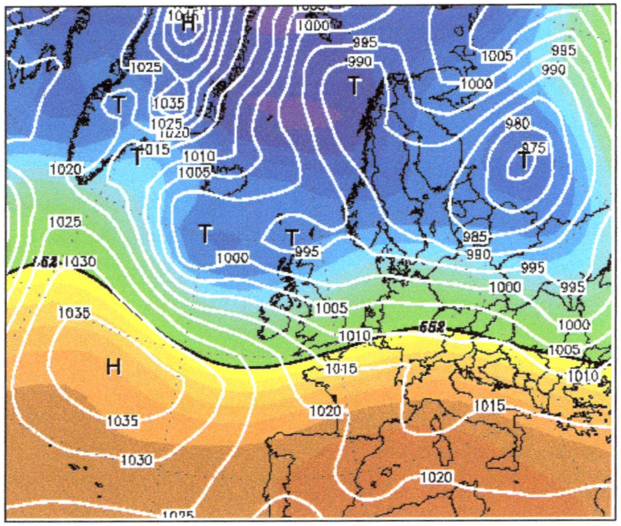

RAIN POTENTIAL

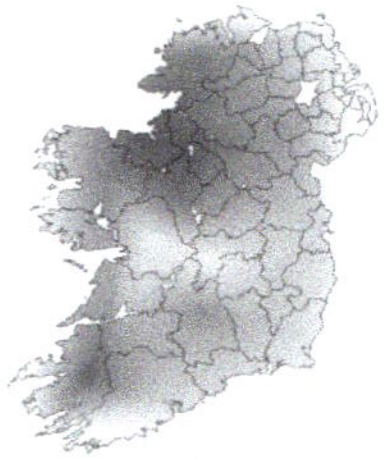

FROST/SNOW

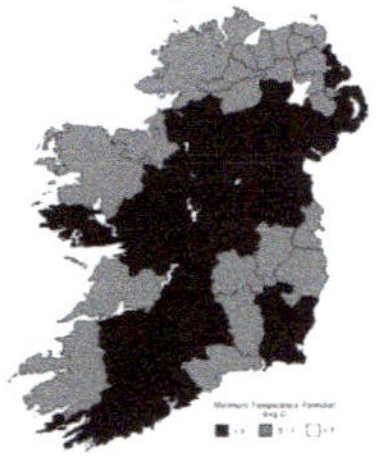

MARCH **15** SUNDAY

High pressure ridges to the west of Ireland while a low pressure trough covers the country. A changeable day throughout, with a mix of sunny and cloudy spells. Chance of isolated overnight wintry showers here and there. Winds generally in a moderate to fresh northerly flow, easing to lighter later. Chance of frost patches overnight and possible brief snow, mostly in the east.

Belfast: Isolated showers, mostly sunny day, very cold, frosts, chance of light snow
Dublin: Partly cloudy, dry, cold, chance of overnight frost patches and light snow
Cork: Partly cloudy, cold, dry, breezy spells
Galway: Partly cloudy, breezy, dry, cold with frosts overnight

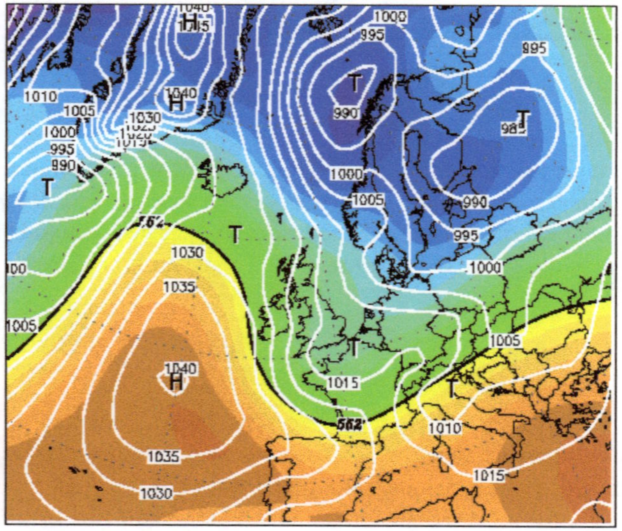

RAIN POTENTIAL

FROST/SNOW

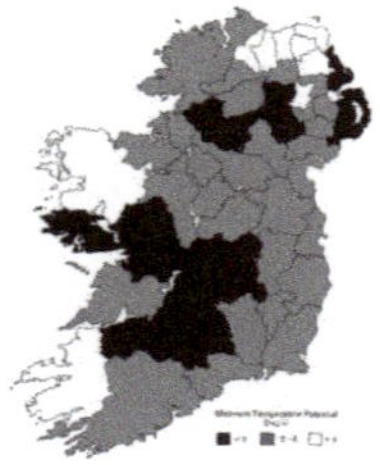

MARCH **16** MONDAY

An anticyclonic system nears the southern coast of Ireland bringing a light and gentle west to southwest flow throughout, although slighlty breezier around the northern coasts. Widespread cloud prevails and a mostly dry day everywhere except for the chance of some isolated showers possible in the far northeast.

Belfast: Cloudy, warmer, dry, moderate breezes
Dublin: Overcast, dry, increasing threats of rain, cool temperatures, light breezes
Cork: Overcast, mostly dry with threats of rain, milder overnight, gentle breezes
Galway: Overcast, isolated showers, cold, light breezes for a time

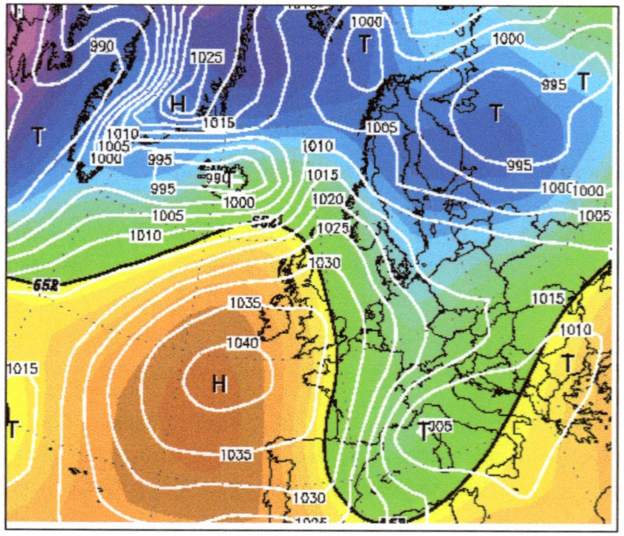

RAIN POTENTIAL

FROST/SNOW

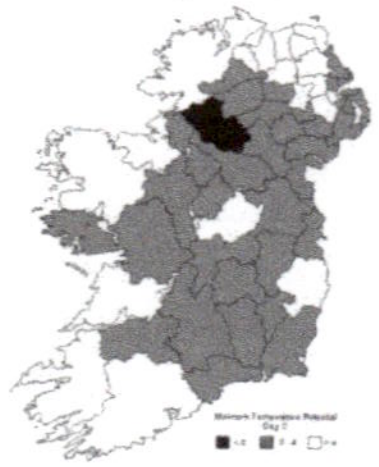

MARCH **17** TUESDAY

High pressure centred to the south of Ireland directs fresh to gusty southwest winds at first, later backing northwesterly later in the day. Wet fronts cross the country bringing outbreaks of wintry rain or drizzle into most districts, some heavier in the west. Chance of thunderstorms later in the day.

Belfast: Overcast, light showers, mild temperatures, fluctuating breezes
Dublin: Unsettled and blustery air, overcast, brief squally showers
Cork: Light showers, dull and gloomy air, unsettled winds
Galway: Overcast, milder temperatures, rain, windy

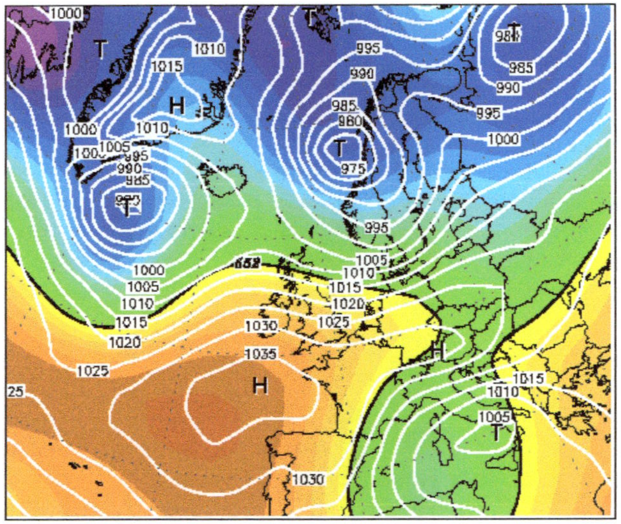

RAIN POTENTIAL

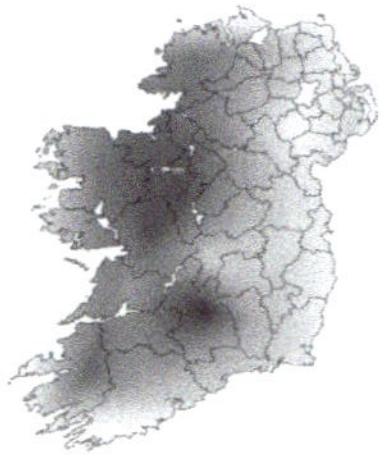

FROST/SNOW

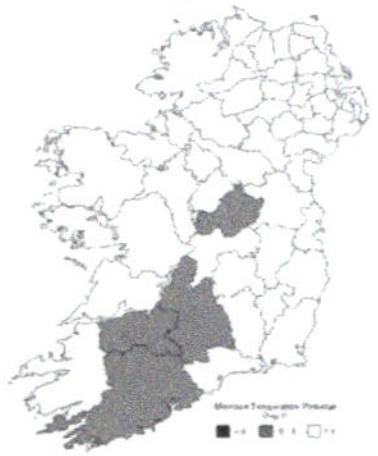

MARCH **18** WEDNESDAY

Low pressure north of Ireland pushes the high pressure ridge to the south of the country, with moderate to strong mostly westerly winds prevailing, more gusty in the north and west. A mixed day as a band of storms cross the country with outbreaks of showers almost everywhere except along the far southern coasts, with sunny spells between the rain bands. Chance of thunderstorms again, possibly with accompanying hail showers.

Belfast: Changeable, mix of sunny spells and light showers, mild day cold night
Dublin: Sunny, light overnight showers, blustery winds at times, cooler
Cork: Partly cloudy, isolated showers, cold overnight, chance of frost patches
Galway: Partly cloudy, rain, blustery, chance of thunderstorms, hail, and gales

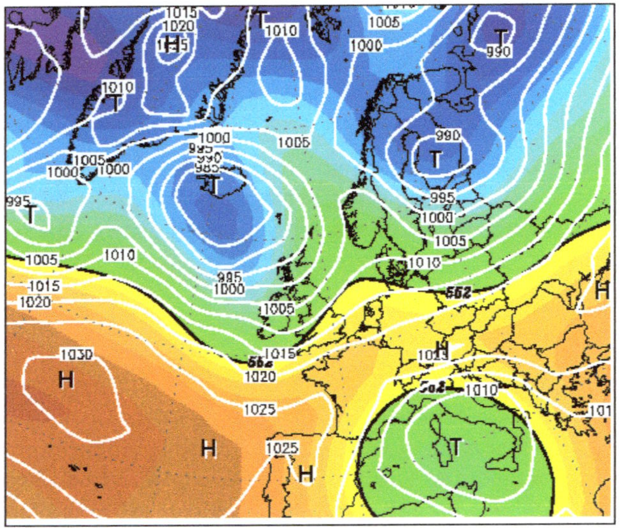

RAIN POTENTIAL

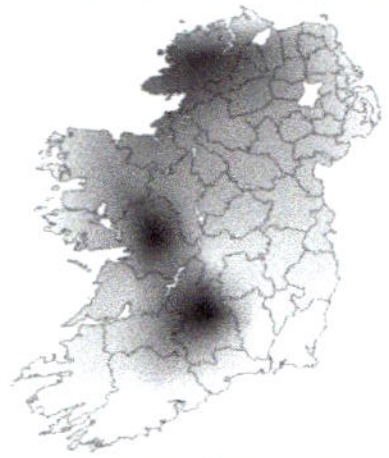

FROST/SNOW

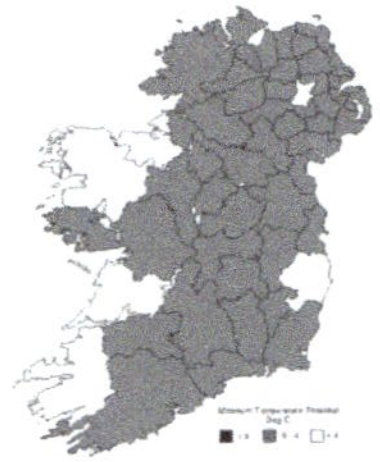

MARCH 19 THURSDAY

A large depression north of Ireland directs very strong and blustery west to southwest winds throughout. Another mixed day of sunny spells and showers everywhere, with heavier rain at times, mostly in central parts. Chance of brief falls of overnight light snow in some western districts.

Belfast: Changeable, mix of showers and sunny spells, colder, blustery winds
Dublin: Stormy, mix of sunny outbreaks and squally showers, milder temperatures
Cork: Mix of light showers and some sunny spells, strong gusty winds
Galway: Storm, squalls, some sunny breakthroughs, gales and gusts likely

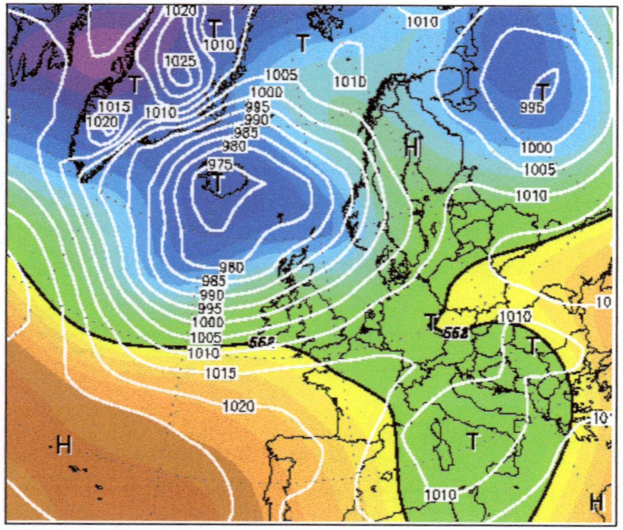

RAIN POTENTIAL

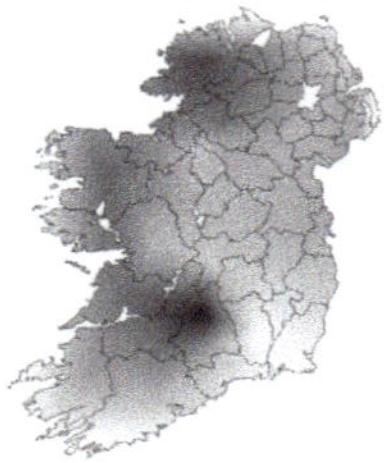

FROST/SNOW

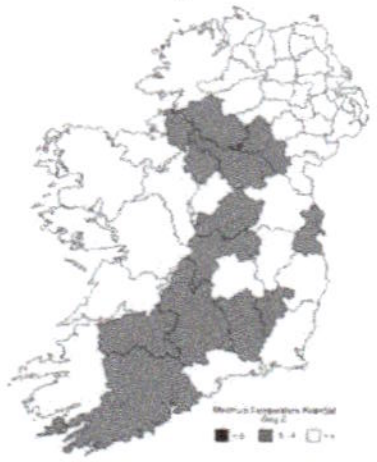

MARCH 20 FRIDAY

The depression centred north of Ireland expands with winds mostly northwestly, moderate for a time, but very unsettled and gusty in the north and later backing southwesterly. A stormy air prevails throughout, with a mix of some sunny spells between the bands of showers. Possible thunderstorms and hail embedded in the rain bands, as well as the chance of overnight snow in some western parts. Rain likely to be heavy for a time in many southern parts of the country.

Belfast: Stormy air, mix of showers and sunny spells, gale gusts possible
Dublin: Unsettled winds, partly cloudy, colder, isolated showers
Cork: Changeable, overnight heavy rain, colder, some sunny spells, strong winds
Galway: Rain, some sunny spells, colder, blustery, chance of overnight snow

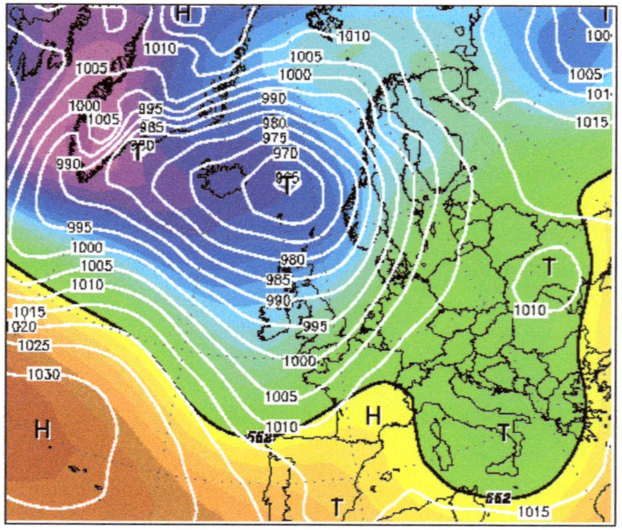

RAIN POTENTIAL

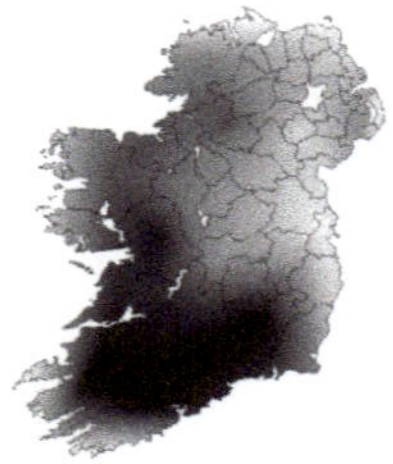

FROST/SNOW

MARCH 21 SATURDAY

The depression that has brought very unsettled conditions over the past few days intensifies as it bears down on Ireland. Winds very strong and stormy, mostly in a south to southwest flow, later turning southeasterly in the north. Very wet and windy day throughout, with only a few brief sunny spells first thing in the day. Widespread rain prevails, as well as the chance of some snow overnight, mostly in the west.

Belfast: Colder, cloudy, light showers, strong winds, gales possible
Dublin: Stormy, cold, cloudy, rain
Cork: Storm air, cloudy, rain, colder, chance of gales
Galway: Cloudy, rain, cold, strong and stormy winds, chance of snow overnight

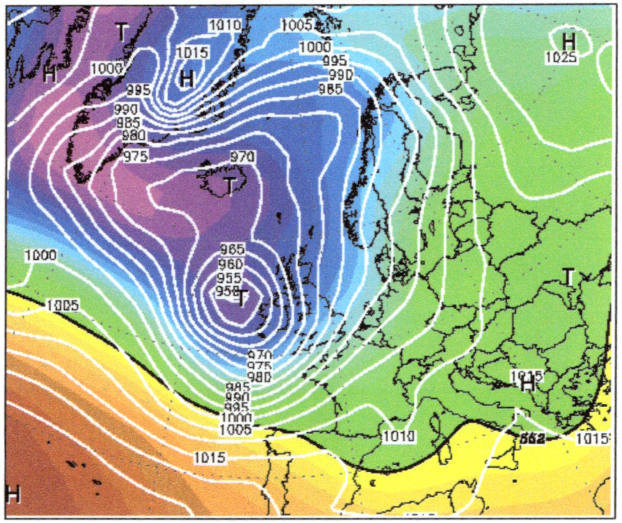

RAIN POTENTIAL

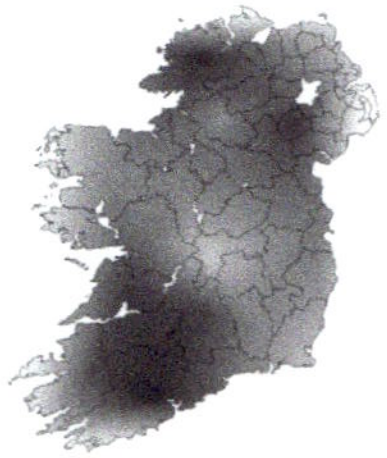

FROST/SNOW

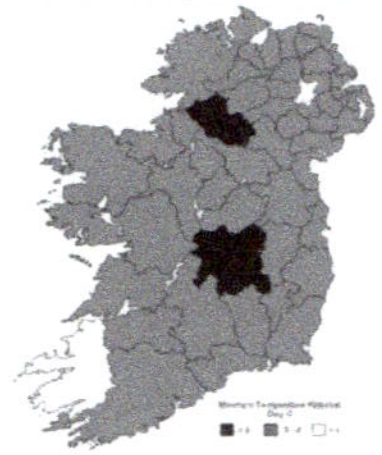

MARCH 22 SUNDAY

Low pressure eases across the country with a northwest change. However, winds remain very strong and gusty throughout, possibly up to gale force along the coasts. Bands of wintry showers cross the country between sunny breakthroughs, with some heavy falls likely in some central areas and in the far northeast, and again, the chance of some snow overnight in western and northern parts.

Belfast: Stormy air, cloudy, rain, cold overnight with freezing rain possible
Dublin: Slightly milder temperatures, blustery, showers, some sunny spells
Cork: Mix of showers and sunny spells, slightly warmer, windy at times
Galway: Scattered showers, brief sunny spells, blustery with gusts, chance of snow

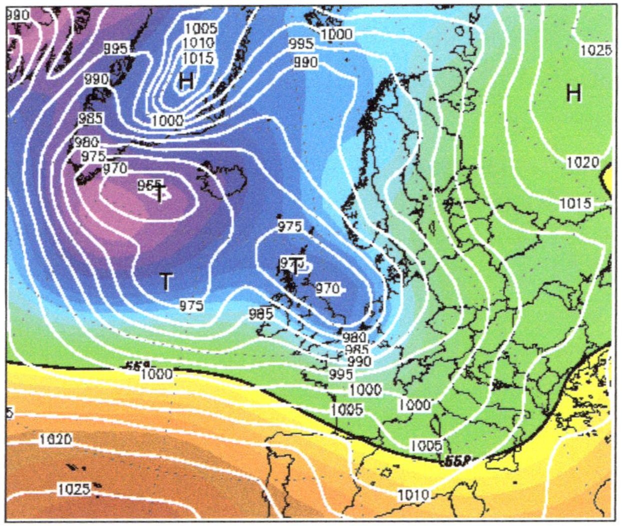

RAIN POTENTIAL

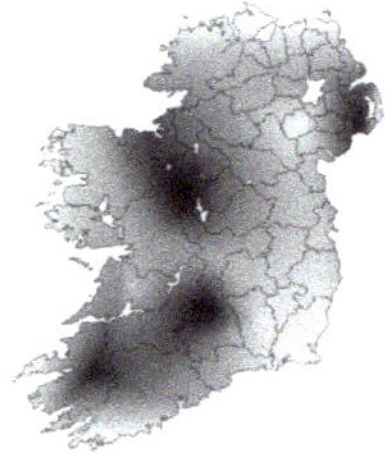

FROST/SNOW

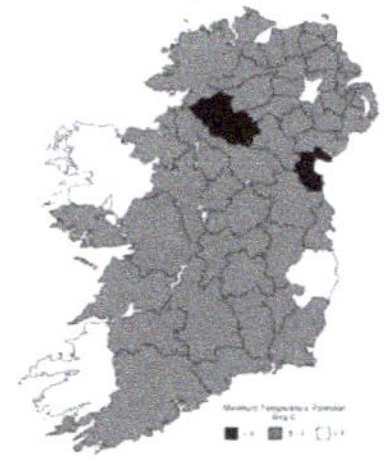

MARCH 23 MONDAY

The depression centred to the immediate north of Ireland reinvigorates the winds again, bringing very strong northwesterly winds, up to storm force at times, particularly in the northern half of the country, before easing later in the day. Another mixed day of sunny spells and bands of squally showers almost everywhere, including some showers possibly falling as hail, sleet, or snow mostly in the west and higher reaches. However, it may be mostly dry around the far southern coastal fringes.

Belfast: Stormy winds, squally showers, occasional sunny spells
Dublin: Mostly sunny, colder, light overnight showers, strong winds for a time
Cork: Sunny, spotty overnight showers, cooler, strong winds, chance of gales
Galway: Strong winds, mix of showers and sunny spells, chance of snow and gales

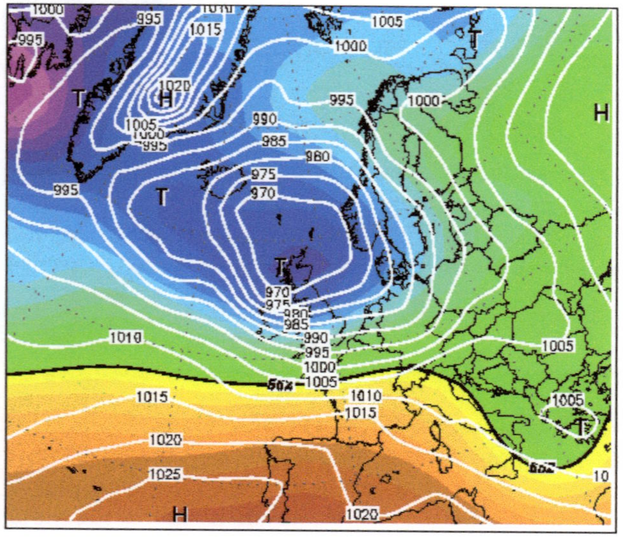

RAIN POTENTIAL

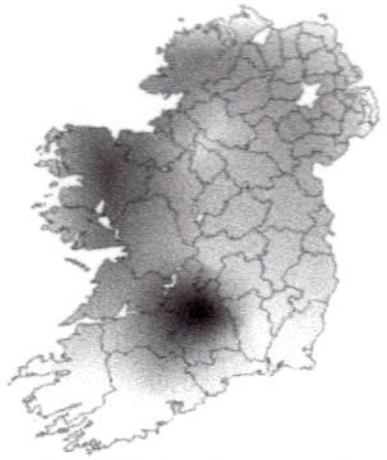

FROST/SNOW

MARCH 24 TUESDAY

The depression that has brought very unsettled conditions across the country over the past few days begins to weaken as it drifts northwards, away from Ireland, while a high pressure ridge builds to the south of the country. Winds ease to generally westerly in flow, mostly light to moderate, but stronger in the far north. Despite the easing of stormy conditions, further bands of lighter rain may still cross the country followed by a few sunny spells. Chance of snow may still prevail mostly in the north and west, while a lighter air may bring some fog patches in the south.

Belfast: Cloudy, colder, light showers, winds easing to moderate
Dublin: Cloudy, brief showers, milder temperatures, winds easing
Cork: Winds easing, overcast, light showers, milder temperatures
Galway: Cloudy and cold, some showers, stormy winds easing later, chance of snow

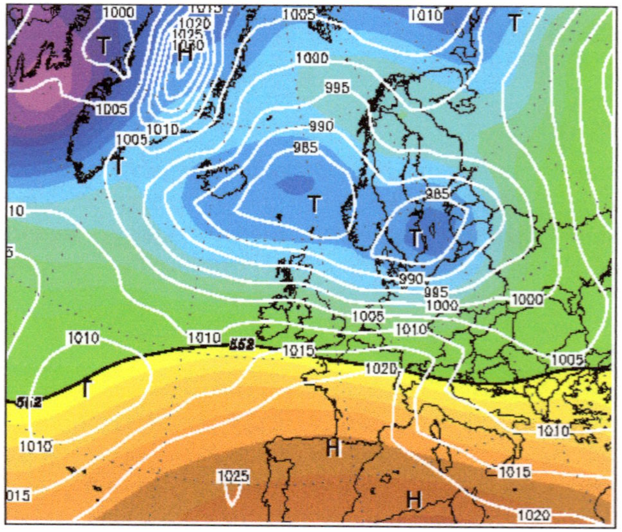

RAIN POTENTIAL

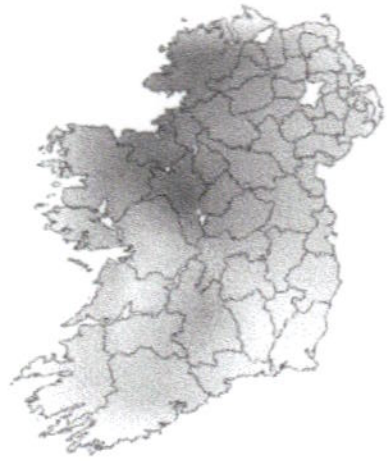

FROST/SNOW

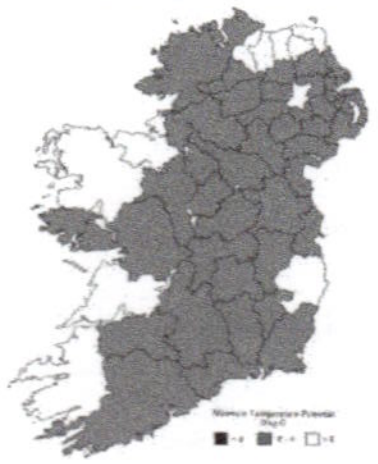

MARCH 25 WEDNESDAY

A complex of high and low pressure centres prevail around Ireland bringing widespread cloud throughout but also lighter northwesterly breezes over the northern half of the country, while southeasterlies may prevail over the southern half. Shower activity likely almost everywhere during the day, with some heavy falls possible for a time mostly in the northwest. In contrast, the gentle air flow across the country may also see patches of fog here and there. Chance of overnight light snow once more mostly in the higher reaches.

Belfast: Overcast, light showers, cold, mostly calm air, chance of fog overnight
Dublin: Overcast, mostly dry, increasing threats of rain, chance of misty patches
Cork: Overcast, scattered showers, light shifting breezes, chance of fog patches
Galway: Overcast, cold, rain, frosts likely, chance of overnight snow, moderate winds

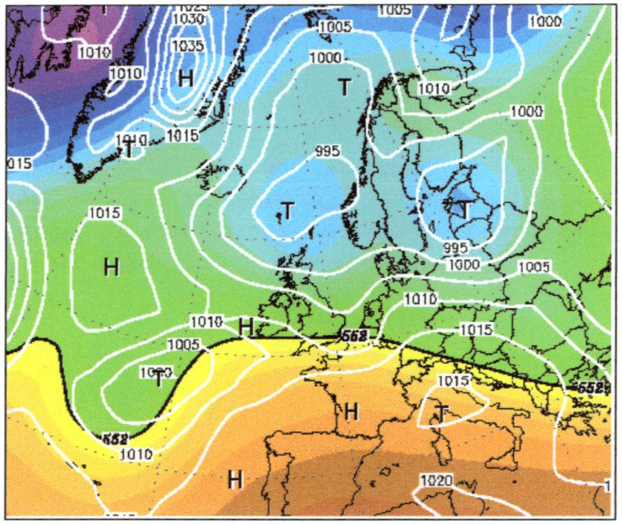

RAIN POTENTIAL

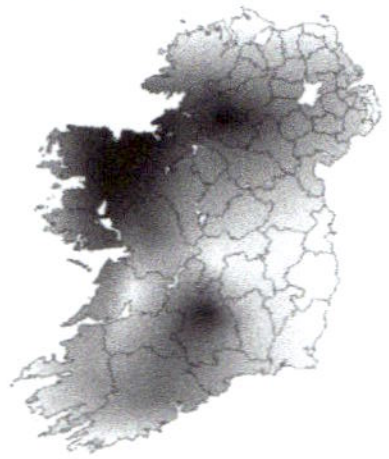

FROST/SNOW

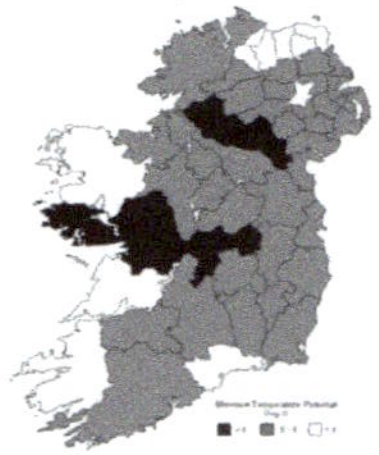

MARCH 26 THURSDAY

A weak depression directs a mostly light south to southeast flow over the country, with the chance of fog overnight in sheltered parts. Widespread cloud prevails. Rain may move into the north of the country at first before spreading quickly into all districts, with the chance of heavier falls and possible localised flooding in some central parts. Milder temperatures prevail.

Belfast: Light breezes to mostly calm air, overcast, rain, chance of fog
Dublin: Overcast, heavy rain, moderate winds
Cork: Overcast, moderate breezes, rain, milder, chance of overnight fog
Galway: Overcast, rain, milder temperatures, fresh breezes at times

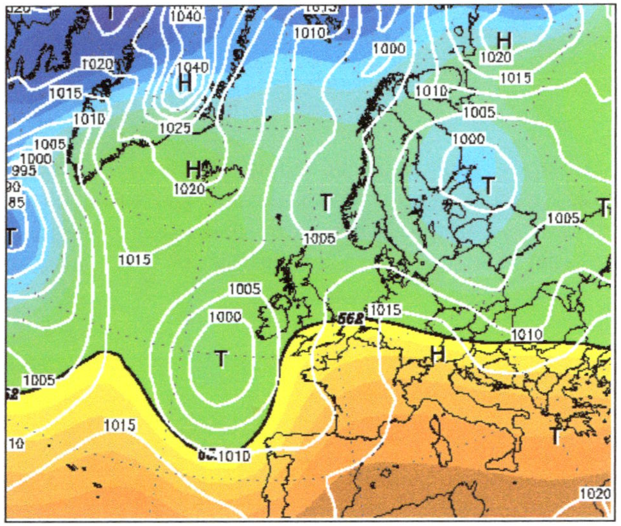

RAIN POTENTIAL

FROST/SNOW

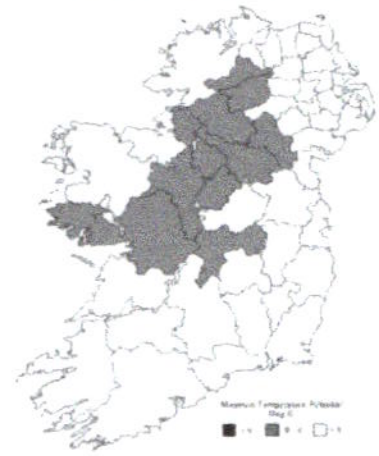

MARCH 27 FRIDAY

A depression crossing Ireland continues to drift away to the east, with a light to moderate northeasterly flow prevailing throughout, and some patches of fog mostly in the north. Cloudy everywhere with some showers still lingering in the east of the country, while mostly dry in the western half of Ireland and in the south.

Belfast: Mostly cloudy, dry, light breezes
Dublin: Cloudy, scattered showers, cool temperatures, moderate breezes
Cork: Overcast, gentle shifting breezes, mostly dry and cooler
Galway: Cloudy and cold, dry, light breezes

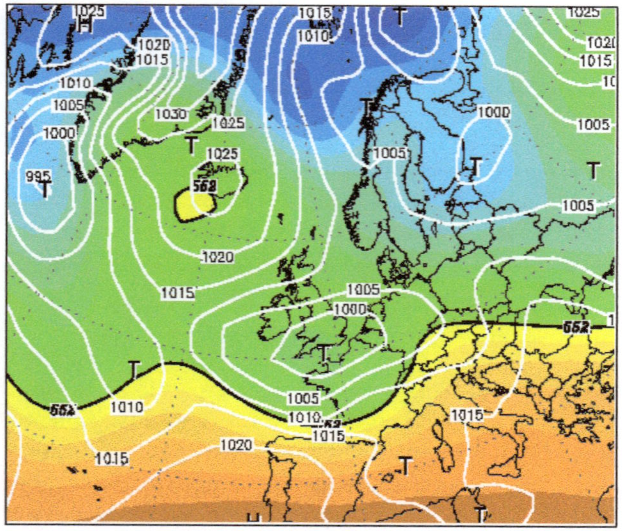

RAIN POTENTIAL

FROST/SNOW

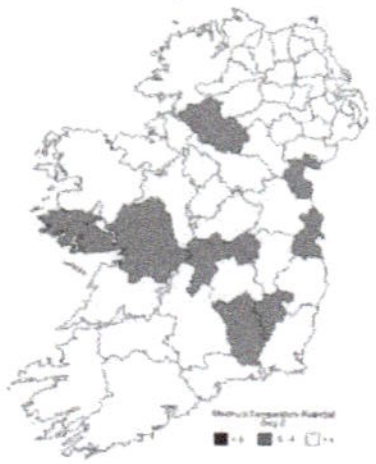

MARCH 28 SATURDAY

A high pressure system to the north of Ireland extends a ridge over the country and a light easterly flow throughout. Colder everywhere overnight with patches of frost and fog likely, particularly in central parts where some frosts may be severe. Mostly dry day everywhere, with good sunny spells prevailing.

Belfast: Dry, sunny, mild day overnight frosts, chance of fog
Dublin: Partly cloudy, dry, cold, light breezes
Cork: Partly cloudy, dry, cold, mostly calm air, fog possible
Galway: Sunny, dry, milder day cold night, gentle breezes, chance of frost overnight

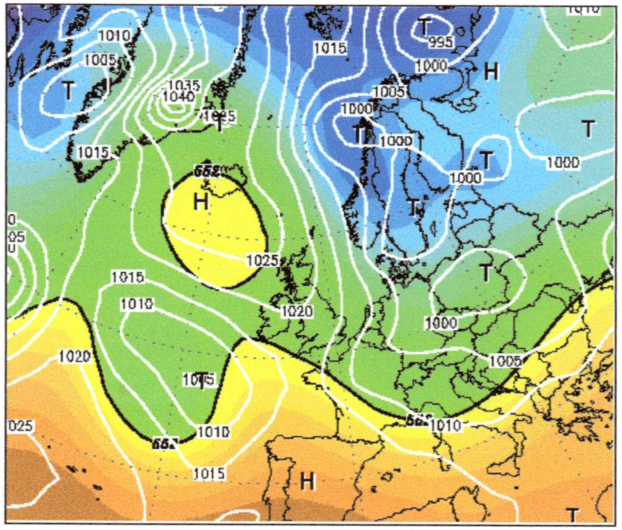

RAIN POTENTIAL

FROST/SNOW

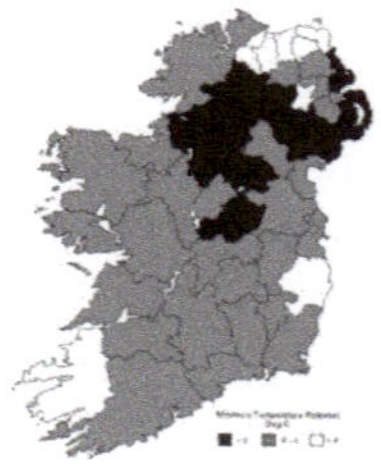

MARCH 29 SUNDAY

High pressure prevails. Light and variable winds prevail throughout, predominately in a northwesterly air. Mostly sunny and dry everywhere, with overnight frost patches in many parts of the country. Chance of snow flurries here and there in higher reaches and in the northwest overnight.

Belfast: Mostly sunny and dry, mild day, frosty overnight, calm and serene air
Dublin: Partly cloudy, cold, frosts, dry, mostly calm air
Cork: Occasional sunny spells, dry, cold, gentle breezes to mostly calm air
Galway: Sunny, dry, overnight frosts likely, light breezes

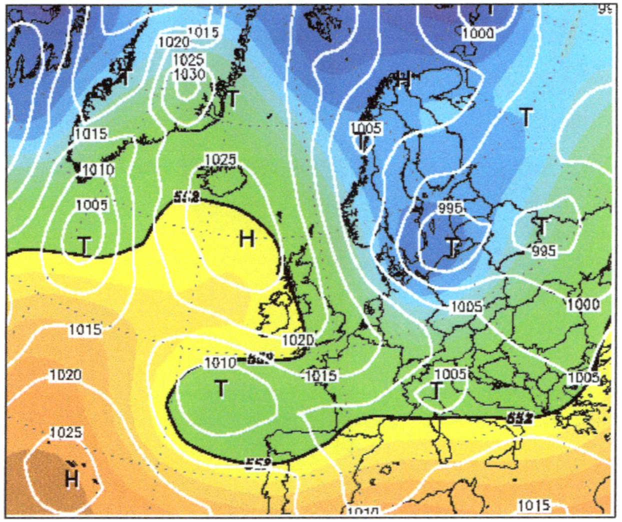

RAIN POTENTIAL

FROST/SNOW

MARCH 30 MONDAY

High pressure prevails with a mostly calm air to light and variable breezes. Sunnier in the south, while cloudier in the north. Patches of frost and fogs in many districts, as well as the chance of some isolated wintry showers in the northwest of the country, possibly falling as light sleet or snow. Dry elsewhere.

Belfast: Cloudy, cold with frost patches likely, dry, calm and misty atmosphere
Dublin: Cold with overnight frosts, cloudy, dry, mostly serene air
Cork: Dry, sunny, mild day cold night, calm and serene air
Galway: Partly cloudy, dry, overnight frosts, mostly calm air, fog possible

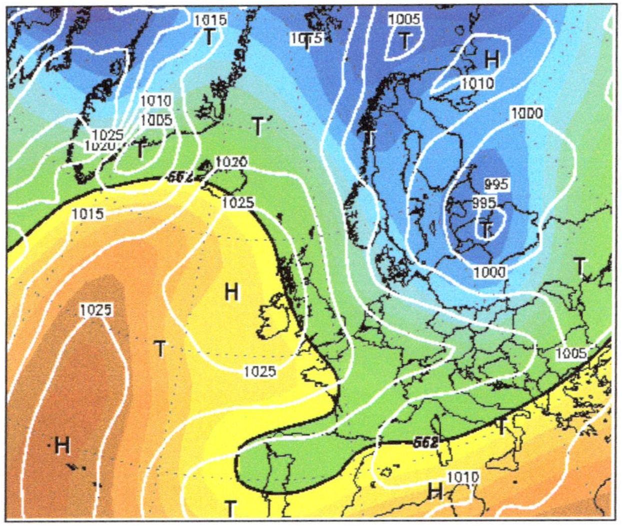

RAIN POTENTIAL

FROST/SNOW

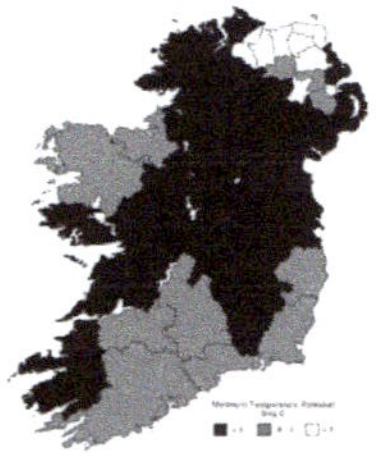

MARCH **31** TUESDAY

The anticyclonic influence of the past couple of days drifts away to the south of the country, while a low pressure system near Iceland nears the north bringing widespread heavy cloud and strengthening northwesterly breezes throughout, with winds becoming strong along the western coasts later in the day. Mostly dry in the south and eastern coastal fringes, while showers develop elsewhere.

Belfast: Overcast, slightly warmer, brief showers, light breezes, chance of fog
Dublin: Overcast, isolated showers, moderate breezes
Cork: Overcast, mostly dry, some threats of rain, milder, gentle breezes
Galway: Overcast and dull, brief showers, milder, moderate to light breezes

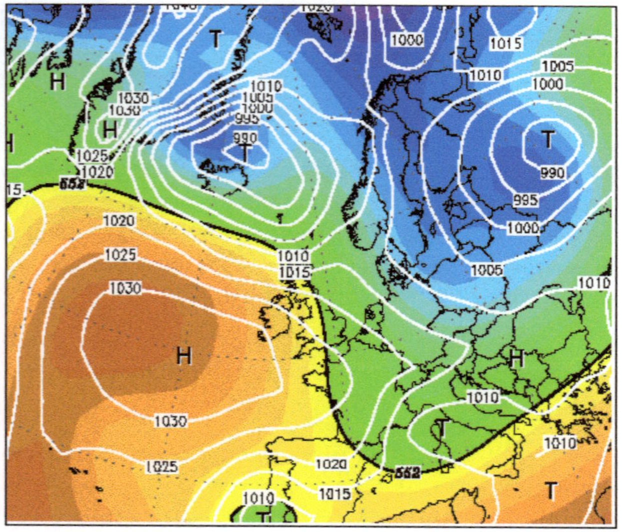

RAIN POTENTIAL

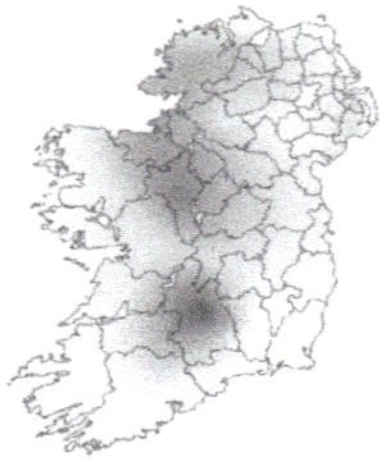

FROST/SNOW

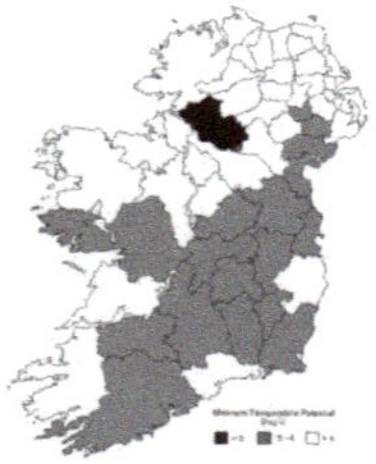

APRIL

PHASES OF THE MOON

2nd	Full Moon
2nd	Crossing Equator
7th	Apogee
8th	Southern Declination
10th	Third Quarter
16th	Crossing Equator
17th	New Moon
19th	Perigee #7
21st	Northern Declination
24th	First Quarter
29th	Crossing Equator

1	2	3	4	5	6	7	8	9	10	11	12	13	14	15	16	17	18	19	20	21	22	23	24	25	26	27	28	29	30
Wed	Thur	Fri	Sat	Sun	Mon	Tue	Wed	Thur	Fri	Sat	Sun	Mon	Tue	Wed	Thur	Fri	Sat	Sun	Mon	Tue	Wed	Thur	Fri	Sat	Sun	Mon	Tue	Wed	Thur
	F XhS					A	V		3Q						XhN	N		P7		^			1Q					XhS	

MONTHLY SUMMARY

1st-11th Low pressure may be re-established to the northeast of Ireland by now, bringing further unsettled conditions throughout, with a cold northerly flow prevailing, and some light rain and occasional thunderstorms and hail activity on most days. A westerly turn midway through this spell may see an increase in daytime temperatures.
1st-10th A large high pressure system to the south of the country brings a rise in air pressure over the country.
2nd Unusually sunny day likely.
2nd-8th Chance of snow during this spell.
5th-7th Chance of widespread frost.
6th Chance of a westerly change in the airflow bringing milder daytime temperatures. An active front system may bring some heavy rain around this time, sometimes accompanied by thunderstorm activity. Despite the rain, it may also be an unusually sunny day.
6th-10th Spell of frequent thunderstorms likely. Heavy rain likely, possibly resulting in some localised flooding, particularly in the east.
8th-11th Chance of very heavy rain in many parts of the country around this time.

Over the next four weeks, low pressure will be the dominant weather, with very little variation in pressure. The unsettled conditions already established this month, may continue to bring further wet spells and showers, some of hail and thunderstorms on numerous days over the next four weeks, while temperatures may be closer to average for this time of year, and sunshine hours above average overall, with the northwest particularly sunnier than average. Ground frosts may be frequent over the next two weeks, before a warmer change.
There may be a wide variation of wetter and drier parts of the country during this outlook period, with some central areas possibly only receiving around half their usual amount of rainfall for this time of year, while other parts in the west up to 30% more rainfall than normal. Overall, there may be between 9 and 17 days on which rain may occur during this outlook period, 8 days with thunderstorm activity, 7 days with hail showers, 1 day with widespread fog, 1 day with gale winds, 5 days with snow and around 20 frost days in inland parts.

12th-14th Milder temperatures and long sunny periods likely.
12th A deep depression nears the north of Ireland which may bring strong winds over the country, with possible gale gusts, particularly in the northwest.
12th-24th A cool and very showery spell across the country, particularly in the north and northwest, as areas of low pressure centred to the north to northwest of Ireland dominate.
13th Chance of widespread fog.
14th Unusually warmer than normal temperatures possible, particularly in the southeast.
15th Cooler change.
16th-18th Spell of wintry showers possible including hail, light at first building to heavier later, occasionally accompanied by thunderstorm activity. Chance of snow.
16th-27th Cooler temperatures may see regular air and ground frosts in numerous districts.
17th-19th Widespread frosts in inland parts likely.
18th Unusually cold with frosts overnight in the midlands and southwest.
21st-22nd Chance of snow.
21st-25th Chance of frequent thunderstorms, heavy rain, and hail showers.
24th-28th Very sunny spell, particularly in the northwest.
25th-28th Anticyclonic spell. Mostly dry and sunny throughout, with light to moderate east to southeast breezes. Chance of widespread ground frosts overnight, while daytime temperatures nearer to normal for this time of year.
26th Chance of severe frosts overnight in inland parts.
27th Very sunny day in the northwest.
29th-30th Low pressure may once again re-establish unsettled conditions throughout the country and persist into the first week to ten days of May. Regular spells of rain or showers likely on most days, sometimes accompanied by thunderstorms.

IRELAND RAINFALL ESTIMATES

April 2026

d=mainly dry, n= nonrecordable, l=light shrs, s=significant shrs, r=rain, h=hvy falls

		ULSTER PROVINCE					CONNAUGHT PROVINCE						LEINSTER PROVINCE				MUNSTER PROVINCE						
		Alderg	Hillsbo	Armag	Loughl	Malin I	Sligo	Ardtari	Belmul	Drums	Galway	Roscoi	Markre	Edende	Kilkeni	Dublin	Shann	Newpo	Tralee	Killarne	Cork	All	Moon
1st	Wed	l	l	l	l	l	l	l	s	l	l	l	l	l	l	l	l	s	s	l	l	40	
2nd	Thur	l	l	n	l	l	s	l	l	s	l	l	s	l	l	d	d	s	l	l	d	38	F XhS
3rd	Fri	l	l	l	l	l	s	l	l	l	l	l	l	l	l	n	n	l	l	l	n	41	
4th	Sat	l	d	l	d	d	l	d	l	d	l	d	d	d	d	l	l	l	l	l	s	22	
5th	Sun	l	l	d	s	l	s	s	s	l	l	s	s	l	d	d	l	l	l	l	l	50	
6th	Mon	l	d	s	l	s	l	l	r	n	l	l	l	d	l	n	l	l	l	s	l	49	
7th	Tues	s	r	l	r	l	h	r	r	r	h	h	l	s	s	l	l	h	r	r	s	171	A
8th	Wed	s	s	s	s	s	r	s	s	r	s	s	s	s	s	s	s	r	s	h	s	141	V
9th	Thur	s	r	s	s	s	s	s	s	s	s	s	s	s	l	l	s	s	r	s	s	132	
10th	Fri	l	n	s	l	l	l	l	d	l	d	n	l	r	h	h	h	l	r	r	r	134	3Q
11th	Sat	l	s	l	l	l	s	l	l	l	l	s	s	l	n	r	l	l	s	s	s	64	
12th	Sun	s	l	s	s	s	s	s	l	l	l	l	l	l	d	l	l	l	d	d	n	45	
13th	Mon	n	d	l	d	d	d	d	l	d	d	d	d	d	d	l	d	l	d	d	d	6	
14th	Tues	d	d	d	d	d	d	d	d	l	d	d	n	d	d	d	d	d	d	d	d	2	
15th	Wed	d	d	d	l	s	l	l	l	l	n	n	l	l	d	l	d	l	d	d	d	13	
16th	Thur	l	d	l	l	l	l	l	l	l	l	l	l	l	d	d	d	l	d	d	d	18	XhN
17th	Fri	l	d	l	l	l	l	l	s	l	l	l	l	d	l	n	d	s	n	l	d	34	N
18th	Sat	l	n	l	l	l	l	l	r	l	l	l	l	l	l	l	l	s	n	l	l	49	
19th	Sun	d	d	n	s	l	l	s	s	d	l	l	l	d	d	d	n	l	l	l	d	30	P7
20th	Mon	l	l	l	l	s	l	l	l	l	l	l	l	l	s	d	l	l	l	l	l	30	
21st	Tues	n	n	d	l	l	l	l	s	s	s	s	s	s	n	n	r	l	s	s	s	65	^
22nd	Wed	l	l	l	s	d	r	s	s	l	l	s	s	l	l	s	l	r	s	l	s	84	
23rd	Thur	l	l	s	l	l	l	l	s	s	s	s	s	s	l	l	l	l	l	l	n	76	
24th	Fri	d	d	d	d	d	n	d	l	l	d	l	l	l	l	s	l	n	d	d	n	13	1Q
25th	Sat	d	d	d	l	d	d	l	l	d	d	d	l	d	d	d	d	d	d	d	d	3	
26th	Sun	d	d	d	d	l	d	d	d	d	d	d	d	d	d	d	d	d	d	d	d	2	
27th	Mon	l	l	d	d	d	d	d	d	d	d	d	d	d	d	d	d	d	d	d	d	2	
28th	Tues	n	l	d	n	l	l	n	d	d	l	d	l	d	d	d	l	d	d	d	n	6	
29th	Wed	d	d	d	l	d	n	l	l	l	d	l	l	d	d	d	d	n	d	d	l	9	XhS
30th	Thur	d	d	d	d	d	l	d	d	l	l	l	n	d	d	d	d	d	d	d	l	8	
	Estimate:	46	47	44	66	64	93	66	104	71	67	80	65	61	50	67	59	98	90	87	56	1378	
	Average:	58	61	58	70	65	78	80	72	67	72	81	78	66	56	54	59	97	110	81	77	1439	
	Trend:	drr	drr	drr	av	av	w tr	drr	w tr	av	av	av	drr	av	av	w tr	av	av	drr	av	drr	av	

IRELAND SUNSHINE ESTIMATES

April 2026

F=fine (8-12 hours of sunshine), pc= partly cloudy (4-7 hours), c=cloudy (1-3 hours), o=overcast (0 hours)

		ULSTER PROVINCE					CONNAUGHT PROVINCE						LEINSTER PROVINCE				MUNSTER PROVINCE						
		Alderg	Hillsbo	Armag	Loughl	Malin I	Sligo	Ardtari	Belmul	Drums	Galway	Roscoi	Markre	Edende	Kilkeni	Dublin	Shannc	Newpo	Tralee	Killarne	Cork	All	Moon
1st	Wed	pc	pc	pc	pc	pc	pc	pc	pc	pc	pc	pc	pc	pc	pc	pc	pc	pc	c	c	pc	120	
2nd	Thur	pc	pc	pc	pc	pc	pc	pc	pc	F	F	pc	pc	F	F	pc	F	F	F	F	F	163	F XhS
3rd	Fri	c	c	c	c	c	pc	c	pc	pc	pc	pc	c	pc	pc	c	pc	pc	c	c	pc	89	
4th	Sat	c	c	c	c	c	c	c	c	c	c	c	c	c	c	pc	c	c	o	o	c	54	
5th	Sun	c	c	c	c	c	c	c	c	c	c	c	c	c	c	o	c	c	c	c	c	50	
6th	Mon	pc	pc	pc	pc	pc	pc	pc	c	pc	pc	pc	pc	pc	F	pc	pc	F	pc	pc	F	127	
7th	Tues	pc	pc	pc	pc	pc	pc	pc	c	pc	pc	pc	pc	pc	pc	F	pc	pc	pc	pc	pc	131	A
8th	Wed	c	c	c	c	c	c	c	pc	c	pc	pc	c	c	pc	c	c	pc	pc	pc	pc	72	V
9th	Thur	c	c	c	c	c	c	c	c	c	c	c	c	c	c	c	pc	pc	c	c	c	65	
10th	Fri	pc	pc	pc	pc	pc	pc	pc	pc	pc	pc	pc	pc	pc	c	pc	c	c	c	c	c	86	3Q
11th	Sat	pc	pc	pc	pc	pc	c	pc	c	pc	c	c	pc	pc	pc	pc	c	pc	o	o	pc	79	
12th	Sun	pc	pc	pc	pc	pc	pc	pc	pc	F	pc	pc	pc	F	F	F	F	F	pc	pc	F	153	
13th	Mon	c	c	c	c	c	c	c	o	pc	c	o	c	pc	pc	pc	c	pc	o	o	F	60	
14th	Tues	pc	pc	pc	pc	pc	c	pc	o	pc	c	c	pc	pc	pc	pc	pc	pc	pc	pc	pc	88	
15th	Wed	pc	pc	pc	pc	pc	4.c	pc	c	pc	c	c	pc	pc	c	pc	c	c	c	c	c	73	
16th	Thur	F	F	F	F	F	pc	F	pc	F	pc	pc	F	F	F	F	F	F	pc	pc	F	174	XhN
17th	Fri	F	F	F	F	F	F	F	pc	F	pc	pc	F	F	F	F	F	F	F	F	F	174	N
18th	Sat	F	F	F	F	F	pc	F	pc	F	pc	pc	F	F	F	F	F	F	F	F	pc	160	
19th	Sun	pc	pc	pc	pc	pc	F	pc	pc	c	pc	pc	pc	c	c	c	c	c	pc	pc	c	91	P7
20th	Mon	pc	pc	pc	pc	pc	F	pc	pc	pc	pc	pc	pc	pc	c	pc	pc	pc	F	F	c	134	
21st	Tues	pc	pc	pc	pc	pc	pc	pc	pc	pc	pc	pc	pc	pc	pc	pc	pc	pc	c	c	F	122	^
22nd	Wed	pc	pc	pc	pc	pc	pc	pc	pc	pc	pc	pc	pc	pc	pc	pc	pc	pc	c	c	pc	122	
23rd	Thur	c	c	c	c	c	c	c	pc	pc	pc	pc	c	pc	pc	c	pc	pc	F	F	F	95	
24th	Fri	F	F	F	F	F	F	F	F	F	F	F	F	F	F	F	F	F	F	F	F	200	1Q
25th	Sat	F	F	F	F	F	F	F	F	F	F	F	F	F	F	F	F	F	F	F	F	231	
26th	Sun	F	F	F	F	F	F	F	F	F	F	F	F	F	F	F	F	F	F	F	F	204	
27th	Mon	F	F	F	F	F	F	F	F	F	F	F	F	F	F	F	F	F	c	c	pc	204	
28th	Tues	pc	pc	pc	pc	pc	F	pc	F	pc	F	F	pc	pc	pc	pc	F	pc	F	F	pc	156	
29th	Wed	pc	pc	pc	pc	pc	F	pc	F	c	pc	pc	pc	c	c	o	pc	pc	pc	pc	c	90	XhS
30th	Thur	c	c	c	c	c	pc	c	F	c	pc	pc	c	c	o	o	o	o	o	o	o	44	
	Estimate:	181	181	181	181	181	179	181	172	184	181	178	181	184	188	182	188	188	165	165	192	3611	
	Average:	151	151	133	152	153	138	140	158	137	149	155	145	151	146	156	148	140	150	150	156	2957	
	Trend:	more	more	more	more	more	more	more	more	more	more	more	more	more	more	more	more	more	more	more	more	more	

APRIL 1 WEDNESDAY

Ireland sits between a large anticyclonic system to the southwest of the country with a depression to the northeast, with a very strong and blustery north to northwest flow prevailing throughout. A very unsettled day everywhere, with a mix of sunny spells and passing light squally showers, some possibly of hail. Cold overnight with chance of some frosts possible mostly in central areas.

Belfast: Blustery air, passing scattered cloud, isolated showers, cold overnight
Dublin: Mostly sunny, squally showers mostly overnight, strong and blustery winds
Cork: Mostly sunny, brief showers, strong and unsettled winds
Galway: Partly cloudy, brief showers, colder, strong gusty winds, chance of gales

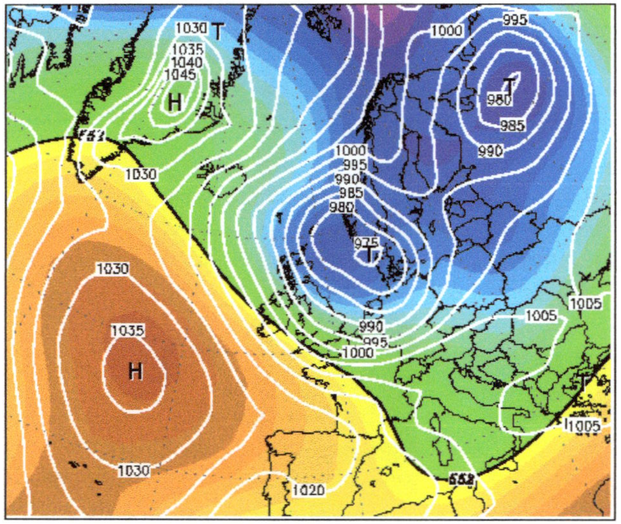

RAIN POTENTIAL

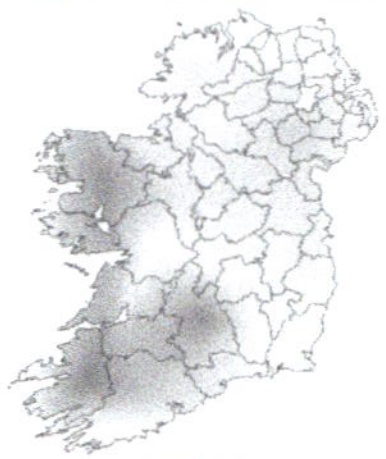

FROST/SNOW

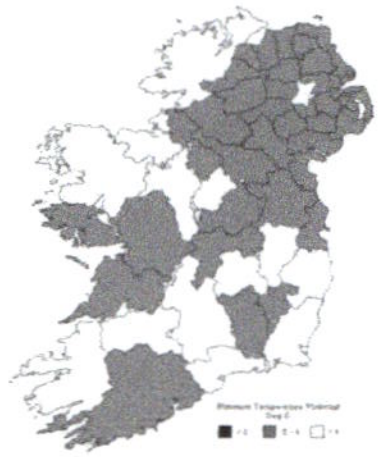

APRIL 2 THURSDAY

The high pressure system moves up to the west of Ireland while a depression drifts southwards to the east of the country, with a prevailing blustery north to northeast flow over most of the day, easing towards evening. Mostly sunny everywhere, with some showers, mostly overnight, persisting in the north, west and central areas, while mostly dry in the east and far south.

Belfast: Mostly sunny, colder, isolated showers, fresh breezes
Dublin: Sunny, colder, dry, strong winds with occasional gusts
Cork: Sunny, dry, cooler, unsettled winds
Galway: Sunny, brief showers, colder, unsettled winds at times

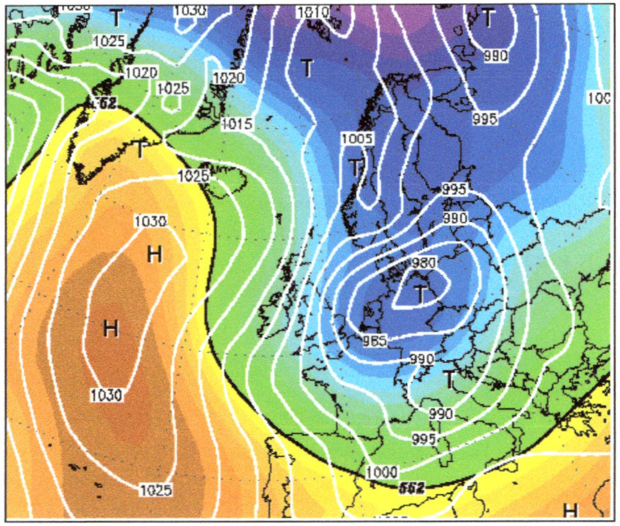

RAIN POTENTIAL

FROST/SNOW

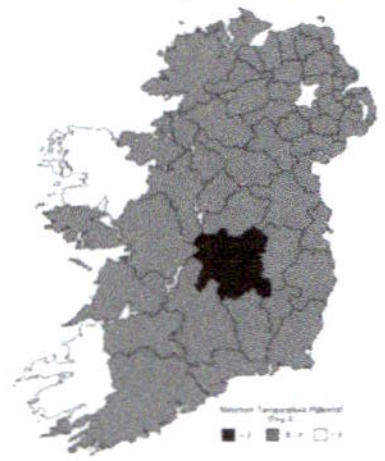

APRIL **3** FRIDAY

A moderate to strong north to northeast flow prevails over the country as high pressure persists to the west and low pressure continues to the east. A mixed day of some sunny spells and passing light showers in most districts for a time, more frequent in the north and west, while more isolated elsewhere. Frosts overnight likely in central parts as well as the chance of snow flurries, for a time in the north.

Belfast: Partly cloudy, colder, light showers, moderate breezes, chance of snow
Dublin: Cloudy and cold, isolated showers, winds easing
Cork: Partly cloudy, isolated showers, milder temperatures, fresh breezes
Galway: Some sunny spells, isolated showers, chance of snow flurries

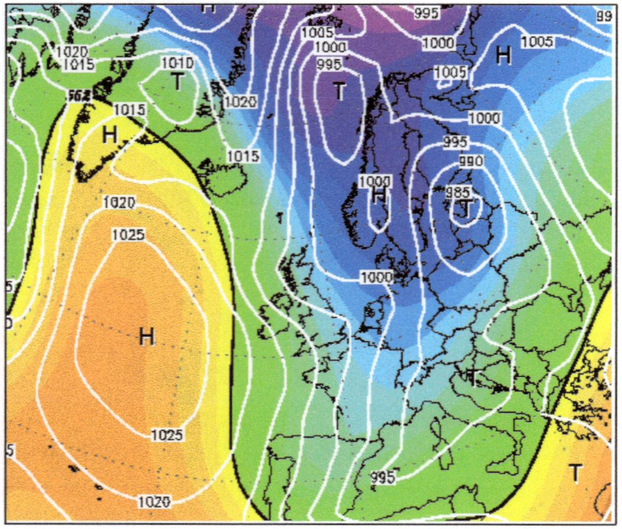

RAIN POTENTIAL

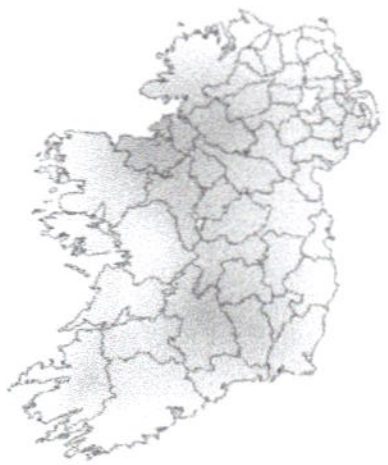

FROST/SNOW

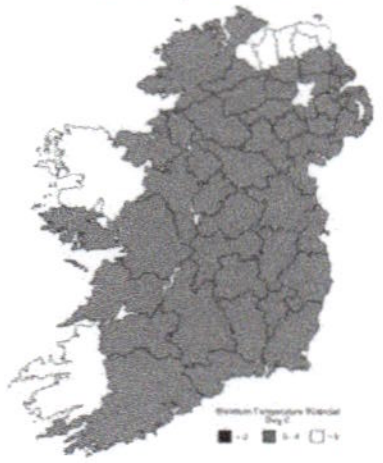

APRIL **4** SATURDAY

Ireland may still sit between high pressure to the west and low pressure to the east, with a lighter northwesterly flow prevailing throughout. Some patches of rain or drizzle overnight in many areas, clearing to mostly dry during the day in the north and east, while shower activity may persist on and off in the south and west. Inland parts may see patches of frost and fog overnight.

Belfast: Light breezes, mostly cloudy, dry and cold
Dublin: Partly cloudy, brief showers, cold northwesterly breezes
Cork: Cloudy, milder temperatures, occasional showers, moderate breezes
Galway: Cloudy, isolated showers, cold northwest breezes

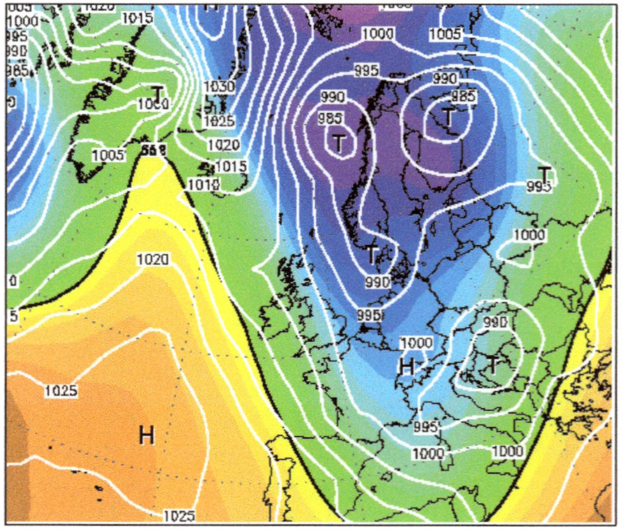

RAIN POTENTIAL

FROST/SNOW

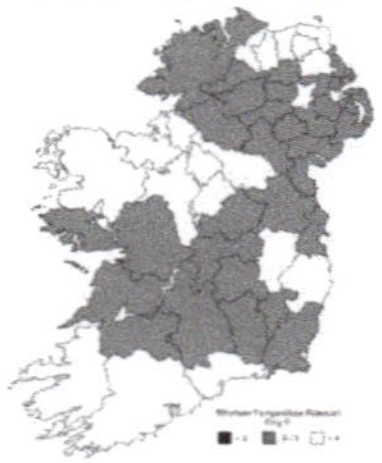

APRIL **5** SUNDAY

A light northwesterly flow prevails as the country sits between high pressure to the west and low pressure to the east. A shower embedded trough brings widespread cloud and light showers and drizzle patches almost everywhere, although mostly dry in the eastern coastal counties. Some patches of fog possible, particularly in northern and central parts, and pockets of frost may affect some western areas. Some brief sunny spells likely later in the day.

Belfast: Cloudy, occasional showers, light breezes, chance of fog patches
Dublin: Overcast, mostly dry, threats of rain, chance of frosts overnight
Cork: Cloudy, light showers, mild temperatures, gentle breezes
Galway: Light showers, frequent cloud, colder, chance of frost overnight

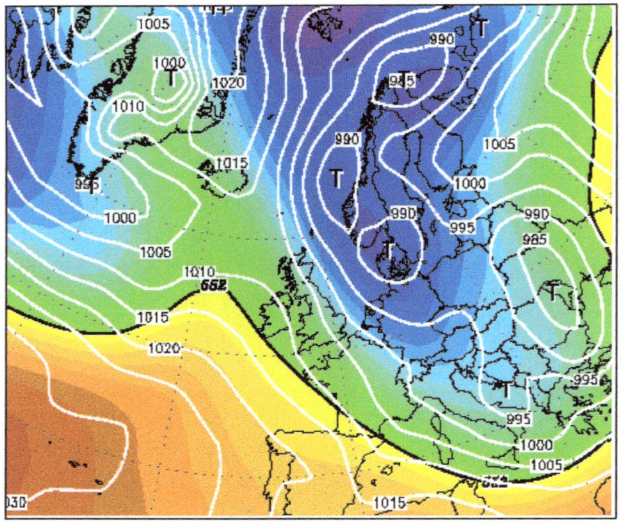

RAIN POTENTIAL

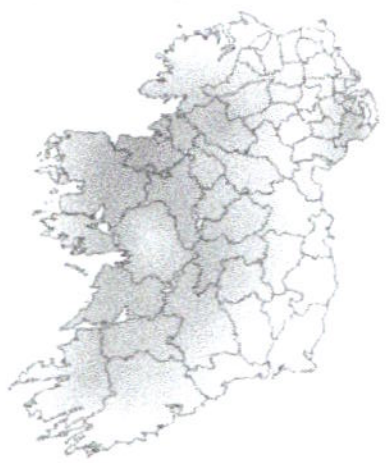

FROST/SNOW

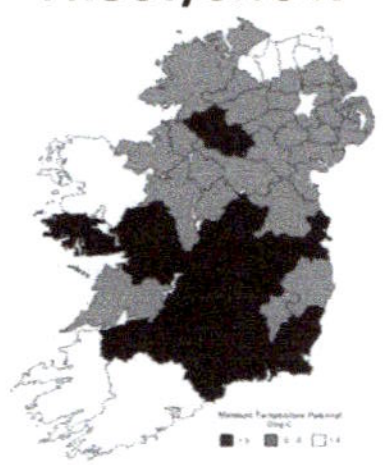

APRIL **6** MONDAY

A low pressure trough associated with a couple of depressions, one to the northwest of Ireland and another to the east, brings a a light west to northwesterly flow over the country. A mixed day with occasional sunny spells and passing bands of showers, some possibly of hail. Chance of patches of fog overnight, particularly in the north and east.

Belfast: Partly cloudy, dry, cold, light breezes, chance of fog overnight
Dublin: Partly cloudy, isolated showers, light breezes
Cork: Sunny and cooler, isolated showers, light breezes, chance of fog overnight
Galway: Partly cloudy, colder, brief showers, light breezes

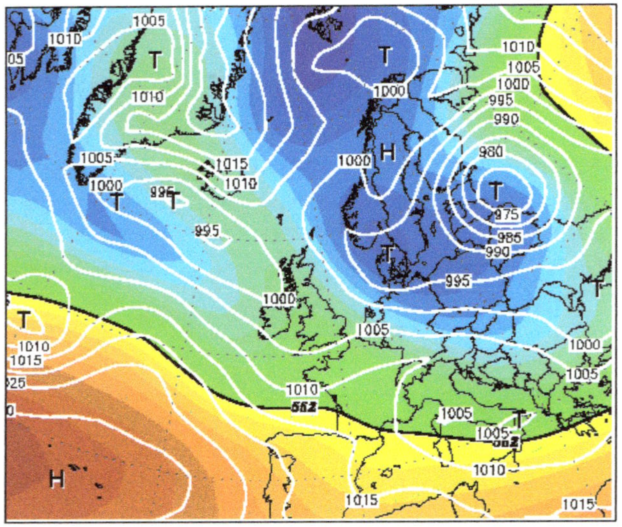

RAIN POTENTIAL

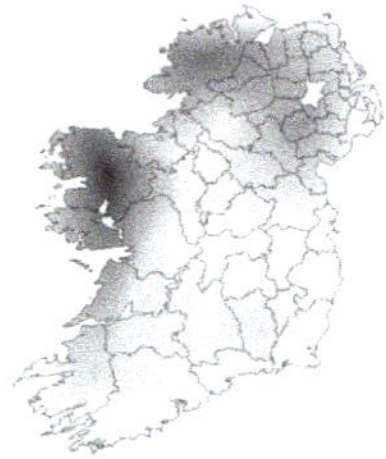

FROST/SNOW

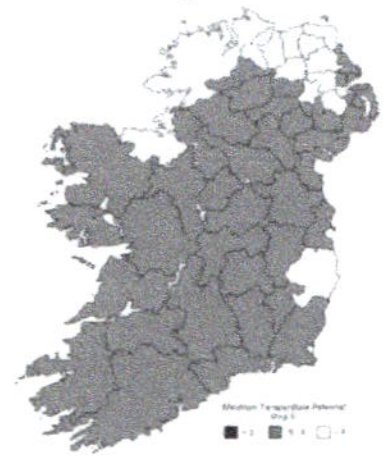

APRIL 7 TUESDAY

A large low pressure trough brings a very light mostly westerly flow over the country with overnight patches of frost. Mostly dry and sunny conditions prevails after the fog lifts in most districts. Later a wet front brings rain into the western areas at first, later spreading into the east. Some precipitation may be particularly heavy for a time in western districts, as well as possibly falling as hail.

Belfast: Changeable, mostly sunny, evening rain, cold overnight, possible fog
Dublin: Light shifting breezes, sunny, cold, brief showers
Cork: Partly cloudy, occasional showers, milder day cold night, gentle breezes
Galway: Mostly cloudy, heavy rain, cold, frosts possible, freshening breezes later

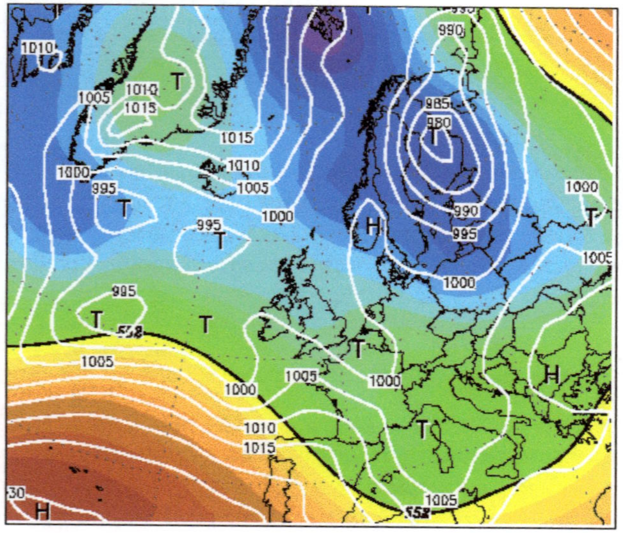

RAIN POTENTIAL

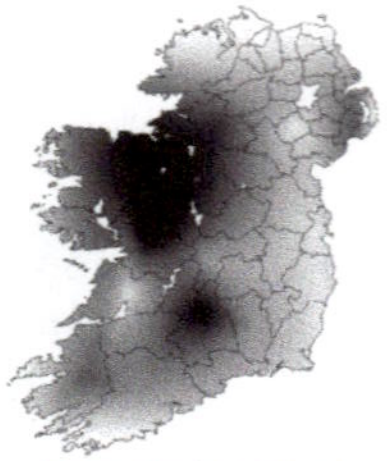

FROST/SNOW

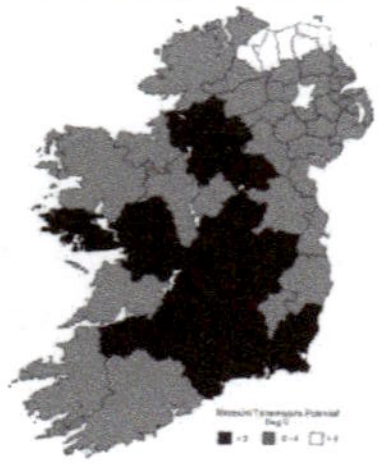

APRIL 8 WEDNESDAY

The depression that was situated to the northwest of Ireland intensifies as it moves onto the country bringing stormy conditions throughout. Strong and unsettled winds develop overnight with west to southwesterly gusts prevailing, sometimes to gale force in places. Cloudy and wet everywhere, with the chance of scattered thunderstorms, hail, or even snow overnight, particularly in the west of the country.

Belfast: Cloudy, milder, scattered showers, blustery at times
Dublin: Cloudy, scattered showers, cold, blustery at times
Cork: Mostly sunny, occasional showers, strong winds
Galway: Rain, sunny spells, strong winds, chance of thunderstorms, hail or snow

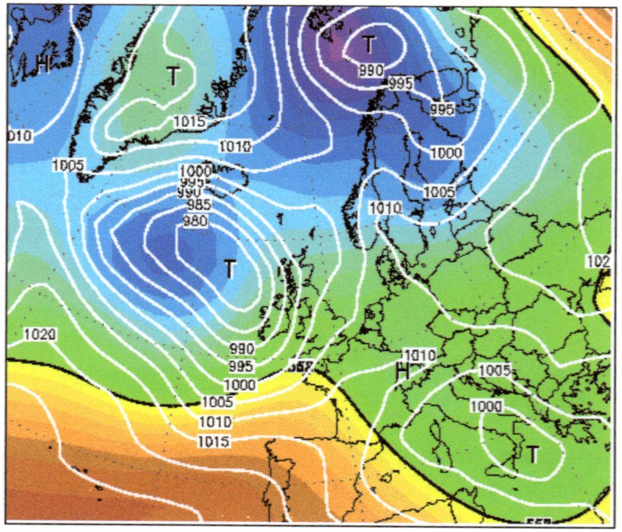

RAIN POTENTIAL

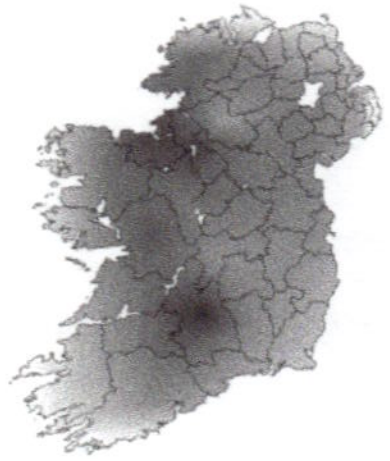

FROST/SNOW

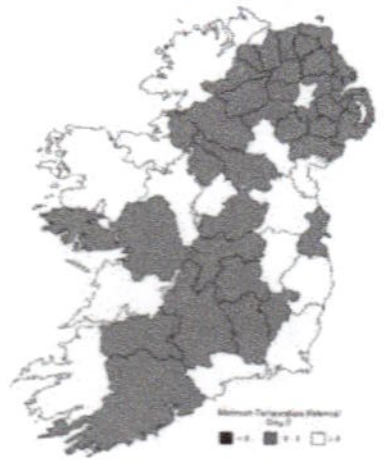

APRIL 9 THURSDAY

Stormy conditions continue as a large depression covers the country. Winds likely to be strong and gusty in a southwest flow. Some brief sunny spells likely in most districts while the passage of wet fronts cross the country bringing frequent showers, the occasional thunderstorm, and possible hail. Chance of further snow overnight mostly in a line from the west, across central areas and into the adjacent east.

Belfast: Cloudy, rain, stormy air, chance of thunderstorms
Dublin: Mostly cloudy, light showers, unsettled winds, chance of thunderstorms
Cork: Partly cloudy, scattered showers, blustery winds, chance of evening snow
Galway: Changeable, rain, some sunny spells, strong winds, chance of thunderstorms

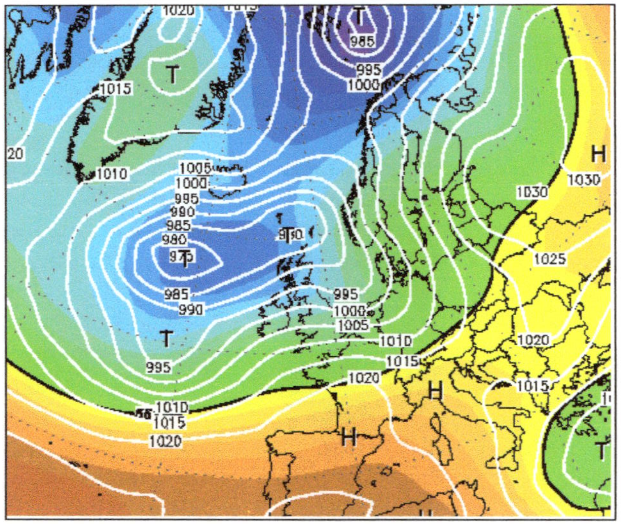

RAIN POTENTIAL

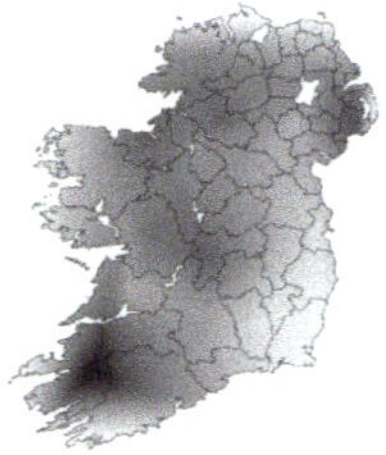

FROST/SNOW

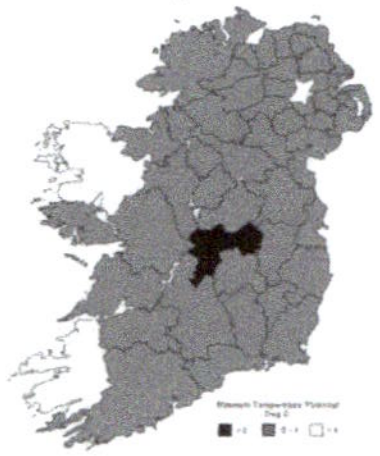

APRIL 10 FRIDAY

A depression over Ireland extends a fresh to strong west to southwest flow across the country. An unsettled day throughout, with a mix of sunny spells and occasional wintry showers almost everywhere; sunnier in the north, cloudier in the south. Thunderstorms likely within the rain bands and possible brief snowfalls overnight, mostly in the higher reaches and in the east. Rain likely to be very heavy for a time in eastern and southern areas, possibly resulting in some pockets of localised flooding.

Belfast: Changeable, partly cloudy, isolated showers, thunderstorms possible
Dublin: Changeable, heavy rain, some sunny spells, chance of thunderstorms
Cork: Cloudy, colder, rain, chance of snow overnight
Galway: Partly cloudy, dry, cold with frosts, fluctuating breezes, static air

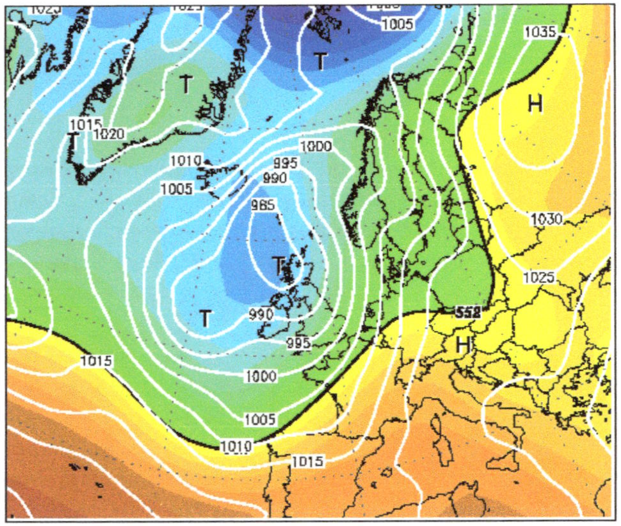

RAIN POTENTIAL

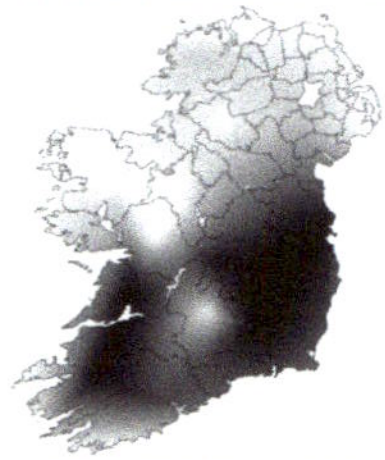

FROST/SNOW

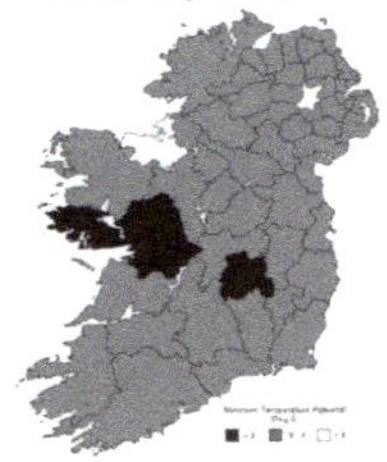

APRIL 11 SATURDAY

A weak depression remains centred over Ireland bringing unsettled and shifting winds throughout, fresher with gusts at times. Another mixed day of some sunny outbreaks between bands of rain. Thunderstorms possible at times, with chance of hail showers. Overnight temperatures still below zero in many parts of the country.

Belfast: Changeable, mix of sunny spells and showers, cold overnight, shifting winds
Dublin: Partly cloudy, evening rain, cold overnight, shifting breezes, static air
Cork: Partly cloudy, occasional showers, milder temperatures, fluctuating breezes
Galway: Cloudy, light showers, shifting breezes, cold with frosts overnight

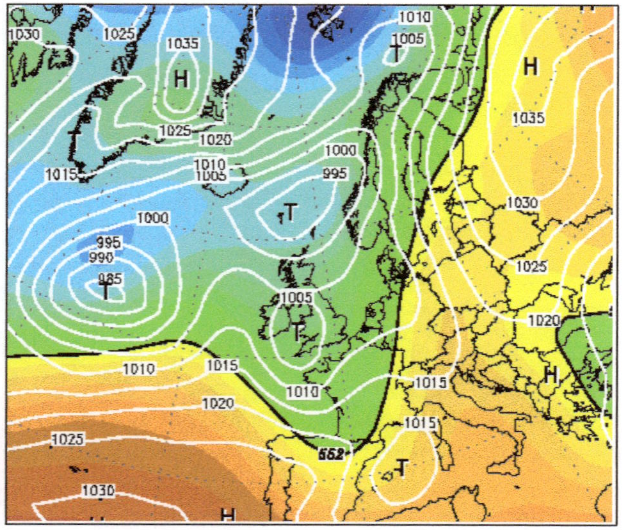

RAIN POTENTIAL

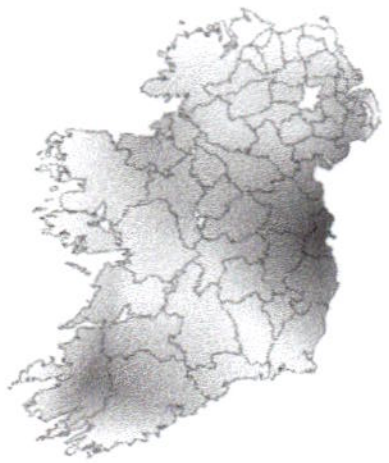

FROST/SNOW

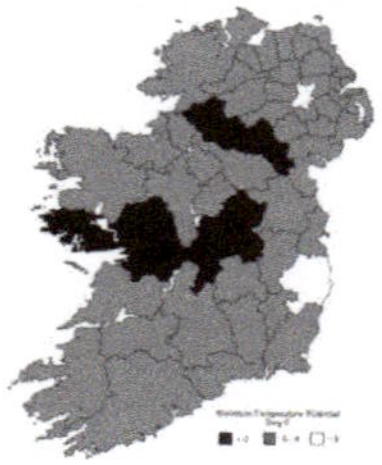

APRIL 12 SUNDAY

The depression over Ireland intensifies as it drifts slowly northwards away from the country. West to southwest winds prevail througout, frequently very strong and blustery, possibly to gale force at times. Further unsettled and changeable conditions prevail, with another day of a mix of sunny spells and bands of scattered showers. Showers may be heavier in the north and west and more isolated in the east and south, possibly mostly dry around the eastern and southern coasts. Temperatures milder than the past few days.

Belfast: Mostly sunny, brief evening showers, milder temperatures, blustery at times
Dublin: Blustery winds, sunny, warmer, isolated evening showers possible
Cork: Sunny, warmer, mostly dry, chance of evening showers, strong winds
Galway: Isolated showers, increasing sunny spells, milder, strong gusty winds

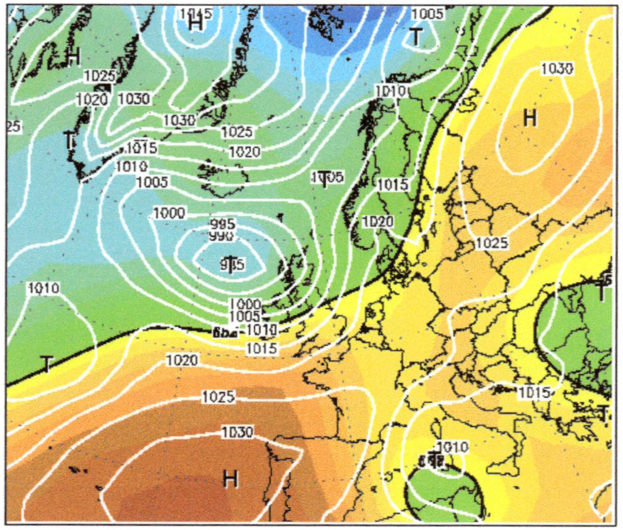

RAIN POTENTIAL

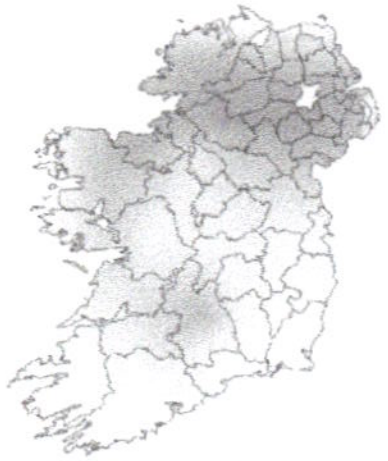

FROST/SNOW

APRIL 13 MONDAY

A high pressure system to the south of Ireland extends a ridge over the country, bringing more settled and drier conditions throughout. Winds generally light, mostly in a southwest flow. Some patches of overnight rain or drizzle likely, mostly confined to the east. Mostly dry elsewhere, with possible patches of fog overnight, and sunniest in the south.

Belfast: Cloudy, mostly dry, threats of rain, light breezes, fog patches possible
Dublin: Partly cloudy, occasional showers, warmer, moderate breezes
Cork: Sunny, warmer, dry, winds easing to light, chance of fog overnight
Galway: Cloudy, dry, fresh breezes, chance of fog overnight

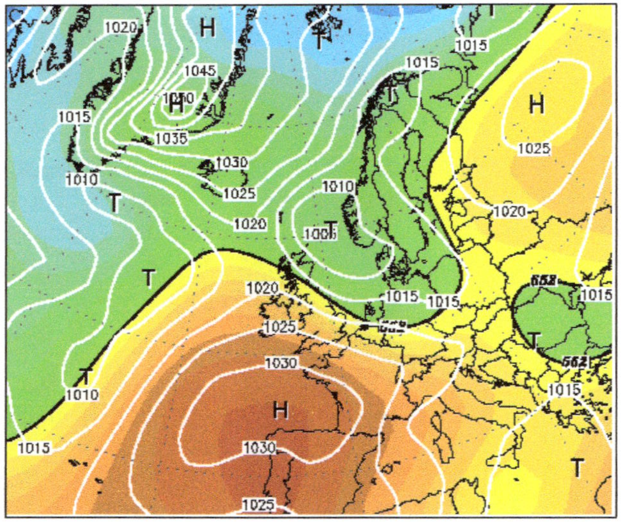

RAIN POTENTIAL

FROST/SNOW

APRIL 14 TUESDAY

Anticyclonic. Partly cloudy and dry throughout the country, with a mostly calm air to a light south to southwest flow prevailing. Patches of fog likely overnight in many districts. Long sunny spells likely in the east of the country, while cloudier to the north and west.

Belfast: Partly cloudy, warmer, dry, gentle breezes to mostly calm air, fog possible
Dublin: Lighter breezes, dry, occasional sunny spells, mild temperatures
Cork: Partly cloudy, dry, mild temperatures, mostly calm air, fog possible
Galway: Cloudy, slightly warmer, dry, light breezes, chance of fog

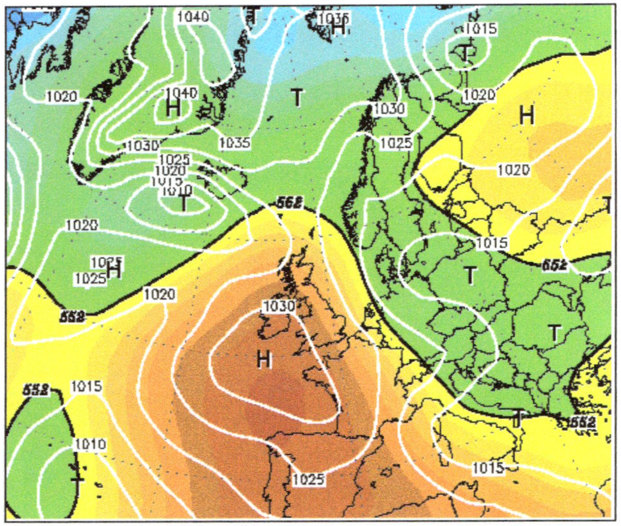

RAIN POTENTIAL

FROST/SNOW

APRIL 15 WEDNESDAY

High pressure continues to dominate Ireland under a gentle west to southwest flow. Partly cloudy everywhere with chance of some misty drizzle patches in the far northwest of the country briefly overnight. Mainly dry elsewhere with chance of patches of morning fog, particularly in the west, south, and central areas. Slightly colder with the return of some light frost patches overnight.

Belfast: Partly cloudy, warmer, dry, mostly calm air, fog patches possible
Dublin: Light breezes, partly cloudy, light showers, cold overnight
Cork: Cloudy, mild temperatures, dry, calm and serene air, fog possible
Galway: Cloudy, isolated showers, gentle shifting breezes

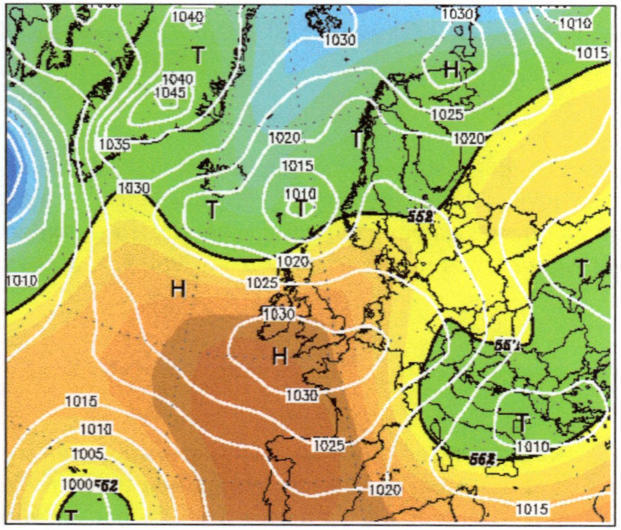

RAIN POTENTIAL

FROST/SNOW

APRIL 16 THURSDAY

High pressure continues over Ireland with northerly breezes prevailing, gentle at first and allowing for scattered patches of fog, but freshening later in the day mostly in the north and northwest. Some brief showers mostly in the north, central and western parts while mostly dry and sunny elsewhere. Showers may be wintry and some of sleet and hail later in the day.

Belfast: Gentle breezes, colder, sunny, mostly dry, chance of brief misty patches
Dublin: Sunny, colder, dry, chance of frost overnight, light breezes to mostly calm air
Cork: Sunny, cooler, dry, light breezes
Galway: Mostly sunny, isolated showers, light breezes, chance of snow flurries

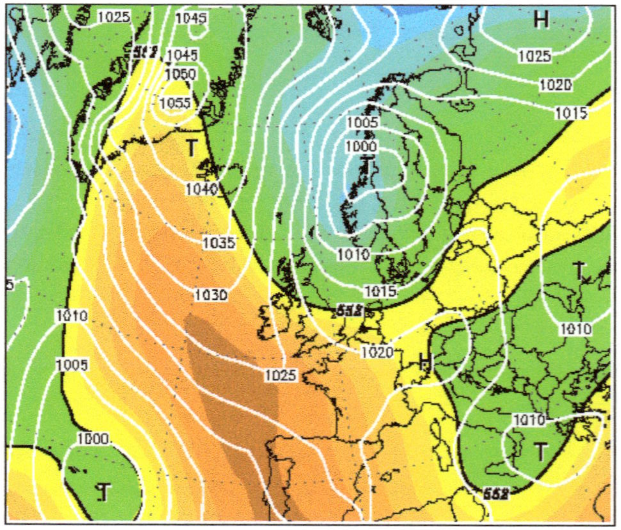

RAIN POTENTIAL

FROST/SNOW

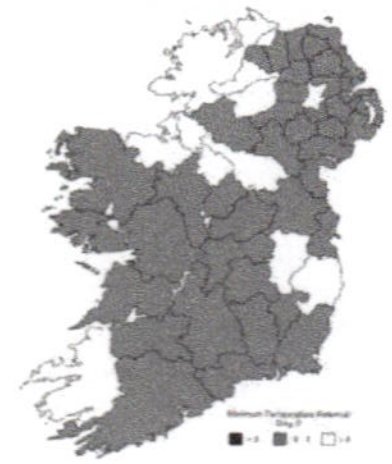

APRIL 17 FRIDAY

Low pressure to the east of Ireland brings fresh to strong northwest winds and a colder air across the country. Very cold with chance of overnight frosts and snowflurries in many parts of the country, accompanying light wintry showers, including some of those of hail and sleet. Mostly sunny day throughout.

Belfast: Breezy, sunny, mostly dry, colder, frost and snowflurries possible
Dublin: Sunny and mostly dry, colder, overight frosts, chance of snow flurries
Cork: Sunny, dry day, breezy spells, overnight frosts or snow flurries possible
Galway: Mostly sunny, light showers, frosts and light snow possible, breezy at times

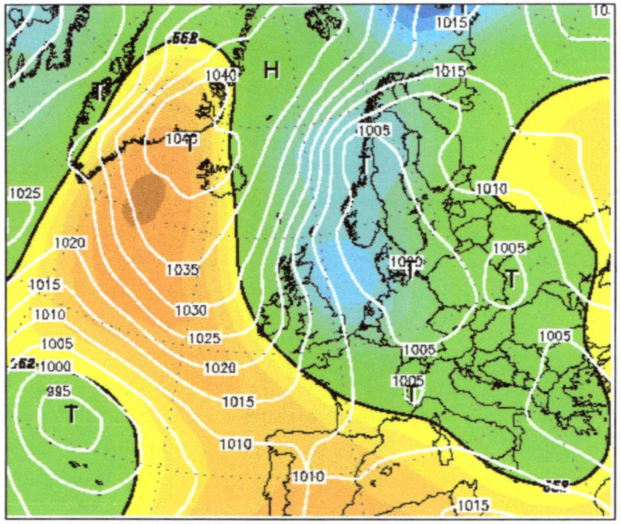

RAIN POTENTIAL

FROST/SNOW

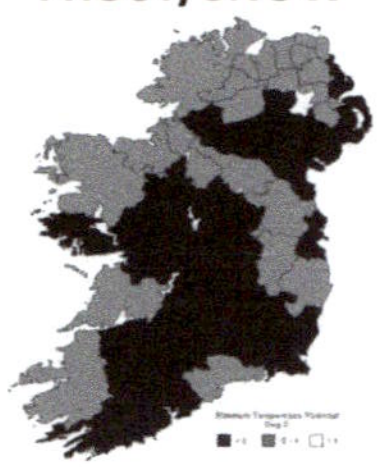

APRIL 18 SATURDAY

High pressure ridges to the west of Ireland while low pressure covers the east of the country, with a fresh to blustery mostly northerly flow prevailing. Some wintry showers likely mostly overnight, as well as the chance of passing thunderstorms, and showers of hail or sleet. Mostly dry and sunny day, except for some afternoon to evening showers likely to return, particularly in the north and west.

Belfast: Sunny, isolated showers, colder, frost possible, chance of snow flurries
Dublin: Sunny, brief overnight showers, cold with frosts, chance of snow
Cork: Mostly sunny, brief showers, overnight frost or light snow possible
Galway: Moderate breezes, partly cloudy, showers, chance of frost and light snow

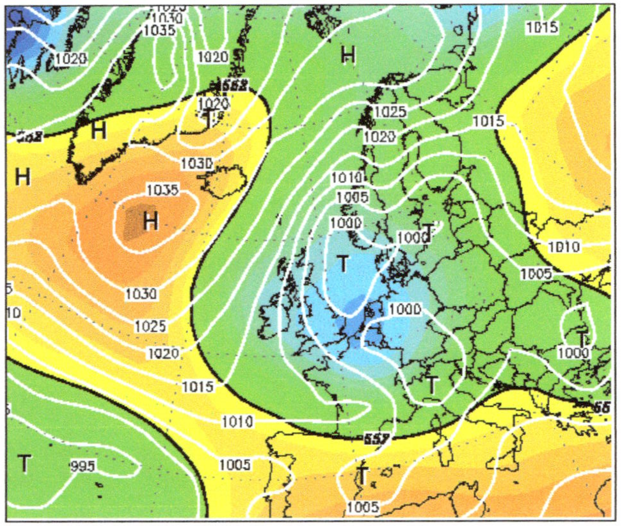

RAIN POTENTIAL

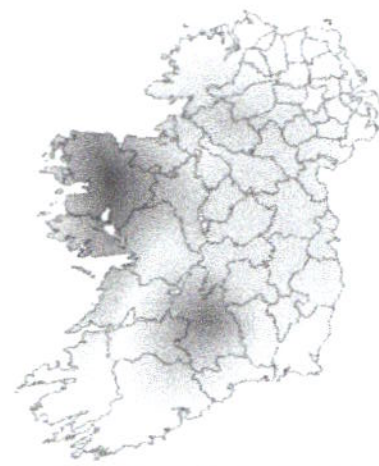

FROST/SNOW

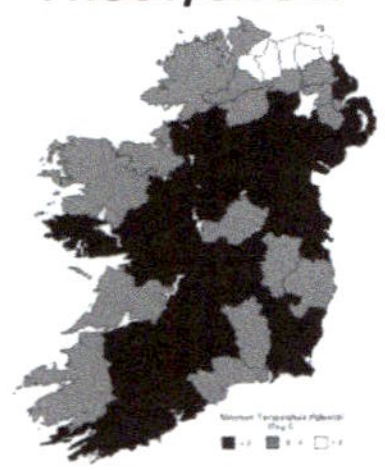

APRIL 19 SUNDAY

Low pressure prevails over the country in a light to moderate northerly flow. A changeable day with a mix of some sunny spells and passing showers, mostly in the western half of the country, where some of the overnight showers may be of sleet and snow in the higher reaches. Chance of fog particularly in the west and north and overnight frost patches in sheltered areas.

Belfast: Partly cloudy, dry, milder temperatures, gentle breezes
Dublin: Cloudy, cold, mostly dry, odd threat of rain, frosts overnight, mostly calm air
Cork: Cloudy, colder, light breezes, dry
Galway: Partly cloudy, light showers, cold, breezy at times

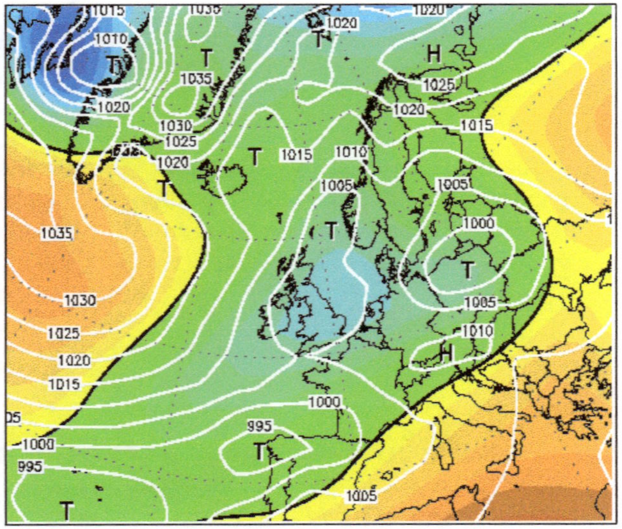

RAIN POTENTIAL

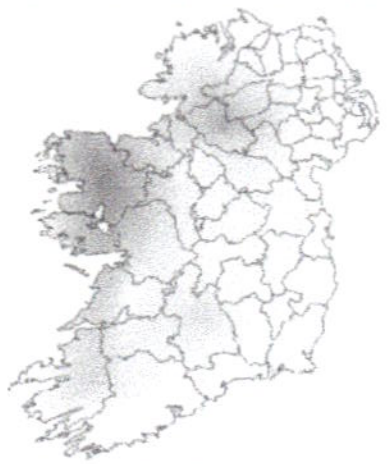

FROST/SNOW

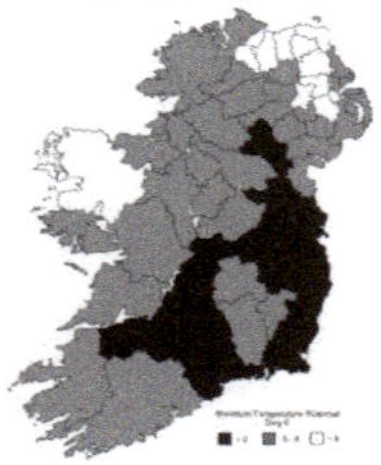

APRIL 20 MONDAY

Low pressure continues to bring a very cold spell throughout the country with ground frosts in many districts as well as some patches of mist or fog here and there. Winds generally very light in a north to northwest flow. Another day with a mix of some light shower activity and occasional sunny spells.

Belfast: Mostly calm and serene air, partly cloudy, isolated showers possible
Dublin: Partly cloudy, dry, mostly calm and serene air, cold
Cork: Calm and serene atmosphere, mostly cloudy, isolated showers
Galway: Brief showers, scattered cloudy spells, cold, frost likely, mostly calm air

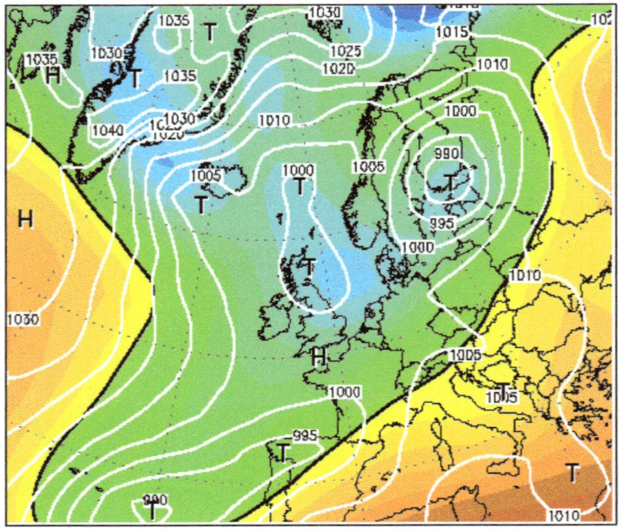

RAIN POTENTIAL

FROST/SNOW

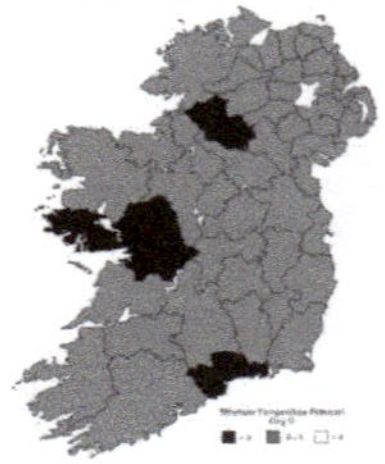

APRIL 21 TUESDAY

Low pressure over Ireland strengthens to the northwest of the country, while winds remain light to moderate in a northwest flow. Some overnight showers in most areas, although mostly dry along the eastern and far northern coasts. Mostly dry and sunny day. Chance of evening thunderstorms arriving late in the day, mostly in the west and southwest.

Belfast: Partly cloudy, isolated showers, gentle breezes to mostly calm air
Dublin: Partly cloudy, isolated showers, gentle breezes to mostly calm air
Cork: Sunny day, overnight showers, cold, gentle breezes to mostly calm air
Galway: Partly cloudy, showers, cold, frosts likely, chance of snow

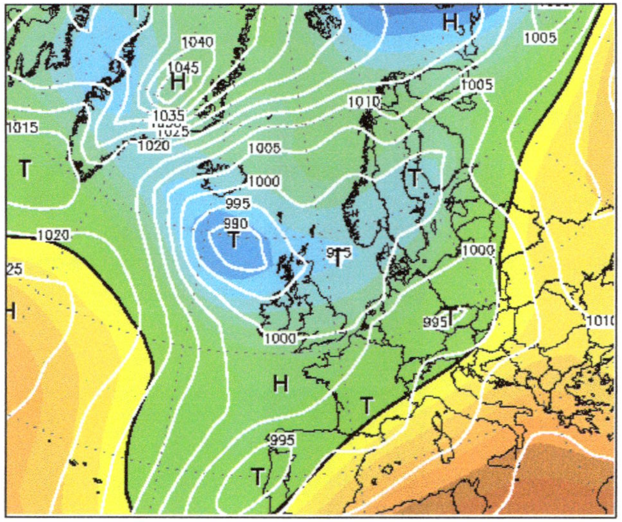

RAIN POTENTIAL

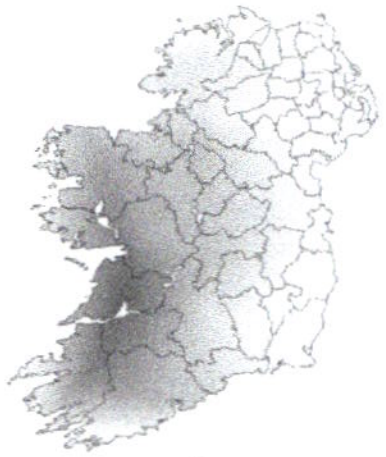

FROST/SNOW

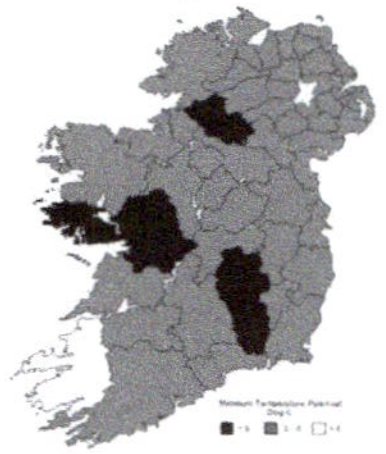

APRIL 22 WEDNESDAY

A weak depression crosses the country bringing a colder north to northwest flow throughout with winds generally light and variable for a time. Some overnight showers in most areas, particularly in the west while drier and with patches of frost mostly in the north. Increased cloud throughout, with the chance of some thunderstorms later in the day, with accompanying wintry showers, including some of hail or sleet.

Belfast: Passing scattered cloud, isolated showers, overnight frost, mostly calm air
Dublin: Changeable, mix of sunny spells and showers, cold overnight, gentle breezes
Cork: Light shifting breezes to mostly calm air, partly cloudy, light showers
Galway: Partly cloudy, light showers, cold, frosts, chance of thunderstorms and hail

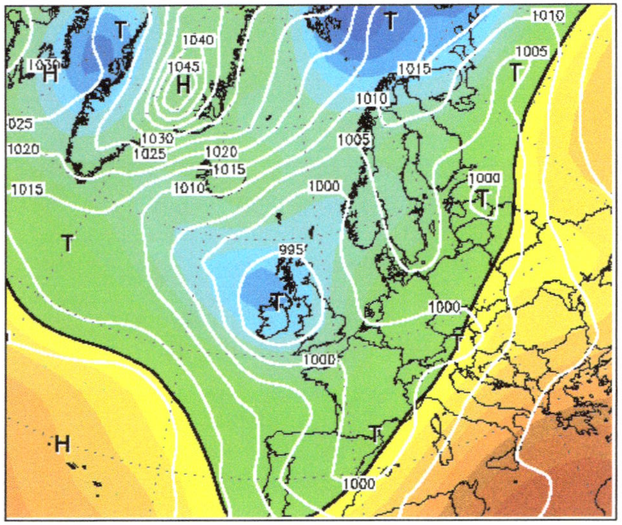

RAIN POTENTIAL

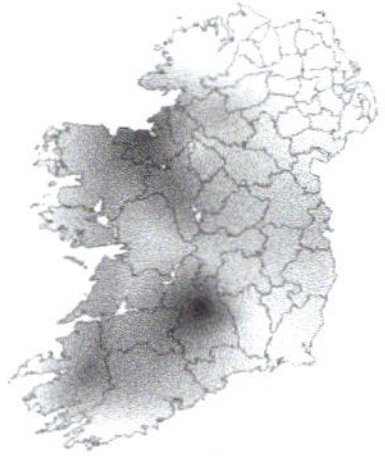

FROST/SNOW

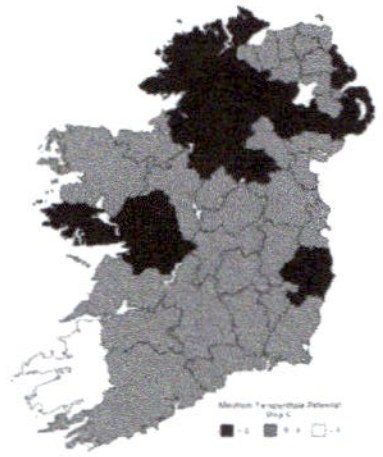

APRIL **23** THURSDAY

The depression over Ireland slowly builds in intensity, with a light to moderate northwest to variable flow prevailing. Overnight showers likely in most parts of the country, being more persistent into the day in the west and north, while clearing to mostly sunny and dry in the south. Chance of afternoon thunderstorms with hail showers once more.

Belfast: Cloudy, light showers, gentle breezes, chance of frost in sheltered pockets
Dublin: Changeable, cloudier, occasional showers, cold, shifting breezes
Cork: Sunny, isolated showers, milder day cold night, light shifting breezes
Galway: Changeable, mix of showers and sunny spells, shifting breezes

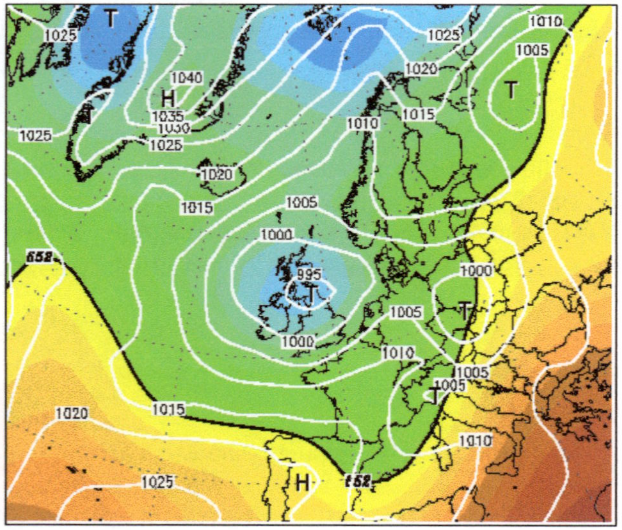

RAIN POTENTIAL

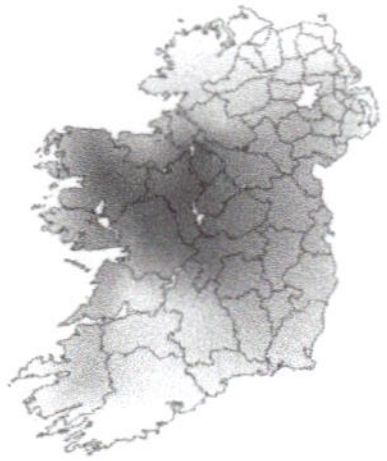

FROST/SNOW

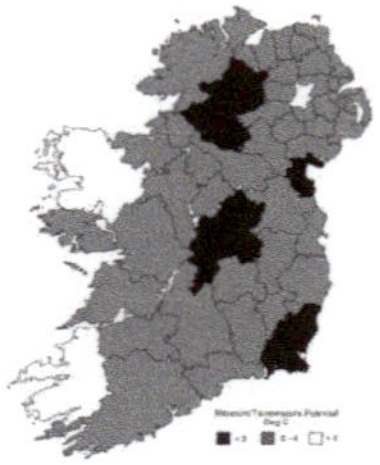

APRIL **24** FRIDAY

As the depression drifts away in an easterly tract, winds ease to mostly light throughout in a gentle north to northwest flow. Some lingering showers in the east of the country and through central parts, while mostly dry and sunny in the north, south, and western coastal areas. Chance of afternoon thunderstorms once more, mostly in the west, however, rainfall amounts may remain light.

Belfast: Sunny, mostly dry, mild day cold night, calm and serene air
Dublin: Showers, sunny, cold overnight, gentle breezes, chance of thunderstorms
Cork: Sunny, isolated showers, mild day cold night, mostly calm, chance of mist
Galway: Sunny, dry, mild day cold night, chance of frost, mostly calm air

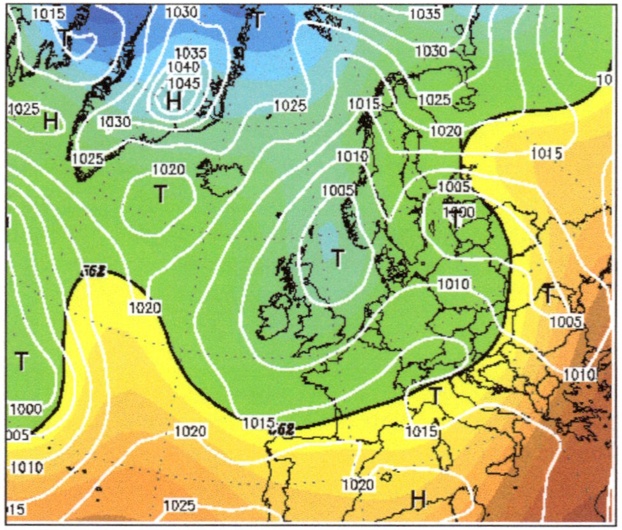

RAIN POTENTIAL

FROST/SNOW

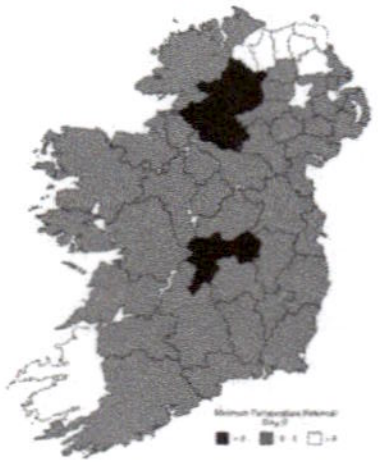

APRIL **25** SATURDAY

A high pressure system to the south of Ireland extends a dry ridge over the country, with a light to mostly calm northerly flow prevailing. Dry and sunny throughout, with scattered misty patches and frost pockets in many districts. Chance of some late evening isolated showers spotting around the west and north.

Belfast: Sunny, dry, warmer day cold night, gentle breezes to mostly calm air
Dublin: Very sunny, dry, mild day cold night, gentle breezes
Cork: Very sunny and dry, mild day cold night, chance of misty patches
Galway: Very sunny and dry, cold, frosts, mostly calm air

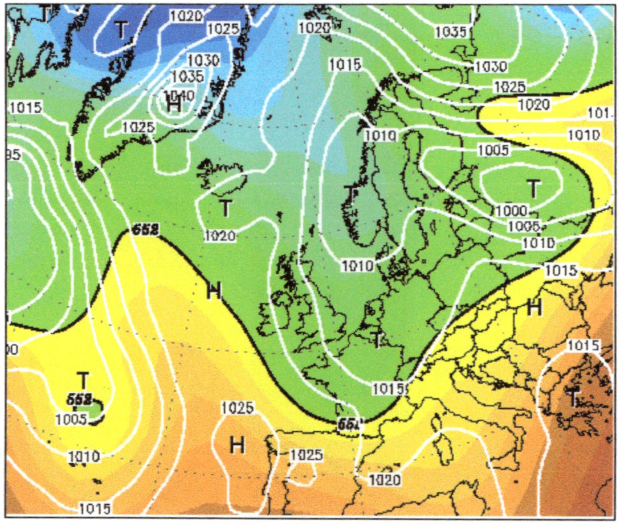

RAIN POTENTIAL

FROST/SNOW

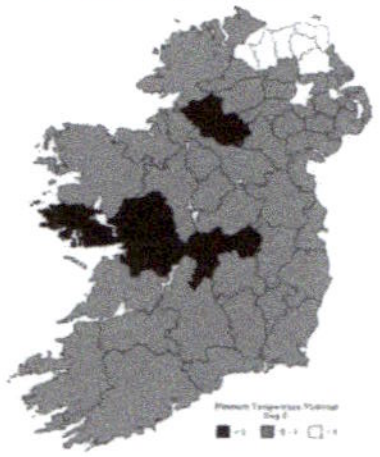

APRIL **26** SUNDAY

Anticyclonic conditions prevail. Winds generally light and variable. Overnight patches of frost in here and there and possible fog briefly mostly in the north to central areas. Dry and sunny day throughout. Chance of some brief showers mostly in the northwest.

Belfast: Sunny, dry, gentle breezes to mostly calm air
Dublin: Sunny and dry, mild day cold night, chance of frost, mostly calm air
Cork: Sunny, dry, gentle shifting breezes
Galway: Sunny, cold and dry, frosts, gentle shifting breezes to mostly calm air

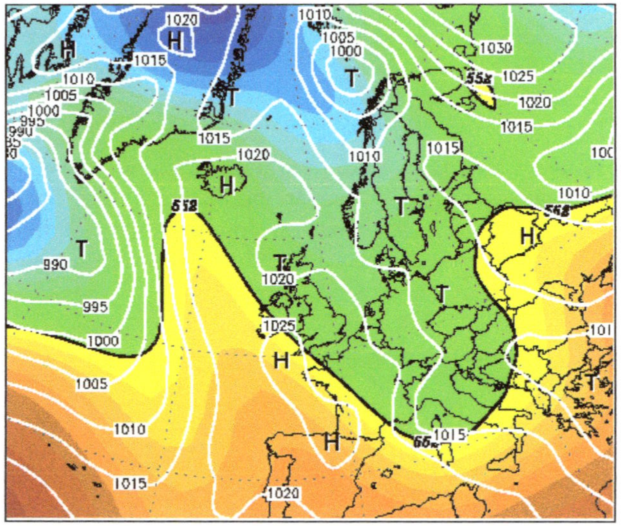

RAIN POTENTIAL

FROST/SNOW

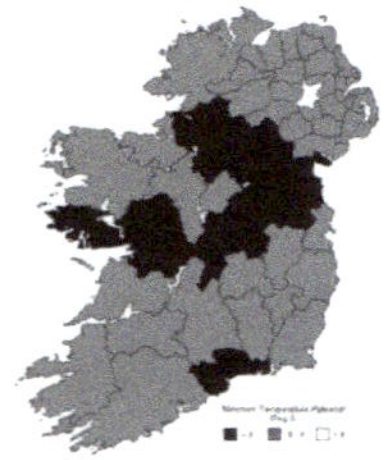

APRIL **27** MONDAY

Anticyclonic conditions continue to bring a mostly dry and sunny day throughout, except for the chance of some isolated showers briefly in the coastal fringes of the northeast. Overnight patches of frost likely, particularly in the north. Winds generally light in a northerly flow, but becoming fresher in the western coasts later in the day.

Belfast: Very sunny, brief overnight showers, chance of frost, light breezes
Dublin: Sunny, colder, dry, frost patches likely, calm and serene air
Cork: Mostly sunny and dry, cooler, light breezes to mostly calm air
Galway: Sunny, dry, mild day overnight frosts, gentle breezes

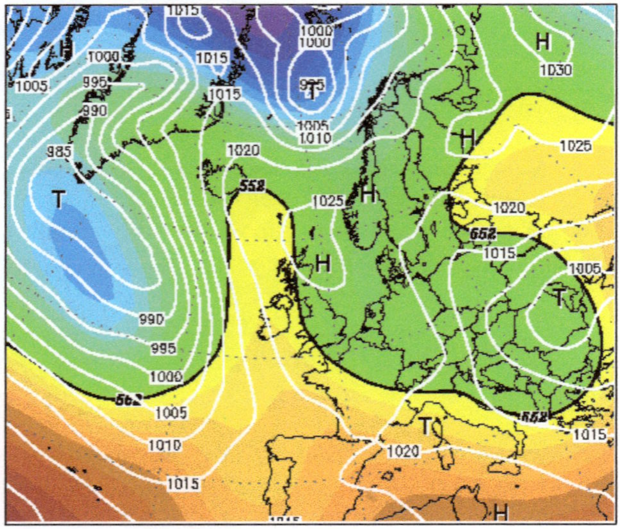

RAIN POTENTIAL

FROST/SNOW

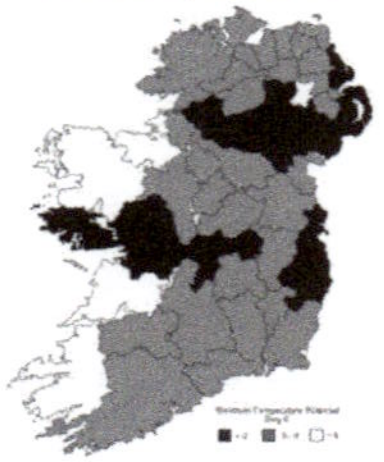

APRIL **28** TUESDAY

The high pressure ridge moves away to the east of Ireland while a large low pressure system moves into the west, with winds strengthening with a southeast change, more gusty at times around the coasts. Some patches of overnight frost still likely, particularly in inland parts. Brief outbreaks of showers likely here and there, but still a mostly dry day with good sunny spells, more cloudy in the east and south.

Belfast: Mostly sunny, light showers, cooler, fresh breezes
Dublin: Partly cloudy, dry, cold, freshening breezes
Cork: Increasing cloudy spells, isolated showers, fresh breezes at times
Galway: Sunny, colder, brief morning showers, windy at times

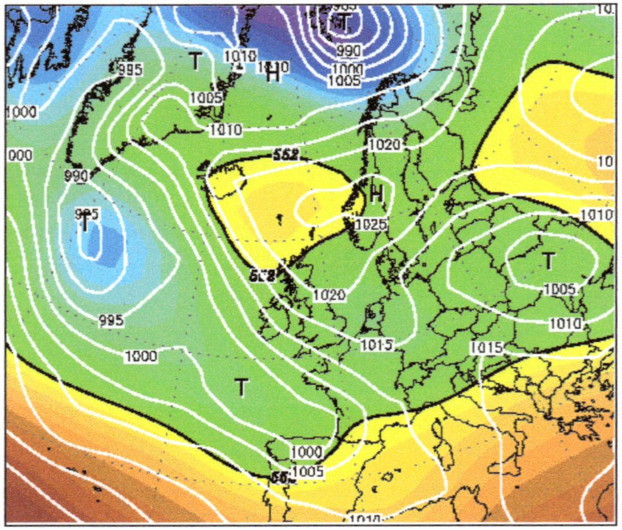

RAIN POTENTIAL

FROST/SNOW

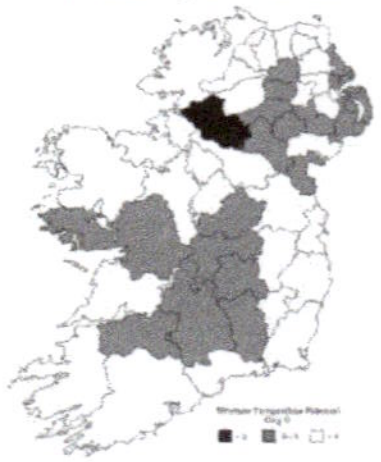

APRIL **29** WEDNESDAY

A weak depression to the west and south of Ireland brings a moderate easterly flow over the country, more blustery at times with possible gale gusts around the coasts. Increased cloud throughout, with passing threats of rain, and some showers likely in inland to western areas, as well as isolated showers in the southeast. Dry elsewhere.

Belfast: Increasing cloud, mostly dry with odd threat of rain, cooler, windy spells
Dublin: Overcast and dull, dry, cool, windy at times
Cork: Cloudy, light showers, fresh breezes
Galway: Unsettled winds, partly cloudy, dry

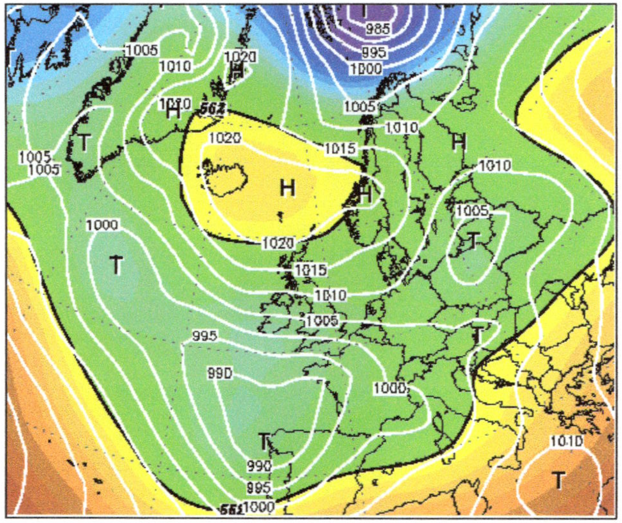

RAIN POTENTIAL

FROST/SNOW

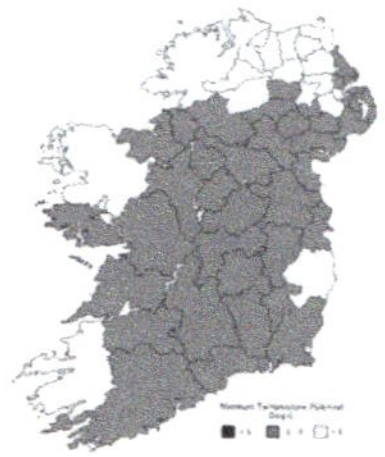

APRIL **30** THURSDAY

A strengthening easterly flow builds over the country as a depression slowly drifts to be centred o the south of Ireland. Mostly cloudy throughout, with some spotty shower activity or drizzle patches mostly in the west to central inland areas and in the southeastern districts. Some longer sunnier spells likely in the far north.

Belfast: Cloudy, mostly dry with odd threat of rain, mild day cold night, windy spells
Dublin: Overcast, mostly dry with odd threat of rain, unsettled winds
Cork: Overcast, brief showers, colder, windy spells
Galway: Partly cloudy, brief showers, cold, strong winds with occasional gusts

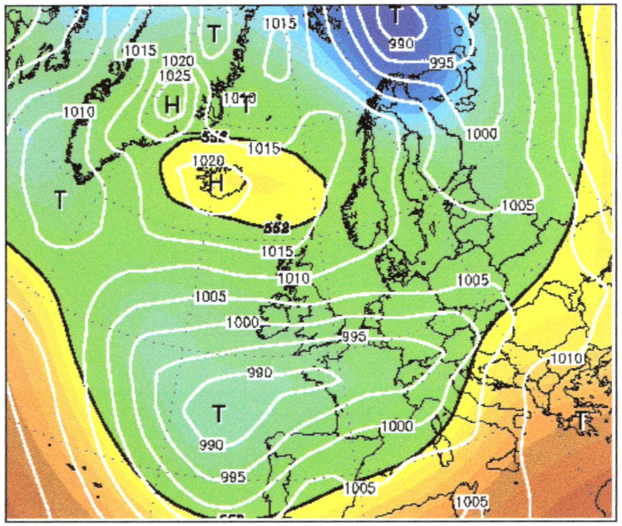

RAIN POTENTIAL

FROST/SNOW

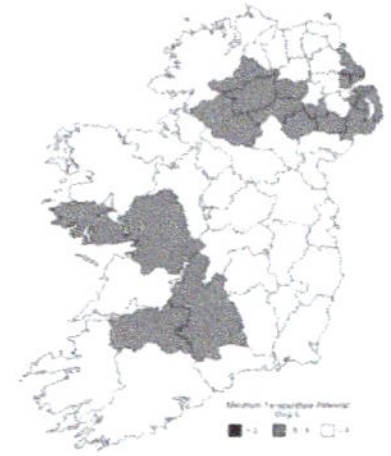

MAY

PHASES OF THE MOON

1st	Full Moon
4th	Apogee
5th	Southern Declination
9th	Third Quarter
13th	Crossing Equator
16th	New Moon
17th	Perigee #3
18th	Northern Declination
23rd	First Quarter
26th	Crossing Equator
31st	Full Moon

1	2	3	4	5	6	7	8	9	10	11	12	13	14	15	16	17	18	19	20	21	22	23	24	25	26	27	28	29	30	31
Fri	Sat	Sun	Mon	Tue	Wed	Thur	Fri	Sat	Sun	Mon	Tue	Wed	Thur	Fri	Sat	Sun	Mon	Tue	Wed	Thur	Fri	Sat	Sun	Mon	Tue	Wed	Thur	Fri	Sat	Sun
F			A	V				3Q				XhN			N	P3	^					1Q			XhS					F

MONTHLY SUMMARY

1st-5th High pressure is replaced by cloudy and changeable weather with frontal systems crossing the country

1st-11th Low pressure re-established at the end of April continues to bring unsettled conditions throughout the country. Regular spells of rain or showers likely on most days, sometimes accompanied by thunderstorms.

3rd The generally easterly flow of the last week may turn south-westerly around this time bringing milder temperatures.

3rd-10th Chance of thunderstorm and heavy rain now and then mostly in the west to midlands areas.

4th Unusually warmer than normal temperatures possible.

5th-11th Chance of heavy rain and thunderstorms.

6th Chance of very heavy rain in the west, particularly around the Maam Valley.

8th-10th Chance of frequent thunderstorms.

10th A northerly change possible around this time, bringing a cooler change.

Over the next four weeks high pressure systems may settle in to be the dominant weather pattern for this outlook period after a brief unsettled start. Anticyclones centred to the north or east of Ireland may bring a prevalent light to moderate easterly flow across the country with plenty dry and sunny days. Overall, many parts of Ireland may only see around half their usual amount of rainfall compared to their averages, with only the south and southwest areas possibly nearer to average, mostly due to some thundery showery events on a number of days. Overall, there may only be around 4-8 wet days in the midlands, east, and north, but up to 10 in the west and southwest. Generally, the prevailing dry and sunny conditions may also result in a warmer than outlook period, with some areas up to 3-4 degrees warmer than what is typical for this time of year, particularly in the south, west and northwest areas. Similarly, overnight temperatures may also be warmer than average, with many areas frost-free and only around 5-6 ground frosts in some inland areas throughout this entire outlook period. Finally, the next four weeks may also be much sunnier than average, with a number of areas possibly seeing around 35-40% more sunshine hours than what is typical for this time of year. In particular, the first half of this outlook period may see regular sunshine levels of up to 10 hours per day in many districts. During this outlook period, there may be 9 days with thunderstorms, only 1 day with hail, 4 days with fog patches, no widespread frosts, but possible snow in some inland parts on 2 days, 3 days with gale winds, and 1 possible tornado.

11th-15th Low pressure with widespread rain and shower activity between sunny spells.

12th Chance of thunderstorms, particularly in the west.

12th-13th Chance of some ground frosts in inland areas and in the southeast.

14th Chance of heavy rain in the southwest and strong winds with possible gale gusts around the north and western coasts.

16th-31st High pressure may become established around this time, with an anticyclone to the east of Ireland slowly moving westward. Some spells of isolated thunderstorms and rain possible, but with most parts of the country mostly dry, warmer, and often very sunny. Mostly light to moderate easterly breezes prevail, and slightly cooler around the northern coasts.

19th Chance of thunderstorms, particularly in the west.

21st-23rd Chance of further thunderstorms in the west.

23rd Hail showers possible in the far northwest.

23rd-26th Very warm and sunny spell.

27th Cooler change with increased cloud.

30th-31st Chance of some ground frosts in some inland areas.

IRELAND RAINFALL ESTIMATES
May 2026

d=mainly dry, n= nonrecordable, l=light shrs, s=significant shrs, r=rain, h=hvy falls

		ULSTER PROVINCE					CONNAUGHT PROVINCE						LEINSTER PROVINCE				MUNSTER PROVINCE						
		Alderg	Hillsbo	Armag	Loughl	Malin H	Sligo	Ardtarı	Belmul	Drums	Galway	Roscoı	Markre	Edende	Kilkenı	Dublin	Shannı	Newpo	Tralee	Killarne	Cork	All	Moon
1st	Fri	n	d	l	d	d	d	d	d	d	d	d	d	l	d	l	d	d	l	d	l	7	F
2nd	Sat	d	d	d	d	d	d	d	d	d	d	d	d	d	d	d	d	d	d	d	d	0	
3rd	Sun	l	s	d	l	d	l	l	l	s	l	s	l	s	d	s	l	l	s	h	s	81	
4th	Mon	l	l	l	s	d	s	s	s	s	s	s	s	l	s	l	s	l	s	s	s	88	A
5th	Tues	l	l	l	r	l	s	r	n	l	s	s	l	l	l	l	l	l	n	l	l	68	V
6th	Wed	l	s	l	l	d	s	l	s	s	l	s	l	l	l	d	l	s	l	s	l	62	
7th	Thur	l	d	l	d	l	d	d	l	d	d	n	l	d	l	d	d	l	s	d	l	16	
8th	Fri	l	l	l	s	s	s	s	l	r	r	l	s	s	l	d	h	s	l	s	s	107	
9th	Sat	l	l	d	n	s	d	n	l	s	n	l	s	d	s	l	l	l	l	d	l	44	3Q
10th	Sun	l	d	l	d	d	d	d	l	l	l	l	s	l	l	d	l	l	d	h	l	45	
11th	Mon	d	d	d	l	s	l	l	l	l	n	l	l	l	d	d	l	l	s	n	d	27	
12th	Tues	d	n	d	l	s	d	l	d	l	d	l	l	l	s	n	l	d	d	l	l	29	
13th	Wed	l	l	s	l	d	l	l	l	l	l	l	s	s	l	d	l	l	l	s	l	45	XhN
14th	Thur	l	l	l	l	d	l	l	d	s	l	s	s	l	s	l	l	l	h	h	r	115	
15th	Fri	l	d	l	d	l	l	d	l	d	d	n	d	d	d	d	n	d	n	d	s	14	
16th	Sat	n	d	l	d	d	d.s	d	d	d	d	d	d	d	d	d	d	d	d	d	n	2	N
17th	Sun	d	d	d	d	d	d	d	d	d	d	d	d	d	d	d	d	d	d	d	d	0	P3
18th	Mon	d	d	d	d	d	d	d	d	d	d	d	d	d	d	d	d	d	d	d	d	0	^
19th	Tues	s	r	d	d	d	l	d	l	l	l	l	d	l	l	n	s	l	l	l	s	44	
20th	Wed	l	l	l	d	l	l	d	l	d	d	d	d	d	d	l	d	d	d	l	d	11	
21st	Thur	l	s	l	d	d	l	d	d	d	d	d	d	d	n	s	d	d	d	d	l	17	
22nd	Fri	l	d	s	l	d	l	l	d	d	d	n	d	n	d	d	d	d	d	d	d	12	
23rd	Sat	l	d	l	d	d	l	d	s	d	l	d	d	d	d	d	l	d	d	s	d	16	1Q
24th	Sun	d	d	d	d	d	d	d	d	d	d	d	d	d	d	d	d	d	d	d	d	0	
25th	Mon	d	d	d	d	d	d	d	d	d	d	d	d	d	d	d	d	d	d	d	d	0	
26th	Tues	d	d	d	d	d	d	d	d	d	d	d	d	d	d	d	d	d	d	d	d	0	XhS
27th	Wed	d	d	d	d	d	d	d	d	d	d	d	d	d	d	d	l	d	d	l	n	4	
28th	Thur	n	l	d	d	d	d	d	d	d	d	d	d	d	d	d	d	d	s	s	s	19	
29th	Fri	d	d	d	d	d	r	d	h	d	d	d	d	d	l	l	d	d	l	l	l	37	
30th	Sat	d	d	d	d	d	l	d	s	d	d	d	d	d	d	d	d	l	d	d	d	9	
31st	Sun	d	d	d	d	d	d	d	d	d	s	d	d	d	d	d	d	n	l	r	d	15	F
	Estimate:	34	41	26	42	33	53	42	51	57	41	38	43	31	34	21	46	35	74	119	73	935	
	Average:	57	59	58	68	58	81	83	70	71	75	82	81	69	60	60	65	95	89	59	82	1423	
	Trend:	drr	drr	drr	drr	drr	drr	drr	drr	drr	drr	drr	drr	drr	drr	drr	drr	drr	drr	wtr	av	drr	

IRELAND SUNSHINE ESTIMATES

May 2026

F=fine (8-12 hours of sunshine), pc= partly cloudy (4-7 hours), c=cloudy (1-3 hours), o=overcast (0 hours)

		ULSTER PROVINCE					CONNAUGHT PROVINCE						LEINSTER PROVINCE				MUNSTER PROVINCE						
		Alderg	Hillsbo	Armag	Loughl	Malin H	Sligo	Ardtarı	Belmul	Drums	Galway	Roscoı	Markre	Edende	Kilkenı	Dublin	Shannc	Newpo	Tralee	Killarne	Cork	All	Moon
1st	Fri	o	o	o	o	o	o	o	o	o	o	o	o	o	o	o	o	o	o	o	o	1	F
2nd	Sat	c	c	c	c	c	c	c	o	c	o	o	c	c	c	pc	o	c	o	o	o	46	
3rd	Sun	pc	pc	pc	pc	pc	c	pc	c	pc	c	c	pc	pc	pc	F	c	pc	pc	pc	c	85	
4th	Mon	F	F	F	F	F	pc	F	pc	F	pc	pc	F	F	pc	F	F	pc	pc	pc	pc	154	A
5th	Tues	pc	pc	pc	pc	pc	pc	pc	F	F	F	F	pc	F	F	F	F	F	F	F	F	176	V
6th	Wed	o	o	o	o	o	o	o	o	o	c	o	o	o	o	o	c	c	o	o	o	8	
7th	Thur	pc	pc	pc	pc	pc	F	pc	F	c	F	F	pc	c	o	o	pc	c	F	F	o	114	
8th	Fri	pc	pc	pc	pc	pc	c	pc	c	pc	c	c	pc	pc	pc	F	c	c	c	c	pc	94	
9th	Sat	c	c	c	c	c	c	c	F	pc	F	F	c	pc	pc	pc	F	F	F	F	pc	127	3Q
10th	Sun	F	F	F	F	F	F	F	pc	F	pc	pc	F	F	F	F	F	F	F	F	pc	172	
11th	Mon	pc	pc	pc	pc	pc	F	pc	F	pc	F	F	pc	pc	pc	o	F	pc	pc	pc	F	139	
12th	Tues	F	F	F	F	F	F	F	F	F	F	F	F	F	F	pc	pc	F	F	F	F	189	
13th	Wed	F	F	F	F	F	F	F	F	pc	F	F	F	pc	pc	F	F	pc	pc	pc	c	157	XhN
14th	Thur	pc	pc	pc	pc	pc	c	pc	c	c	c	c	pc	c	c	pc	c	c	o	o	o	60	
15th	Fri	F	F	F	F	F	F	F	F	F	F	F	F	F	F	F	F	F	F	F	F	202	
16th	Sat	F	F	F	F	F	F	F	F	F	F	F	F	F	F	F	F	F	pc	pc	c	226	N
17th	Sun	F	F	F	F	F	F	F	F	F	F	F	F	F	F	F	F	F	pc	pc	F	216	P3
18th	Mon	F	F	F	F	F	F	F	pc	F	pc	pc	F	F	F	F	pc	pc	F	F	pc	205	^
19th	Tues	pc	pc	pc	pc	pc	pc	pc	pc	pc	c	pc	pc	pc	pc	F	c	pc	o	o	o	108	
20th	Wed	o	o	o	o	o	c	o	pc	c	pc	pc	o	c	c	o	c	c	pc	pc	pc	53	
21st	Thur	o	o	o	o	o	o	o	c	o	c	c	o	o	o	o	c	c	pc	pc	o	28	
22nd	Fri	pc	pc	pc	pc	pc	pc	pc	F	pc	pc	pc	pc	pc	pc	F	o	c	pc	pc	pc	121	
23rd	Sat	F	F	F	F	F	pc	F	F	F	pc	pc	F	F	F	F	pc	F	pc	pc	F	167	1Q
24th	Sun	F	F	F	F	F	F	F	F	F	F	F	F	F	F	F	F	F	F	F	F	265	
25th	Mon	F	F	F	F	F	F	F	F	F	F	F	F	F	F	F	F	F	F	F	F	285	
26th	Tues	F	F	F	F	F	F	F	F	F	F	F	F	F	F	F	F	F	F	F	F	273	XhS
27th	Wed	F	F	F	F	F	F	F	F	F	F	F	F	F	pc	F	F	pc	F	F	o	178	
28th	Thur	c	c	c	c	c	c	c	F	c	pc	F	c	c	c	c	pc	pc	pc	pc	c	80	
29th	Fri	pc	pc	pc	pc	pc	pc	pc	pc	pc	c	c	pc	pc	pc	F	o	c	pc	pc	c	97	
30th	Sat	pc	pc	pc	pc	pc	pc	pc	pc	pc	pc	pc	pc	pc	pc	c	pc	pc	F	F	pc	120	
31st	Sun	pc	pc	pc	pc	pc	pc	pc	F	pc	pc	pc	pc	pc	pc	pc	c	c	c	c	o	99	F
	Estimate:	220	220	220	220	220	204	220	235	209	219	227	220	209	198	235	201	200	206	206	160	4245	
	Average:	189	189	165	191	192	189	188	190	153	180	183	172	176	170	189	179	168	180	180	187	3611	
	Trend:	more	more	more	more	more	more	more	more	more	more	more	more	more	more	more	more	more	more	more	less	more	

MAY 1 FRIDAY

A depression centred to the south of Ireland directs a fresh to strong east to northeasterly flow over the country, easing to moderate later in the day. Widespread cloud througout with patchy showers in the southeast, and isolated showers in central parts to the adjacent eastern coasts. Mostly dry elsewhere.

Belfast: Dull and gloomy, dry, mild temperatures, unsettled winds at times
Dublin: Overcast, brief showers, cold, windy at times
Cork: Overcast, light showers, moderate breezes
Galway: Fresh breezes, overcast, cold, dry with passing threats of rain

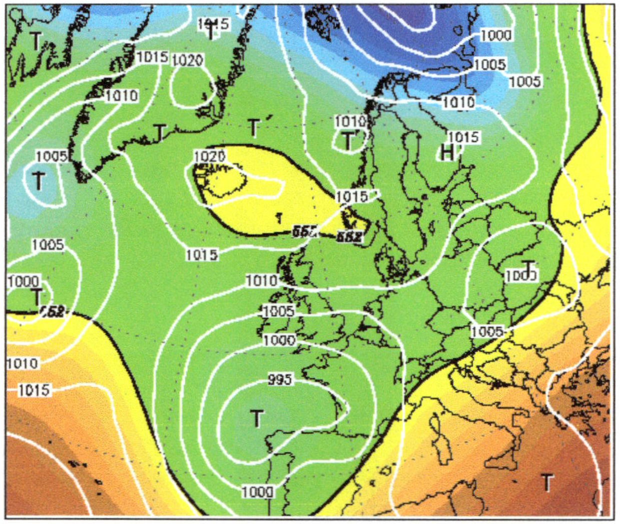

RAIN POTENTIAL

FROST/SNOW

MAY 2 SATURDAY

A weak low pressure influence maintains a light to variable easterly flow across the country. Mostly dry throughout, with widespread cloud over the south and west, with brief sunnier spells in the north and east. Chance of fog patches overnight in the south and west and possible brief outbreaks of drizzle in northern parts. Cold overnight with patches of frost likely mostly in inland parts.

Belfast: Partly cloudy, dry, mild temperatures, light shifting breezes
Dublin: Partly cloudy, milder, dry, light breezes
Cork: Overcast and dull, dry, milder temperatures, mostly calm air
Galway: Overcast, slightly warmer, dry, light breezes

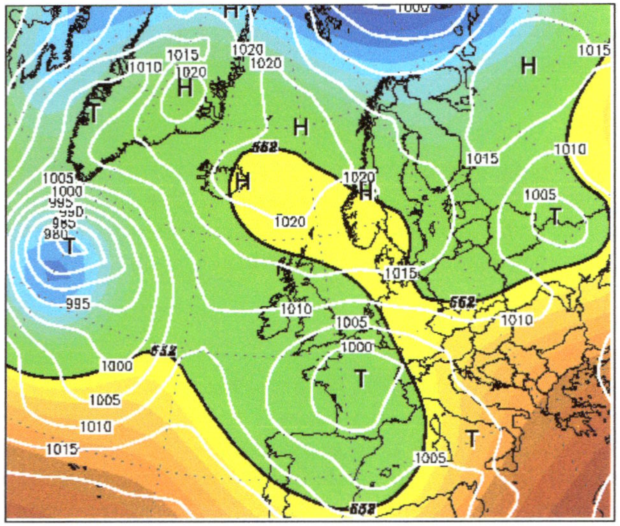

RAIN POTENTIAL

FROST/SNOW

MAY 3 SUNDAY

A large depression to the west of Ireland directs a light southeasterly flow throughout, freshening later with a southwesterly change. A changeable day as a frontal trough crosses the country, first in the west and moving eastward, bringing a mix of sunny spells and occasional showers in the north and east, while cloudier with passing showers elsewhere. Chance of fog overnight, mostly in the east and south.

Belfast: Changeable, occasional showers, passing sunny breakthroughs, light breezes
Dublin: Sunny, warmer day colder night, occasional showers, moderate breezes
Cork: Cloudy, scattered showers, light breezes, chance of fog patches overnight
Galway: Cloudy, light showers, moderate breezes

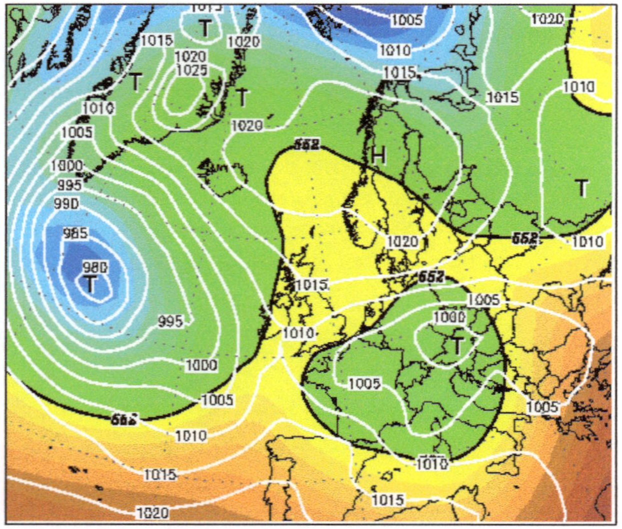

RAIN POTENTIAL

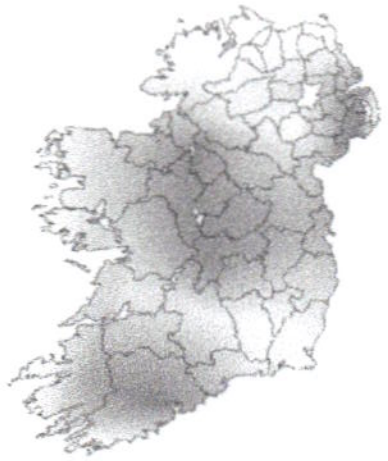

FROST/SNOW

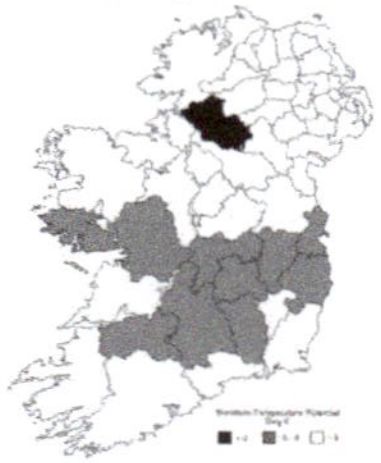

MAY 4 MONDAY

The depression to the west of Ireland drifts nearer to the country, with light southeasterly breezes prevailing, becoming fresher as the day progresses. A changeable day, mostly sunny and dry at first with a band of showers later crossing the country from the southwest at first and spreading northeastwards as the day progresses with only the far northern coastal fringes remaining mostly dry. Chance of fog patches overnight mostly in the north, east, and south.

Belfast: Sunny, warmer, brief showers, light breezes
Dublin: Sunny, isolated showers, warmer, fresh breezes at times
Cork: Changeable, mix of sunny spells and showers, light breezes, fog possible
Galway: Mix of showers and sunny spells, mild day cold overnight, moderate breezes

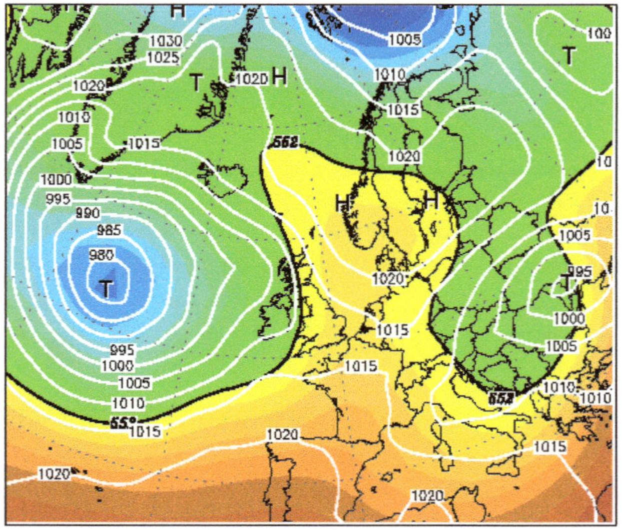

RAIN POTENTIAL

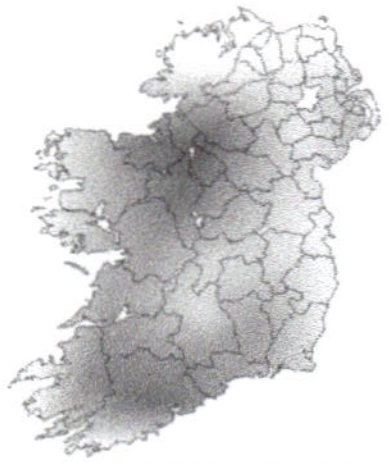

FROST/SNOW

MAY 5 TUESDAY

The depression to the west of Ireland maintains a light to moderate west to southwest flow across the country. Some outbreaks of rain overnight, clearing to a mostly sunny and dry conditions until later in the afternoon when another front may cross the country bearing further showers and possible thunderstorms, and including the chance of hail showers at times.

Belfast: Mostly sunny, light showers, warmer, chance of thunderstorms
Dublin: Sunny, mild temperatures, light showers, moderate breezes
Cork: Sunny, brief showers, moderate breezes at times
Galway: Sunny, cooler, occasional showers overnight, moderate winds

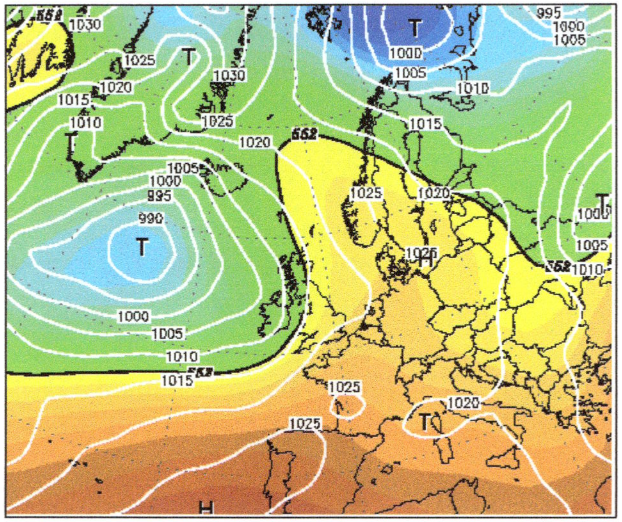

RAIN POTENTIAL

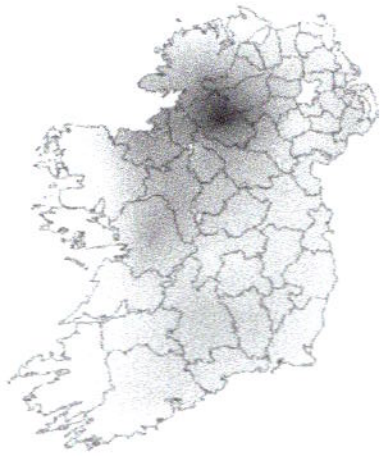

FROST/SNOW

MAY 6 WEDNESDAY

The depression to the west of Ireland begins to drift further northwest of the country allowing for a high pressure ridge to move up over Ireland. Despite the advancing high pressure, widespread cloud and patches of showers may prevail everywhere, before clearing to a mostly dry and cloudy afternoon. Chance of fog settling in towards evening in the south and east. Winds generally light in a southwesterly flow, but gustier for a time mostly along the western coasts.

Belfast: Overcast, occasional showers, windy spells
Dublin: Overcast, mostly dry, threats of rain, mild temperatures, moderate breezes
Cork: Overcast, occasional showers, light breezes, chance of misty patches
Galway: Cloudy, fresh winds at times, brief showers, mild temperatures

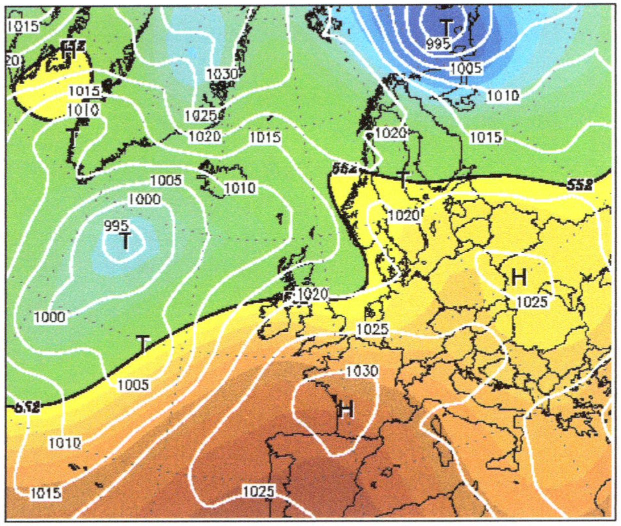

RAIN POTENTIAL

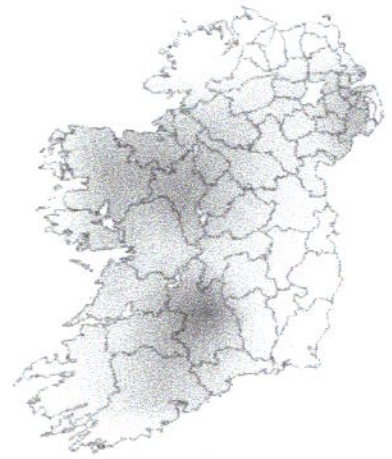

FROST/SNOW

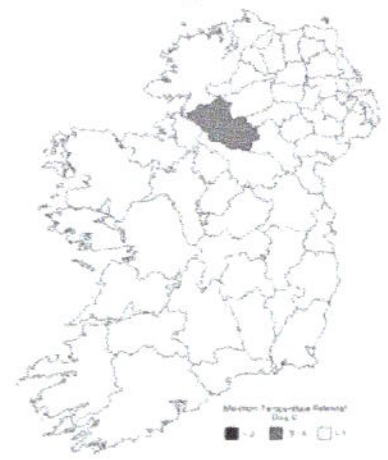

MAY **7** THURSDAY

A large depression to the west of Ireland moves closer to the country directing stronger south to southwesterly winds throughout. Outbreaks of rain and drizzle likely mostly in the south and west, with more isolated showers elsewhere.

Belfast: Unsettled winds, dry and mild, occasional sunny spells
Dublin: Overcast, dry, mild temperatures, breezy at times, chance of fog overnight
Cork: Overcast, light showers, warmer, fresh breezes at times
Galway: Sunny, dry, mild temperatures, windy at times

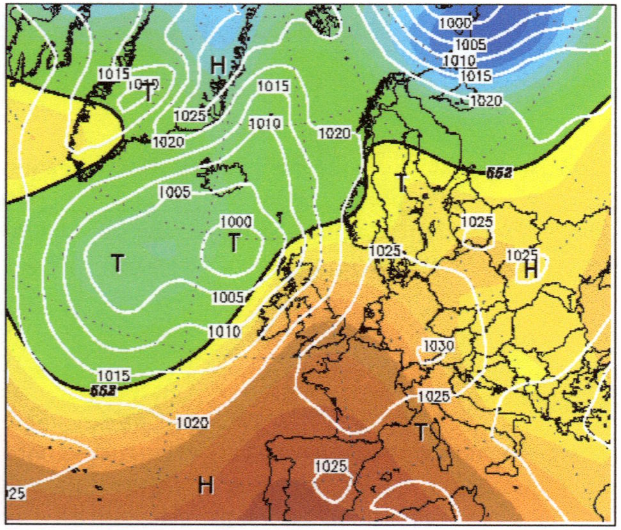

RAIN POTENTIAL

FROST/SNOW

MAY **8** FRIDAY

The depression to the north of Ireland drifts deeper into the North Atlantic towards Iceland, with winds easing to light in a mostly southwesterly flow, more variable later in the day. A low pressure trough covers the country, bringing a mix of sunny spells and patches of rain throughout, heavier at times in the west of the country and possibly accompanied by thunderstorm activity. Chance of fog overnight in central to southern parts.

Belfast: Partly cloudy, light showers, warmer day colder night, light breezes
Dublin: Sunny, static air, cooler, mostly dry, chance of overnight misty patches
Cork: Partly cloudy, outbreaks of rain, slightly cooler, chance of overnight fog
Galway: Cloudy, rain, light breezes

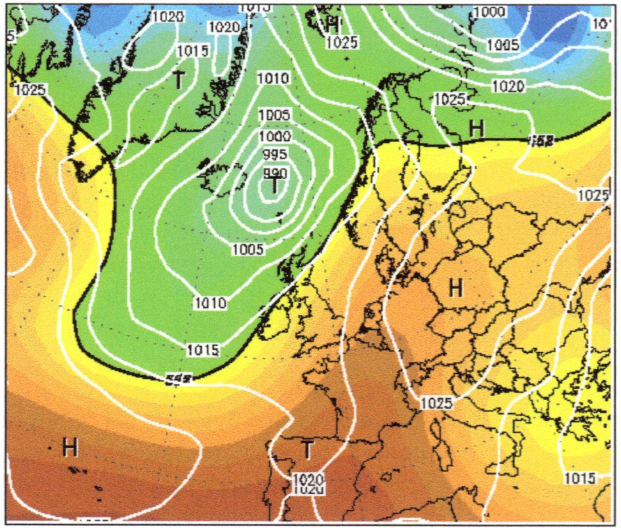

RAIN POTENTIAL

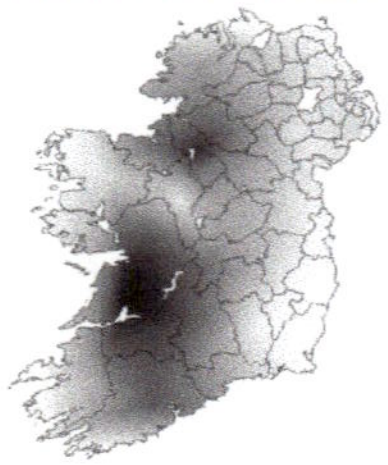

FROST/SNOW

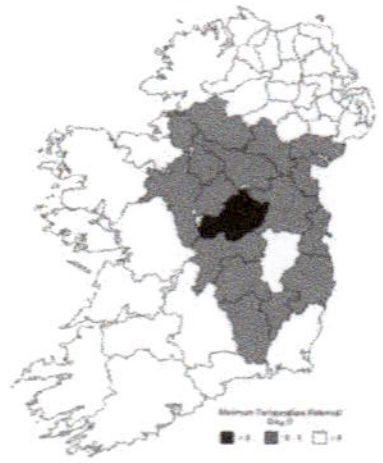

MAY 9 SATURDAY

A weak low pressure centre forms over Ireland bringing light to moderate west to southwest breezes, backing northwest later in the day. Widespread rain or drizzle almost everywhere overnight, with more frequent showers during the day, occasionally accompanied by thunderstorm activity, between pleasant outbreaks of sunshine. Clearer by evening with the chance of some patches of ground frost developing, mostly in inland areas.

Belfast: Mostly cloudy, light showers, mild day colder night, gentle shifting breezes
Dublin: Light breezes to mostly calm air, partly cloudy, isolated showers
Cork: Partly cloudy, light showers, gentle shifting breezes
Galway: Sunny, isolated evening showers, mild day cold night, gentle shifting breezes

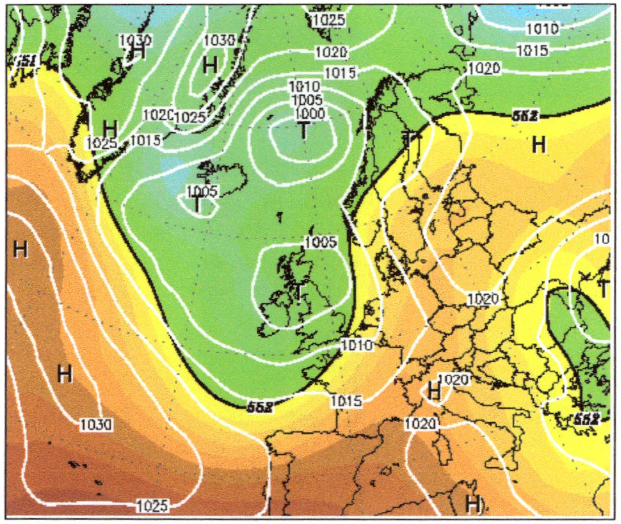

RAIN POTENTIAL

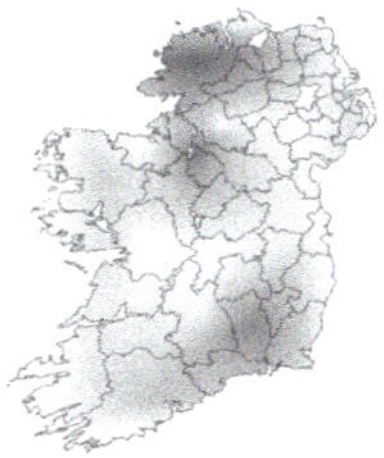

FROST/SNOW

MAY 10 SUNDAY

A low pressure trough between two depression centres sits over the country, bringing a mostly calm and stable air throughout, with patches of frost overnight mostly in the west to inland parts, and overnight fog patches here and there. Mostly sunny day everywhere, with some isolated showers possible mostly in the south and west. Chance of afternoon thunderstorms.

Belfast: Very sunny, colder, dry, serene air, chance of overnight misty patches
Dublin: Sunny, dry, mild day cold night, light breezes to mostly calm air
Cork: Partly cloudy, light showers, mostly calm air
Galway: Partly cloudy, light showers, shifting breezes, chance of frost overnight

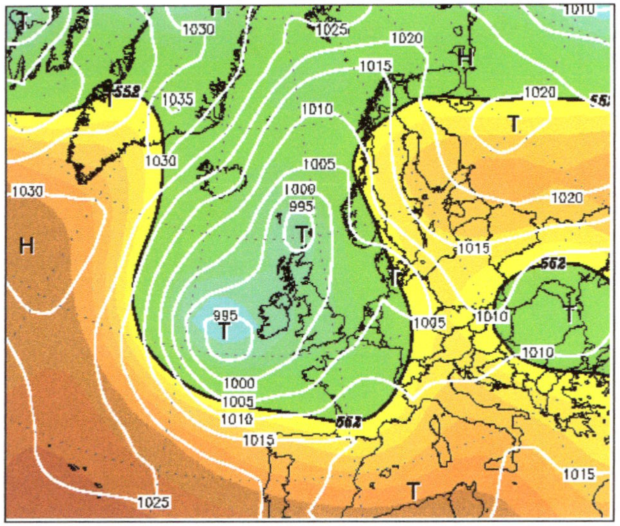

RAIN POTENTIAL

FROST/SNOW

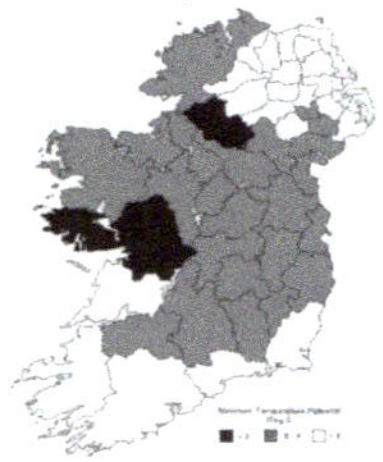

MAY 11 MONDAY

Low pressure strengthens over the country with winds building to fresh at times in a northeasterly flow, backing northwesterly later in the day. Dull and gloomy but mostly dry in the east, with passing threats of rain, with sunnier spells elsewhere. Chance of showers developing later in the day, mostly in the west of the country from the northwest to the southwest, while frost pockets may form overnight in some inland parts.

Belfast: Partly cloudy, mostly dry, warmer day, cool night, freshening breezes
Dublin: Overcast, dry, fluctuating breezy spells
Cork: Sunny, dry, mild temperatures, fluctuating and shifting breezes
Galway: Sunny, isolated showers, fluctuating winds

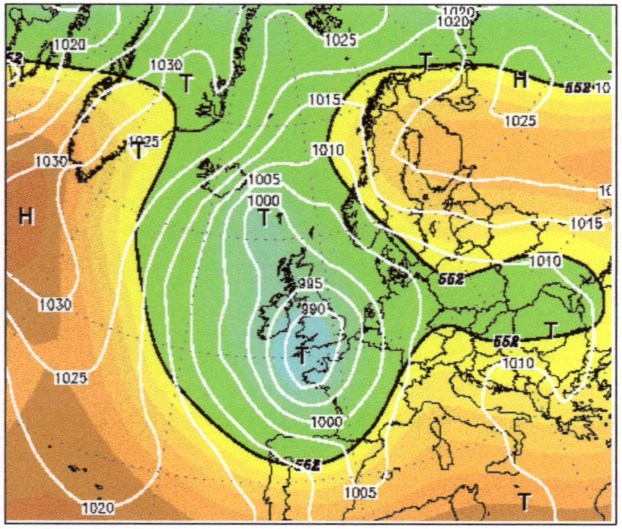

RAIN POTENTIAL

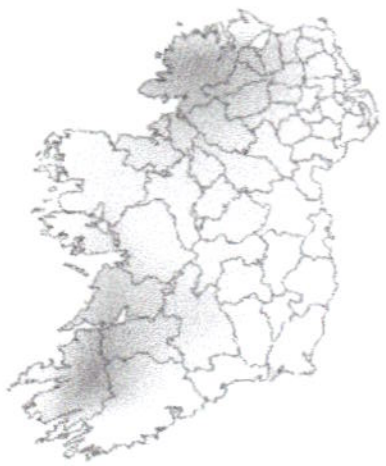

FROST/SNOW

MAY 12 TUESDAY

The depression centred over Ireland begins to weaken, with winds easing to light in a northwesterly flow. Possible patches of fog overnight mostly in the north to central areas, and frost patches in inland areas. Mostly dry with good sunny spells throughout the day. Chance of evening showers for a brief time in most districts, occasionally accompanied by thunderstorms.

Belfast: Sunny, slightly cooler, isolated showers, gentle breezes, chance of fog
Dublin: Mostly sunny and dry, chance of isolated evening showers, shifting breezes
Cork: Sunny, brief evening showers, cold overnight, light shifting breezes
Galway: Dry, sunny, mild day cold night, chance of frost, light shifting breezes

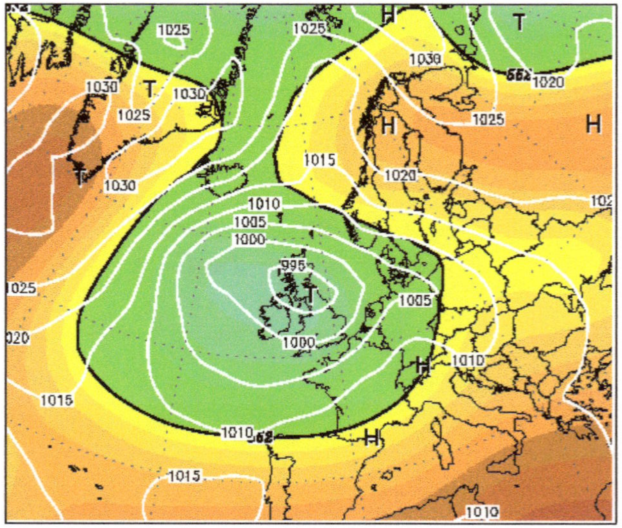

RAIN POTENTIAL

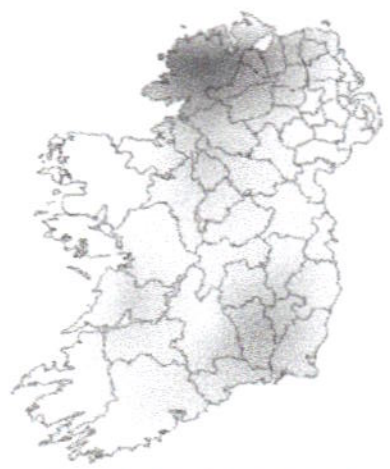

FROST/SNOW

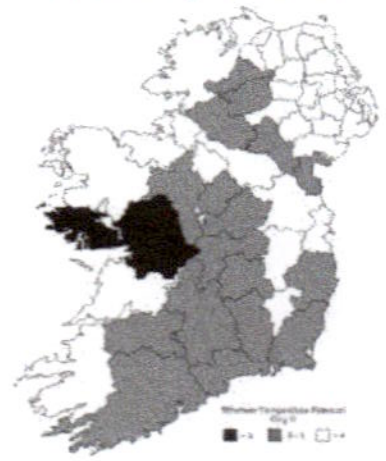

MAY **13** WEDNESDAY

The depression over Ireland continues to weaken as it drifts off to the west of the country, with winds generally light to moderate in a south to southwesterly flow. Patches of overnight fog likely in sheltered areas, clearing to a mostly dry and sunny day. Chance of a weak front later in the day spreading light showers from the south and west at first and later moving into the north and east. Some shower activity may be accompanied by thunderstorms and hail showers.

Belfast: Sunny, brief showers, slightly warmer day, cool night, gentle breezes
Dublin: Sunny, mostly dry, mild day cold night, gentle breezes
Cork: Mostly cloudy, brief showers, moderate breezes, chance of fog overnight
Galway: Sunny, evening showers, overnight frost patches possible, light breezes

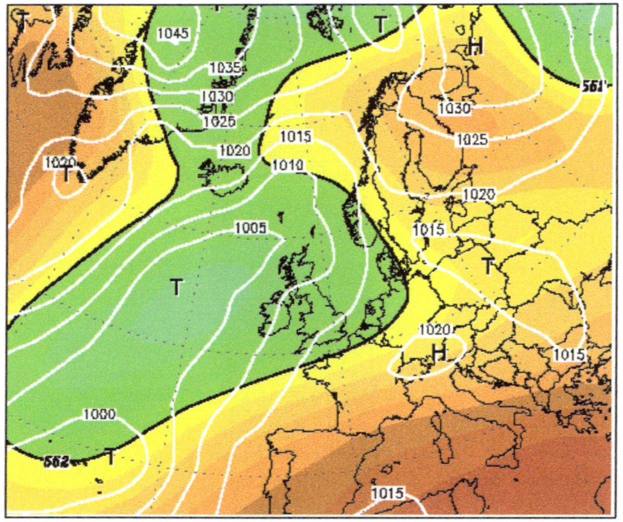

RAIN POTENTIAL

FROST/SNOW

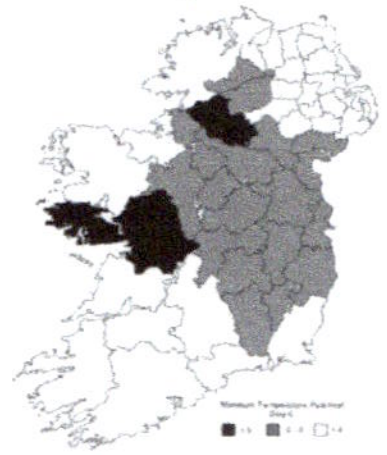

MAY **14** THURSDAY

A depression to the west of Ireland brings strong southerly winds up through the country, with increased cloud throughout. A rain embedded trough may move into southern parts at first and spread shower activity northward almost everywhere, although remaining mostly dry in the northwest. Some outbreaks of rain may be heavy for a time in southern areas.

Belfast: Partly cloudy, light showers, warmer, windy spells
Dublin: Partly cloudy, light showers, warmer, freshening breezes
Cork: Overcast, rain, milder temperatures, strengthening breezes
Galway: Cloudy, brief showers, milder, windy with occasional gusts at times

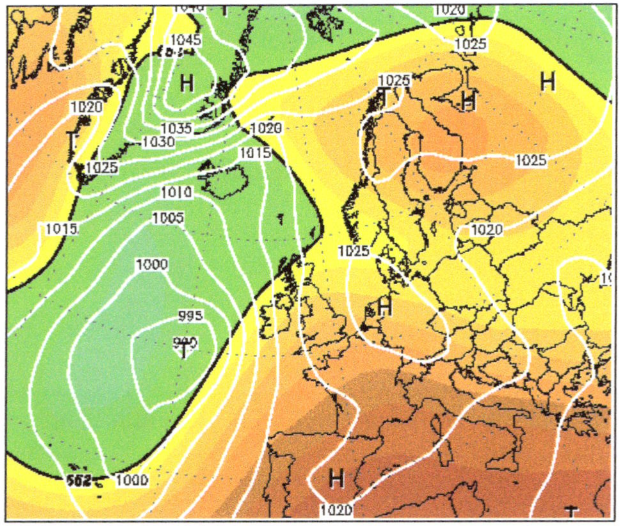

RAIN POTENTIAL

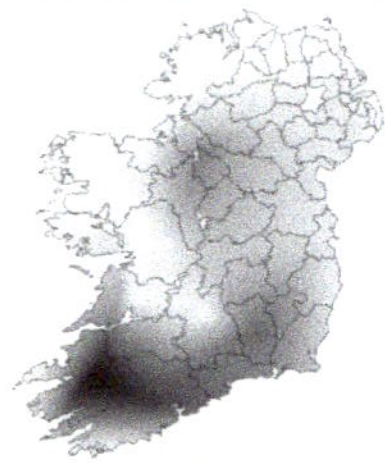

FROST/SNOW

MAY **15** FRIDAY

Low pressure continues to prevail over Ireland with fresh to strong southerly winds for a time, easing later in the day with the chance of evening patches of fog developing. Some light showers likely at first, mostly in the southeast and more isolated in the west and north, before a mostly sunny and dry day develops throughout.

Belfast: Sunny, dry, warmer, blustery at times
Dublin: Sunny, mostly dry, warmer, windy
Cork: Very sunny, occasional showers, warmer, unsettled breezes at times
Galway: Sunny, mild temperatures, dry, strong winds with occasional gusts

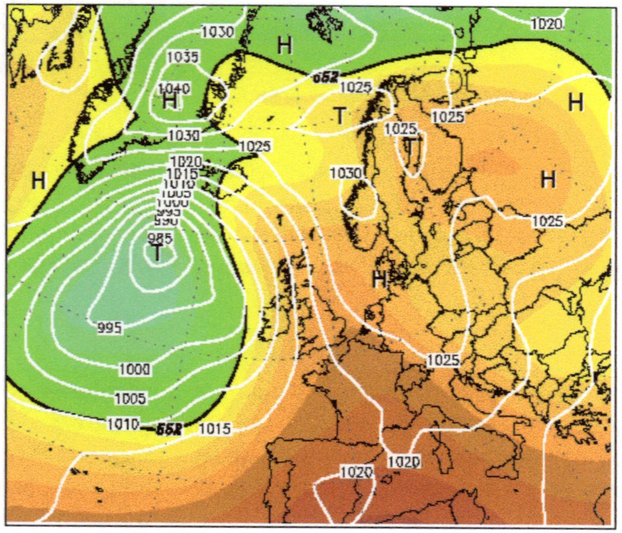

RAIN POTENTIAL

FROST/SNOW

MAY **16** SATURDAY

High pressure moves up over the country behind the departing depression. A mostly calm to gentle southerly breezes prevails throughout, with possible patches of fog overnight in many parts of the country. Very sunny, warm, and dry almost everywhere, but with increased cloud around southern parts. Chance of some drizzle briefly in the northeastern areas for a time, while low lying fog may persist for a lengthy time in eastern coastal parts.

Belfast: Very sunny, warmer, dry, gentle breezes, chance of fog patches overnight
Dublin: Very bright and sunny, dry, cooler, calm and serene air, chance of fog
Cork: Cloudy, calm air, isolated showers, cooler, fog patches possible
Galway: Sunny, warmer day colder night, dry, winds easing to light

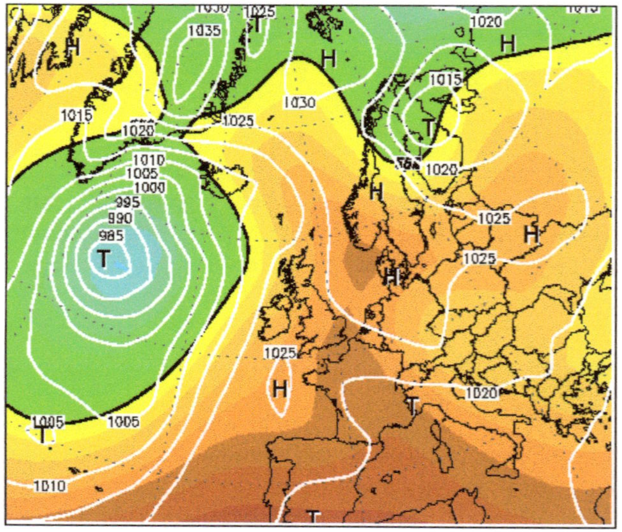

RAIN POTENTIAL

FROST/SNOW

MAY **17** SUNDAY

High pressure maintains a light southerly flow over the country. Patches of mist and fog likely in many districts, lifting to a mostly dry and sunny day throughout, except for some increased scattered cloud likely in the southwest for a time.

Belfast: Sunny, dry, warmer day, cooler night, gentle breezes
Dublin: Very sunny, dry, warmer, mostly calm air, fog patches possible
Cork: Dry, sunny, warmer, gentle and pleasant air, fog possible
Galway: Warm, sunny, dry, light breezes

RAIN POTENTIAL

FROST/SNOW

MAY **18** MONDAY

An anticyclonic air prevails. Winds light and generally southerly in flow. Mostly sunny and dry throughout the country, with slightly more cloud lingering around the southwest and south coasts for a time, after morning fog patches clear. Mild temperatures in the south, while very warm elsewhere.

Belfast: Very sunny, warm and dry, cool overnight, gentle breezes to mostly calm air
Dublin: Very sunny and dry, warm day cool night, mostly calm atmosphere
Cork: Partly cloudy, dry, mostly serene and pleasant air, chance of fog
Galway: Warm, scattered cloudy spells, dry, gentle breezes

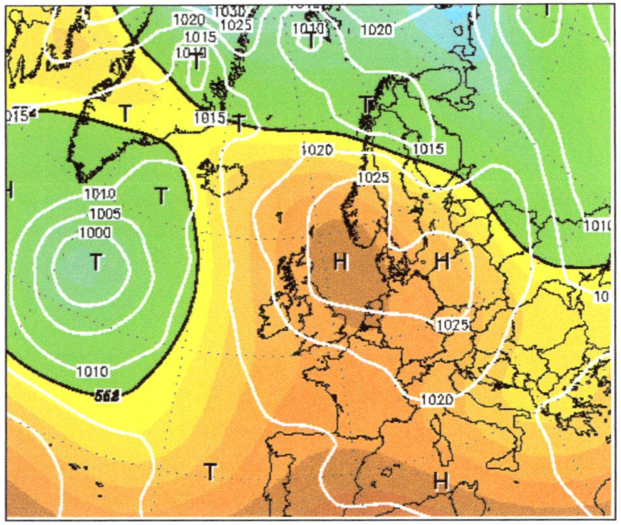

RAIN POTENTIAL

FROST/SNOW

MAY 19 TUESDAY

The high pressure ridge moves to the east of Ireland, while a low pressure trough associated with a weak depression centred just south of Greenland moves into the west of the country. A light southeasterly flow prevails. Mostly dry overnight before a wet front may move into the west and south of the country at first, increasing cloud and bringing outbreaks of drizzle later turning to showers as the front spreads throughout the country in a northeastward tract.

Belfast: Warm and mostly sunny day, overnight rain, gentle breezes
Dublin: Sunny, mostly dry, chance of isolated showers, gentle breezes
Cork: Overcast, occasional showers, gentle breezes, cooler, fog patches possible
Galway: Warm, mostly cloudy, light showers, chance of thunderstorms

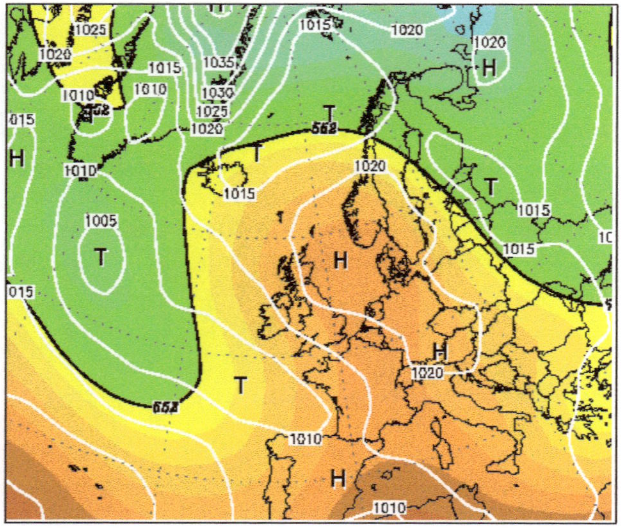

RAIN POTENTIAL

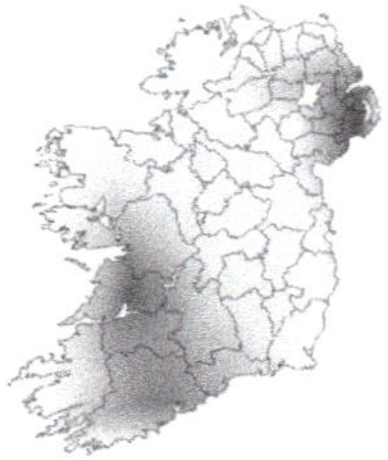

FROST/SNOW

MAY 20 WEDNESDAY

A low pressure trough maintains a weak southeasterly flow over Ireland, backing westerly later in the day. Generally dull and gloomy in the east and north with drizzle patches and light showers, while mostly dry with some sunny spells in the west, sunniest and dry in the south.

Belfast: Overcast, light showers, gentle breezes, warm day mild night
Dublin: Overcast and cooler, light showers, gentle breezes, chance of fog patches
Cork: Gentle shifting breezes, dry, passing scattered cloud
Galway: Partly cloudy, cooler, dry, gentle breezes

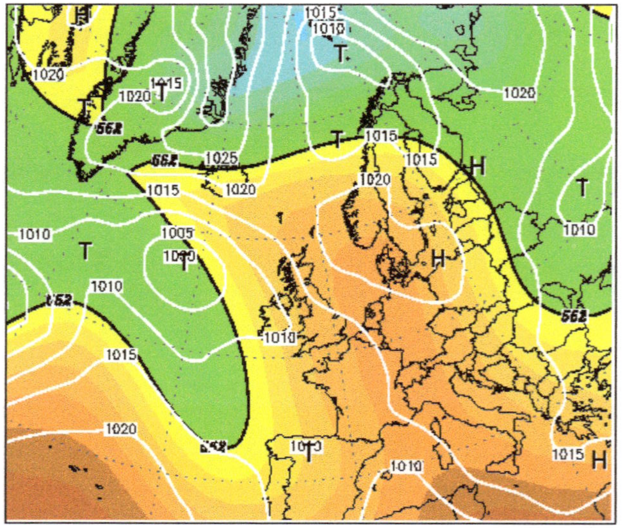

RAIN POTENTIAL

FROST/SNOW

MAY **21** THURSDAY

A light southwesterly flow to a mostly calm air prevails with widespread cloud throughout the country. Some pockets of fog likely overnight in sheltered areas. Showers in the northeast, east, and south, while mostly dry in the northwest, west and southwest. Chance of a band of thunderstorms crossing the east overnight and moving into the northeast later.

Belfast: Overcast, occasional showers, cooler, serene air, chance of fog overnight
Dublin: Overcast, scattered showers, calm atmosphere, chance of thunderstorms
Cork: Overcast, brief showers, warmer, calm and pleasant air, chance of fog
Galway: Cloudy, dry, mild temperatures, calm and serene atmosphere

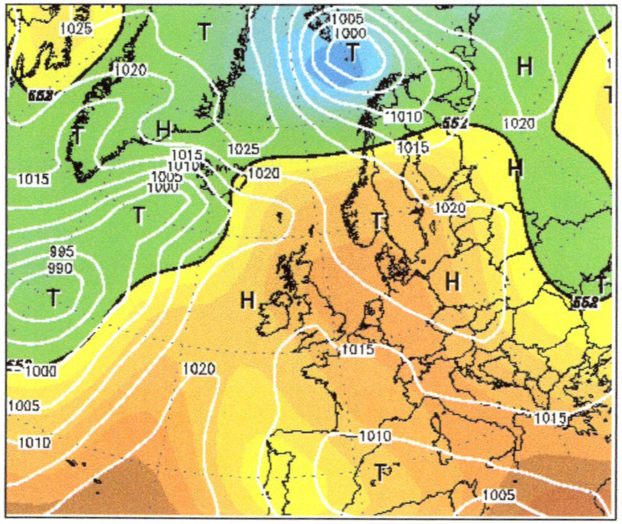

RAIN POTENTIAL

FROST/SNOW

MAY **22** FRIDAY

A weak high pressure flow maintains a gentle south to southwesterly air over the country, with patches of fog likely overnight in many districts, some dense and slow to clear. Some showers in the northern to northeast areas, possibly accompanied by thunderstorms. Mostly dry and sunny elsewhere.

Belfast: Partly cloudy, dry, mild temperatures, calm and pleasant air
Dublin: Sunny, warmer day cooler night, dry, calm atmosphere, fog possible
Cork: Warmer, dry, partly cloudy, mostly calm air, chance of fog overnight
Galway: Partly cloudy, dry, mild temperatures, calm and pleasant air

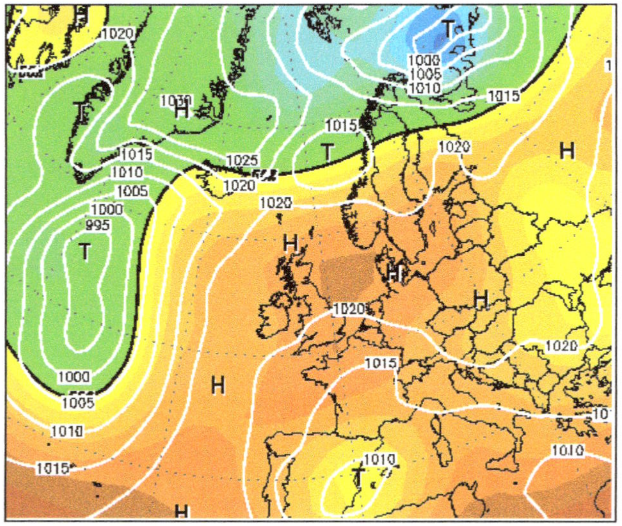

RAIN POTENTIAL

FROST/SNOW

MAY **23** SATURDAY

An anticyclonic system to the north of Ireland directs a gentle easterly flow across the country. Some patches of mist and fog overnight. Mostly sunny everywhere and dry at first before some showers arrive in the west and northeast later in the day, possibly accompanied by thunderstorm activity. Dry elsewhere.

Belfast: Sunny, warm, dry, serene atmosphere
Dublin: Dry, sunny, mild day cool night, calm and serene air
Cork: Very sunny, dry, warm day cool night, calm air, chance of fog overnight
Galway: Mostly sunny, brief showers, gentle breezes, chance of thunderstorm

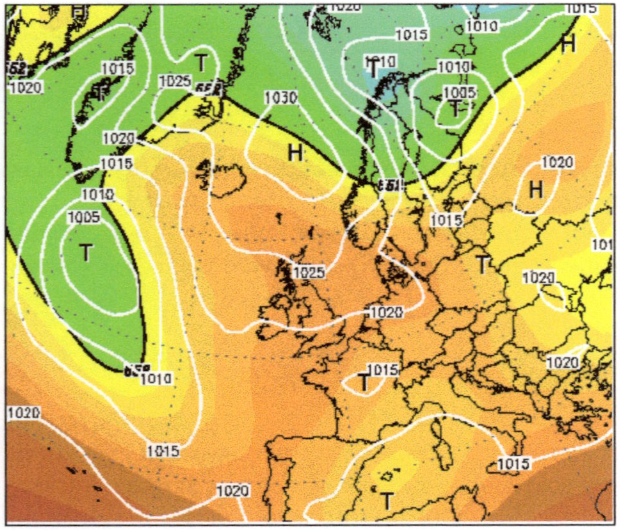

RAIN POTENTIAL

FROST/SNOW

MAY **24** SUNDAY

High pressure maintains a dry and light east to southeasterly flow over the country, with long periods of sunshine everywhere. Chance of patches of fog overnight, mostly in the midlands.

Belfast: Very sunny and warm day, cool overnight, dry, gentle shifting breezes
Dublin: Very bright and sunny, cooler, dry, gentle breezes
Cork: Very bright and sunny, warm, dry, calm and pleasant air
Galway: Very sunny, dry and warm, gentle breezes

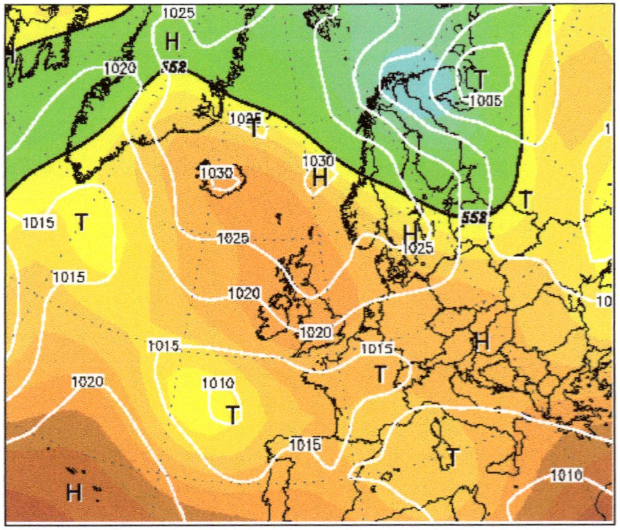

RAIN POTENTIAL

FROST/SNOW

MAY **25** MONDAY

Anticyclonic conditions continue to prevail throughout the country with a very sunny and dry day throughout. Winds generally very light in a easterly flow. Temperatures may be unusually warmer than what is normal for this time of year.

Belfast: Very sunny, dry, moderate temperatures, gentle breezes
Dublin: Very bright and sunny, dry, mild day cool night, gentle breezes
Cork: Bright and sunny, dry, warm, calm and serene atmosphere
Galway: Very sunny, mild temperatures, dry, mostly calm and pleasant air

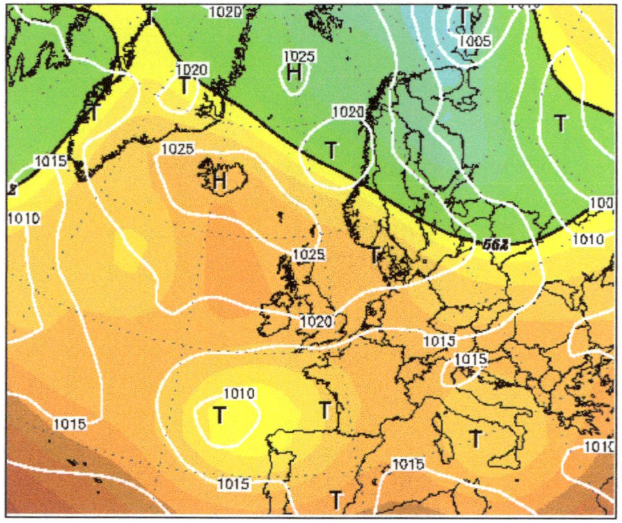

RAIN POTENTIAL

FROST/SNOW

MAY **26** TUESDAY

High pressure weakens with a low pressure trough building to the northwest of Ireland. Winds continue to be very light in a east to northeast flow, with slightly cooler temperatures returning. Very sunny and dry throughout.

Belfast: Very sunny, dry, pleasant atmosphere
Dublin: Pleasant and very sunny air, gentle breezes, dry, cold overnight
Cork: Mostly calm and pleasant air, very sunny, dry, cooler
Galway: Mild temperatures, very sunny, dry, light breezes to mostly calm air

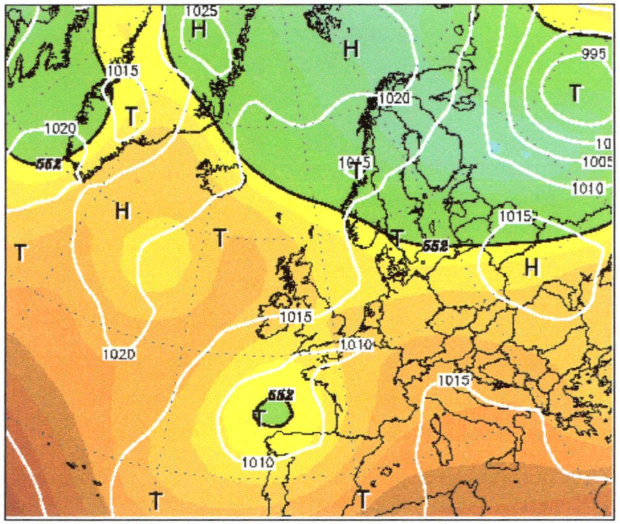

RAIN POTENTIAL

FROST/SNOW

MAY 27 WEDNESDAY

The high pressure ridge drifts to the west of Ireland, while a weak low pressure trough moves across Ireland. Winds remain very light in a generally northerly to variable flow. Cloudy with the chance of some isolated showers overnight in the south of the country. Sunny and dry elsewhere with the chance of fog patches for a time in the west.

Belfast: Sunny, dry, mild temperatures, calm atmosphere
Dublin: Sunny, dry, mild day cold night, calm atmosphere
Cork: Overcast and dull, isolated showers, cooler, mostly calm air
Galway: Sunny, dry, mild temperatures, calm and pleasant air, chance of fog patches

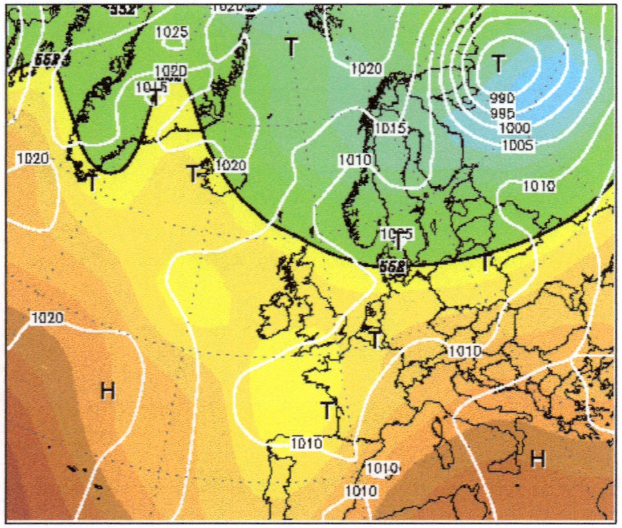

RAIN POTENTIAL

FROST/SNOW

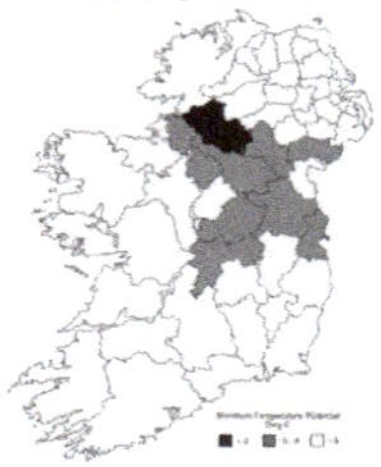

MAY 28 THURSDAY

Low pressure stabilises over Ireland increasing cloud throughout but still maintaining very light breezes in an easterly flow. Some showers over the southern areas of the country for a time, before they ease to patches of mist or drizzle that may linger around most of the day. Mostly cloudy but dry elsewhere, sunnier for a time in the west. Chance of some isolated coastal showers in the northeast of Ireland later in the day.

Belfast: Cloudy, isolated showers, mild temperatures, gentle shifting breezes
Dublin: Mostly cloudy, dry, cold overnight, gentle shifting breezes
Cork: Cloudy, occasional showers, mostly calm air
Galway: Gentle breezes, mostly sunny, cooler and dry

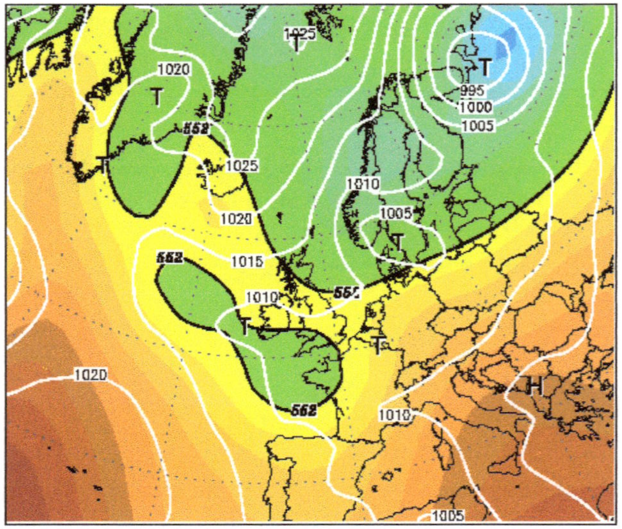

RAIN POTENTIAL

FROST/SNOW

MAY 29 FRIDAY

Low pressure dominates Ireland with a very light easterly flow still continuing throughout. A wet front may bring some rain and drizzle patches in many areas overnight, and with heavier falls, possibly accompanied by thunderstorm activity, in the west later in the day. Only the north of the country may remain mostly dry and sunny.

Belfast: Partly cloudy, cooler, dry, mostly calm atmosphere
Dublin: Sunny, brief morning showers possible, cold overnight, mostly calm air
Cork: Cloudy, spotty showers, cooler, calm and serene air
Galway: Cloudy, dry, mild temperatures, calm atmosphere

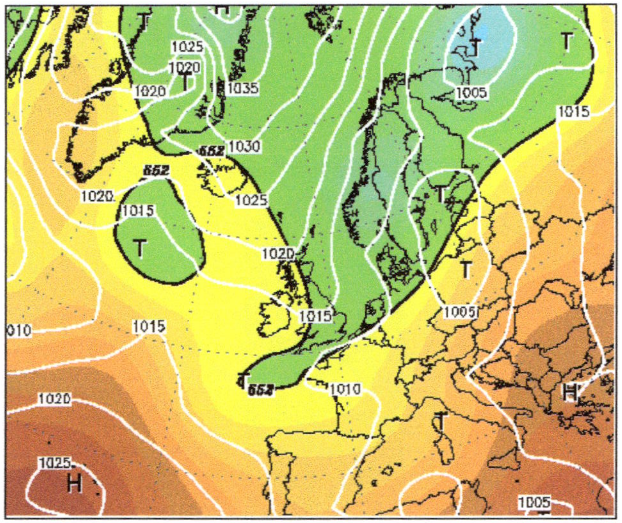

RAIN POTENTIAL

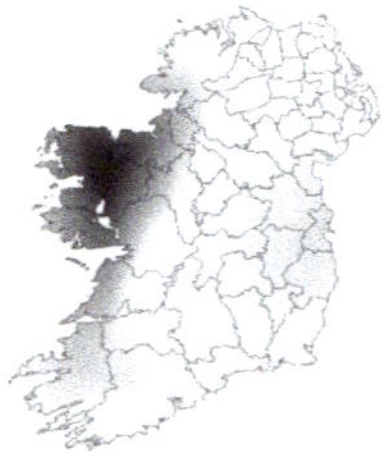

FROST/SNOW

MAY 30 SATURDAY

A depression forms to the west of Ireland, while a weak trough still maintains a very light easterly flow over the country, later becoming more gusty with a southeast change. Chance of inland frost patches. A changeable day with a mix of sunny breakthroughs and the odd isolated shower, possibly accompanied by thunderstorma activity mostly in the west and south, while remaining mostly dry in the north and east.

Belfast: Partly cloudy, dry, mild day cold night, serene air
Dublin: Mostly cloudy, dry, mild day cold night, calm and pleasant air
Cork: Mostly calm air, partly cloudy, dry, slightly warmer
Galway: Light shifting breezes, mild temperatures, dry, occasional cloudy spells

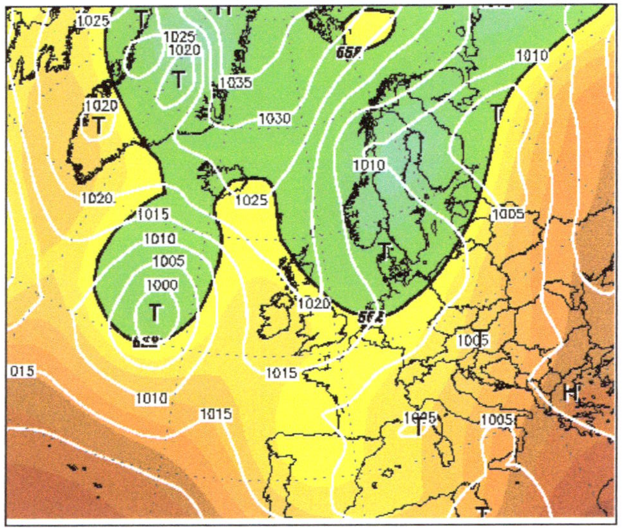

RAIN POTENTIAL

FROST/SNOW

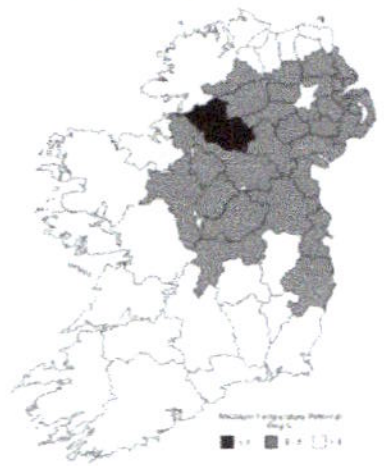

MAY 31 SUNDAY

The depression west of Ireland slow nears the country, while a weak high pressure ridge remains almost stalled to the east. Winds generally light to moderate in a south to southeasterly flow. Some lingering spots of showers mostly in the inland western districts. Mostly dry elsewhere with widespread cloud in the south and more frequent sunny spells in the north and east.

Belfast: Partly cloudy, dry, mild temperatures, occasional breezy spells
Dublin: Partly cloudy, dry, mild day cold night, light breezes
Cork: Overcast, cooler, mostly dry, increasing threats of rain, freshening winds
Galway: Mix of sunny spells and showers, freshening breezes, slightly cooler

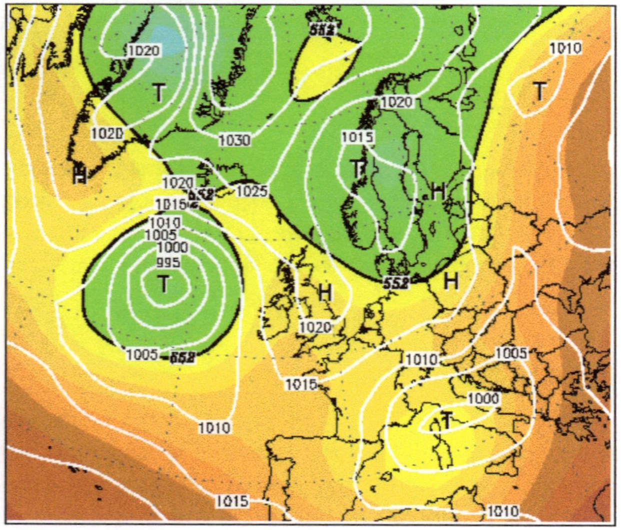

RAIN POTENTIAL

FROST/SNOW

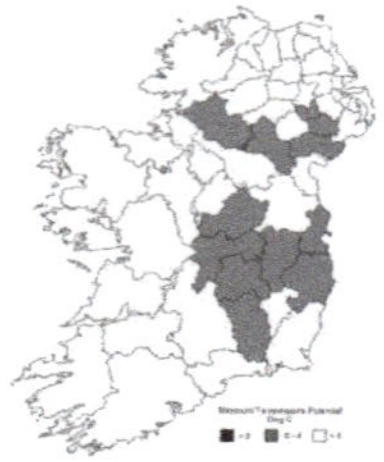

JUNE

PHASES OF THE MOON

1st	Apogee
1st	Southern Declination
8th	Third Quarter
9th	Crossing Equator
14th	Perigee #2
15th	New Moon
15th	Northern Declination
21st	First Quarter
22nd	Crossing Equator
28th	Apogee
28th	Southern Declination
29th	Full Moon

1	2	3	4	5	6	7	8	9	10	11	12	13	14	15	16	17	18	19	20	21	22	23	24	25	26	27	28	29	30
Mon	Tue	Wed	Thur	Fri	Sat	Sun	Mon	Tue	Wed	Thur	Fri	Sat	Sun	Mon	Tue	Wed	Thur	Fri	Sat	Sun	Mon	Tue	Wed	Thur	Fri	Sat	Sun	Mon	Tue
A V							3Q	XhN					P2	N ^						1Q	XhS						A V	F	

MONTHLY SUMMARY

1st-11th Low pressure may return bringing bands of rain, occasionally heavy and accompanying thunderstorm activity, particularly in the west. Temperatures may still be above normal almost everywhere. Winds may be fresh to strong in a south-easterly flow at first, later backing to easterly and decreasing.

1st Very heavy rain likely in the south of the country.

1st-3rd A band of heavy rain may cross the country in an easterly flow, with thunderstorms possible mostly in the west.

4th-7th Further rain bands may cross the east and south of the country.

5th-6th Strong north-easterly winds to gale force possible for a time before easing and becoming variable.

6th Unusually bright and sunny in the northwest, particularly around Malin Head.

10th-11th Chance of intense thunderstorms and heavy showers in the west of the country.

11th Chance of thunderstorms and showers, particularly in the west, where unsettled winds may see the formation of a funnel cloud system.

The next four weeks ahead may see the return of unsettled conditions becoming the dominant weather patterns. Overall, the first couple of weeks may still see fairly average to above average air pressures, before dropping to low pressure areas situated mostly to the west or northwest of Ireland being the dominant feature for the last two weeks. Generally, the next four weeks may be wetter than normal almost everywhere, with sunshine and temperatures nearer to normal. Rain activity may occur on up to 15-20 days during this outlook period, occasionally heavy, and with frequent thunderstorm activity, with most parts of the country receiving above average rain, and some parts of the south and southeast almost twice as much recorded rainfall than what is normal for this time of year. It may be warm and very sunny at first during this outlook period, before temperatures return to closer to average values and sunshine levels drop to average. This may result in air temperatures being slightly above normal overall, but slightly cooler in some inland parts and sunnier in the south and west, with average sunshine overall in the north and east. Despite the increased warmth during this period, there may be some ground frosts noted at times in some inland areas, mostly in the first couple of weeks ahead. During this outlook period there may be at least 3 days with thunderstorms, 3 days with hail, widespread fogs on 3 days but no air frosts but 5 days with ground frosts in some inland areas, 1 day with gales, but no snow.

12th-17th Low pressure systems near to Ireland prevail bringing warm and humid conditions. Despite this, it may be sunnier than normal for this time of year.

12th-22nd Warmer than average temperatures likely.

12th Chance of widespread thunderstorms. Fog patches may be widespread overnight.

12th-13th Very warm spell.

14th Light south to southwest winds may prevail, freshening later with a westerly change. Very sunny day likely.

15th A deep depression near the northwest of Ireland may bring an active frontal system with heavy rain and cooler temperatures in many areas of the country with thunderstorms possible mostly in the west.

16th Cooler change with the chance of hail showers and possible frost patches in inland parts.

16th-18th Possibly the sunniest spell of summer.

18th-27th An anticyclonic influence to the southwest of Ireland may dominate the country bringing a mostly dry but cloudy spell with very little observed rain activity.

19th-20th Lighter air with chance of widespread fog patches overnight.

20th Very warm day.

21st High pressure may peak around now.

22nd A brief wet and cooler respite possible, as a band of rain may cross the country bringing some light showers, mostly around the coasts. Winds may take a northerly change.

26th Chance of frost patches overnight in inland parts.

28th-11th July Unsettled conditions may return as deep depressions become established near to the country.

29th-30th Chance of thunderstorms and heavy rain, mostly in the western half of the country.

30th Unusually sunny day possible.

IRELAND RAINFALL ESTIMATES

June 2026

d=mainly dry, n= nonrecordable, l=light shrs, s=significant shrs, r=rain, h=hvy falls

		ULSTER PROVINCE					CONNAUGHT PROVINCE						LEINSTER PROVINCE				MUNSTER PROVINCE						
		Alderg	Hillsbo	Armag	Loughl	Malin H	Sligo	Ardtar	Belmul	Drums	Galway	Roscoi	Markre	Edende	Kilkenı	Dublin	Shann	Newpo	Tralee	Killarn	Cork	All	Moon
1st	Mon	l	s	d	l	d	l	l	s	l	s	s	l	s	s	l	s	s	h	h	h	142	A V
2nd	Tues	s	s	d	r	l	l	r	s	r	l	s	r	l	l	r	s	r	l	l	l	106	
3rd	Wed	l	l	l	d	n	l	d	l	l	d	l	l	d	d	d	s	l	s	d	l	23	
4th	Thur	d	d	d.l	d	d	d	d	d	d	d	d	d	d	d	d	d	d	l	d	d	2	
5th	Fri	d	d	d	d	d	d	d	d	d	d	d	d	d	d	n	d	d	d	d	s	4	
6th	Sat	d	d	d	d	d	d	d	d	d	d	d	d	d	r	l	d	d	d	d	l	13	
7th	Sun	l	l	d	d	d	d	d	d	d	d	d	d	l	d	s	d	d	d	d	l	13	
8th	Mon	l	d	l	d	d	l	d	d	d	d	d	d	d	d	d	d	d	d	d	d	2	3Q
9th	Tues	d	d	d	l	d	d	l	d	d	l	d	d	d	n	d	d	l	l	l	l	8	XhN
10th	Wed	d	d	d	n	d	n	n	l	d	d	d	l	l	l	d	s	l	s	l	l	24	
11th	Thur	d	d	n	d	d	d	d	d	r	d	l	n	d	d	d	d	d	d	d	d	16	
12th	Fri	d	d	d	d	d	d	d	d	s	l	l	d	d	d	d	r	d	d	d	d	21	
13th	Sat	d	d	d	s	n	d	s	d	l	s	l	l	l	d	d	s	d	d	d	d	24	
14th	Sun	d	d	d	s	l	n	s	l	l	r	l	s	n	l	d	d	r	l	r	d	65	P2
15th	Mon	l	s	d	l	r	s	l	h	s	r	r	l	s	s	r	s	l	h	d	s	140	N ^
16th	Tues	l	d	s	h	d	l	h	l	r	r	s	r	l	s	l	d	l	d	l	d	114	
17th	Wed	l	n	l	d	l	l	d	d	l	d	s	d	l	l	d	d	d	l	d	l	19	
18th	Thur	d	d	d	d	d	d	d	d	d	d	n	d	d	d	d	l	n	l	l	d	6	
19th	Fri	d	d	d	d	d	d	d	d	d	d	d	d	d	d	d	l	d	d	d	l	4	
20th	Sat	d	d	d	d	d	d	d	d	d	d	d	d	d	d	d	d	d	d	d	d	1	
21st	Sun	d	d	d	d	d	n	d	l	d	d	d	l	l	d	d	d	d	d	d	d	5	1Q
22nd	Mon	l	l	d	l	l	l	l	l	l	d	l	l	s	s	l	l	l	l	s	l	48	XhS
23rd	Tues	l	d	l	l	l	l	l	d	n	r	n	l	d	d	d	d	d	d	d	d	16	
24th	Wed	d	d	d	d	l	d	d	n	l	d	d	l	d	d	d	d	l	d	l	d	9	
25th	Thur	d	d	d	d	l	l	d	l	s	s	n	l	l	n	d	d	l	d	d	d	19	
26th	Fri	d	d	d	d	l	d	d	d	l	d	l	d	d	d	n	l	n	l	d	d	7	
27th	Sat	n	d	l	l	d	n	l	d	l	s	d	d	s	d	d	d	l	l	n	d	23	
28th	Sun	l	l	d	s	s	s	s	s	l	s	s	l	s	s	s	l	s	l	r	s	95	A V
29th	Mon	s	s	s	l	l	s	l	l	n	s	l	l	h	h	r	h	d	h	r	h	171	F
30th	Tues	l	l	n	l	l	l	l	l	s	l	s	l	s	d	s	d	l	d	d	d	37	
	Estimate:	23	29	17	64	34	29	64	42	74	95	64	57	66	61	60	74	52	109	78	85	1175	
	Average:	62	66	58	71	70	82	84	72	74	80	82	82	74	61	67	70	90	108	72	81	1505	
	Trend:	drr	drr	drr	av	drr	drr	drr	drr	av	wtr	drr	drr	av	av	av	av	drr	av	av	av	drr	

IRELAND SUNSHINE ESTIMATES

June 2026

F=fine (8-12 hours of sunshine), pc= partly cloudy (4-7 hours), c=cloudy (1-3 hours), o=overcast (0 hours)

		ULSTER PROVINCE					CONNAUGHT PROVINCE						LEINSTER PROVINCE				MUNSTER PROVINCE						
		Alderg	Hillsbo	Armag	Loughl	Malin I	Sligo	Ardtarı	Belmul	Drums	Galway	Roscoı	Markre	Edend(	Kilkenı	Dublin	Shann(	Newpo	Tralee	Killarn(	Cork	All	Moon
1st	Mon	c	c	c	c	c	o	c	o	c	o	o	c	c	c	c	o	o	c	c	o	24	A V
2nd	Tues	o	o	o	o	o	o	o	c	c	c	c	o	c	c	o	c	c	pc	pc	pc	38	
3rd	Wed	pc	pc	pc	pc	pc	pc	pc	pc	pc	pc	pc	pc	pc	pc	pc	pc	pc	c	c	pc	103	
4th	Thur	F	F	F	F	F	pc	F	F	F	F	F	F	F	F	F	F	F	F	F	F	201	
5th	Fri	F	F	F	F	F	F	F	F	F	F	F	F	F	pc	pc	F	pc	F	F	pc	199	
6th	Sat	F	F	F	F	F	F	F	F	pc	F	F	F	pc	pc	pc	pc	pc	pc	pc	c	168	
7th	Sun	c	c	c	c	c	c	c	pc	o	c	pc	c	c	o	o	o	o	o	o	o	35	
8th	Mon	o	o	o	o	o	o	o	pc	c	c	c	o	c	c	o	c	c	c	c	c	30	3Q
9th	Tues	pc	pc	pc	pc	pc	pc	pc	F	pc	pc	pc	pc	pc	c	pc	pc	pc	pc	pc	o	110	XhN
10th	Wed	pc	pc	pc	pc	pc	pc	pc	pc	pc	pc	pc	pc	pc	pc	pc	c	pc	c	c	c	99	
11th	Thur	F	F	F	F	F	F	F	F	F	F	F	F	F	F	F	pc	F	pc	pc	F	216	
12th	Fri	F	F	F	F	F	F	F	F	F	F	F	F	F	F	F	F	F	c	c	F	190	
13th	Sat	F	F	F	F	F	pc	F	pc	F	pc	pc	F	F	F	F	c	pc	F	F	F	157	
14th	Sun	F	F	F	F	F	F	F	F	F	F	F	F	F	F	F	F	F	F	F	F	267	P2
15th	Mon	c	c	c	c	c	pc	c	F	pc	F	F	pc	pc	pc	c	F	pc	F	F	F	127	N ^
16th	Tues	pc	pc	pc	pc	pc	F	pc	o	pc	pc	pc	pc	pc	pc	o	F	F	F	F	F	131	
17th	Wed	pc	pc	pc	pc	pc	pc	pc	F	F	F	F	pc	F	F	F	F	F	F	F	F	196	
18th	Thur	F	F	F	F	F	F	F	pc	F	pc	pc	F	F	F	F	F	F	pc	pc	F	193	
19th	Fri	c	c	c	c	c	c	c	pc	c	pc	pc	c	c	c	c	pc	c	o	o	o	55	
20th	Sat	F	F	F	F	F	pc	F	o	F	o	o	F	F	F	F	o	pc	o	o	F	120	
21st	Sun	F	F	F	F	F	pc	F	pc	F	pc	pc	F	F	F	F	F	F	F	F	F	199	1Q
22nd	Mon	o	o	o	o	o	o	o	o	c	o	o	o	c	c	c	o	o	o	o	c	13	XhS
23rd	Tues	pc	pc	pc	pc	pc	pc	pc	pc	pc	pc	pc	pc	pc	pc	pc	c	pc	pc	pc	F	108	
24th	Wed	pc	pc	pc	pc	pc	F	pc	F	pc	pc	pc	pc	pc	pc	pc	c	pc	o	o	c	117	
25th	Thur	pc	pc	pc	pc	pc	pc	pc	F	pc	F	F	pc	pc	pc	pc	F	F	F	F	c	156	
26th	Fri	pc	pc	pc	pc	pc	pc	pc	pc	pc	pc	pc	pc	pc	pc	c	pc	pc	F	F	pc	114	
27th	Sat	c	c	c	c	c	c	c	c	pc	c	c	c	pc	pc	pc	c	pc	pc	pc	pc	81	
28th	Sun	c	c	c	c	c	c	c	pc	c	pc	pc	c	c	pc	pc	c	c	o	o	c	67	A V
29th	Mon	c	c	c	c	c	pc	c	F	c	pc	F	c	c	c	c	c	c	c	c	c	79	F
30th	Tues	F	F	F	F	F	F	F	F	F	F	F	F	F	F	F	F	F	F	F	F	202	
	Estimate:	187	187	187	187	187	180	187	219	189	199	210	187	190	193	191	177	185	181	181	194	3794	
	Average:	167	167	142	169	171	162	162	161	137	152	159	155	158	155	180	154	146	152	154	172	3173	
	Trend:	more	more	more	more	more	more	more	more	more	more	more	more	more	more	more	more	more	more	more	more	more	

JUNE **1** MONDAY

A depression centred to the west of Ireland directs a fresh southeasterly flow across the country, with occasional gusts at times. Mostly cloudy to overcast throughout, with showers in most districts. Rain may be particularly heavy for a time in the south and southwest.

Belfast: Cloudy, passing showers, strong winds at times
Dublin: Mostly cloudy, light showers, windy spells
Cork: Fresh breezes at times, overcast, heavy rain, cool
Galway: Overcast, scattered showers, strong and unsettled winds, chance of gales

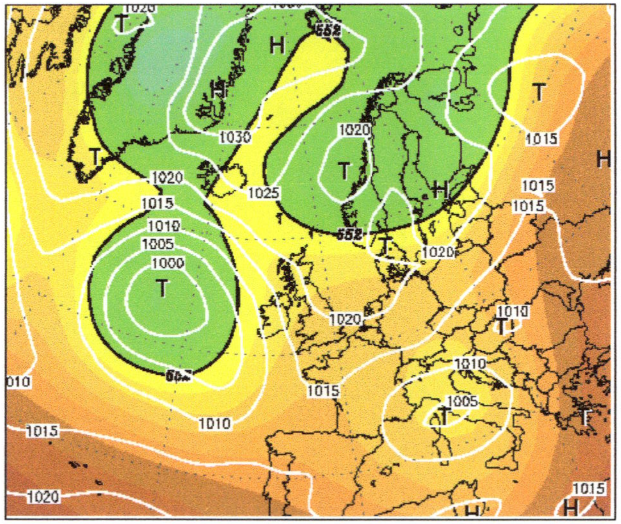

RAIN POTENTIAL

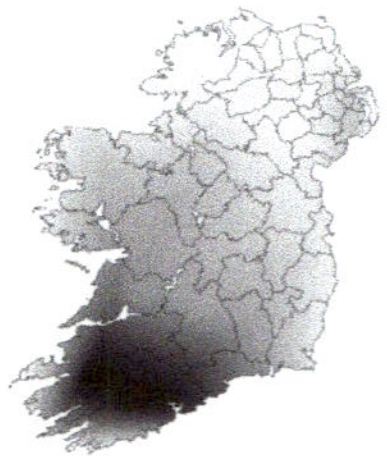

FROST/SNOW

JUNE **2** TUESDAY

A depression continues to sit to the west of Ireland directing a light to moderate south to southeast flow throughout. Cloudy and dull with passing showers in most parts of the country, although mostly dry and sunny in the far south. Chance of thunderstorms for a time in the west and possible patches of fog in the south.

Belfast: Overcast and dull, scattered showers, windy spells
Dublin: Overcast, rain, cool temperatures, breezy spells
Cork: Mostly sunny, slightly warmer, brief showers, light breezes, chance of fog
Galway: Cloudy, occasional showers, strong winds, chance of thunderstorms

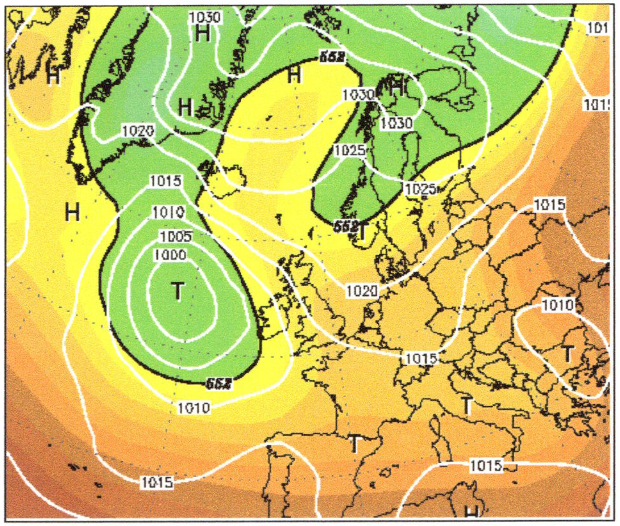

RAIN POTENTIAL

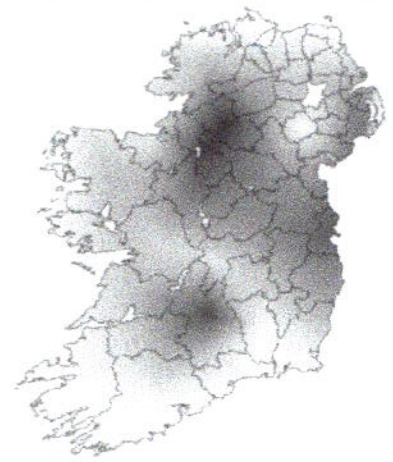

FROST/SNOW

JUNE **3** WEDNESDAY

The dpression to the west of Ireland still maintains a moderate south to southeast flow over the country, light breezes overnight with the chance of fog patches in the north and west. Mostly dry with increasing sunny spells except for some cloudier spells and brief showers likely in southern and western parts for a time. Chance of thunderstorms once more in the west.

Belfast: Isolated showers, increasing sunny spells, winds easing, chance of fog
Dublin: Dry, occasional sunny spells, warmer day cooler night, gentle breezes
Cork: Partly cloudy, light showers, gentle breezes
Galway: Partly cloudy, mostly dry, odd threat of rain, mild day cool night, static air

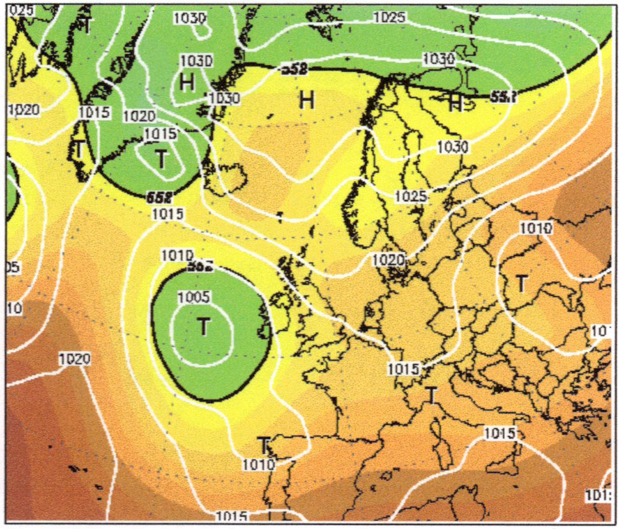

RAIN POTENTIAL

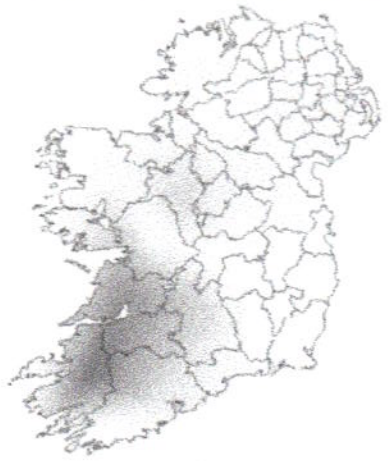

FROST/SNOW

JUNE **4** THURSDAY

The depression that has brought changeable conditions over the past few days slips to be centred to the south of the country, with light winds taking a easterly change. Mostly dry and sunny throughout. Some blustery spells likely around the northern coasts.

Belfast: Sunny, warmer, dry, moderate breezes
Dublin: Sunny, mild temperatures, dry, moderate winds
Cork: Very sunny, mostly dry, warmer, moderate breezes
Galway: Very sunny, dry, warmer, fresh breezes

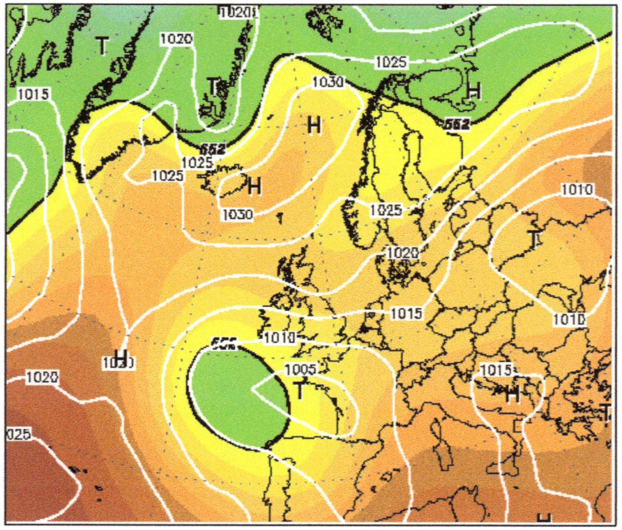

RAIN POTENTIAL

FROST/SNOW

JUNE 5 FRIDAY

A fresher northeasterly flow may develop over the country around this time as a high pressure system ridges to the north and low pressure lingers to the south, with winds more blustery at times in southern and western areas. Some brief showers may occur in the southeast later in the day, possibly accompanying thunderstorms and some of hail, while mostly dry and sunny elsewhere.

Belfast: Sunny and dry, mild day cool night, blustery at times
Dublin: Mostly sunny, isolated evening showers, blustery at times
Cork: Mostly sunny, light showers, slightly cooler, blustery at times
Galway: Sunny, warm day cool night, dry, unsettled winds, chance of gales and gusts

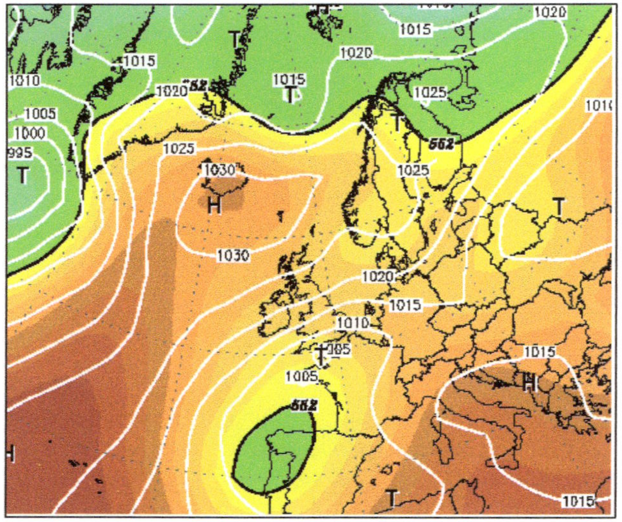

RAIN POTENTIAL

FROST/SNOW

JUNE 6 SATURDAY

High pressure prevails over the west and northern half of the country bringing mostly sunny and dry conditions, while a weak depression south of Ireland brings some showers and increased cloud into the southeast. Winds generally moderate to strong in a northeastery flow, more gusty at times around the coasts.

Belfast: Dry, mild day cool night, long sunny spells, moderate winds, occasional gusts
Dublin: Unsettled winds at times, partly cloudy, brief showers
Cork: Cloudy, brief showers, strong winds, occasional gusts, mild day cool night
Galway: Sunny, dry, strong winds, chance of gales

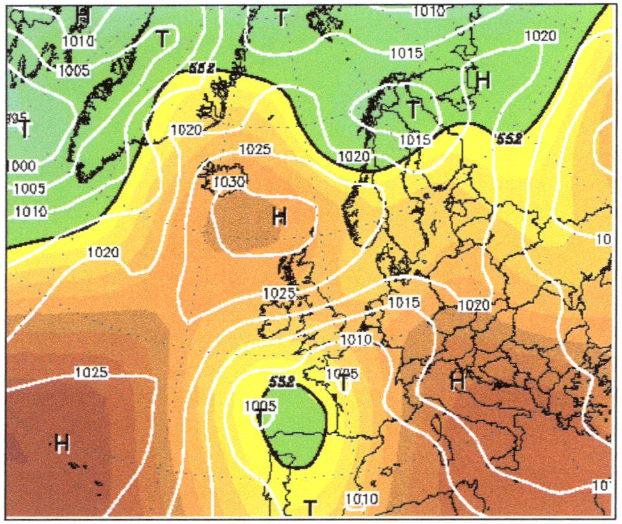

RAIN POTENTIAL

FROST/SNOW

JUNE **7** SUNDAY

The high pressure ridge maintains a mostly cloudy but dry day over the north and west of the country, with moderate to fresh northeasterly winds prevailing. In contrast, it may be very dull and gloomy with brief showers and squally winds at times in the east and southeast. Chance of thunderstorms in some central and eastern areas, with possible showers of hail.

Belfast: Cloudy, light showers, mild day cool night, fresh breezes
Dublin: Overcast and dull, colder, scattered showers, windy spells
Cork: Overcast, blustery at times, occasional squally showers, colder
Galway: Mostly cloudy, dry, strong and unsettled winds

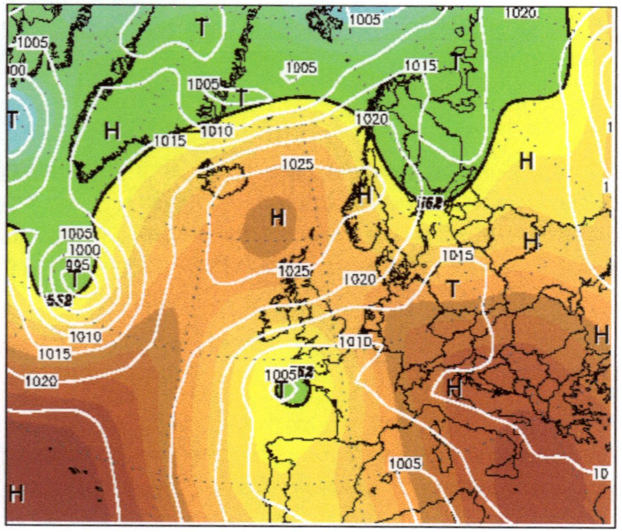

RAIN POTENTIAL

FROST/SNOW

JUNE **8** MONDAY

A low pressure trough brings a lighter easterly flow across the country, with widespread cloud but mostly dry conditions throughout. Chance of some misty patches overnight, mostly in the north, west and central parts.

Belfast: Overcast, cooler, dry, threats of rain, lighter winds, misty patches possible
Dublin: Overcast, milder, dry with odd threat of rain, moderate breezes
Cork: Mostly cloudy, warmer, dry, light breezes
Galway: Partly cloudy, dry, mild day cool night, windy spells

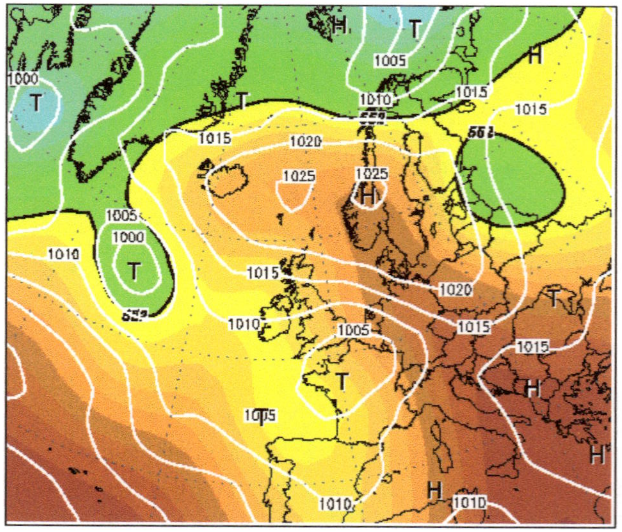

RAIN POTENTIAL

FROST/SNOW

JUNE 9 TUESDAY

A weak low pressure influence settles over the country bringing very gentle generally easterly breezes to a mostly calm air throughout, with patches of fog likely in many districts, particularly in the north and central parts. Mostly dry throughout with increasing sunny spells as the morning mist patches clear. Chance of some isolated showers or brief drizzle patches for a time in the southern parts of the country.

Belfast: Partly cloudy, dry, gentle breezes to mostly calm air, chance of fog
Dublin: Warmer, partly cloudy, dry, gentle shifting breezes
Cork: Overcast, spotty showers, colder, gentle breezes, chance of fog patches
Galway: Mostly sunny, brief evening showers, mostly calm air, warm day cold night

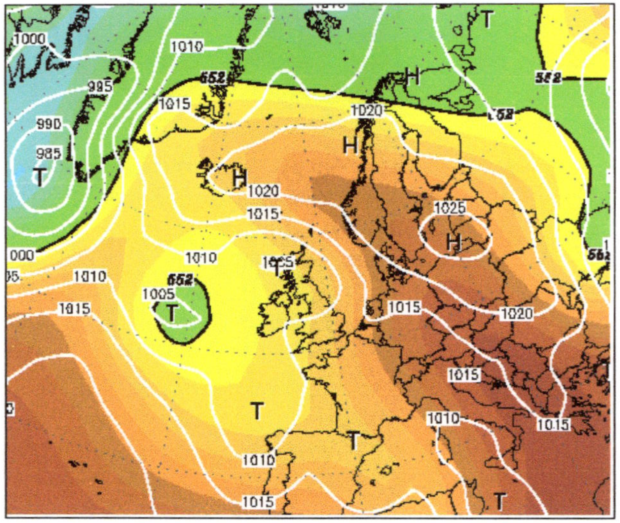

RAIN POTENTIAL

FROST/SNOW

JUNE 10 WEDNESDAY

A weak low pressure system brings a mostly calm and settled day throughout the country with chance of fog patches in central and southern parts. Sunniest in the north and east with increasing cloudy spells in the west and south. Some showers may briefly affect the west and midlands at first, and later into southern parts, possibly accompanied by thunderstorms. Mostly dry in the north and east.

Belfast: Warm and dry, pleasant sunny spells, mostly calm air
Dublin: Mostly sunny, warm day cool night, dry, gentle breezes to mostly calm air
Cork: Mostly cloudy, brief showers, mostly calm air, chance of fog
Galway: Partly cloudy, dry, mostly calm air

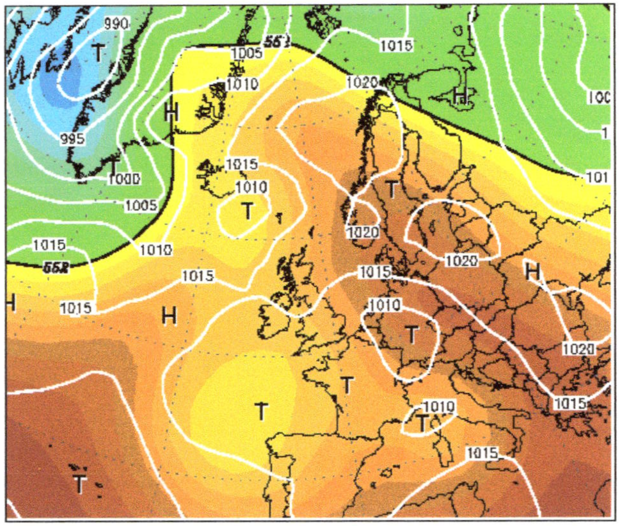

RAIN POTENTIAL

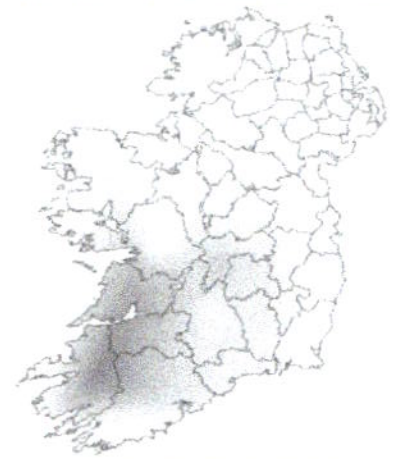

FROST/SNOW

JUNE 11 THURSDAY

Mostly calm and serene air throughout the country with scattered fogs in many districts, some quite dense and slow to clear, particularly in the south. Mostly dry and sunny everywhere after the fog lifts, although chance of some thundery showers possible briefly in the afternoon in central parts.

Belfast: Sunny, dry, warmer, calm and pleasant air
Dublin: Calm and pleasant warm atmosphere, very bright and sunny, dry
Cork: Sunny, warmer, dry, calm atmosphere, misty fog patches possible
Galway: Sunny, warm and dry, calm and pleasant air

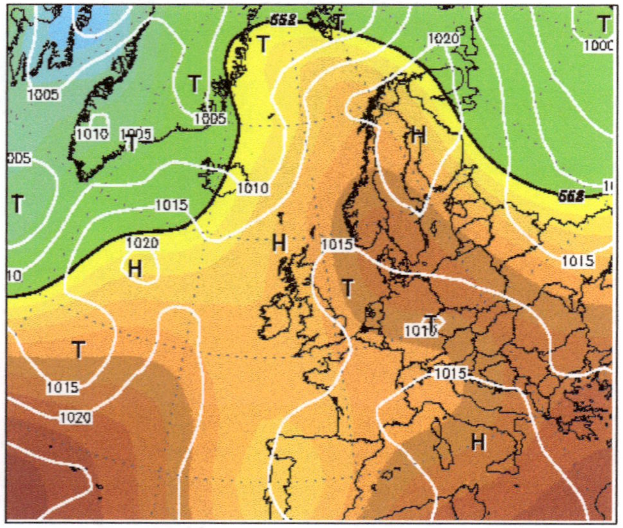

RAIN POTENTIAL

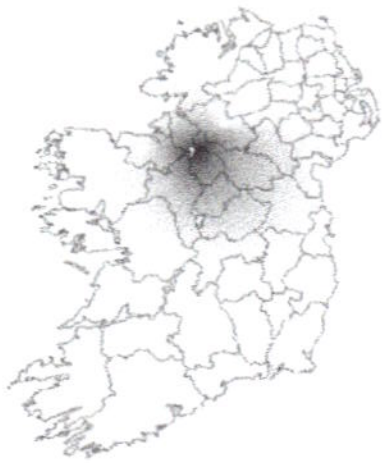

FROST/SNOW

JUNE 12 FRIDAY

A high pressure system centred to the north of Ireland extends a weak ridge over the country and a very gentle easterly air prevails. Chance of some thunderstorms and showers mostly in the west of the country, while mostly sunny and dry elsewhere, with overnight patches of fog mostly in the south.

Belfast: Pleasant and serene atmosphere, warm, dry
Dublin: Sunny, slightly milder, dry, calm and pleasant air, cool overnight
Cork: Sunny, warm day cool night, dry, calm and serene air, fog likely
Galway: Sunny, warm, evening showers, calm air, chance of thunderstorms

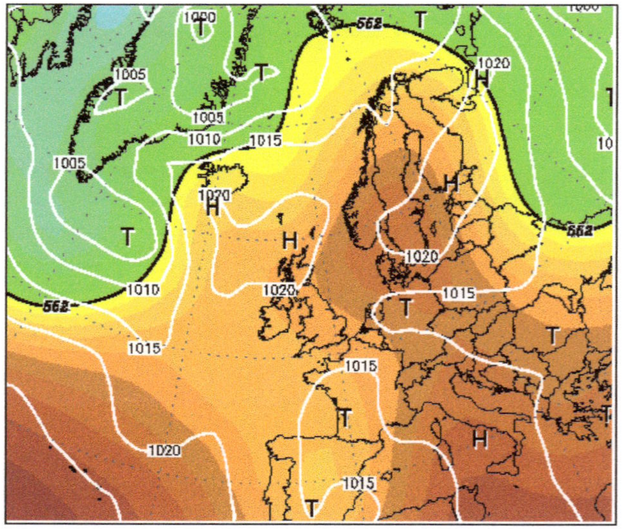

RAIN POTENTIAL

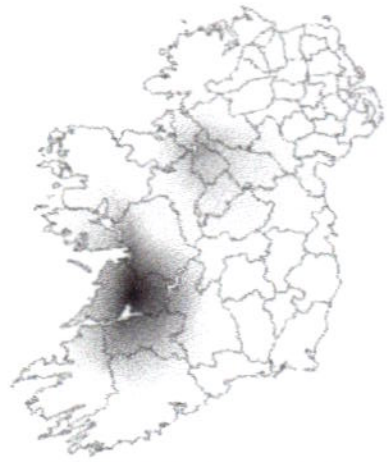

FROST/SNOW

JUNE **13** SATURDAY

The high pressure ridge drifts away northwards allowing a low pressure trough to flow over the country in a very gentle and variable flow. Widespread patches of mist and fog overnight, some slow to clear in the morning. Some brief showers and scattered cloud in western to central parts, while sunny and dry elsewhere. Chance of further thunderstorms mostly in the north, west, and midlands.

Belfast: Sunny, warm, dry, mostly calm air, chance of brief evening patches of fog
Dublin: Sunny, dry, mild day cool night, calm and pleasant air
Cork: Sunny, slightly cooler, dry, mostly calm air, fog possible
Galway: Mix of showers and sunny outbreaks, cooler, light shifting breezes, static air

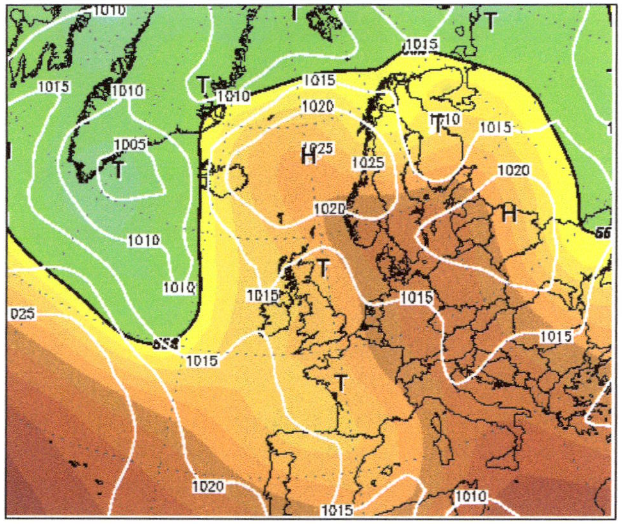

RAIN POTENTIAL

FROST/SNOW

JUNE **14** SUNDAY

Low pressure centred to the northwest of Ireland directs a light west to southwest flow over the country. Some outbreaks of showers and possible thunderstorms overnight in the northwest at first and down through central and inland districts onto the southeast later. Despite this, most of the day may be generally dry and with long sunny spells. Showers may return again in the evening.

Belfast: Very bright and sunny, warm, dry, moderate shifting breezes
Dublin: Very bright and sunny, dry, light breezes at times
Cork: Very bright and sunny, mild day cool night, dry, gentle shifting breezes
Galway: Sunny, cooler, overnight rain, chance of thunderstorms, windy spells

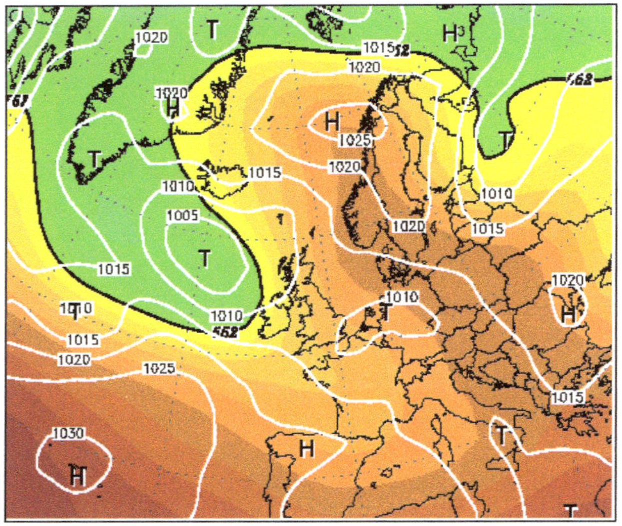

RAIN POTENTIAL

FROST/SNOW

JUNE 15 MONDAY

A depression to the northwest of Ireland intensifies with moderate south to southwest winds prevailing, becoming more blustery later in the day. An unsettled day throughout with a band of rain crossing the country, in the west at first and spreading eastwards during the day, with wet conditions reaching all districts by the end of the day. Some sunny spells develop behind the passage of the wet front. Rain may be very heavy at times in western areas.

Belfast: Changeable, cooler, mix of sunny spells and showers, freshening winds
Dublin: Cloudy, rain, cooler, fluctuating breezes
Cork: Changeable, sunny, overnight showers, mild day cool night, breezy spells
Galway: Sunny, overnight rain, unsettled winds

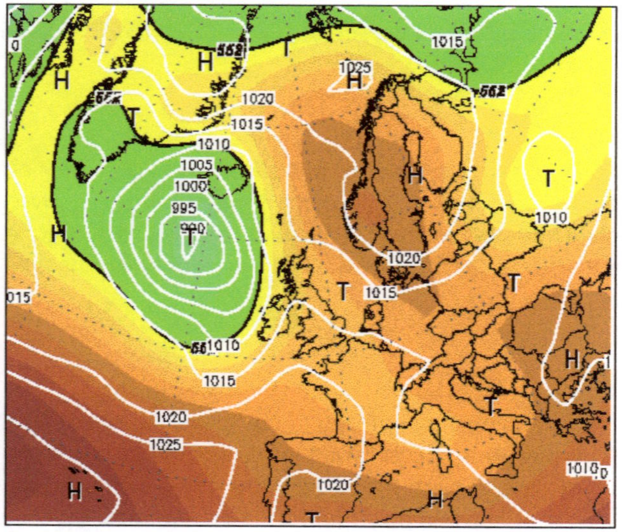

RAIN POTENTIAL

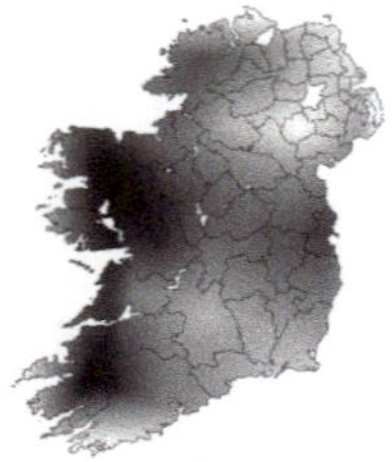

FROST/SNOW

JUNE 16 TUESDAY

A depression to the northwest of Ireland drifts further away allowing for a very light westerly air to prevail, becoming mostly calm almost everywhere except along the northern coasts. Some shower activity in most districts, except around the coastlines of the far north, west and south. Patches of fog overnight in many districts likely, as well as the chance of some patches of ground frost in the midlands. Possible thunderstorms and heavy showers for a time in the west. Mostly dry and sunny in the south.

Belfast: Partly cloudy, cooler, dry, mostly calm air, chance of overnight fog
Dublin: Colder, overcast, light overnight showers, mostly calm air
Cork: Bright and sunny, dry, mild day cold night, calm atmosphere
Galway: Changeable, partly cloudy, rain, cold overnight, moderate breezes

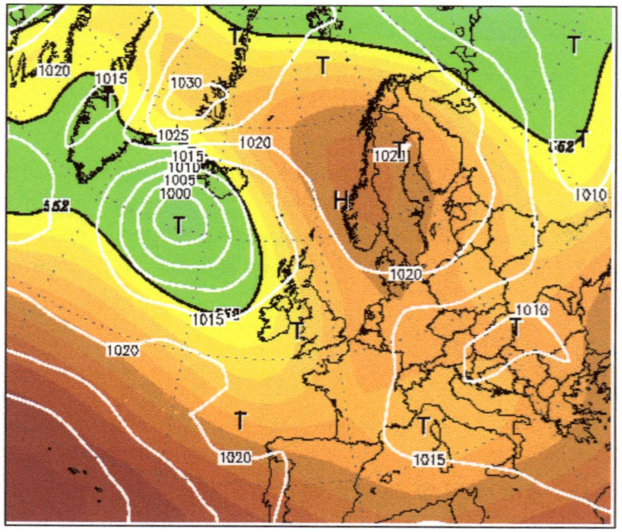

RAIN POTENTIAL

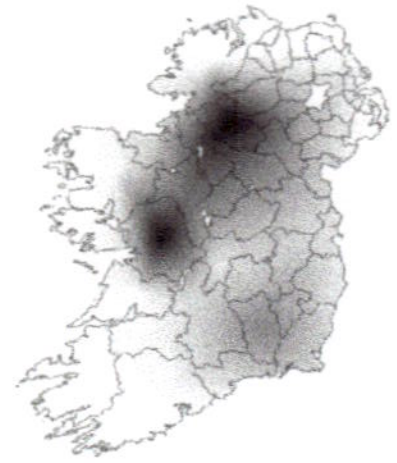

FROST/SNOW

JUNE 17 WEDNESDAY

A high pressure system to the southwest of Ireland extends a weak ridge over the country and a light to mostly calm westerly air. Mostly sunny everywhere, with patches of morning fog in many districts. Some brief overnight showers possible in the eastern half of the country, while remaining mostly dry in the west.

Belfast: Mostly sunny, isolated overnight showers, calm and pleasant air
Dublin: Sunny, dry, warmer, mostly calm air
Cork: Mostly sunny, mild day cool night, light breezes, chance of light showers
Galway: Sunny, dry, mild day cold night, light breezes

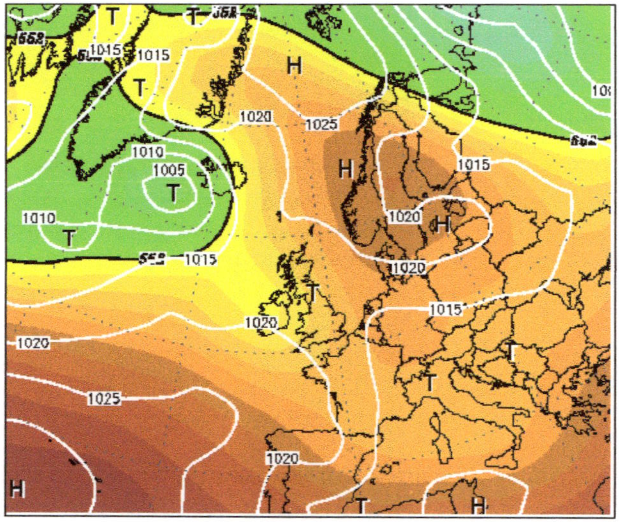

RAIN POTENTIAL

FROST/SNOW

JUNE 18 THURSDAY

A high pressure ridge maintains a very light to moderate westerly flow over the country bringing a pleasant, mostly dry and sunny day throughout. Some fog likely overnight in many districts, quick to clear. Chance of some brief outbreaks of light showers or drizzle patches in the southwest.

Belfast: Sunny, dry, mild temperatures, light breezes to mostly calm air
Dublin: Very sunny, dry, warm day cool night, calm atmosphere, chance of fog
Cork: Sunny, mostly dry, odd threat of rain overnight, mild day cool night, calm air
Galway: Partly cloudy, dry, mostly calm air

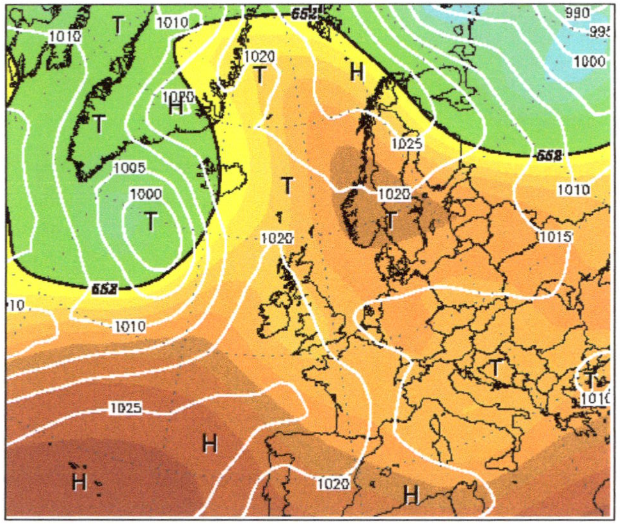

RAIN POTENTIAL

FROST/SNOW

JUNE 19 FRIDAY

High pressure continues to bring a very light humid and westerly air over the country, with chance of patchy showers mostly overnight in the southwest. Mostly dry and cloudy elsewhere. Overnight fog patches likely once again.

Belfast: Mostly calm and pleasant air, cloudy, warm, dry
Dublin: Calm and cloudy atmosphere, dry
Cork: Overcast, passing showers, slightly warmer, mostly calm air
Galway: Mostly calm and pleasant air, warm, dry, passing scattered cloud

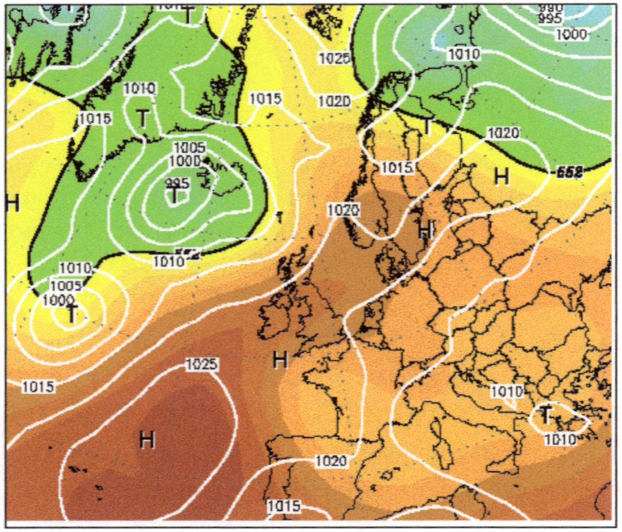

RAIN POTENTIAL

FROST/SNOW

JUNE 20 SATURDAY

High pressure builds over the country with very light to moderate southwesterly breezes prevailing, fresher at times mostly around the northern coasts. Sunny, warm, and dry everywhere. Chance of fog patches in many areas and some brief coastal drizzles overnight.

Belfast: Sunny, warm, dry, light breezes
Dublin: Sunny, warm, dry, serene atmosphere
Cork: Warm, sunny, dry, calm atmosphere, overnight fog patches possible
Galway: Overcast, slightly cooler, dry, gentle breezes, odd threat of rain

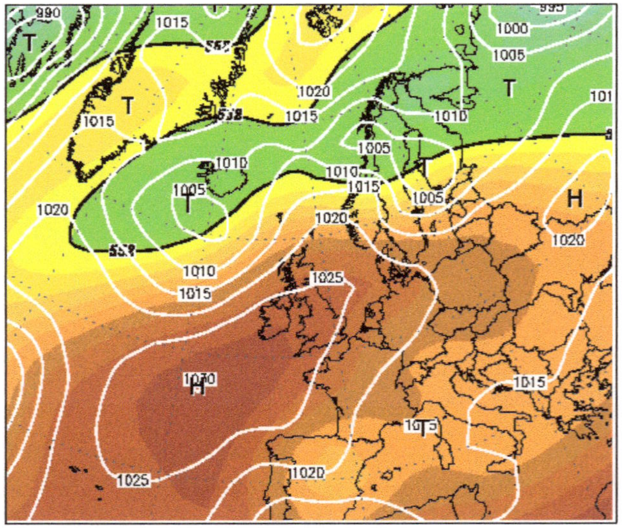

RAIN POTENTIAL

FROST/SNOW

JUNE **21** SUNDAY

Anticyclonic air brings a gentle west to northwest flow and mild temperatures throughout the country. Mostly sunny and dry everywhere, except for the chance of some brief drizzle patches overnight in inland parts. Overnight patches of fog likely in the north and east.

Belfast: Warmer, sunny, dry, gentle breezes
Dublin: Warm and sunny, dry, light breezes to mostly calm air
Cork: Bright and sunny, dry, warm, light breezes to mostly calm air
Galway: Partly cloudy, dry, gentle breezes

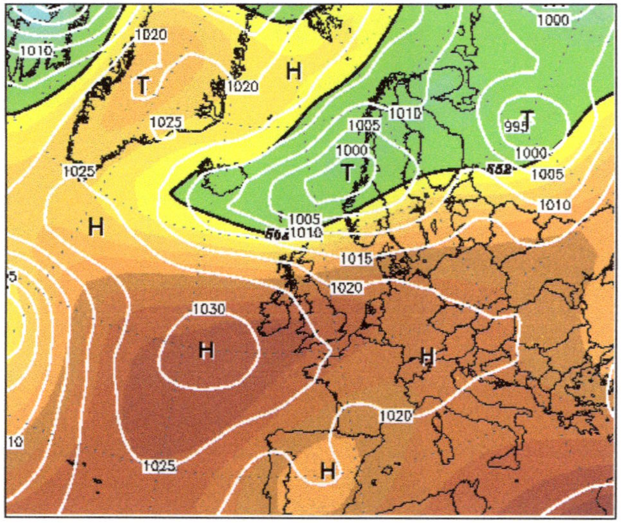

RAIN POTENTIAL

FROST/SNOW

JUNE **22** MONDAY

High pressure maintains a gentle west to northwest flow over the country. Increased cloud everywhere, with showers likely to develop later in the day, arriving in the north at first and spreading graduallly southwards. Chance of thunderstorms accompanying some of the showers in northern to central parts.

Belfast: Mostly calm air, overcast, light morning showers, cooler
Dublin: Cloudy, brief showers, slightly cooler, gentle breezes
Cork: Cloudy and warm, brief spotty showers, gentle breezes
Galway: Overcast, cooler, dry with increasing threats of rain, light breezes

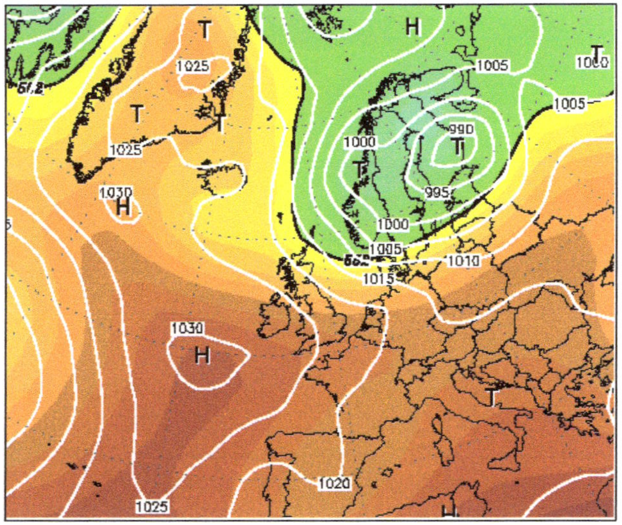

RAIN POTENTIAL

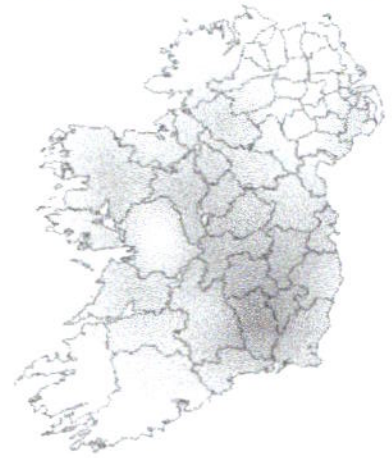

FROST/SNOW

JUNE **23** TUESDAY

A high pressure ridge sits to the west of Ireland while a low pressure system builds to the east. Winds generally light to moderate in a north to northwesterly flow. Some brief outbreaks of showers and thunderstorms possible overnight in the west, while fog prevails in the east. Mostly dry and sunny day everywhere.

Belfast: Gentle breezes, colder, dry, scattered cloudy spells
Dublin: Light breezes to mostly calm air, partly cloudy, cool and dry
Cork: Sunny, dry, cooler, moderate breezes
Galway: Partly cloudy, rain, cold overnight, moderate breezes

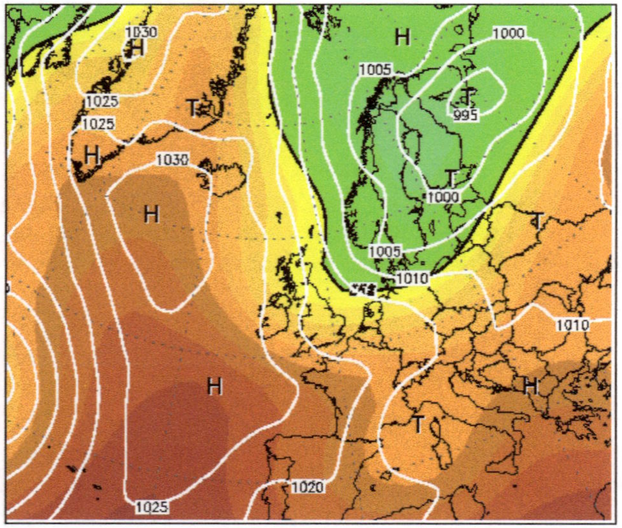

RAIN POTENTIAL

FROST/SNOW

JUNE **24** WEDNESDAY

High pressure remains almost stalled to the northwest of Ireland blocking the low pressure to the northeast from advancing nearer to Ireland, with a light northerly air prevailing. A few showers may be noted overnight mostly in the northwest and southeast. Mostly dry day everywhere with a mix of sunny and cloudy spells, cloudiest in the south.

Belfast: Partly cloudy, dry, mild temperatures, light breezes
Dublin: Partly cloudy, dry, mild day cool night, calm atmosphere
Cork: Cloudy, dry, cool, mostly calm air
Galway: Partly cloudy, dry, mild day cool night, moderate breezes

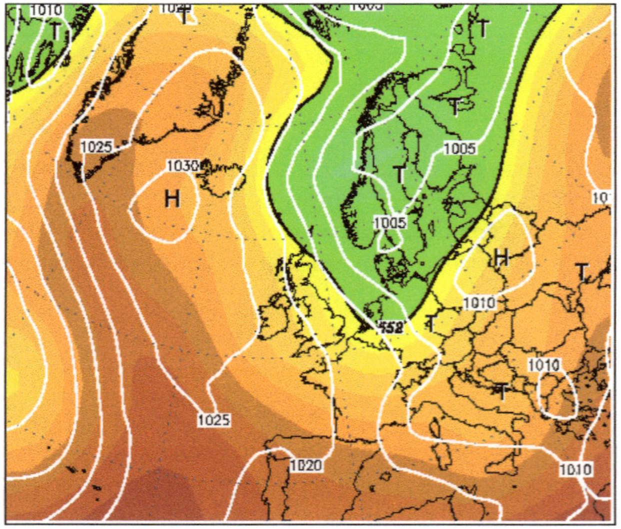

RAIN POTENTIAL

FROST/SNOW

JUNE 25 THURSDAY

The high and low pressure systems almost remain stationary with a light northerly flow persisting throughout the country. Some showers overnight in the north and west, clearing to a mostly dry day. Cloudy in the south, while sunnier elsewhere.

Belfast: Partly cloudy, dry, milder day colder night, gentle breezes
Dublin: Mostly sunny, dry, mild day cold night, gentle breezes
Cork: Mostly cloudy, dry, mild day cool night, gentle breezes
Galway: Changeable, sunny, morning rain, fresh breezes, cold overnight

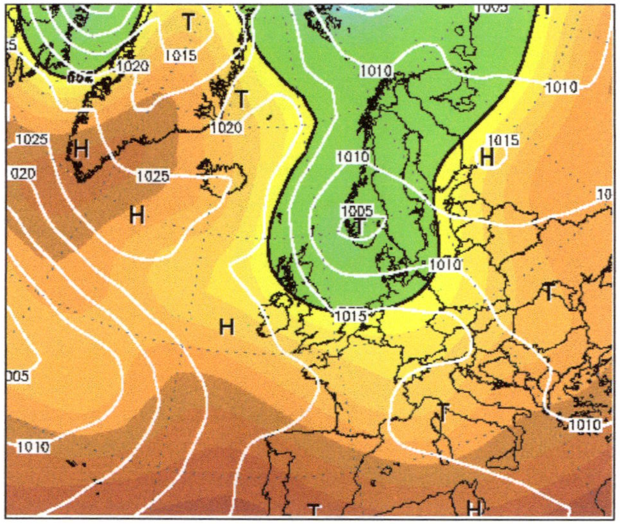

RAIN POTENTIAL

FROST/SNOW

JUNE 26 FRIDAY

Some slight movement in the weather systems around this time sees the high pressure system slip further away to the west, while low pressure nears the south and another low to the east. Despite this, a light northerly flow persists bringing a mostly dry and calm day throughout. Chance of some isolated showers and increased cloud in the northwest and in the central west. Breezes becoming fresher for a time, mostly along the eastern and southern coastal fringes.

Belfast: Increasing cloudy spells, dry, mild day cool night, light shifting breezes
Dublin: Mostly cloudy, isolated showers, mild day cold night, gentle breezes
Cork: Partly cloudy, dry, mild day cool night, shifting breezes
Galway: Dry, partly cloudy, mild day cold night, shifting breezy spells

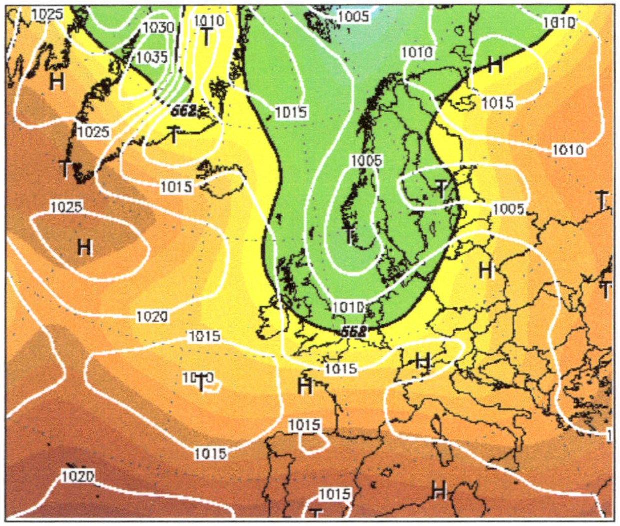

RAIN POTENTIAL

FROST/SNOW

JUNE **27** SATURDAY

The high pressure ridge leaves the west of the country while a low pressure trough moves onto Ireland with a light north to northwesterly flow prevailing. Mostly cloudy in the north and west with some longer sunnier spells in the east and south. Chance of brief showers moving into the west of the country during the day mostly sprinkling inland areas. Dry around the coasts, with possible isolated showers in adjacent coastal areas.

Belfast: Mostly cloudy and dry, cooler, light shifting breezes
Dublin: Partly cloudy and dry, mild day cold overnight, gentle breezes
Cork: Partly cloudy, dry, mild day cool night, moderate breezes
Galway: Cloudy, occasional showers, fresh breezes

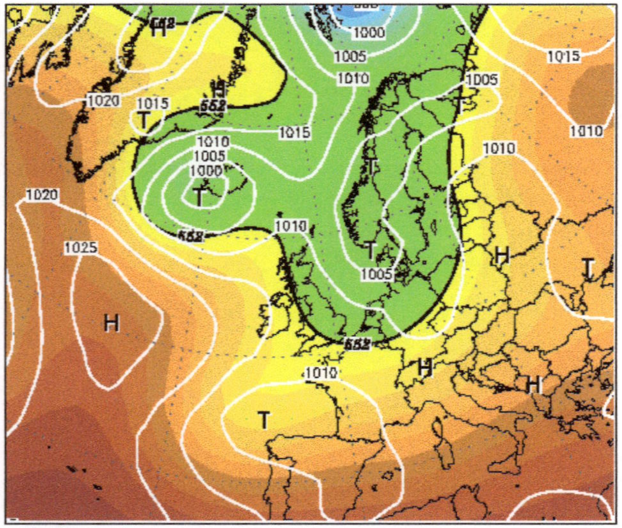

RAIN POTENTIAL

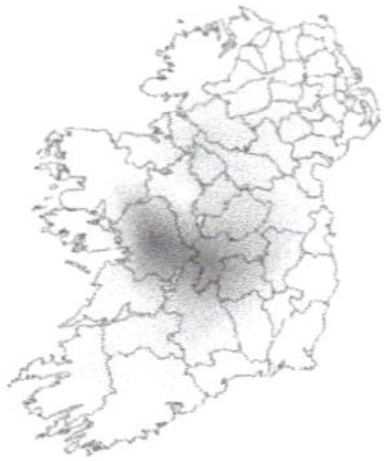

FROST/SNOW

JUNE **28** SUNDAY

A depression north of Ireland may bear down onto Ireland around this time with winds becoming fresh to strong at times, in a west to southwest flow, with the chance of gale gusts for a time in the north. Mostly cloudy everywhere with passing showery spells.

Belfast: Freshening winds, mostly cloudy, light showers
Dublin: Warmer, mix of showers and sunny spells, unsettled breezes
Cork: Cloudy, occasional showers, light breezes freshening later in the day
Galway: Mix of showers and sunny outbreaks, freshening winds

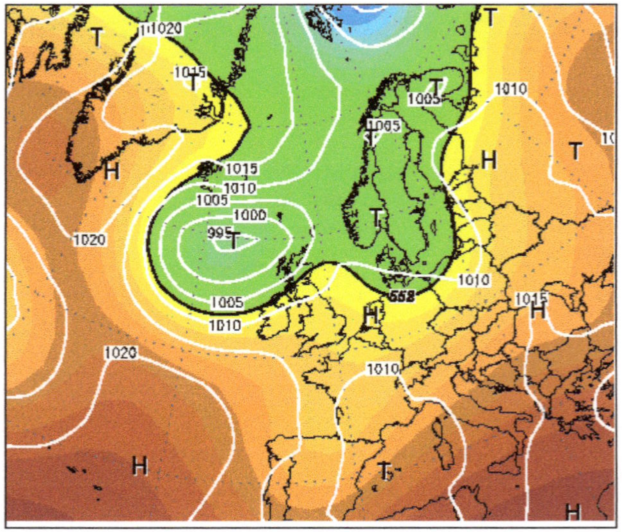

RAIN POTENTIAL

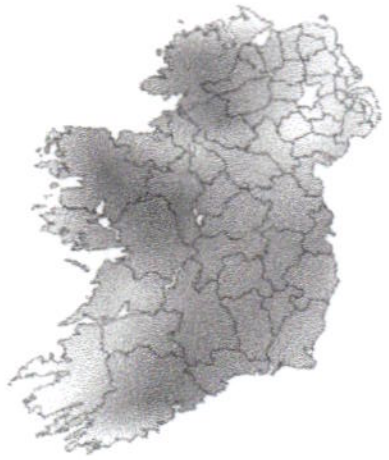

FROST/SNOW

JUNE **29** MONDAY

A depression moves onto the country bringing stormy conditions throughout, with strong and blustery southwesterly winds prevailing, up to gale force at times. Cloudy everywhere, with showers at first in all districts. Later rain may turn heavy, particularly in the southern half of the country, and possibly accompanied by thunderstorm activity.

Belfast: Unsettled and blustery winds at times, mostly cloudy, scattered showers
Dublin: Cloudy, cooler, rain, unsettled winds
Cork: Cloudy, heavy rain, colder, blustery
Galway: Mix of showers and sunny spells, strong and unsettled winds

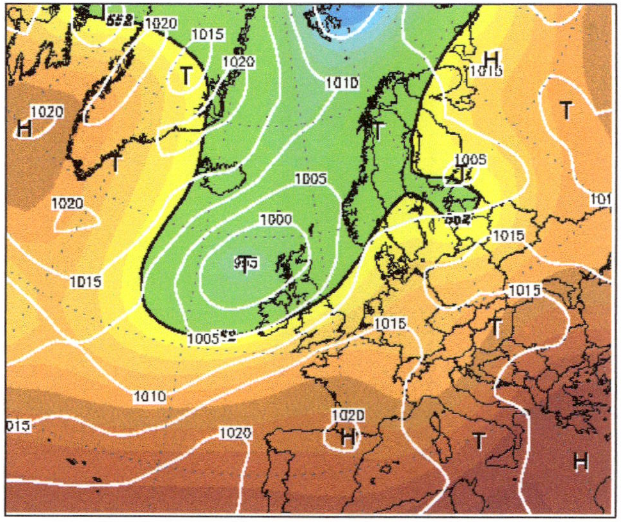

RAIN POTENTIAL

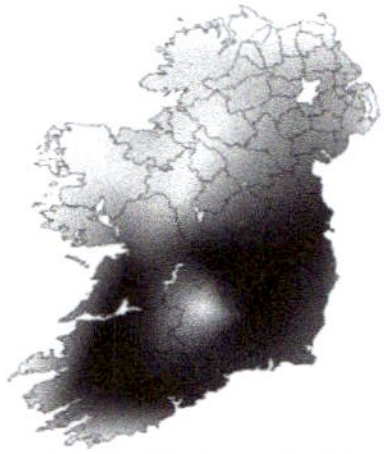

FROST/SNOW

JUNE **30** TUESDAY

The depression over Ireland intensifies producing a stormy air throughout. Winds generally in a west to northwesterly flow, unsettled and very blustery at times. A changeable day with a mix of sunny spells and bands of showers, mostly in the north and east, while clearing to mostly dry and sunny in the south and west. Chance of thunderstorms.

Belfast: Stormy air with unsettled winds, isolated showers, long sunny spells
Dublin: Sunny, occasional showers, milder, unsettled winds with occasional gusts
Cork: Sunny, warmer day cooler night, dry, unsettled and blustery winds at times
Galway: Sunny, brief showers, mild day cold night, strong and unsettled winds

RAIN POTENTIAL

FROST/SNOW

JULY

PHASES OF THE MOON

7th	Third Quarter
7th	Crossing Equator
12th	Northern Declination
13th	Perigee #4
14th	New Moon
20th	Crossing Equator
21st	First Quarter
25th	Apogee
26th	Southern Declination
29th	Full Moon

1	2	3	4	5	6	7	8	9	10	11	12	13	14	15	16	17	18	19	20	21	22	23	24	25	26	27	28	29	30	31
Wed	Thur	Fri	Sat	Sun	Mon	Tue	Wed	Thur	Fri	Sat	Sun	Mon	Tue	Wed	Thur	Fri	Sat	Sun	Mon	Tue	Wed	Thur	Fri	Sat	Sun	Mon	Tue	Wed	Thur	Fri
						3Q XhN					^	P4	N						XhS	1Q				A	V			F		

MONTHLY SUMMARY

1st-11th Unsettled conditions may continue as deep depressions are established near to the country.
1st-3rd A deep depression may bring very wet conditions as a band of heavy rain and scattered thunderstorms and some hail showers slowly crosses the country; particularly wet in the south and southeast. Chance of localised flooding.
1st Unusually sunny day possible.
3rd Winds may be particularly strong and with gale gusts at times.
4th-11th Generally cloudy and dull spell.
7th Chance of thunderstorms and heavy rain.
10th Chance of hail showers.
11th Heavy rain possible, particularly in western parts.

The next four weeks may be very wet and very dull almost everywhere; cooler over the first couple of weeks, while warmer over the second half of the outlook period. Low pressure systems may be the dominant weather patterns, particularly over the first and fourth weeks with frequent rain or showers, while higher pressure during the second and third weeks may bring more settled conditions throughout. Overall, this outlook period may be wetter than what is typical for this time of year with up to twice as much rainfall in the south and southeast, while nearer to average in the north and northwest. Overall, there may be up to 12-16 days with rain activity, which is slightly more than what is normal for this time of year. Sunshine hours may be slightly higher than average in the north and northwest, while below average elsewhere, and particularly duller than normal for this time of year in the southwest. Temperatures may be average to slightly warmer than normal in most parts of the country, except in some parts of the east and south which may be cooler than average. Overnight minimum temperatures may be unusually cooler than normal for this time of year in a brief spell over the third week ahead. During this outlook period there may be thunderstorms on at least 5 days, 1 day with hail showers, 24 days when fog is observed, 10 days when winds may reach gale force for a time, but no days with frost or snow.

12th-13th Intense low pressure prevails. Chance of gale gusts.
12th-17th Chance of spells of heavy rain.
12th-21st A spell of changeable days with cooler temperatures and a mix of sunny spells and of rain or showers likely as areas of low pressure linger to the southwest of Ireland. Winds generally south to southwesterly at first, later backing northwesterly and strengthening.
13th Chance of widespread thunderstorm activity.
13th-16th Very sunny spell likely.
14th-16th Chance of a brief spell of heavy rain.
15th Chance of some isolated patches of light frost overnight mostly in inland areas.
21st Chance of hail showers.
22nd-7th August Spell of high pressure as an anticyclonic influence becomes established to the southwest of Ireland, later moving over the country in a northeastwardly flow. Some spells of light rain now and then.
24th-26th Chance of fog and isolated patches of light frost overnight.
27th-28th A front may cross the country bringing a brief spell of heavy rain, mostly over northern areas.
29th Chance of widespread shower activity.
30th More frequent sunny spells may develop around this time, as well as warmer temperatures.

IRELAND RAINFALL ESTIMATES
July 2026

d=mainly dry, n= nonrecordable, l=light shrs, s=significant shrs, r=rain, h=hvy falls

		ULSTER PROVINCE					CONNAUGHT PROVINCE						LEINSTER PROVINCE				MUNSTER PROVINCE						
		Alderg	Hillsbo	Armag	Loughl	Malin H	Sligo	Ardtara	Belmul	Drums	Galway	Roscom	Markre	Edende	Kilkenn	Dublin	Shannc	Newpo	Tralee	Killarne	Cork	All	Moon
1st	Wed	l	s	d	d	d	d	d	d	l	r	l	l	s	h	d	d	l	d	r	d	60	
2nd	Thur	r	h	n	h	r	s	h	r	h	h	r	s	h	h	h	h	h	h	h	h	400	
3rd	Fri	s	n	r	s	r	h	s	h	l	d	s	r	l	d	l	l	s	l	l	d	130	
4th	Sat	n	l	d	l	n	d	l	n	l	d	l	s	l	d	l	l	l	l	l	d	25	
5th	Sun	l	l	n	r	s	l	r	s	s	s	s	r	l	l	l	l	h	r	r	r	129	
6th	Mon	l	l	l	s	l	s	s	s	l	r	s	s	s	l	n	l	s	l	r	l	91	
7th	Tues	s	l	s	l	l	s	l	l	s	r	l	l	s	h	s	h	s	h	h	h	179	3Q XhN
8th	Wed	l	s	l	l	l	l	l	l	l	h	l	l	l	l	n	l	s	s	d	l	64	
9th	Thur	l	l	l	l	l	l	l	d	l	l	l	l	l	l	l	l	l	d	l	d	24	
10th	Fri	l	l	n	l	l	n	l	n	l	l	l	s	l	d	l	l	l	n	d	s	26	
11th	Sat	l	l	l	s	l	r	s	h	s	r	s	d	l	s	d	d	r	r	r	d	133	
12th	Sun	l	l	l	s	l	s	s	r	l	l	l	s	s	l	n	s	s	r	h	r	118	^
13th	Mon	l	s	n	s	h	l	s	l	n	d	s	l	h	h	l	s	d	s	l	r	134	P4
14th	Tues	s	r	s	d	r	l	d	d	s	d	r	d	r	s	r	d	d	l	s	l	96	N
15th	Wed	r	s	h	d	n	r	d	l	l	r	s	l	h	r	l	l	l	s	r	h	139	
16th	Thur	s	s	l	l	s	l	l	n	r	l	s	l	s	r	h	r	l	s	s	s	130	
17th	Fri	s	s	r	l	d	s	l	l	l	l	l	s	s	r	s	s	d	s	l	h	112	
18th	Sat	l	n	s	l	l	l	l	d	l	d	n	s	d	d	l	d	d	n	l	n	28	
19th	Sun	n	n	d	d	l	n	d	l	l	l	d	s	d	l	d	l	l	s	l	s	32	
20th	Mon	s	s	s	h	l	l	h	l	s	h	s	h	r	l	s	r	s	s	s	l	186	XhS
21st	Tues	s	d	r	l	s	s	l	l	d	d	d	d	s	l	r	s	l	l	r	l	65	1Q
22nd	Wed	d	d	n	n	l	n	n	d	n	d	l	l	d	n	d	d	l	d	d	d	7	
23rd	Thur	d	d	d	d	n	n	d	l	d	d	d	d	d	d	l	l	d	d	d	l	5	
24th	Fri	d	d	d	d	d	d	d	n	d	d	d	d	d	d	d	n	l	l	l	d	7	
25th	Sat	d	d	d	l	l	n	l	l	n	l	d	n	d	d	d	d	l	l	d	d	10	A
26th	Sun	d	d	d	l	l	n	l	l	d	d	d	l	d	d	d	d	n	l	d	d	8	V
27th	Mon	l	l	n	s	r	l	s	l	l	d	l	s	n	d	n	d	n	d	d	d	44	
28th	Tues	l	d	l	l	h	l	l	n	l	l	l	s	l	d	l	l	l	d	d	d	46	
29th	Wed	l	l	d	s	s	l	s	l	l	d	l	s	l	l	d	l	l	s	d	l	45	F
30th	Thur	l	d	s	l	l	l	l	l	d	l	n	l	d	d	d	l	d	l	s	d	21	
31st	Fri	l	d	l	d	n	l	d	d	n	l	d	d	d	d	d	d	d	d	d	d	4	
Estimate:		90	84	95	118	146	104	118	114	107	164	97	121	159	149	118	105	126	142	170	174	2497	
Average:		71	77	63	86	81	93	95	79	73	87	86	93	71	59	56	66	101	103	75	79	1593	
Trend:		wtr	wtr	wtr	wtr	wtr	wtr	wtr	wtr	wtr	wtr	wtr	wtr	wtr	wtr	wtr	wtr	wtr	wtr	wtr	wtr	wtr	

IRELAND SUNSHINE ESTIMATES
July 2026

F=fine (8-12 hours of sunshine), pc= partly cloudy (4-7 hours), c=cloudy (1-3 hours), o=overcast (0 hours)

		ULSTER PROVINCE					CONNAUGHT PROVINCE						LEINSTER PROVINCE				MUNSTER PROVINCE						
		Alderg	Hillsbo	Armag	Loughl	Malin I	Sligo	Ardtara	Belmul	Drums	Galway	Roscoi	Markre	Edende	Kilkenı	Dublin	Shannc	Newpo	Tralee	Killarne	Cork	All	Moon
1st	Wed	F	F	F	F	F	F	F	F	F	F	F	F	F	F	F	F	F	F	F	pc	255	
2nd	Thur	o	o	o	o	o	o	o	o	o	c	o	o	o	o	o	c	c	o	o	o	8	
3rd	Fri	pc	pc	pc	pc	pc	c	pc	c	pc	pc	c	pc	pc	F	pc	F	F	F	F	F	119	
4th	Sat	c	c	c	c	c	pc	c	pc	c	c	pc	c	c	c	c	o	c	c	c	pc	67	
5th	Sun	o	o	o	o	o	o	o	o	o	o	o	o	o	o	o	o	o	c	c	o	11	
6th	Mon	F	F	F	F	F	F	F	F	F	F	F	F	F	F	F	F	F	pc	pc	F	182	
7th	Tues	c	c	c	c	c	c	c	F	c	pc	pc	c	c	c	c	c	c	o	o	c	58	3Q XhN
8th	Wed	o	o	o	o	o	o	o	o	o	o	o	o	o	c	o	o	o	o	o	c	11	
9th	Thur	c	c	c	c	c	c	c	o	c	c	c	c	c	c	c	c	c	o	o	c	40	
10th	Fri	F	F	F	F	F	F	F	F	F	F	F	F	F	pc	F	pc	pc	pc	pc	pc	170	
11th	Sat	c	c	c	c	c	0.6	c	c	c	c	c	c	c	c	pc	c	c	o	o	c	40	
12th	Sun	c	c	c	c	c	pc	c	pc	c	pc	pc	c	c	c	c	pc	pc	pc	pc	pc	92	^
13th	Mon	pc	pc	pc	pc	pc	pc	pc	F	pc	F	F	pc	pc	c	pc	pc	pc	F	F	c	130	P4
14th	Tues	c	c	c	c	c	pc	c	F	pc	F	F	c	pc	pc	o	F	F	F	F	F	132	N
15th	Wed	F	F	F	F	F	F	F	pc	pc	pc	pc	F	pc	pc	F	pc	pc	o	o	o	125	
16th	Thur	pc	pc	pc	pc	pc	pc	pc	o	pc	c	o	pc	pc	pc	pc	c	pc	o	o	pc	77	
17th	Fri	pc	pc	pc	pc	pc	o	pc	c	pc	c	c	pc	pc	pc	F	c	c	c	c	c	77	
18th	Sat	c	c	c	c	c	c	c	c	pc	c	c	c	pc	pc	pc	pc	pc	o	o	pc	74	
19th	Sun	pc	pc	pc	pc	pc	c	pc	c	pc	pc	pc	pc	pc	pc	pc	F	pc	c	c	pc	109	
20th	Mon	o	o	o	o	o	o	o	o	c	o	o	o	c	c	o	c	c	o	o	c	15	XhS
21st	Tues	c	c	c	c	c	c	c	pc	c	pc	pc	c	c	pc	c	c	c	c	c	F	61	1Q
22nd	Wed	c	c	c	c	c	c	c	c	c	c	c	c	c	c	c	c	c	pc	pc	c	54	
23rd	Thur	c	c	c	c	c	c	c	pc	c	pc	pc	c	c	c	c	pc	pc	c	c	c	62	
24th	Fri	pc	pc	pc	pc	pc	o	pc	o	pc	o	o	pc	pc	pc	F	c	pc	o	o	c	86	
25th	Sat	c	c	c	c	c	o	c	o	c	o	o	c	c	c	c	o	c	o	o	c	23	A
26th	Sun	pc	pc	pc	pc	pc	pc	pc	pc	pc	pc	pc	pc	pc	pc	F	pc	pc	o	o	c	122	V
27th	Mon	o	o	o	o	o	o	o	o	c	o	o	o	c	c	o	c	c	c	c	c	18	
28th	Tues	o	o	o	o	o	o	o	c	c	c	c	o	c	c	o	o	c	o	o	pc	23	
29th	Wed	c	c	c	c	c	c	c	o	c	o	o	c	c	pc	c	o	c	o	o	pc	34	F
30th	Thur	F	F	F	F	F	F	F	F	F	F	F	F	F	F	F	pc	F	pc	pc	F	186	
31st	Fri	F	F	F	F	F	F	F	F	F	F	F	F	F	F	F	F	F	pc	pc	F	200	
	Estimate:	131	131	131	131	131	115	131	152	138	141	147	131	138	145	146	129	137	106	106	142	2661	
	Average:	152	152	135	143	133	121	122	133	129	144	157	142	155	144	167	142	121	142	140	167	2841	
	Trend:	less	less	av	less	av	av	av	more	av	av	less	less	less	av	less	less	more	less	less	less	less	

JULY 1 WEDNESDAY

A depression near the northwest of Scotland and another depression to the southwest of Ireland forms a low pressure trough over the country and a moderate southeasterly flow up throughout. Mostly dry and sunny day throughout with the arrival of some showers in the northwest and outbreaks of rain at times in central parts to the southeast of Leinster towards the evening.

Belfast: Bright and sunny day, breezy spells, occasional showers, mild temperatures
Dublin: Very bright and sunny, mild day cool night, dry, breezy at times
Cork: Fresh breezes at times, partly cloudy, cooler, dry, increasing threats of rain
Galway: Partly cloudy, mild temperatures, light breezes, storm threatening

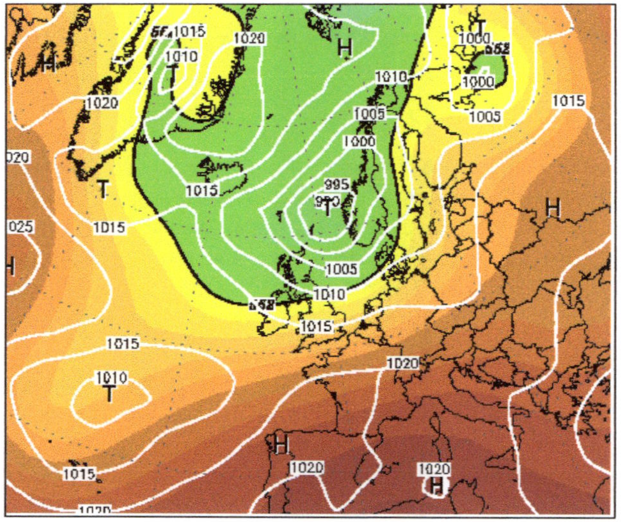

RAIN POTENTIAL

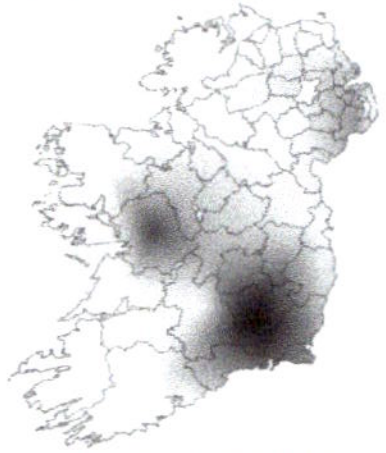

FROST/SNOW

JULY 2 THURSDAY

A trough between two depressions, one to the northeast of Ireland and another to the southwest, directs a light southeasterly flow over the country at first before turning stormy with a northwesterly change as the depression to the northeast draws nearer. Mostly dry with some sunny spells at first before widespread stormy conditions develop bringing heavy rain and strong and gusty winds throughout, with gale gusts around the coasts.

Belfast: Heavy rain, mild day cool night, dull, gloomy and stormy air later in the day
Dublin: Partly cloudy, light winds strengthening later in the day, brief showers
Cork: Overcast, winds strengthening, stormy later in the day with heavy rain
Galway: Changeable, very sunny, mild day cool night, breezy, rain developing later

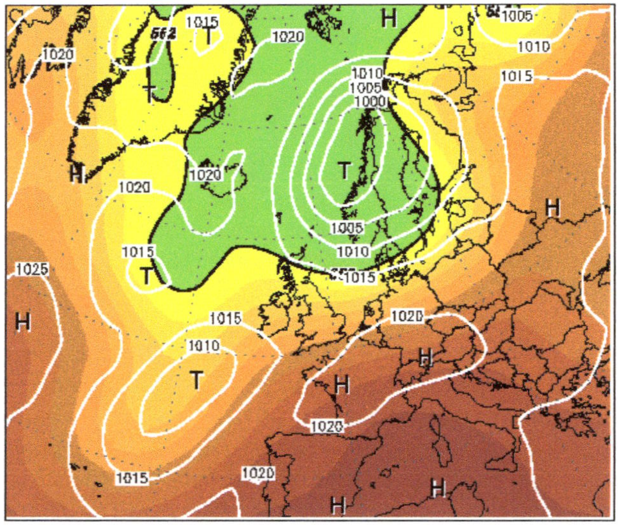

RAIN POTENTIAL

FROST/SNOW

JULY 3 FRIDAY

Stormy conditions prevail throughout Ireland overnight as a depression crosses the country, with winds up to gale force at times, in variable directions as the storm passes. Widespread rain activity throughout, easing later in the day particularly in the southern half of the country, where increasing sunny spells may develop, while cloud and occasional drizzle patches may persist over the northern half of Ireland for most of the day.

Belfast: Stormy air easing, partly cloudy, cooler, isolated showers and drizzle patches
Dublin: Overcast, heavy rain, stormy air
Cork: Clearing, strong and gusty winds, mostly sunny and dry
Galway: Heavy rain, cloudy, cooler, strong winds easing briefly

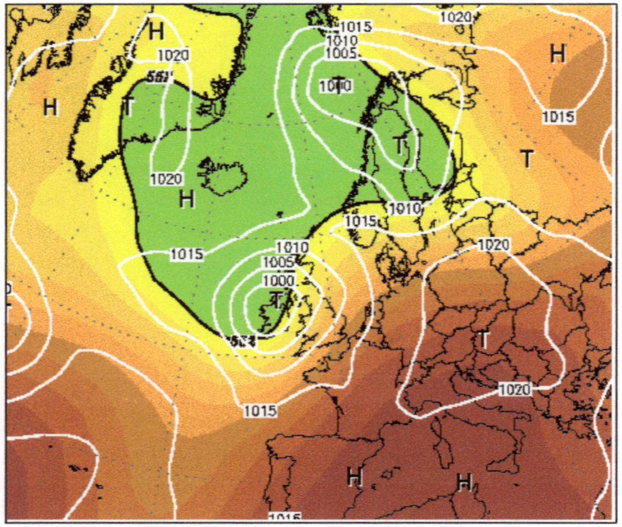

RAIN POTENTIAL

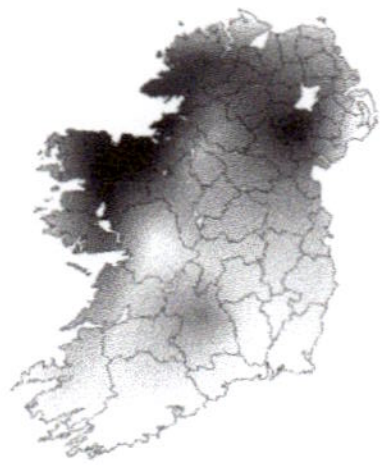

FROST/SNOW

JULY 4 SATURDAY

The stormy conditions of the past couple of days ease throughout as a low pressure trough between two depressions lingers, with one depression having drifted away to the northeast, and the other remains lingering to the southwest of the country, Mostly cloudy skies may persist throughout the country, with the odd spotty showers likely, mostly in the eastern half of Ireland, while mostly dry in the west and south. Winds generally light and in a southeasterly flow, mostly directed from a high pressure ridge south of Ireland.

Belfast: Mostly cloudy, spotty showers, moderate breezes
Dublin: Cloudy, showers, winds easing
Cork: Partly cloudy, dry, passing threats of rain, unsettled breezes easing
Galway: Mostly cloudy, dry, warmer day cooler night, light breezes

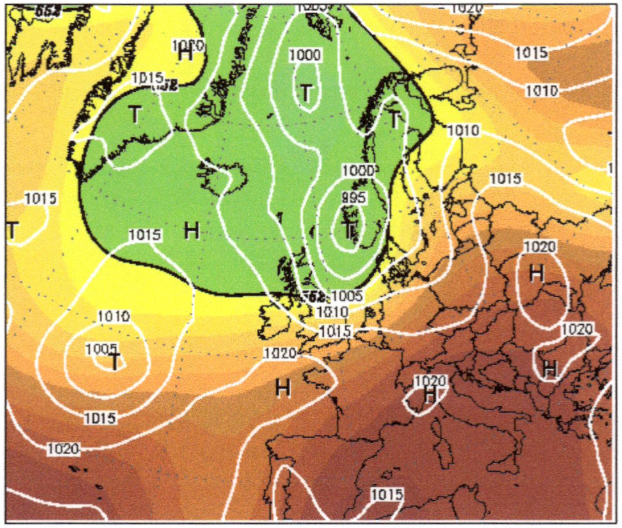

RAIN POTENTIAL

FROST/SNOW

JULY **5** SUNDAY

The depression that has lingered to the southwest of Ireland strengthens as it moves nearer to the western coasts of the country, while the weak high pressure ridge drifts away to the east of the country, with a light southerly flow at first, later becoming more strong and gusty. Dull and gloomy almost everywhere, with outbreaks of showers in most districts, while storm conditions build and threaten later in the day.

Belfast: Overcast, brief showers, light breezes strengthening later in the day
Dublin: Overcast, spotty showers, light breezes strengthening later in the day
Cork: Overcast, colder, rain, light winds strengthening later in the day
Galway: Overcast, cooler, scattered showers, storm threatening

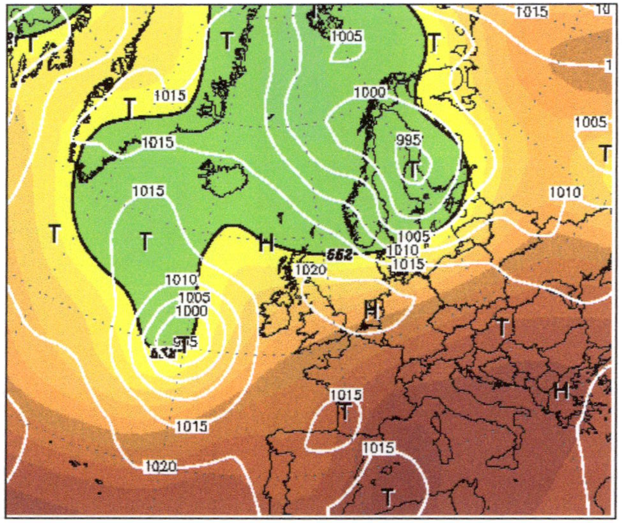

RAIN POTENTIAL

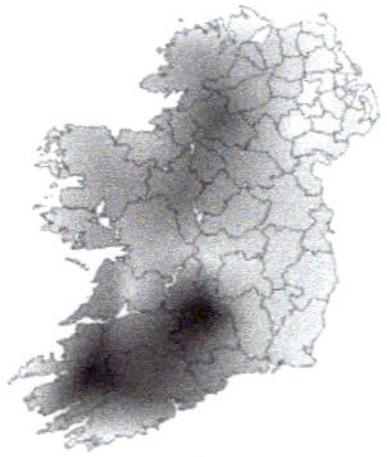

FROST/SNOW

JULY **6** MONDAY

A strong depression over Ireland directs unsettled and blustery south to southwest winds throughout, with winds up to gale force possible at times. An unstable day likely with a mix of sunny spells interspersed by squally showers, more frequent in the west while isolated to the east.

Belfast: Sunny, light showers, cooler, strong and unsettled winds
Dublin: Blustery winds, sunny, warmer, isolated showers
Cork: Stormy air, showers, warmer, long sunny spells
Galway: Stormy air, overnight rain, long sunny spells, chance of gales

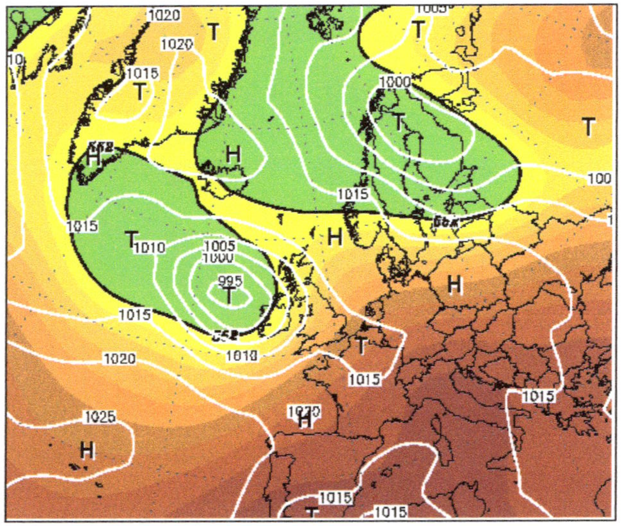

RAIN POTENTIAL

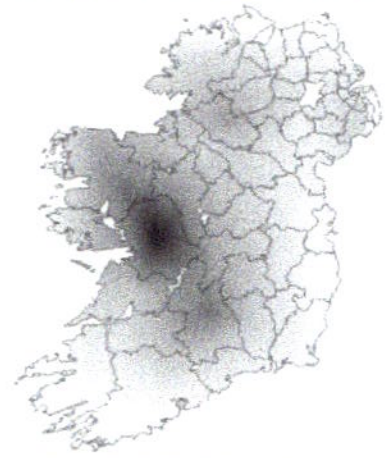

FROST/SNOW

JULY **7** TUESDAY

A high pressure ridge to the south of Ireland slowly pushes the stormy depression northwards with a fresh to moderate mostly westerly flow prevailing. A trough embedded with thunderstorms on the edge of the high pressure ridge, may bring further widespread rain throughout Ireland, with very heavy falls likely in the southern half of the country, including in the midlands.

Belfast: Cloudy, light showers, slightly warmer, blustery at times
Dublin: Stormy and colder air, cloudy, squally showers
Cork: Stormy, heavy rain, mostly cloudy and cooler, chance of thunderstorms
Galway: Partly cloudy, rain, strong winds, gales possible

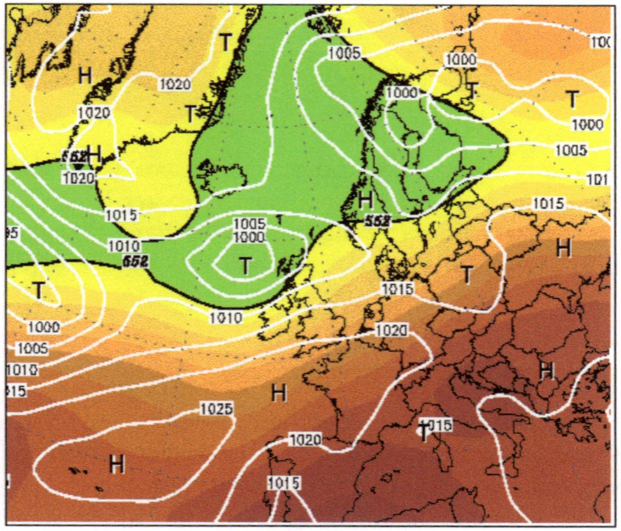

RAIN POTENTIAL

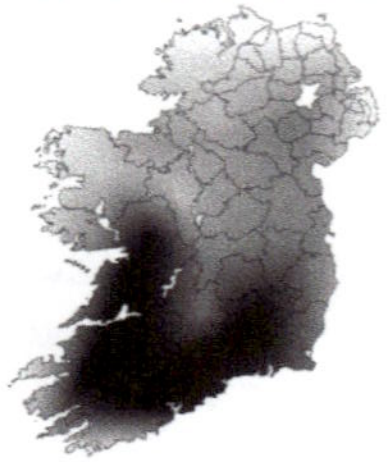

FROST/SNOW

JULY **8** WEDNESDAY

As the depression to the north of Ireland disipates, another strengthens to the west of the country, while the high pressure ridge still may linger to the south of Ireland. Winds generally light to moderate throughout in a west to southwest flow, but sometimes more gusty around the northern coasts. Patches of rain or drizzle likely almost everywhere, with only a few brief sunny spells.

Belfast: Unsettled breezes, overcast, scattered showers
Dublin: Overcast, warmer, isolated showers, unsettled breezes
Cork: Cloudy, windy, warmer, passing showers
Galway: Overcast, heavy rain, mild day cool night, blustery at times

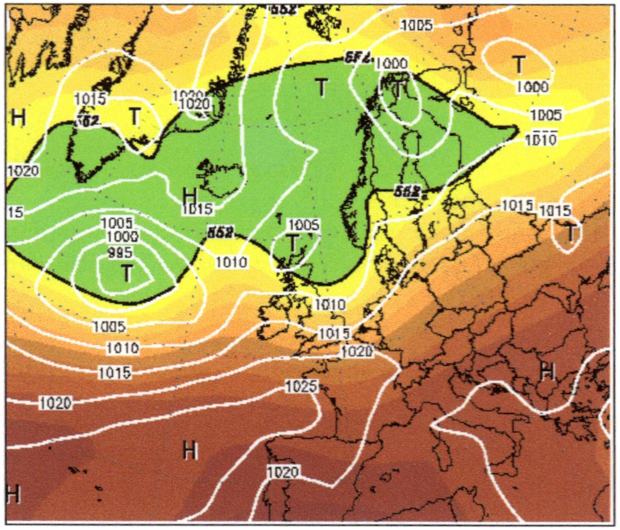

RAIN POTENTIAL

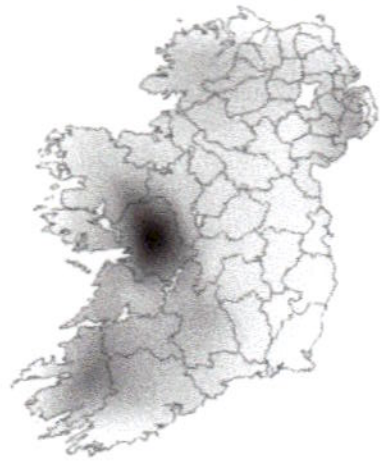

FROST/SNOW

JULY **9** THURSDAY

A low presure trough associated with a weakening depression centred to the northwest of Ireland maintains a light to moderate west to southwest flow, fresher at times along the northern and western coasts. A mostly cloudy day throughout, with patches of drizzle likely, particularly in the north to central regions, while mostly dry around the western, southern, and eastern coasts.

Belfast: Cloudy, light showers, fresh to strong breezes
Dublin: Fresh and breezy air, cloudy, light showers
Cork: Mostly cloudy, dry, mild temperatures, windy spells, passing threats of rain
Galway: Cloudy, brief showers, mild temperatures, strong winds

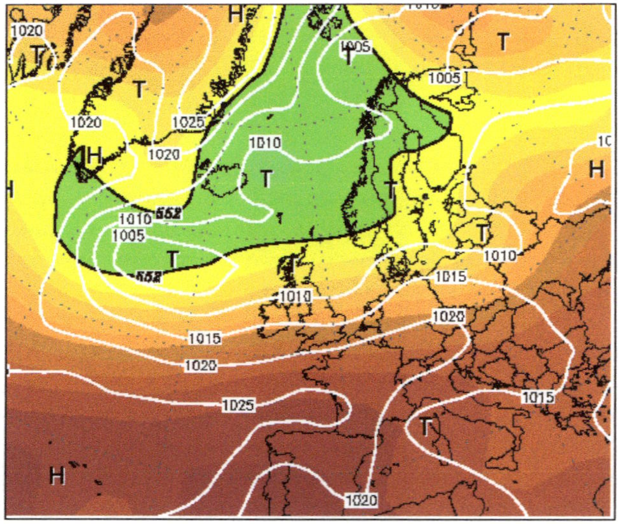

RAIN POTENTIAL

FROST/SNOW

JULY **10** FRIDAY

A weak low pressure system still influences the country as it continues to prevail under a light to moderate west to southwest flow, more fresher around the coasts, particularly in the north and west. A mixed day of some sunny spells and spotty showers or drizzle patches likely throughout, sunniest in the west and north.

Belfast: Sunny, isolated evening showers, mild temperatures, fresh breezes
Dublin: Windy spells, sunny, light evening showers
Cork: Partly cloudy, occasional showers, mild temperatures, windy at times
Galway: Sunny, brief evening showers, strong winds

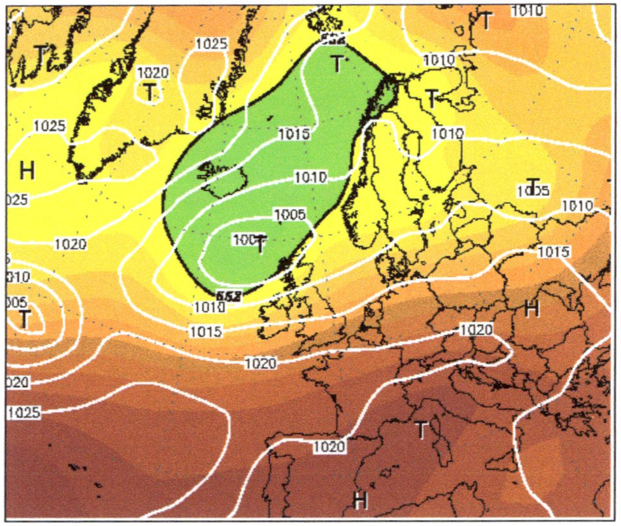

RAIN POTENTIAL

FROST/SNOW

JULY **11** SATURDAY

A depression to the west of Ireland strengthens as it nears the western, while the low pressure to the north lingers and a high pressure ridge may also remain almost stalled to the south. Winds generally light in a southwesterly flow for much of the day, but later strengthening and becoming very gusty as the depression to the west advances onto Ireland. Mostly cloudy throughout, with rain developing almost everywhere, with falls particularly heavy at times in the west.

Belfast: Moderate breezes, cloudy, light showers
Dublin: Partly cloudy, dry, winds easing to moderate
Cork: Cloudy, dry, mild day cool night, moderate breezes
Galway: Cloudy, rain, strong winds at times

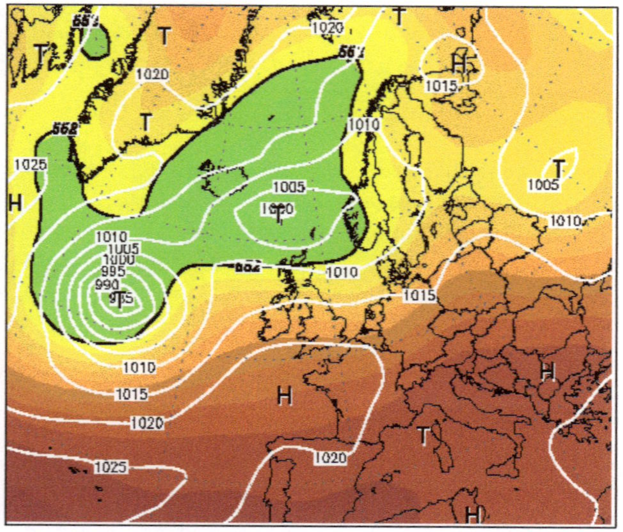

RAIN POTENTIAL

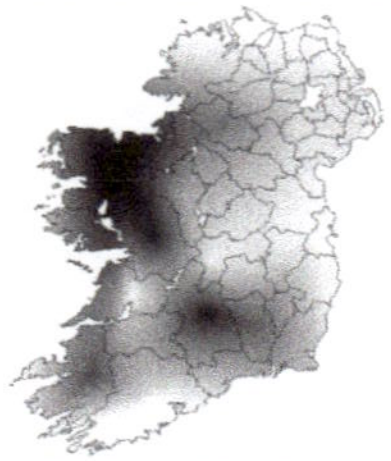

FROST/SNOW

JULY **12** SUNDAY

A large depression to the west of Ireland directs a very strong and gusty southwest flow over the country, and including up to gale force later with a southeasterly change. Stormy conditions may prevail throughout, with widespread showers prevailing, some possibly accompanying thunderstorm activity. Despite this, some brief sunny outbreaks may occur, more frequently in the east and north.

Belfast: Strong and unsettled winds, partly cloudy, light showers, slightly warmer
Dublin: Cloudy, isolated showers, mild temperatures, strong winds with gusts
Cork: Strong winds, rain, passing sunny outbreaks, possible thunderstorms
Galway: Partly cloudy, light showers, stormy and static air, strong winds and gusts

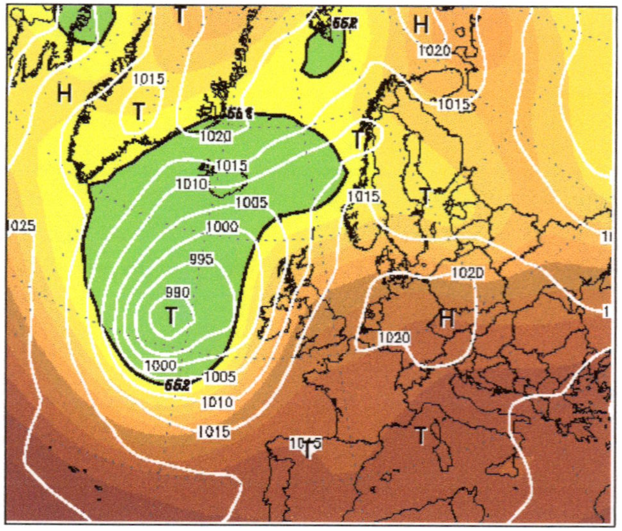

RAIN POTENTIAL

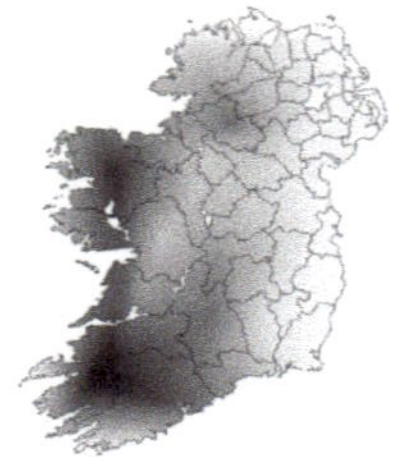

FROST/SNOW

JULY **13** MONDAY

The depression centred to the northwest of Ireland intensifies strengthening the southwest winds up to gale force for a time, before easing later in the day when turning southerly. A changeable day likely as storm conditions prevail throughout, with a mix of some sunny spells between bands of rain, more heavy in the southeast and northwest, and possibly accompanied by thunderstorm activity.

Belfast: Stormy air, mix of showers and sunny spells, slightly cooler, chance of gales
Dublin: Changeable, mix of showers and sunny spells, strong winds, possible gales
Cork: Cloudy, mild temperatures, rain, strong winds, chance of thunderstorms
Galway: Sunny, dry, strong blustery winds easing later in the day, cold overnight

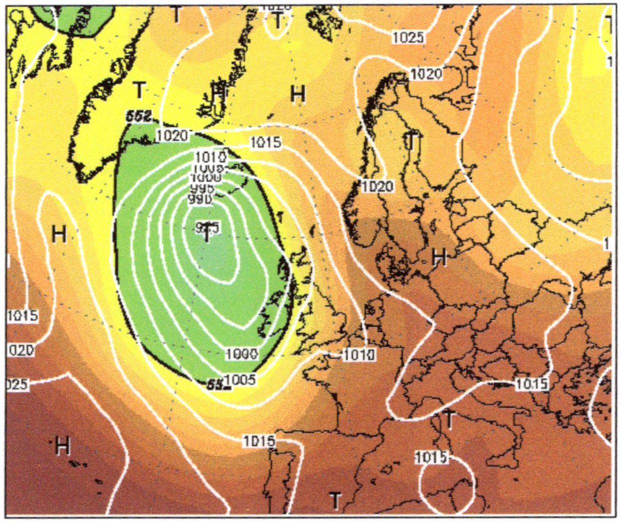

RAIN POTENTIAL

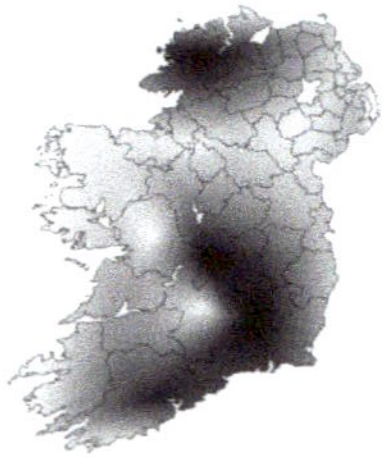

FROST/SNOW

JULY **14** TUESDAY

The depression centred to the northwest of Ireland weakens allowing for winds to become light in a variable to westerly flow, mostly calm later in the day. Mostly cloudy with outbreaks of rain still may persist in the north and east, while sunnier and a drier day develops in the south and west. Chance of thunderstorms still in the north, while possible fog patches may develop overnight in central parts.

Belfast: Cloudy, rain, winds easing to light, chance of thunderstorms
Dublin: Overcast, rain, colder, winds easing, chance of misty patches or drizzle
Cork: Light winds, brief overnight showers, mostly sunny day
Galway: Sunny, dry, mild day cold night, gentle breezes to mostly calm air

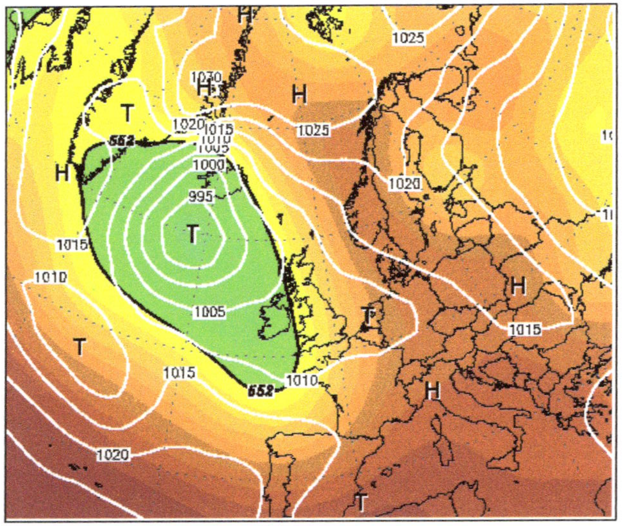

RAIN POTENTIAL

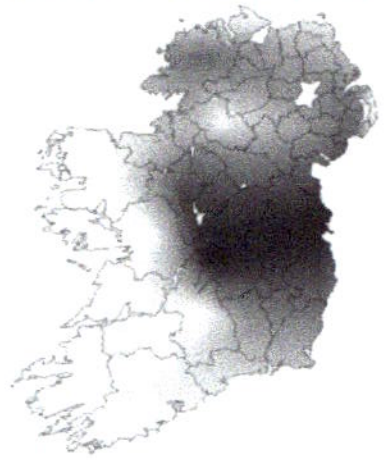

FROST/SNOW

JULY 15 WEDNESDAY

A couple of low pressure centres linger to the west of Ireland, while a low pressure trough prevails over the country. Winds generally light to moderate throughout, mostly in an easterly flow. Chance of fog patches for a time overnight in sheltered districts. Mostly sunny and dry in the northwest, while cloudier with outbreaks of rain elsewhere. It may be particularly dull and gloomy with heavy rain for a time in the southeast.

Belfast: Mostly calm air, cooler, sunny, overnight showers, chance of fog patches
Dublin: Light breezes to mostly calm, sunny, warmer day cooler night, brief showers
Cork: Overcast, heavy rain, cooler, gentle breezes
Galway: Partly cloudy, overnight rain, mild day cold night, mostly calm air

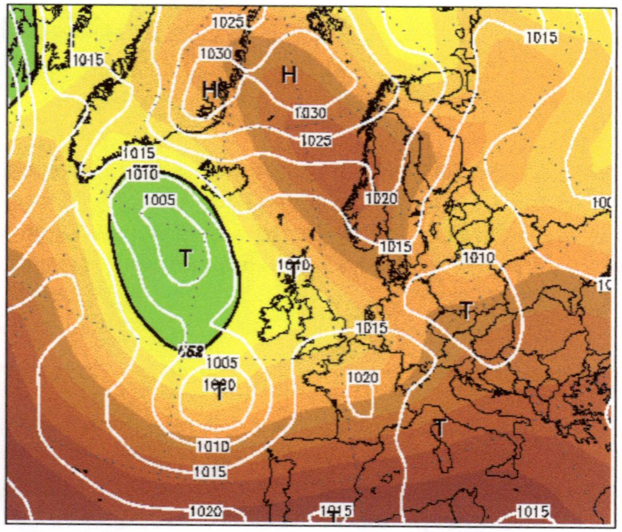

RAIN POTENTIAL

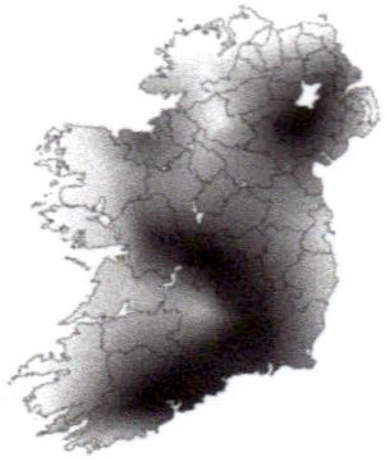

FROST/SNOW

JULY 16 THURSDAY

A weak depression moves down onto Ireland bringing another stormy day throughout, with winds moderate to fresh for a time in a westerly flow, later becoming strong and gusty with a southeasterly change. A changeable day throughout, with a mix of some sunny spells, interspersed with outbreaks of heavy rain or showers everywhere, occasionally accompanied by thunderstorm activity.

Belfast: Changeable, warmer, mix of sunny breakthroughs and overnight rain
Dublin: Stormy, strong and gusty winds, heavy rain, some sunny spells
Cork: Stormy, rain, some sunny spells, warmer, chance of thunderstorms
Galway: Stormy, gale gusts possible, cloudy, squalls

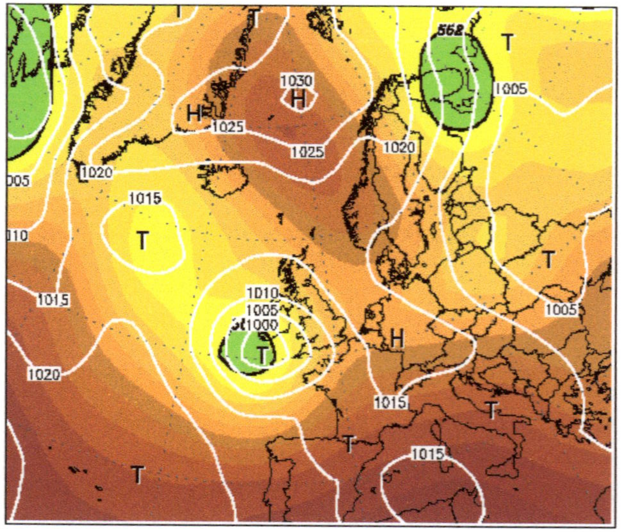

RAIN POTENTIAL

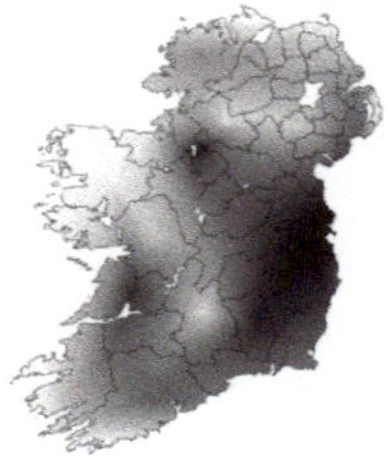

FROST/SNOW

JULY 17 FRIDAY

A depression lingers over Ireland bringing fluctuating winds mostly in a northeasterly flow. Mostly cloudy at first with widespread squally rain activity, some heavier and associated with thunderstorm activity particularly in the south. Rain may be heaviest for a time in the southeast of the country, while lighter and more scattered in the northwest.

Belfast: Stormy, partly cloudy, occasional showers, unsettled winds
Dublin: Sunny day, overnight rain, mild temperatures, strong fluctuating winds
Cork: Cloudy, heavy rain, slightly cooler, strong and changeable winds, static air
Galway: Cloudy, brief squally showers, strong and unsettled winds, chance of gales

RAIN POTENTIAL

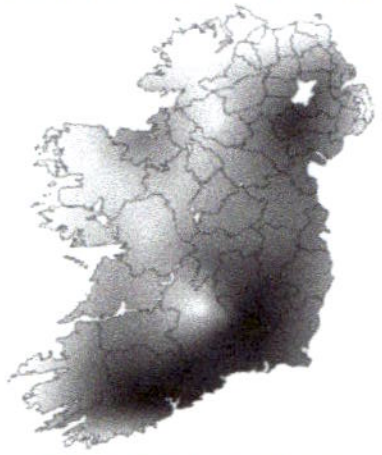

FROST/SNOW

JULY 18 SATURDAY

A depression over Ireland weakens with winds becoming light to moerate in a north to northeast flow. Some lingering overnight thunderstorms and showers may prevail for a time, mostly in the north, before the day clearing to mostly dry with passing scattered cloud throughout. Chance of fog in central parts for a time.

Belfast: Mostly cloudy, cooler, isolated showers, moderate breezes
Dublin: Partly cloudy, mild day cool night, isolated showers, winds easing
Cork: Moderate breezes, partly cloudy, isolated showers
Galway: Cooler, cloudy, dry, chance of gale gusts and thunderstorms overnight

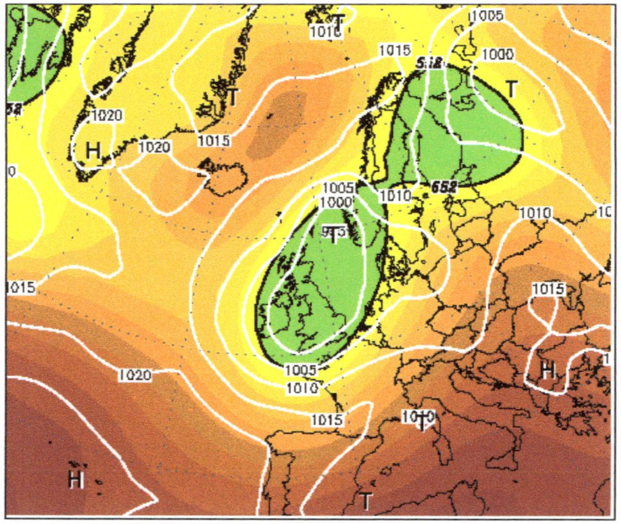

RAIN POTENTIAL

FROST/SNOW

JULY **19** SUNDAY

A weak low pressure trough may see winds easing to light to moderate in a generally northwesterly flow over the country. Mostly sunny during the day, with isolated showers overnight in the north and more showery spells in the southern half of the country. Chance of lingering thunderstorms mostly in the west overnight.

Belfast: Partly cloudy, isolated showers, shifting windy spells
Dublin: Mostly sunny and dry, fresh breezes at times
Cork: Partly cloudy, scattered showers, mild day cool night, unsettled winds
Galway: Partly cloudy, light showers, blustery spells, chance of thunderstorms

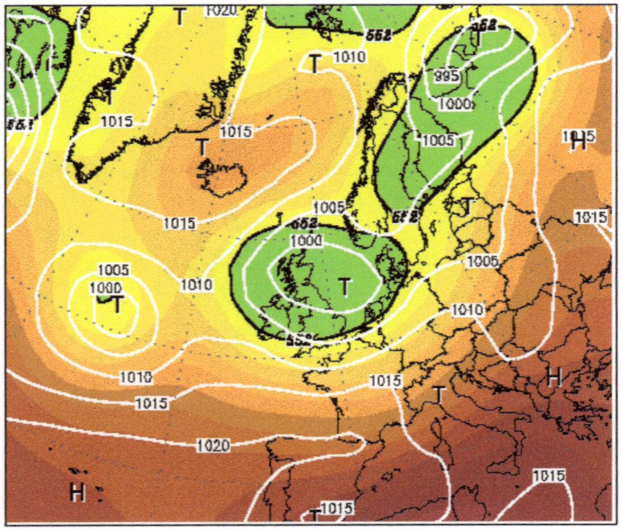

RAIN POTENTIAL

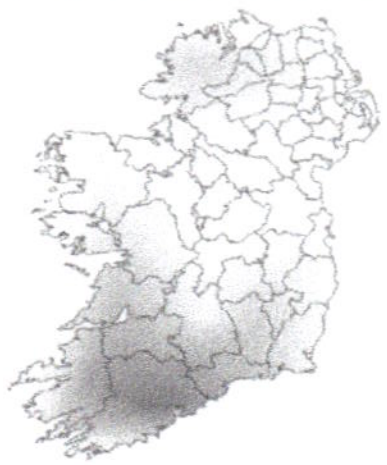

FROST/SNOW

JULY **20** MONDAY

A weak depression to the west of Ireland directs a light to moderate mostly southerly flow over the country. Mostly cloudy and dull throughout, with showers and misty drizzles returning to most districts. Winds may be fresher for a time around the western and southern coasts. Chance of fog overnight in sheltered parts in the southern half of the country.

Belfast: Overcast, scattered showers, winds easing to light
Dublin: Overcast, fluctuating breezes occasional showers, cooler
Cork: Cloudy, brief showers, fluctuating breezes, chance of misty drizzles
Galway: Overcast, heavy rain, warmer temperatures, blustery at times

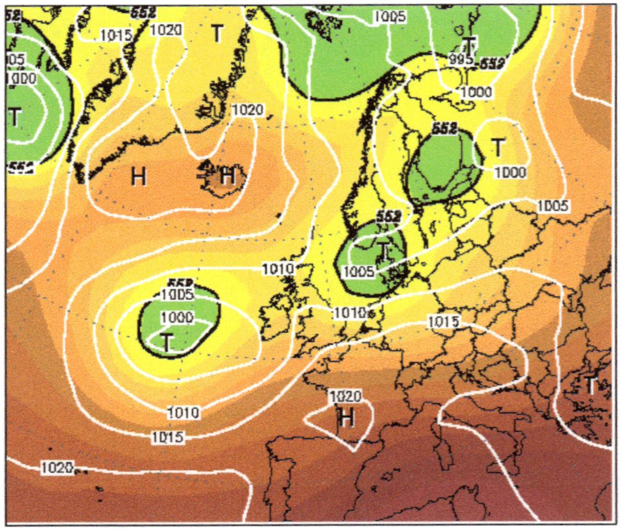

RAIN POTENTIAL

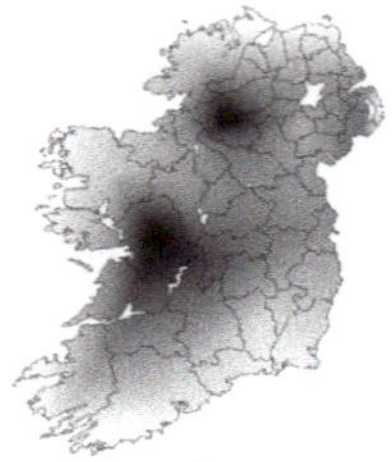

FROST/SNOW

JULY **21** TUESDAY

A depression continues to linger over Ireland with variable winds prevailing, mostly light to moderate, but fresher at times to the north. Mostly cloudy and calm with patches of fog likely in the north and east, while sunnier in the south and west. Chance of a band of rain crossing the country during the day, bringing some light showers in most districts for a time, some heavier mostly in the east around the coasts.

Belfast: Cloudy, dry, mild temperatures, light shifting breezes, chance of fog
Dublin: Cloudy, rain, light breezes with chance of misty patches overnight
Cork: Sunny, warmer, light showers, moderate breezes
Galway: Partly cloudy, cooler, dry, shifting breezy spells

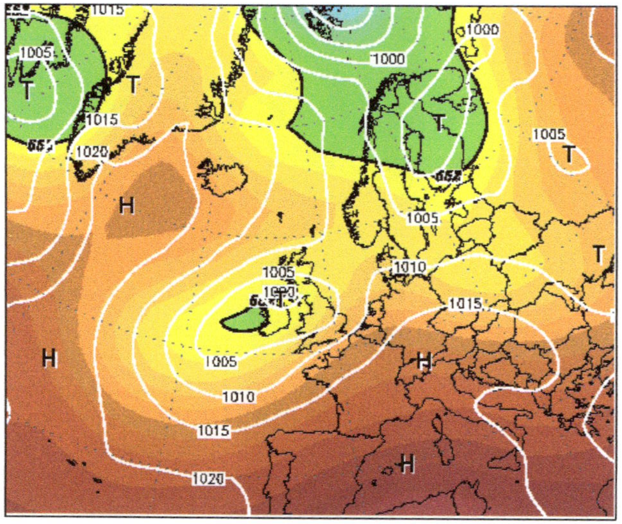

RAIN POTENTIAL

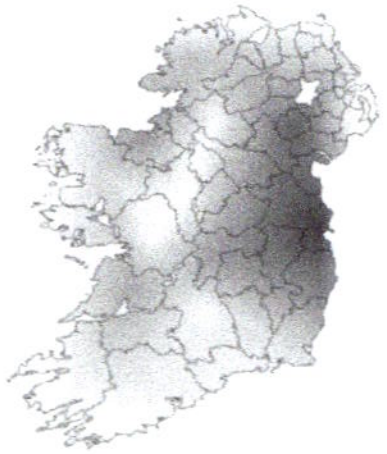

FROST/SNOW

JULY **22** WEDNESDAY

Low pressure continues to linger over Ireland while slowly drifting away to the northeast of Ireland, with light to moderate northerly breezes prevailing, slightly fresher along the northern coastal fringes. Mostly cloudy and dry throughout, with patches of fog likely in the north and central areas overnight. Chance of some isolated showers briefly in the northwest.

Belfast: Cooler, cloudy, dry, fluctuating winds at times, chance of fog overnight
Dublin: Moderate breezes, cloudy, mostly dry with passing threat of rain, cooler
Cork: Dry, mostly cloudy, cooler, shifting breezy spells
Galway: Cloudy, cooler, dry, fresh and shifting breezes

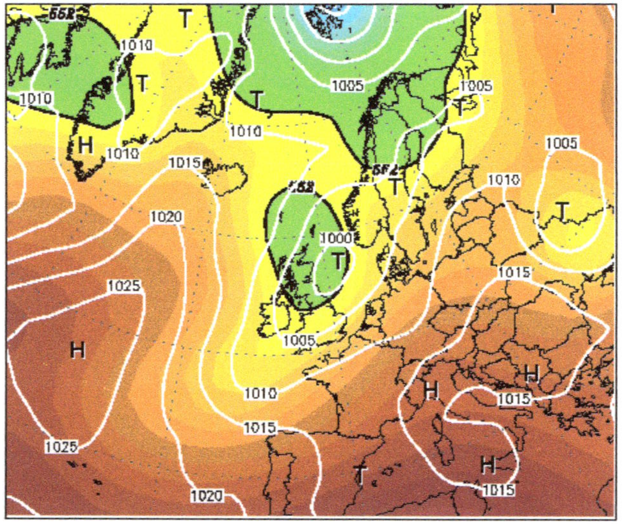

RAIN POTENTIAL

FROST/SNOW

JULY **23** THURSDAY

A high pressure ridge sits to the southwest of Ireland, while low pressure continues to prevail over the country. Winds generally light throughout in a northwesterly flow. A changeable day likely with a mix of passing sunny spells, more freqhent in the south and west, and some spotty showers mostly in the east and south. Chance of fog patches overnight mostly in inland parts.

Belfast: Cloudy, dry, moderate breezes
Dublin: Mostly cloudy, light showers, coolish, moderate breezes
Cork: Partly cloudy, brief showers, mild temperatures, light breezes
Galway: Partly cloudy, dry, mild day cool night, moderate breezes

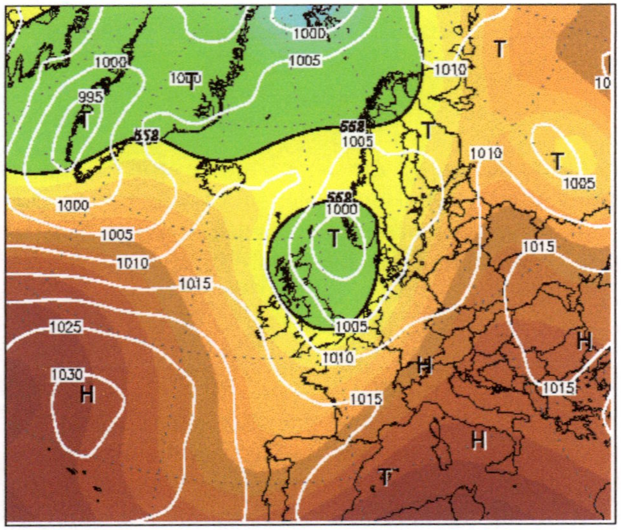

RAIN POTENTIAL

FROST/SNOW

JULY **24** FRIDAY

A weak high pressure ridge nears the south of the country while low pressure continues to linger to the north. Winds generally very light in a variable to westerly flow to mostly calm throughout. Mostly dry with good sunny spells in the eastern half of the country and in the north, while cloudier to overcast in the south and west, with the chance of some outbreaks of spotty showers or drizzle. Patches of fog likely mostly in the east.

Belfast: Partly cloudy, dry, mild day cool night, calm and pleasant air
Dublin: Very sunny, warm and dry, calm and pleasant air
Cork: Cloudy, mostly dry, threats of rain, calm air, chance of misty patches
Galway: Overcast, dry, increasing threats of rain, mostly calm air, chance of fog

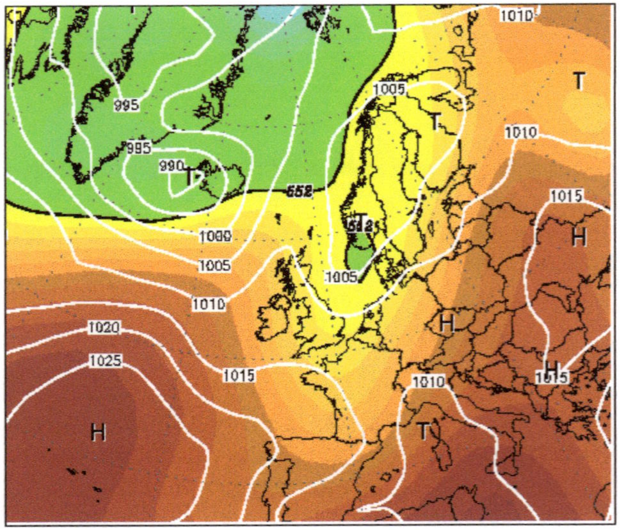

RAIN POTENTIAL

FROST/SNOW

JULY 25 SATURDAY

The high pressure ridge to the south of the country slowly builds as it nears Ireland, with a light westerly flow prevailing throughout. Chance of some passing drizzle, mist, or fog patches overnight, mostly in the southern half of the country, before clearing to a mostly dry and cloudy day throughout.

Belfast: Cloudy and warm, dry, moderate breezes
Dublin: Cloudy, dry, warm, moderate breezes
Cork: Cloudy, warm, mostly dry and calm air, chance of fog patches
Galway: Overcast, light showers, chance of fog patches, warmer

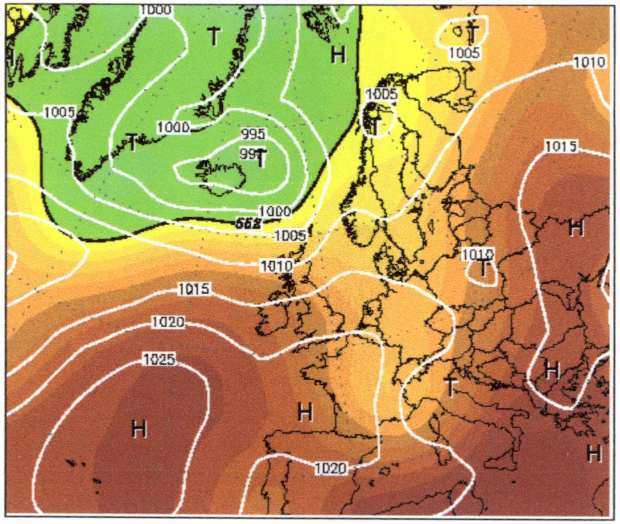

RAIN POTENTIAL

FROST/SNOW

JULY 26 SUNDAY

A high pressure ridge crosses the country with winds freshening for a time in a mostly west to northwest flow, more gusty for a time along the northern and western coasts. Some showers likely overnight in the northwest, with patches of fog and mist elsewhere. Mostly dry day throughout, with good sunny spells in the north and east, while cloudier in the south and west.

Belfast: Mostly sunny, dry, warmer, fresh breezes
Dublin: Sunny and warm, dry, occasional breezy spells
Cork: Warm, cloudy, dry, moderate breezes
Galway: Partly cloudy, dry, slightly warmer, fresh breezes, chance of gale gusts

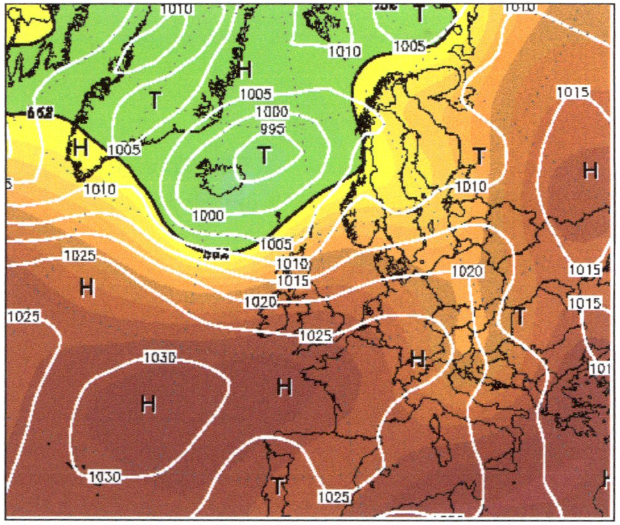

RAIN POTENTIAL

FROST/SNOW

JULY **27** MONDAY

A high pressure system centred to the southwest of Ireland maintains a ridge over the country, with light to moderate northwest breezes prevailing throughout, fresher with gusts at times in the north. Mostly cloudy throughout with rain in the northwest of the country, some brief drizzle patches elsewhere in the northern half of the country, and mostly dry with threats of rain elsewhere.

Belfast: Overcast, cooler, brief showers, fresh breezes
Dublin: Overcast, isolated showers, drizzle patches possible, fresh breezes at times
Cork: Cooler, cloudy, dry, light to moderate breezes
Galway: Overcast, cooler, dry with increasing threats of rain, windy at times

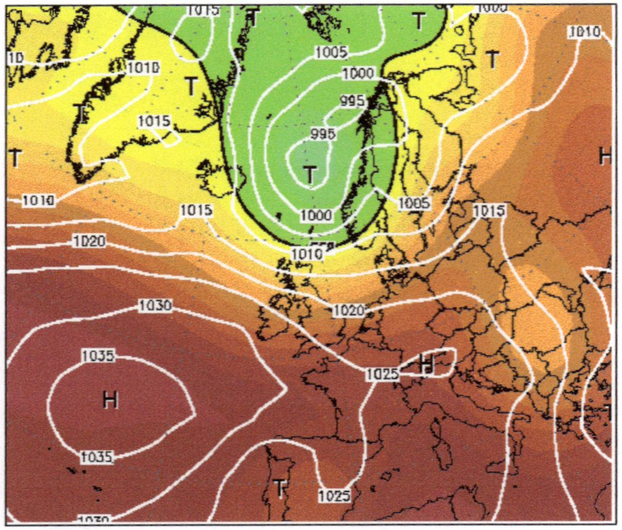

RAIN POTENTIAL

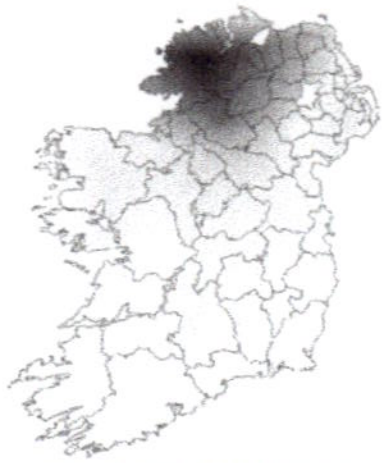

FROST/SNOW

JULY **28** TUESDAY

The high pressure ridge slips to the south of Ireland while a low pressure trough moves down onto the country from the north. Winds generally northwesterly, light to moderate in most areas, but fresher at times in the north and west. Partly cloudy and dry in the south, while cloudier to dull with some showers elsewhere. Showers may be heavier in the northwest parts of the country.

Belfast: Mostly dull and gloomy, dry, cooler, moderate breezes
Dublin: Overcast, moderate breezes, coolish, spotty showers
Cork: Partly cloudy and mostly dry, mild temperatures, moderate breezes
Galway: Cloudy, light showers, mild temperatures, fresh breezes

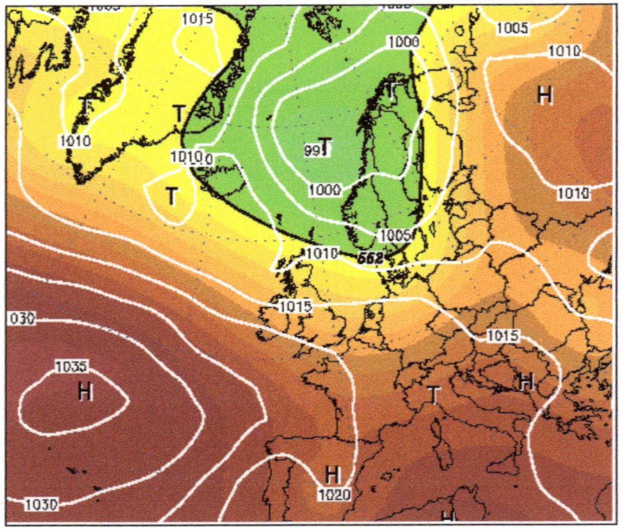

RAIN POTENTIAL

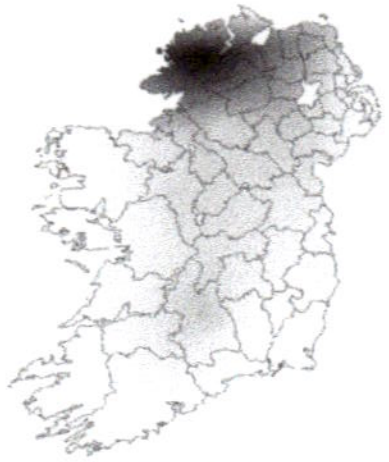

FROST/SNOW

JULY **29** WEDNESDAY

The high pressure ridge remains almost stalled to the south of Ireland while low pressure sits to the north, with moderate northwesterly breezes prevailing, fresher with occasional gusts mostly around the western and northern coastal fringes. Mostly dry and cloudy throughout the day with patches of showers or drizzle in most parts of the country overnight. Sunniest in the south.

Belfast: Cloudy, mild temperatures, light showers, moderate breezes
Dublin: Cloudy, slightly warmer, mostly dry, odd threat of rain, moderate breezes
Cork: Partly cloudy, warm, isolated showers, moderate breezes
Galway: Overcast, mostly dry with chance of spotty showers, fresh breezes

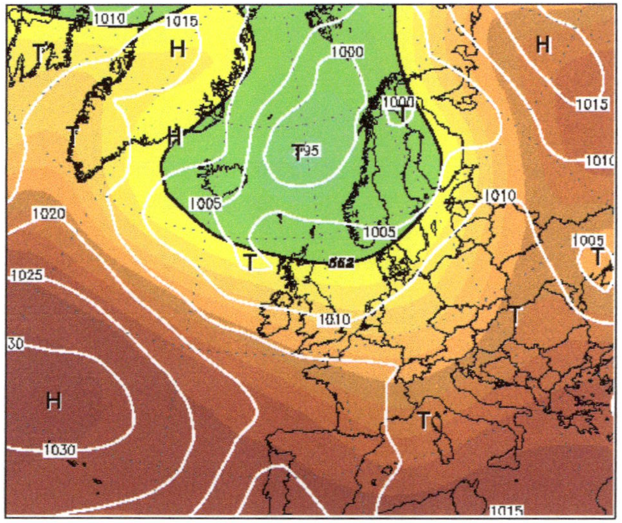

RAIN POTENTIAL

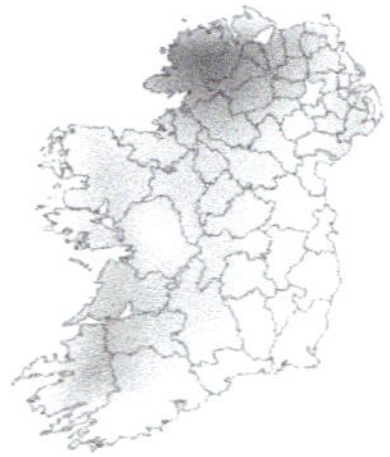

FROST/SNOW

JULY **30** THURSDAY

A depression to the northeast of Ireland directs a fresh to strong northwesterly flow, with some outbreaks of rain or drizzle overnight mostly over the northern half of the country. Mostly dry and sunny day throughout. Winds may be up to gale force for a time in the west and northern coastal regions.

Belfast: Sunny, dry, mild temperatures, windy with occasional gusts
Dublin: Sunny, mild temperatures, mostly dry, fluctuating windy spells
Cork: Sunny, dry, slightly cooler, fresh breezes
Galway: Brief showers, sunny, warmer day cooler night, windy, chance of gales

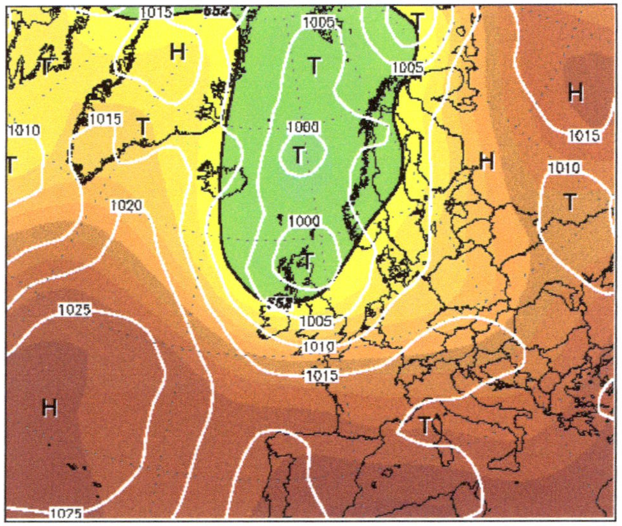

RAIN POTENTIAL

FROST/SNOW

JULY 31 FRIDAY

A high pressure ridge moves up over the country behind the departing depression, now centred to the northeast of Ireland. A light to moderate northwesterly air prevails bringing a mostly dry and sunny day throughout. Chance of some patchy drizzle or misty patches overnight, mostly in the northeast and midlands.

Belfast: Sunny, mostly dry, mild temperatures, chance of misty patches overnight
Dublin: Sunny, dry, mild day cool night, moderate breezes
Cork: Dry and sunny, mild temperatures, moderate breezes
Galway: Sunny, mild day cold night, brief showers, fresh breezes

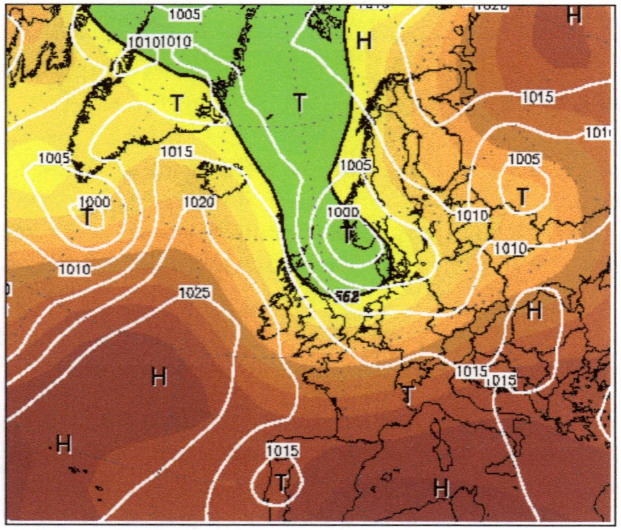

RAIN POTENTIAL

FROST/SNOW

AUGUST

PHASES OF THE MOON

3rd	Crossing Equator
6th	Third Quarter
8th	Northern Declination
12th	New Moon
16th	Crossing Equator
20th	First Quarter
22nd	Apogee
22nd	Southern Declination
28th	Full Moon
30th	Crossing Equator

1	2	3	4	5	6	7	8	9	10	11	12	13	14	15	16	17	18	19	20	21	22	23	24	25	26	27	28	29	30	31
Sat	Sun	Mon	Tue	Wed	Thur	Fri	Sat	Sun	Mon	Tue	Wed	Thur	Fri	Sat	Sun	Mon	Tue	Wed	Thur	Fri	Sat	Sun	Mon	Tue	Wed	Thur	Fri	Sat	Sun	Mon
		XhN			3Q		^		P8		N				XhS				1Q		A V						F		XhN	

MONTHLY SUMMARY

1st-7th Spell of high pressure continues from the anticyclonic influence that became established to the southwest of Ireland over the previous week. Some spells of light rain now and then.
1st Chance of unusually low minimum temperatures overnight, with possible light ground frost in some inland places.
1st-8th A wet week when some rain can be expected every day in many parts of the country despite some long sunny spells.
4th-8th Possibly the warmest spell of summer.
8th-11th Return of very unsettled weather associated with slow moving depressions centred near the southwest of Ireland, bringing heavy rain and some thunderstorms possible. Chance of flooding in some southern inland parts and more aggressive thunderstorms in the midlands.
11th-12th A burst of rain may be particularly heavy in southern parts of the country, possibly bringing some localised or flash flooding. Chance of thunderstorms in some parts of the southwest.

The next four weeks may be cooler than what is normal for this time of year despite the summer level temperatures, as well as very wet and gloomy. Depressions near Ireland may bring a succession of Atlantic frontal systems bearing some considerable amounts of rainfall across the country at times. The first two weeks of this outlook period may be dominated by depressions and low pressure. In contrast, the third and fourth weeks may see a return to higher pressure. As such, the first half of this look-ahead period may see considerable rain, while during the second half of this outlook period, while still seeing almost some daily rain activity, most of it may be much lighter and sparser than the first two weeks. During this outlook period there may be up to 20-24 days with rain compared to the average of 12-14 days, which may also see sunshine levels exceptionally lower than normal over the next four weeks. This unusually heavy rain during the first half of this outlook period may cause damage to crops and farming lands affecting harvesting, particularly in the east and southern areas. Overall, overnight minimum temperatures may be up to two degrees warmer than average, while maximum day temperatures may be below average almost everywhere. During this outlook period there may be 10 days with thunderstorms, 1 day with hail, 14 days with brief fog patches, but no days with gale gusts, snow, or frosts.

12th-29th A wet spell as Atlantic depressions crossing or near to Ireland may be the dominant weather feature, with a series of very dull days and wet fronts that may bring spells of persistent rain, heavy at times. Thunderstorms may accompany many of these wet spells. There may also be occasions when flash flooding may occur, and local catchments may see their river levels rise above flood level at times. Daytime temperatures may be cooler than normal, but overnight temperatures milder than average. Winds may be generally light to moderate in a southwest flow.
12th Chance of widespread heavy rain across the country.
12th-14th Chance of thunderstorms.
17th Widespread rain and scattered thunderstorms likely throughout, with very heavy falls at times in western parts. Chance of hail showers. Heavy downpours at this time may result in some flooding in the west of the country.
20th-21st A wet front may bring further widespread rain and possible thunderstorms across the country as well as stronger winds. Rainfall may be particularly heavy in the east of the country, where a risk of severe flooding may develop.
22nd-23rd Further widespread rain likely almost everywhere.
23rd An intense depression to the immediate north of Ireland may bring heavy rain throughout.
24th Very heavy rain in the west, particularly in the Mayo county region.
25th Unusually cool but very sunny day across the country.
26th-27th Chance of significant heavy rain occurring in the River Barrow catchment region, which may possibly cause some significant flooding.
27th Chance of heavy rain and scattered thunderstorms in many parts of the country.
27th-29th Strong and unsettled winds likely, particularly around the northwest.
30th-11th September Atlantic depressions may drift further northwards away from Ireland allowing for higher pressure to become established over the country bringing a more settled period, despite some scattered rain activity may still prevail almost daily. During this spell, rainfall amounts may be much lighter and sparser than the previous couple of weeks.
31st Chance of thunderstorms.

IRELAND RAINFALL ESTIMATES
August 2026

d=mainly dry, n= nonrecordable, l=light shrs, s=significant shrs, r=rain, h=hvy falls

		ULSTER PROVINCE					CONNAUGHT PROVINCE						LEINSTER PROVINCE				MUNSTER PROVINCE						
		Alderg	Hillsbo	Armag	Loughl	Malin H	Sligo	Ardtarı	Belmul	Drums	Galway	Roscoı	Markre	Edende	Kilkenı	Dublin	Shann	Newpo	Tralee	Killarne	Cork	All	Moon
1st	Sat	l	l	l	l	d	l	l	l	d	d	d	l	d	d	d	d	s	d	d	d	16	
2nd	Sun	d	d	d	d	d	l	d	l	d	d	d	d	d	d	d	d	n	l	d	d	5	
3rd	Mon	d	d	d	d	d	l	d	l	d	d	d	d	l	n	d	d	l	d	d	d	4	XhN
4th	Tues	d	d	d	d	d	l	d	l	n	n	d	d	d	n	d	d	n	d	d	n	4	
5th	Wed	d	d	d	l	s	d	l	d	l	n	l	l	n	l	l	l	d	d	d	d	14	
6th	Thur	d	d	d	d	l	d	d	d	d	d	d	d	d	d	d	d	d	l	l	d	3	3Q
7th	Fri	d	d	d	d	d	d	d	d	d	d	d	d	d	l	d	d	l	d	s	d	10	
8th	Sat	l	s	d	l	d	d	l	d	s	l	l	l	h	s	d	l	l	s	r	r	82	^
9th	Sun	s	r	s	l	d	s	l	l	r	s	s	l	s	s	s	r	s	h	h	h	196	
10th	Mon	r	h	l	d	l	s	d	r	d	s	d	n	h	h	r	d	d	s	d	s	137	P8
11th	Tues	r	s	r	r	h	s	r	l	h	r	s	r	s	s	r	h	r	l	h	r	251	
12th	Wed	r	h	s	s	r	s	s	l	s	s	r	s	r	l	s	r	h	d	l	s	156	N
13th	Thur	s	l	r	l	l	r	l	r	s	l	l	r	l	d	l	l	l	l	n	l	83	
14th	Fri	l	l	l	l	r	l	l	l	l	l	n	l	d	n	l	s	s	l	l	d	49	
15th	Sat	l	n	l	d	d	l	d	d	d	l	d	l	l	s	d	l	n	l	s	s	34	
16th	Sun	s	s	l	s	s	l	s	l	s	s	l	l	s	s	s	s	l	l	l	s	85	XhS
17th	Mon	s	r	l	d	s	l	d	l	r	h	l	l	h	s	h	h	d	s	l	s	214	
18th	Tues	s	d	s	d	d	s	d	l	d	d	d	s	d	l	s	d	d	d	n	l	29	
19th	Wed	s	r	l	s	d	l	s	d	s	h	r	r	r	r	n	d	r	s	r	l	145	
20th	Thur	s	s	s	s	s	s	s	r	h	s	s	r	h	d	s	h	r	h	s	r	302	1Q
21st	Fri	s	s	s	l	l	s	l	s	l	r	l	d	l	l	l	s	r	l	s	l	90	
22nd	Sat	h	h	l	r	s	s	r	r	h	s	h	r	h	h	r	r	r	r	r	r	295	A V
23rd	Sun	r	l	h	l	r	r	l	r	l	s	n	l	h	s	h	s	r	h	h	s	206	
24th	Mon	s	s	s	h	d	h	h	sl	h	r	r	h	s	l	l	s	h	s	s	r	274	
25th	Tues	s	d	h	l	l	h	l	r	l	l	d	l	d	d	d	d	l	d	l	r	86	
26th	Wed	s	r	n	h	s	s	h	r	h	h	h	h	r	h	l	s	h	r	r	s	296	
27th	Thur	r	h	s	s	r	r	s	h	s	h	l	h	h	s	h	s	l	s	l	r	278	
28th	Fri	h	l	ll	l	l	h	l	s	l	r	l	s	l	s	l	l	l	s	s	s	142	F
29th	Sat	l	l	l	d	l	l	d	l	l	l	l	l	s	l	s	r	s	h	r	r	95	
30th	Sun	l	d	s	n	d	l	n	d	l	d	l	s	l	l	l	n	l	l	l	d	26	XhN
31st	Mon	d	d	d	d	d	d	d	d	l	r	l	l	s	l	d	l	d	l	s	r	50	
	Estimate:	181	193	167	110	121	170	110	173	206	225	127	174	303	163	214	210	207	193	208	204	3656	
	Average:	84	87	76	101	95	102	104	102	88	108	109	102	86	72	73	82	133	106	77	97	1882	
	Trend:	w tr	w tr	w tr	av	w tr	w tr	av	w tr	w tr	w tr	w tr	w tr	w tr	w tr	w tr	w tr	w tr	w tr	w tr	w tr	w tr	

IRELAND SUNSHINE ESTIMATES
August 2026

F=fine (8-12 hours of sunshine), pc= partly cloudy (4-7 hours), c=cloudy (1-3 hours), o=overcast (0 hours)

		ULSTER PROVINCE					CONNAUGHT PROVINCE						LEINSTER PROVINCE				MUNSTER PROVINCE						
		Alderg	Hillsbo	Armag	Loughl	Malin H	Sligo	Ardtarr	Belmul	Drums	Galway	Roscom	Markre	Edende	Kilkenn	Dublin	Shannc	Newpo	Tralee	Killarne	Cork	All	Moon
1st	Sat	pc	pc	pc	pc	pc	c	pc	c	pc	c	c	pc	pc	pc	pc	c	c	o	o	c	63	
2nd	Sun	o	o	o	o	o	c	o	o	c	c	o	o	c	c	o	c	c	c	c	c	26	
3rd	Mon	pc	pc	pc	pc	pc	c	pc	pc	pc	pc	pc	pc	pc	pc	pc	pc	pc	F	F	c	104	XhN
4th	Tues	c	c	c	c	c	pc	c	pc	pc	pc	pc	c	pc	pc	c	pc	pc	pc	pc	pc	90	
5th	Wed	c	c	c	c	c	o	c	pc	c	pc	pc	c	c	c	c	c	c	pc	pc	c	53	
6th	Thur	pc	pc	pc	pc	pc	c	pc	pc	pc	pc	pc	pc	pc	pc	pc	pc	pc	c	c	pc	91	3Q
7th	Fri	F	F	F	F	F	pc	F	pc	pc	pc	pc	F	pc	pc	F	pc	pc	pc	pc	pc	139	
8th	Sat	pc	pc	pc	pc	pc	c	pc	c	pc	c	c	pc	pc	pc	F	o	c	o	o	o	97	^
9th	Sun	pc	pc	pc	pc	pc	pc	pc	pc	pc	c	pc	pc	pc	pc	pc	o	c	c	c	o	96	
10th	Mon	pc	pc	pc	pc	pc	F	pc	pc	pc	F	pc	pc	pc	pc	pc	F	F	F	F	F	151	P8
11th	Tues	o	o	o	o	o	o	o	o	o	c	c	o	o	o	o	c	o	o	o	o	6	
12th	Wed	pc	pc	pc	pc	pc	pc	pc	pc	pc	pc	pc	pc	pc	F	pc	F	F	pc	pc	F	135	N
13th	Thur	pc	pc	pc	pc	pc	pc	pc	F	pc	F	F	pc	pc	F	pc	F	F	c	c	F	136	
14th	Fri	pc	pc	pc	pc	pc	pc	pc	c	pc	c	c	pc	pc	pc	pc	pc	pc	pc	pc	pc	110	
15th	Sat	c	c	c	c	c	c	c	pc	c	c	pc	c	c	c	pc	o	c	o	o	o	55	
16th	Sun	o	o	o	o	o	o	o	o	o	o	o	o	o	o	o	c	o	o	o	o	4	XhS
17th	Mon	c	c	c	c	c	o	c	c	c	c	c	c	c	pc	c	o	c	c	c	pc	46	
18th	Tues	o	o	o	o	o	o	o	pc	c	c	c	o	c	c	o	o	c	c	c	c	27	
19th	Wed	pc	pc	pc	pc	pc	pc	pc	pc	pc	pc	pc	pc	pc	pc	pc	pc	pc	c	c	pc	93	
20th	Thur	c	c	c	c	c	c	c	c	c	c	c	c	c	c	c	o	c	o	o	c	35	1Q
21st	Fri	pc	pc	pc	pc	pc	pc	pc	pc	pc	pc	pc	pc	pc	pc	pc	pc	pc	c	c	pc	98	
22nd	Sat	c	c	c	c	c	o	c	pc	c	c	c	c	c	c	c	o	o	o	o	c	26	A V
23rd	Sun	o	o	o	o	o	o	o	o	c	c	c	o	c	c	c	c	c	pc	pc	pc	41	
24th	Mon	c	c	c	c	c	o	c	o	c	o	o	c	c	pc	pc	o	c	pc	pc	c	52	
25th	Tues	pc	pc	pc	pc	pc	c	pc	F	pc	F	F	pc	pc	F	F	F	F	F	F	F	152	
26th	Wed	c	c	c	c	c	o	c	o	c	o	o	c	c	c	c	o	o	o	o	o	18	
27th	Thur	c	c	c	c	c	pc	c	pc	c	pc	pc	c	c	c	c	c	c	F	F	pc	85	
28th	Fri	c	c	c	c	c	c	c	o	c	o	o	c	c	c	c	c	c	c	c	o	28	F
29th	Sat	c	c	c	c	c	o	c	o	c	c	o	c	c	c	c	c	c	c	c	o	22	
30th	Sun	c	c	c	c	c	c	c	c	c	c	c	c	c	c	o	c	c	o	o	pc	42	XhN
31st	Mon	c	c	c	c	c	c	c	F	c	pc	pc	c	c	c	c	c	c	o	o	c	66	
	Estimate:	106	106	106	106	106	85	106	118	114	110	115	106	115	123	126	101	112	104	104	119	2186	
	Average:	146	146	133	140	133	127	127	144	122	140	152	136	148	140	158	138	124	140	140	159	2793	
	Trend:	less	less	less	less	less	less	less	less	av	less	less	less	less	less	less	less	less	less	less	less	less	

AUGUST **1** SATURDAY

A high pressure ridge maintains a light to moderate northwesterly flow over the country becoming more westerly later in the day. Mostly dry with passing scattered cloud throughout, except for the chance of some patches of fog here and there, and chance of some light showers overnight mostly in the in the west and southeast, more isolated in the northeast.

Belfast: Partly cloudy, brief showers, mild day cool night, light breezes, chance of fog
Dublin: Partly cloudy, warmer day colder night, dry, northwesterly breezes
Cork: Cloudy, dry, mild day cool night, light breezes to mostly calm air
Galway: Light breezes to mostly calm air, cloudy, dry, mild day cold night

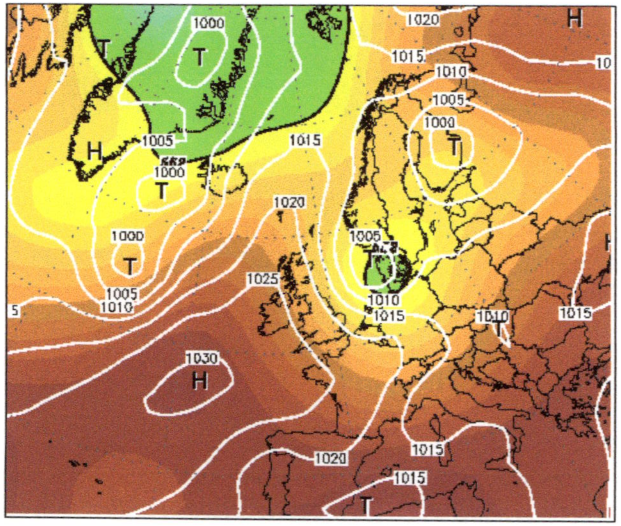

RAIN POTENTIAL

FROST/SNOW

AUGUST **2** SUNDAY

A strong high pressure ridge maintains a light westerly flow and increased cloud throughout, very dull and gloomy over the northern half of the country. Mostly dry except for some patches of light drizzle for a time in the west and along the northern coasts. Chance of fog patches overnight.

Belfast: Overcast, dry, some threat of rain, mild temperatures, chance of mist
Dublin: Overcast, dry, warmer, fresh breezes
Cork: Mostly cloudy, dry, mostly calm air, chance of fog overnight
Galway: Cloudy, warmer, dry, fresh breezes

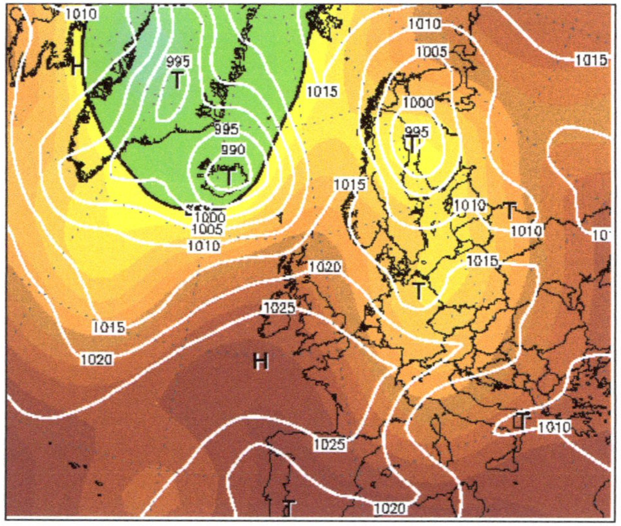

RAIN POTENTIAL

FROST/SNOW

AUGUST 3 MONDAY

A mild and dry high pressure influence dominates the country bringing a mostly dry, calm and serene air throughout with patches of fog likely in many districts overnight, and pleasant sunny spells during the day. Chance of some thunderstorms off the western coasts, which may briefly produce some light showers later in the day.

Belfast: Warm, dry, partly cloudy, light air, chance of overnight fog patches
Dublin: Partly cloudy, mild and dry, mostly calm air, fog possible
Cork: Cloudy, slightly cooler, dry, calm and serene air, chance of fog
Galway: Partly cloudy, pleasant, dry, warm, static atmosphere possible

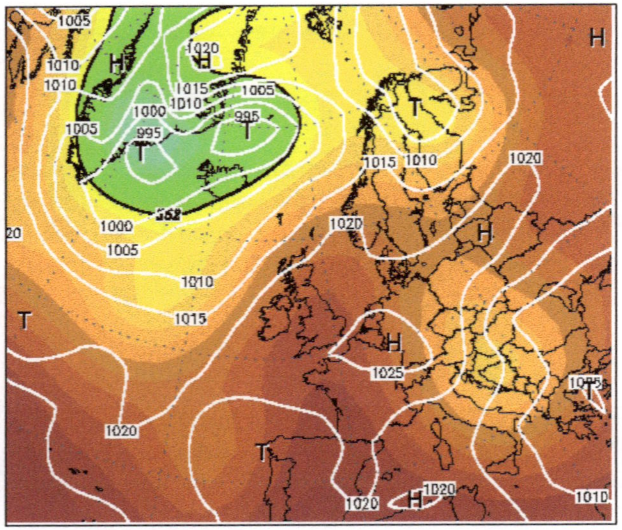

RAIN POTENTIAL

FROST/SNOW

AUGUST 4 TUESDAY

A high pressure system centred to the northeast of Ireland extends a light southeasterly flow throughout, while the low pressure trough lingers to the south of the country. Mostly dry and sunny in the west and south, while cloudier elsewhere. Chance of some overnight showers likely here and there, some possibly accompanying thunderstorm activity. Chance of fog patches for a time in central to eastern parts.

Belfast: Warm and mostly dry, scattered cloudy spells, light breezes
Dublin: Cloudy, warm, dry, light breezes, chance of fog
Cork: Light breezes, warm, partly cloudy, mostly dry, chance of isolated showers
Galway: Warmer, partly cloudy, isolated showers, chance of overnight thunderstorm

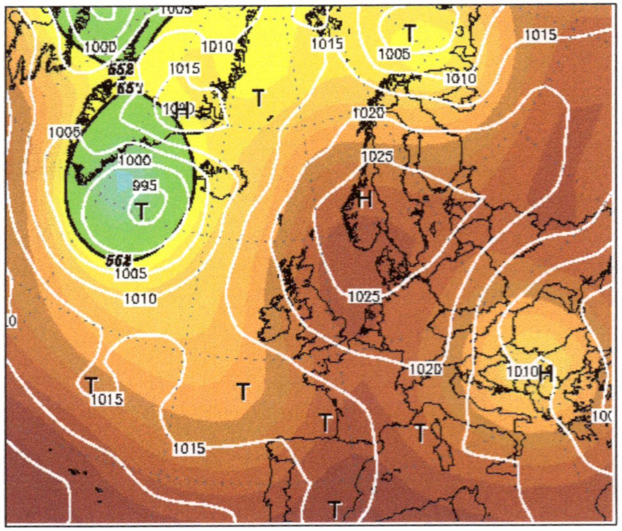

RAIN POTENTIAL

FROST/SNOW

AUGUST 5 WEDNESDAY

The high pressure ridge contracts northeasterly away from the country while a depression to the northwest extends a cloudy trough over Ireland with light to moderate southeasterly breezes prevailing, later backing southwesterly and freshening. Humid with some showers mostly over the northern half of the country, while mostly dry in the south. Chance of further thunderstorms mostly in the northwest.

Belfast: Mostly cloudy, dry, warm, moderate breezes
Dublin: Cloudy, moderate breezes, warm, brief showers
Cork: Dry, mostly cloudy, warm and pleasant air
Galway: Warm, mostly dry, partly cloudy, chance of isolated evening showers

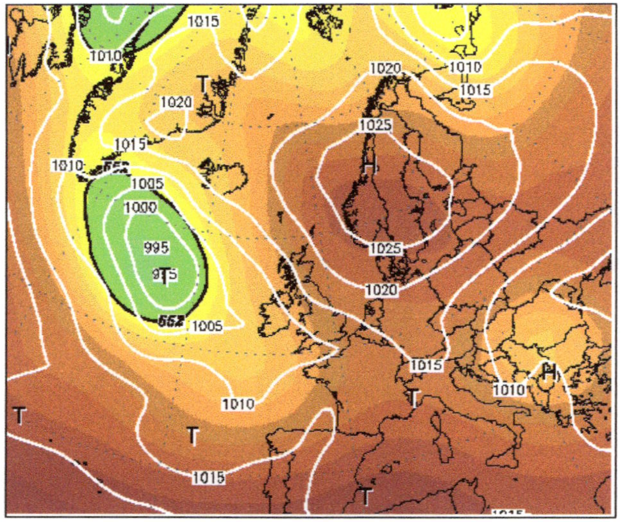

RAIN POTENTIAL

FROST/SNOW

AUGUST 6 THURSDAY

A depression centred to the northwest of Ireland intensifies as it drifts nearer to Greenland, while high pressure lingers to the northeast, allowing for a large low pressure trough to cover Ireland in a very gentle and variable, mostly calm air. Scattered patches of fog likely in many parts of the country, clearing to a mostly dry day with good sunny spells throughout, except for the chance of some isolated showers possible some northern and southwestern areas.

Belfast: Calm and warm air, dry, occasional cloudy spells, chance of fog patches
Dublin: Partly cloudy, warm, dry, calm and serene air
Cork: Partly cloudy, dry, mostly calm air, chance of fog
Galway: Mostly calm and warm air, partly cloudy, dry

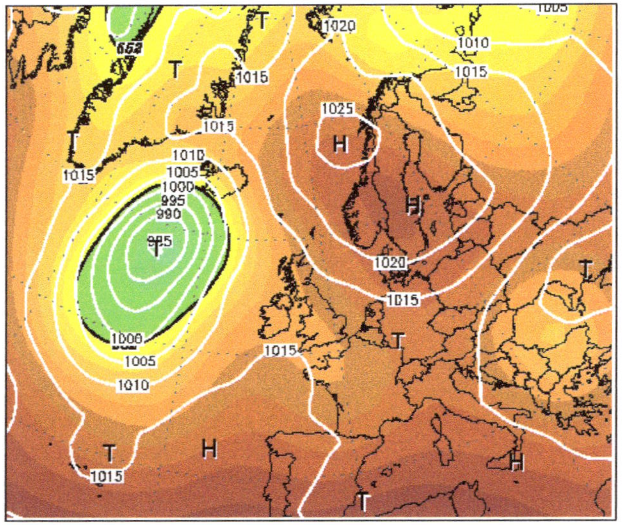

RAIN POTENTIAL

FROST/SNOW

AUGUST 7 FRIDAY

The low pressure trough nestled between a depression to the northwest of Ireland and a high to the northeast continues to maintain a very pleasant and serene air over the country. Mostly sunny and dry throughout with patches of fog in many inland parts. Chance of some isolated showers mostly in the southeast. Cloud may thicken with increasing threats of rain in the west and south later in the evening.

Belfast: Sunny, dry, warmer, calm and pleasant air, chance of fog patches overnight
Dublin: Warm day cool night, sunny, dry, mostly calm air
Cork: Calm and pleasant air, mild temperatures, partly cloudy, dry
Galway: Calm and pleasant air, partly cloudy, dry, warm

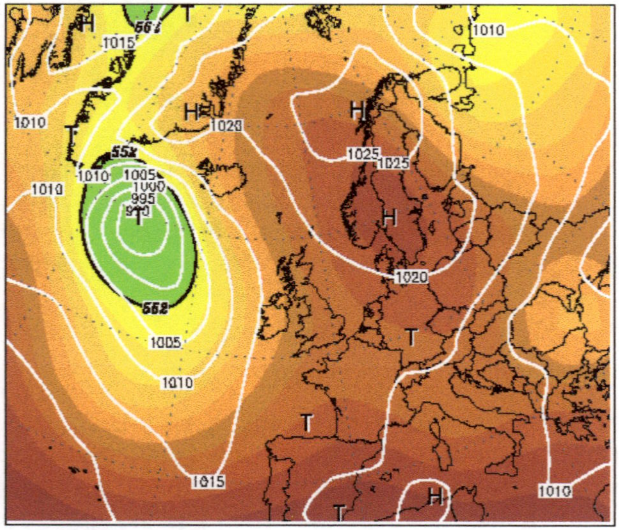

RAIN POTENTIAL

FROST/SNOW

AUGUST 8 SATURDAY

The low pressure trough persists over the country with a very light to moderate flow prevailing, variable in direction. Some overnight patches of mist here and there, particularly in inland parts. A changeable day throughout, with a mix of sunny spells and showers in most areas, although very dull in the south, and mostly dry along the western and northern coasts. Chance of a band of thunderstorms moving into the southeast in the evening.

Belfast: Warm day cool night, mix of showers and sunny spells, misty patches
Dublin: Sunny, mostly dry, warm, calm and pleasant air
Cork: Overcast, rain, cooler, light breezes
Galway: Warm, cloudy, brief evening showers, mostly calm air

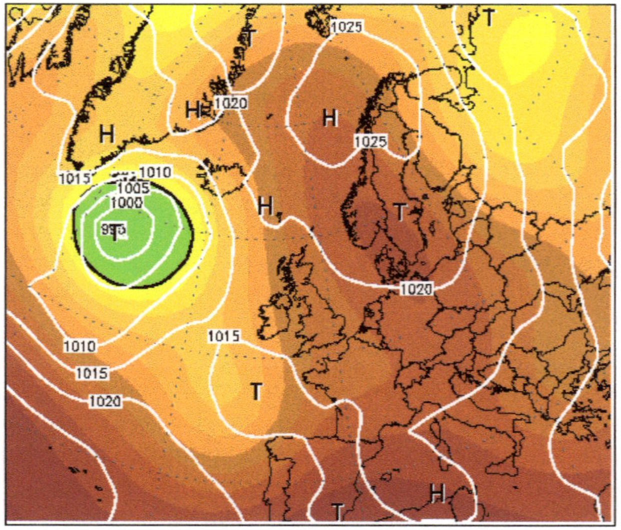

RAIN POTENTIAL

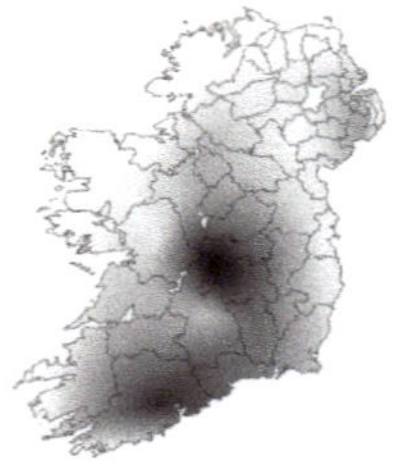

FROST/SNOW

AUGUST 9 SUNDAY

A weak low pressure centre forms within the trough situated over Ireland, bringing fresher breezes, variable in direction. Another mixed day of some sunny breakthroughs between bands of rain crossing the country, with more persistent cloud, drizzle, and spells of heavy rain likely in southern areas, possibly accompanied by thunderstorm activity.

Belfast: Partly cloudy, hot, rain, fluctuating winds, static air
Dublin: Mostly sunny, occasional showers, warm, gentle shifting breezes
Cork: Overcast, cooler, heavy rain, breezy spells, chance of drizzles or fog patches
Galway: Cloudy and warm, occasional showers, shifting breezy spells

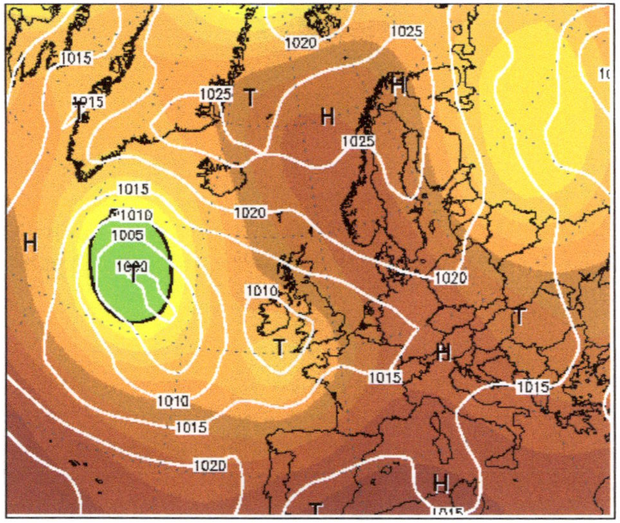

RAIN POTENTIAL

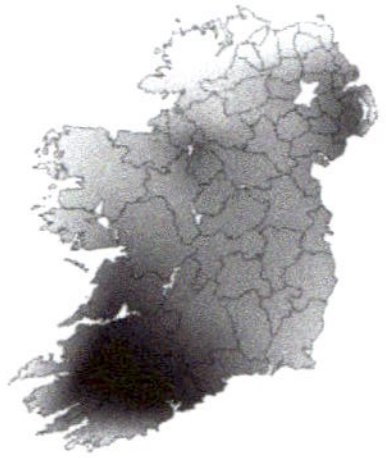

FROST/SNOW

AUGUST 10 MONDAY

A depression to the west of Ireland directs a moderate southerly flow over the country, strong with occasional gusts along the northwestern coasts at times. Some sunny spells during the day between widespread outbreaks of rain and drizzle at both ends of the day. Rain may be particularly heavy in the southeast and up later into the northeast.

Belfast: Warm, partly cloudy, heavy rain, unsettled winds
Dublin: Cloudy, rain, warm temperatures, unsettled windy spells
Cork: Sunny, warmer, evening rain, windy
Galway: Changeable, sunny day, cooler, evening showers, windy spells

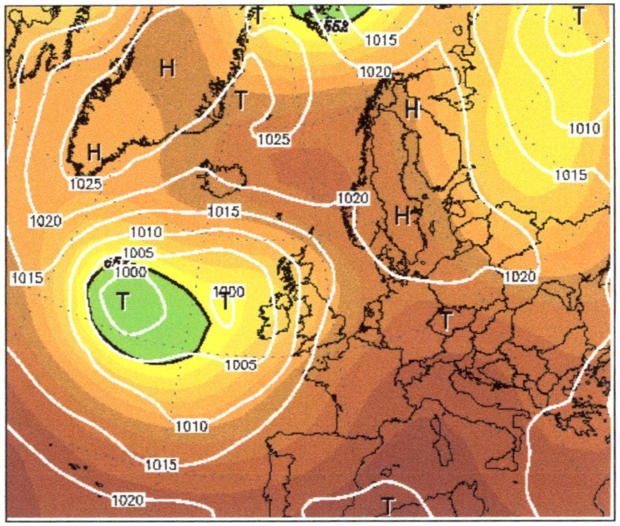

RAIN POTENTIAL

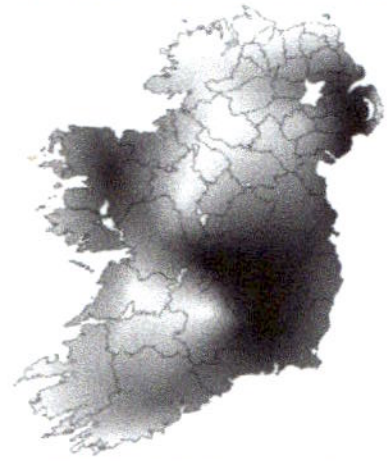

FROST/SNOW

AUGUST **11** TUESDAY

A depression immediately to the west of Ireland maintains a moderate mostly southerly flow over the country, fresher at times particularly along the coasts. Dull and gloomy with rain and drizzle everywhere, easing for a time, before another stormy band of rain and accompanying thunderstorms again crosses the country, starting in the east and moving westward, and may be heavy enough in some areas to cause some localised flooding, particularly in the southeast and northwest areas.

Belfast: Overcast, warm and wet, rain, fresh winds
Dublin: Overcast, rain, cooler, windy spells
Cork: Overcast, rain, cooler, windy
Galway: Cloudy, rain, strong breezes, mild day cool night

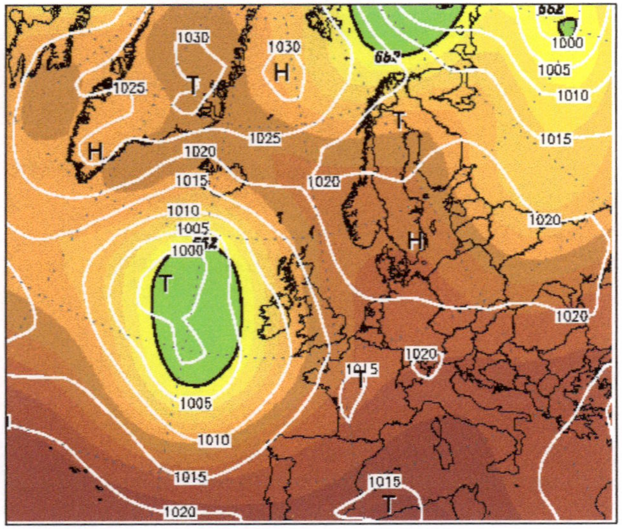

RAIN POTENTIAL

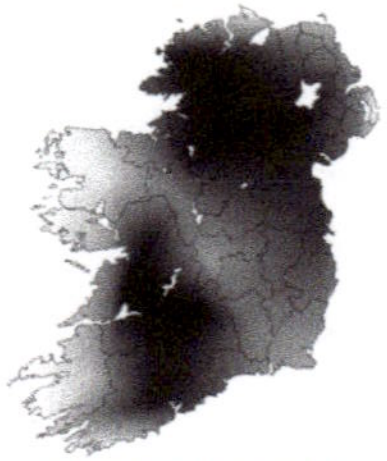

FROST/SNOW

AUGUST **12** WEDNESDAY

A weak depression continues to linger over the country with winds generally light and variable in direction, later backing more southwesterly. An unsettled day throughout with bands of rain and occasional sunny spells. Rain activity may be particularly heavy at times in the northeast and midlands, possibly associated with further thunderstorm activity.

Belfast: Fluctuating winds, heavy rain, sunny spells, warm, chance of thunderstorms
Dublin: Partly cloudy, mild temperatures, overnight rain, shifting windy spells
Cork: Overnight showers, sunny and dry day, warmer, shifting breezy spells
Galway: Shifting windy spells, mix of showers and sunny spells

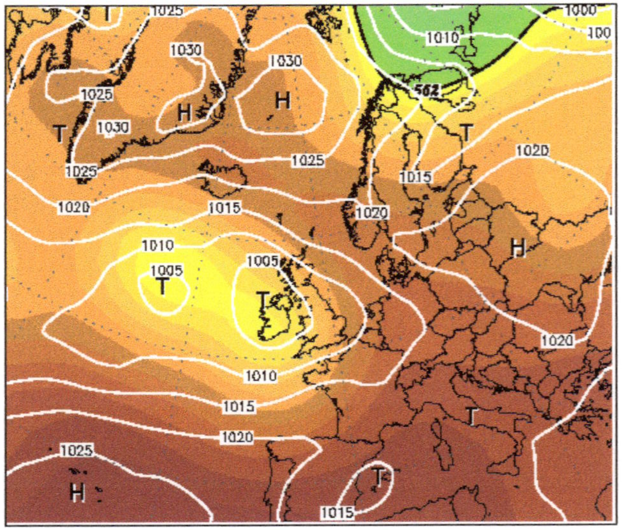

RAIN POTENTIAL

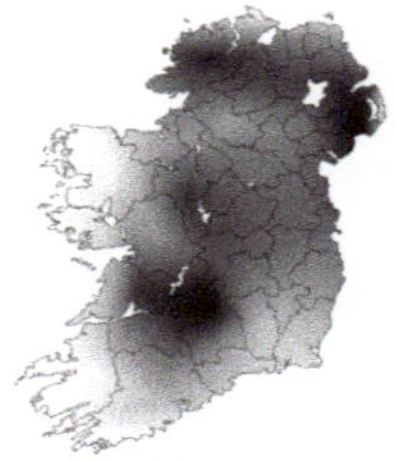

FROST/SNOW

AUGUST **13** THURSDAY

A depression centred over Ireland weakens with a light southwest to westerly change. Another mixed day of sunny spells and passing showers, with heavier showers at times in the west and northeast, while lighter to more isolated in elsewhere.

Belfast: Showers, warm, some sunny spells, shifting winds
Dublin: Partly cloudy, light showers, warmer, moderate breezes
Cork: Sunny, light overnight showers, mild temperatures, windy at times
Galway: Sunny, light overnight showers, mostly dry day, moderate breezes

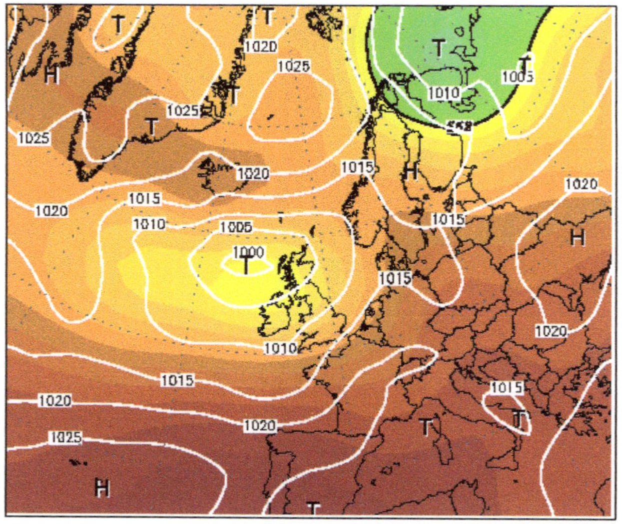

RAIN POTENTIAL

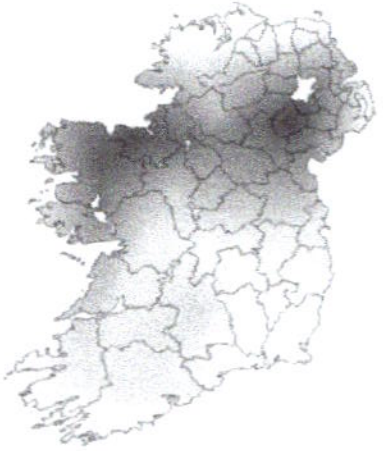

FROST/SNOW

AUGUST **14** FRIDAY

The depression that has lingered over Ireland the past few days slowly drifts away to be centred to the north of the country, while maintaining a trough of low pressure crossing the country in a light to moderate westerly flow throughout. Changeable in the north and west with a mix of scattered showers and sunny spells, while mostly dry and sunny in the far south and eastern coastal areas. Showers may be heavy for a time in the far northwest, possibly accompanied by thunderstorms.

Belfast: Warm, partly cloudy, light showers, moderate breezes
Dublin: Partly cloudy, spotty showers, fresh breezes at times
Cork: Mostly sunny and dry, mild temperatures, blustery at times
Galway: Mostly cloudy, occasional showers, fresh breezes and some gusts

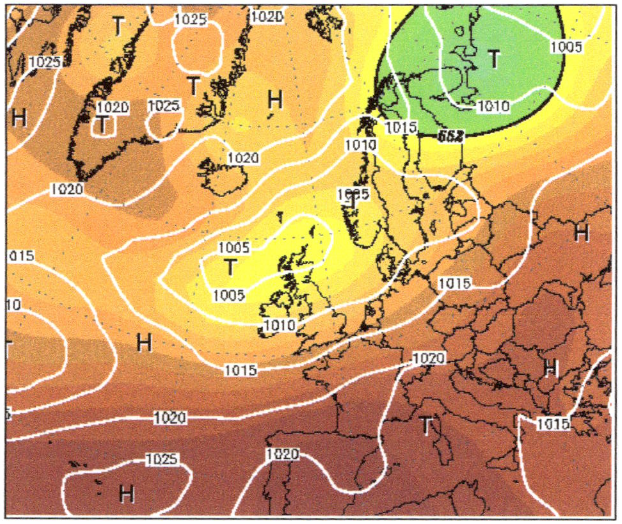

RAIN POTENTIAL

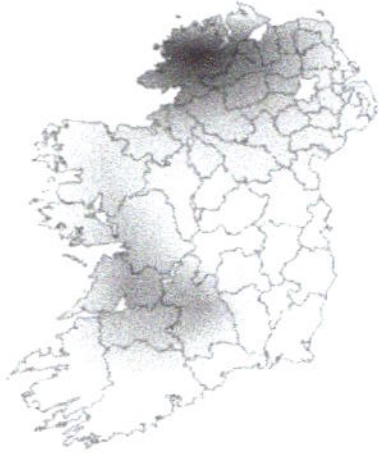

FROST/SNOW

AUGUST **15** SATURDAY

Low pressure continues to bring changeable conditions throughout the country. Winds generally light in a west to northwest flow. Mostly cloudy everywhere, with chance of fog patches overnight in northern parts. Some showers likely in almost all districts except for a mostly dry day around the northwestern coasts, which should also see the most sunshine.

Belfast: Slightly cooler, mostly cloudy, isolated showers,, chance of fog overnight
Dublin: Partly cloudy, mostly dry with increasing threats of rain, moderate breezes
Cork: Overcast, scattered showers, cooler, breezy spells
Galway: Cloudy, light showers, moderate breezes

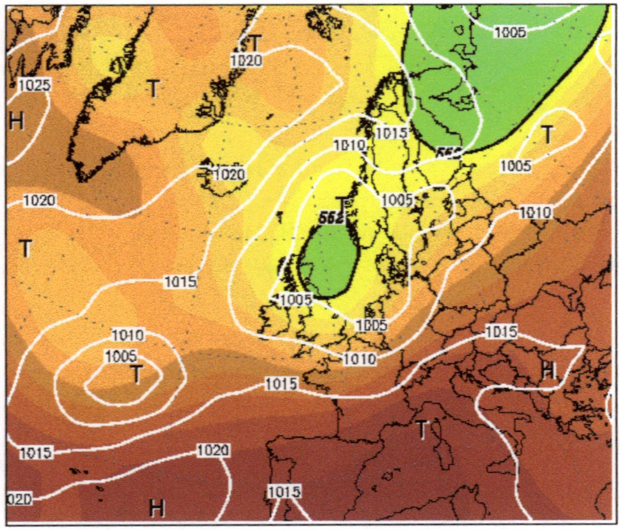

RAIN POTENTIAL

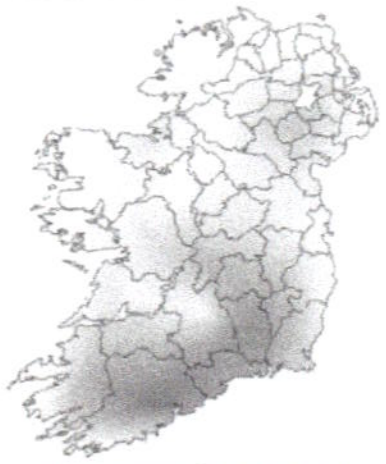

FROST/SNOW

AUGUST **16** SUNDAY

A depression to the southwest of Ireland directs a fresh south to southeast flow over the country, with occasional gusts at times. Dull and gloomy everywhere, with a humid atmosphere, and outbreaks of showers throughout. Bands of rain may cross the southwest of Ireland at first, and then spread northwards to cover the rest of the country during the day.

Belfast: Unsettled air, overcast, scattered showers
Dublin: Overcast, scattered showers, shifting breezy spells
Cork: Unsettled winds, overcast, scattered showers, chance of misty drizzles
Galway: Overcast, scattered showers, strong winds with occasional gusts

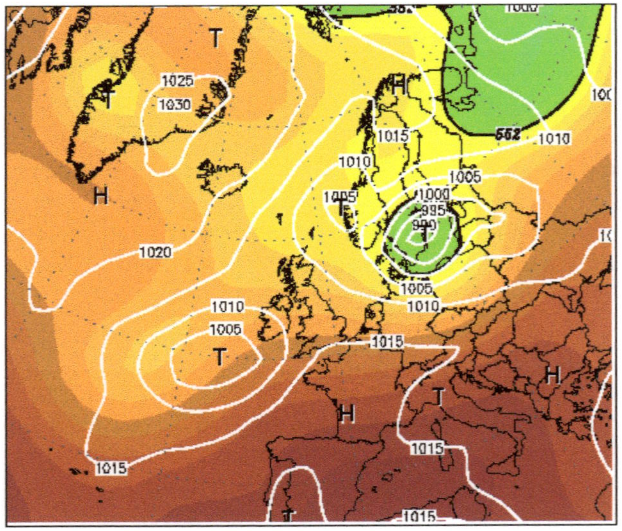

RAIN POTENTIAL

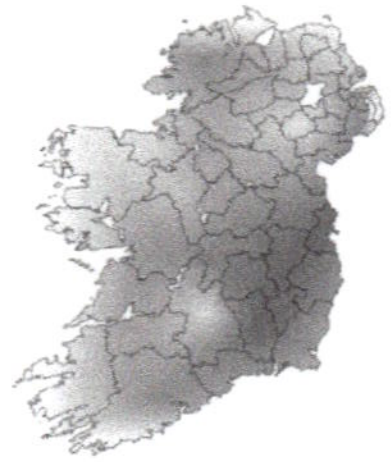

FROST/SNOW

AUGUST **17** MONDAY

Low pressure brings further widespread rain activity throughout, and with lighter winds in most districts in a southwesterly flow, a little fresher around the southwestern coasts. Mostly cloudy everywhere, with rain activity in all districts, including the likelihood of thunderstorms and heavier rain, particularly in a line from the central west coast to the central east coast. Slightly more frequent sunny spells in southern parts.

Belfast: Cloudy, rain, cooler, light breezes, chance of fog
Dublin: Warmer, heavy rain, brief sunny breakthroughs, moderate winds
Cork: Changeable, warmer, mix of sunny spells and showers, moderate breezes
Galway: Cloudy, heavy rain, slightly warmer, fluctuating winds

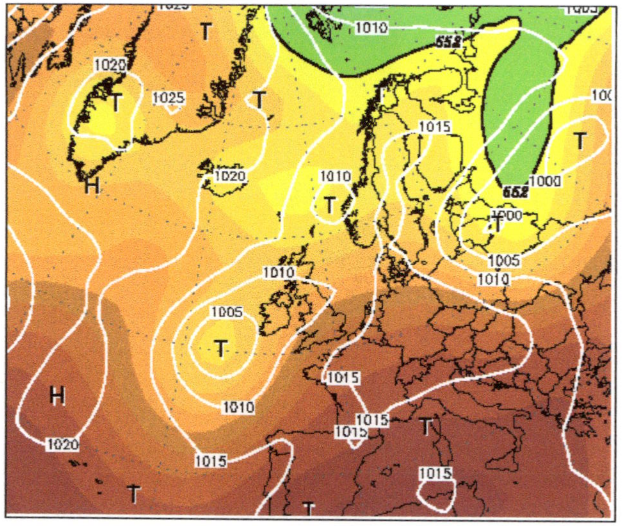

RAIN POTENTIAL

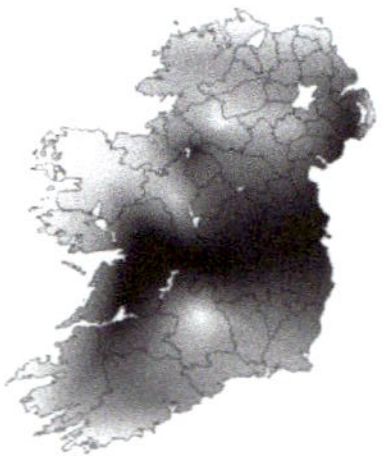

FROST/SNOW

AUGUST **18** TUESDAY

A weak depression sits over Ireland bringing unsettled and changeable winds, but predominately northwesterly in direction. Humid, dull and gloomy almost everywhere, but slightly sunnier in the south. Mostly dry around the southwestern coasts, with some showers or drizzle patches elsewhere, more frequent in the northeast.

Belfast: Overcast, mostly dry, threats of rain, slightly warmer, unsettled winds
Dublin: Overcast, occasional showers, cooler, unsettled windy spells at times
Cork: Mild temperatures, light showers, windy, some sunny breakthroughs
Galway: Cloudy, cooler, dry, shifting windy spells

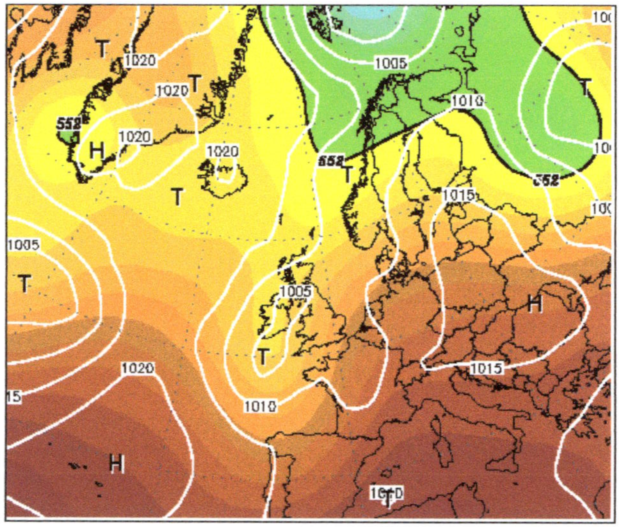

RAIN POTENTIAL

FROST/SNOW

AUGUST **19** WEDNESDAY

Low pressure continues to linger over Ireland while a high pressure ridge sits to the far southwest of the country. Winds generally light in a north to northwesterly flow, a little fresher at times around the western coasts. Mostly dry and sunny around the northwest coastal areas, with increased cloud and outbreaks of rain elsewhere. Rain may be particularly heavy for a time in the midlands and in the far northeast.

Belfast: Rain, cooler, some sunny spells, windy spells at times
Dublin: Partly cloudy, isolated showers, coolish, static air, breezy at times
Cork: Partly cloudy, spotty showers, mild day cool night, moderate breezes
Galway: Fresh to strong breezes, partly cloudy, rain, mild day cool night

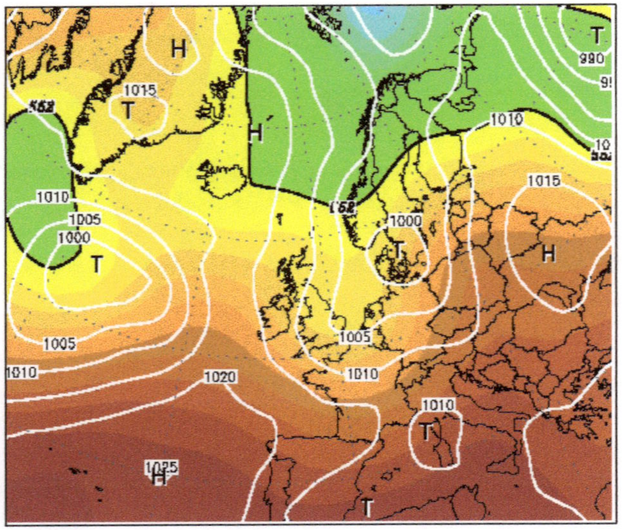

RAIN POTENTIAL

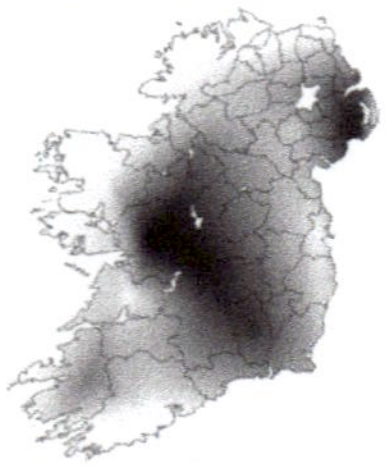

FROST/SNOW

AUGUST **20** THURSDAY

A depression to the northwest of Ireland bears down onto the country around this time, with winds becoming strong and blustery with a southwest change, possibly to gale force for a time especially in the western half of the country. Stormy conditions prevail everywhere, with widespread cloud, some thunderstorm activity and heavy rain in many areas.

Belfast: Cloudy, scattered showers, slightly warmer, strong and unsettled winds
Dublin: Cloudy, warm, chance of thunderstorms, heavy rain, possible flooding
Cork: Cloudy, rain, blustery
Galway: Stormy air, strong and gusty winds, chance of gales, cloudy, showers

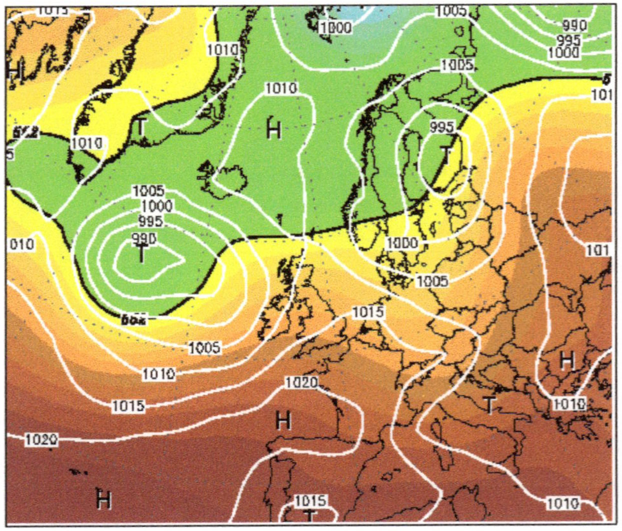

RAIN POTENTIAL

FROST/SNOW

AUGUST **21** FRIDAY

A depression over Ireland continues to bring very unsettled west to southwest winds across the country, possibly up to gale force at times. A stormy air prevails throughout, with widespread spells of rain, more intense in the west, but also occasional sunny spells. Only the far southern coasts and southeastern coastal areas may be mostly dry throughout the day.

Belfast: Warm, wet, and windy, mix of showers and sunny spells
Dublin: Partly cloudy, light showers, cooler, strong and gusty winds, chance of gales
Cork: Partly cloudy, brief showers, strong and blustery winds, chance of gales
Galway: Storm, rain, some sunny breakthroughs, slightly cooler

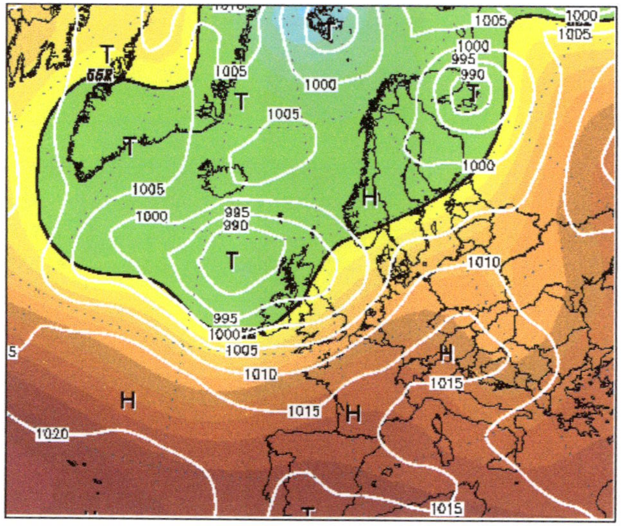

RAIN POTENTIAL

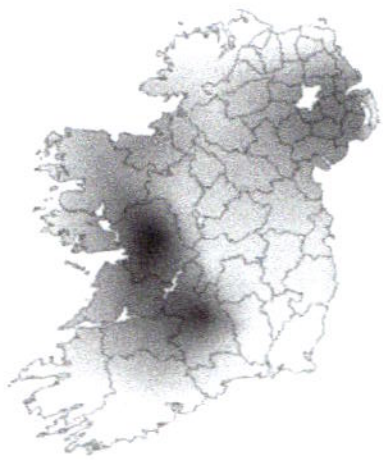

FROST/SNOW

AUGUST **22** SATURDAY

A depression continues to bring widespread stormy conditions throughout the country with winds easing to moderate in a southwesterly flow. Widespread cloud and bands of rain cross the country throughout the day, in the southwest and first and spreading northwards into all areas as the day progresses. Falls may be heavy at times, particularly in inland parts and in the northeast.

Belfast: Mostly overcast, heavy rain, stormy air
Dublin: Cloudy, stormy air, rain
Cork: Cloudy, rain, slightly cooler, strong winds with gusts likely
Galway: Cloudy, rain, cooler, strong winds with gusts, chance of gales for a time

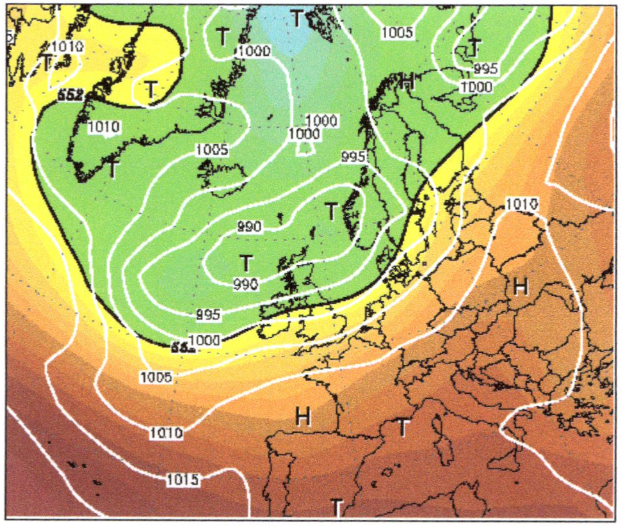

RAIN POTENTIAL

FROST/SNOW

AUGUST **23** SUNDAY

Low pressure persists over the country with winds easing to light and variable in direction. Mostly cloudy throughout with the chance of patches of fog overnight in sheltered areas. Further bands of rain may once again cross the country throughout the day, with some heavy falls likely at times particularly in the northeast and into the midlands.

Belfast: Light breezes, mostly overcast, brief showers, chance of fog
Dublin: Cloudy, heavy rain, colder, winds easing to light for a time
Cork: Partly cloudy, cooler, passing showers, light breezes for a time, chance of fog
Galway: Winds easing to mostly calm air, cloudy, scattered showers, slightly warmer

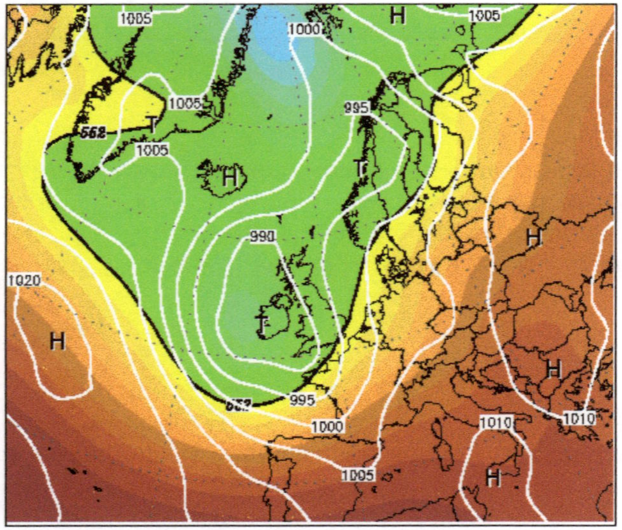

RAIN POTENTIAL

FROST/SNOW

AUGUST **24** MONDAY

Low pressure may continue to persist bringing further widespread rain activity throughout the country, lighter in the east and far northwest. Bands of rain, including scattered thunderstorms may occur in many districts, possibly bringing some torrential downpours particularly in the western half of the country. It may be mostly dry and sunny in the far northwest coastal fringes. Winds generally in a moderate north to northwest flow, but stronger with gusts along the southwest coasts.

Belfast: Cloudy, colder, occasional showers, unsettled and fluctuating winds
Dublin: Partly cloudy, light showers, unsettled and changeable winds
Cork: Mostly cloudy, rain, mild day cool night, stormy air, strong winds and gusts
Galway: Overcast, rain, cooler, strong winds, chance of gales and gusts

RAIN POTENTIAL

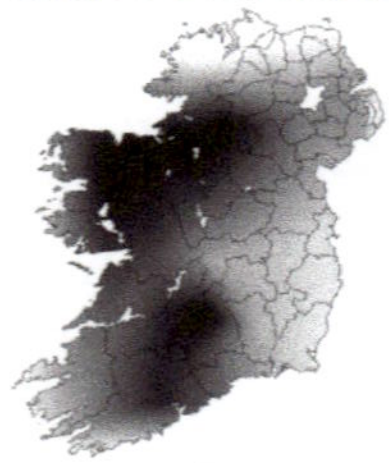

FROST/SNOW

AUGUST **25** TUESDAY

Low pressure still maintains a moderate west to southwest flow over the country. Overnight rain with some heavy falls at times likely in the west and south, with some showers elsewhere, clearing to a mostly dry day with increasing sunny spells.

Belfast: Partly cloudy, mostly dry, threats of rain, slightly warmer, breezy at times
Dublin: Dry, sunny, moderate breezes
Cork: Unsettled winds, mix of rain and sunny spells
Galway: Unsettled air, showers, sunny day, warmer day cool night, strong breezes

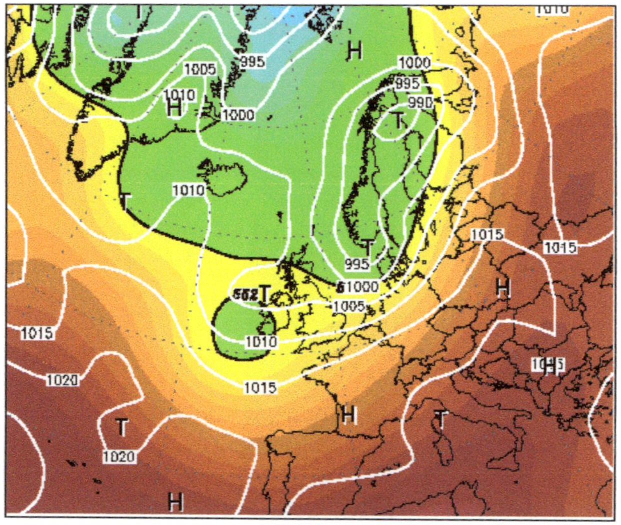

RAIN POTENTIAL

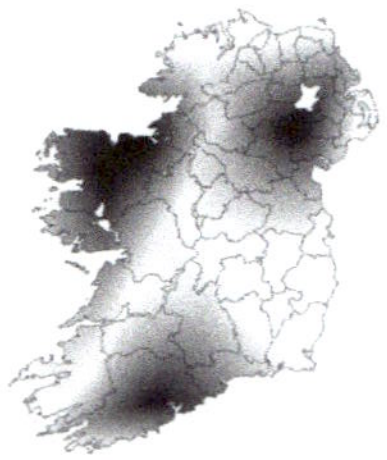

FROST/SNOW

AUGUST **26** WEDNESDAY

A weak high pressure influence moves over the country, bringing widespread cloud and outbreaks of showers everywhere, sometimes heavy in inland and western parts. Winds generally light in variable directions, with the occasional gust possible mostly around the southwest coasts.

Belfast: Cloudy, rain, gentle breezes, chance of misty patches
Dublin: Cloudy, cooler, isolated showers, light breezes to mostly calm air
Cork: Overcast, rain, cooler, light breezes
Galway: Overcast, heavy rain, cooler, light breezes strengthening later in the day

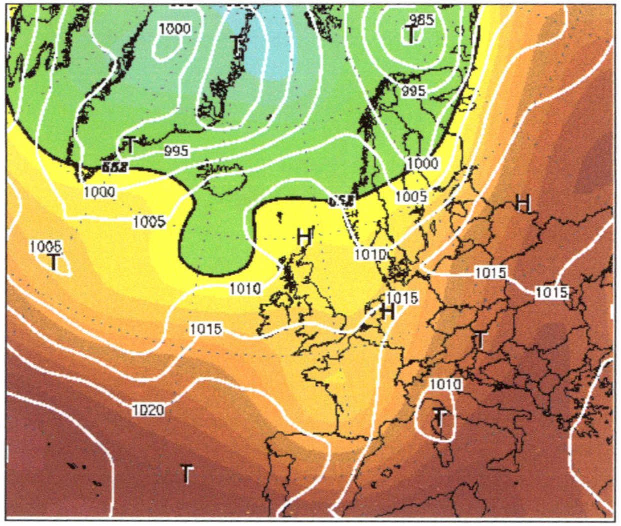

RAIN POTENTIAL

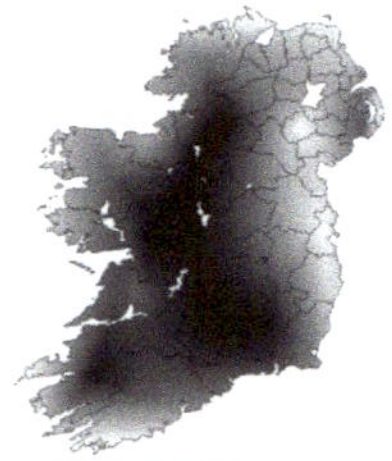

FROST/SNOW

AUGUST 27 THURSDAY

Stormy conditions return throughout the country as a depression crosses Ireland once again. Winds generally southerly, fresh to strong at times, and particularly gusty in southern areas. An unsettled day everywhere, with a mix of some sunny spells between bands of rain and possible scattered thunderstorms. Rain may be particularly heavy at times in central areas and in the far north.

Belfast: Stormy air, heavy rain, brief sunny outbreaks, windy with gusts
Dublin: Stormy, cloudy, heavy rain, cooler, thunderstorms possible, wind gusts
Cork: Partly cloudy, rain, slightly warmer, windy at times
Galway: Partly cloudy, mild day cold night, stormy air, heavy rain, gusty winds

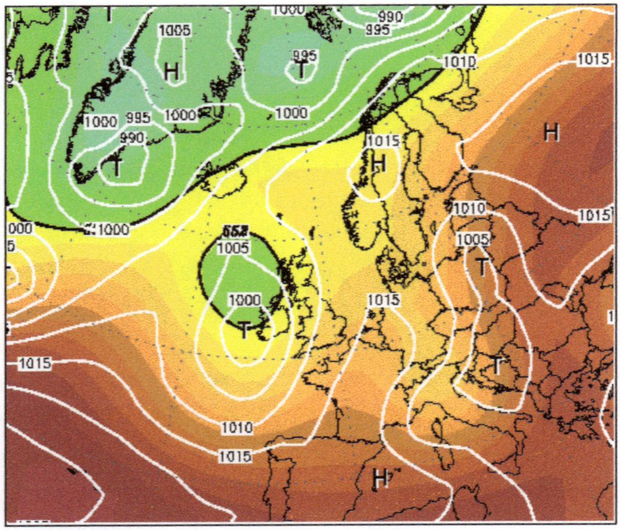

RAIN POTENTIAL

FROST/SNOW

AUGUST 28 FRIDAY

The depression that brought stormy conditions over the past few days may have now drifted to away to the west of Ireland, easing winds throughout to generally light in a west to southwest flow. Overnight rain may persist, while later clearing to a mostly dry afternoon with some sunny spells, particularly in the north. However, further bands of rain may return again by evening bringing more widespread scattered showers.

Belfast: Cloudy, cooler, light showers, unsettled and changeable winds
Dublin: Cloudy, light showers, slightly warmer, strong and unsettled winds, static air
Cork: Overcast, slightly cooler, rain, fluctuating and shifting winds
Galway: Overcast, rain, unsettled and fluctuating winds

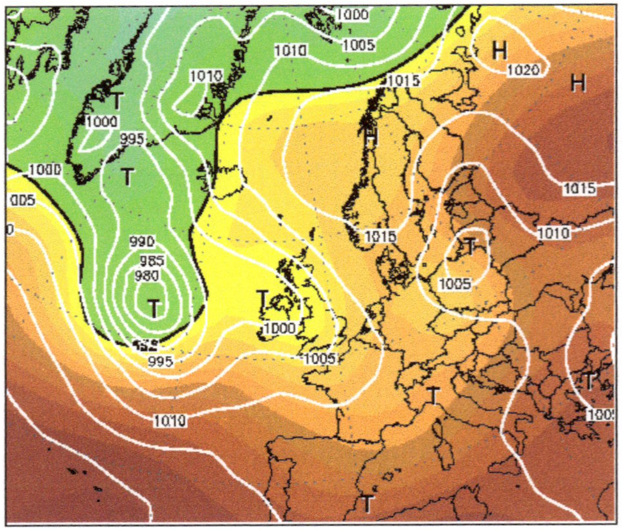

RAIN POTENTIAL

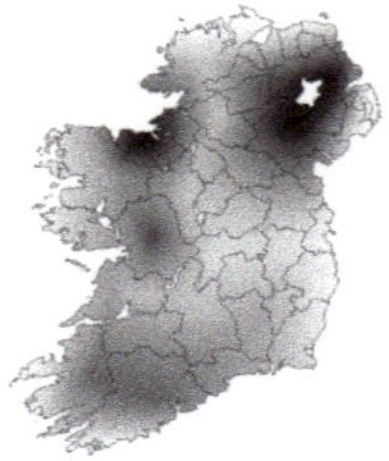

FROST/SNOW

AUGUST **29** SATURDAY

The depression to the west of Ireland may once again drift back down over the country bringing a return to moderate to strong southerly winds, backing northwesterly later in the day. A stormy air may return with widespread cloud, strong gusty winds at times, and scatered showers throughout. Rain may be heavy at times in southern parts. Chance of thunderstorms and gale gusts.

Belfast: Cloudy, light showers, warmer, strong blustery winds at times
Dublin: Cloudy, scattered showers, strong and unsettled winds with gusts
Cork: Strong gusty winds, overcast, stormy air, rain
Galway: Cloudy, slightly warmer, brief showers, strong winds, chance of gales

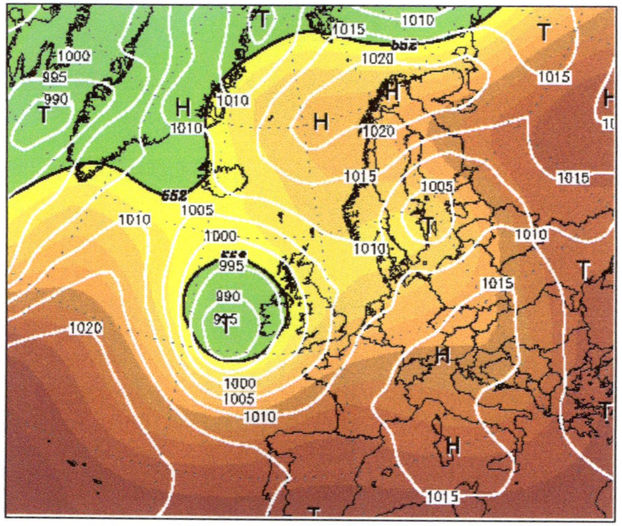

RAIN POTENTIAL

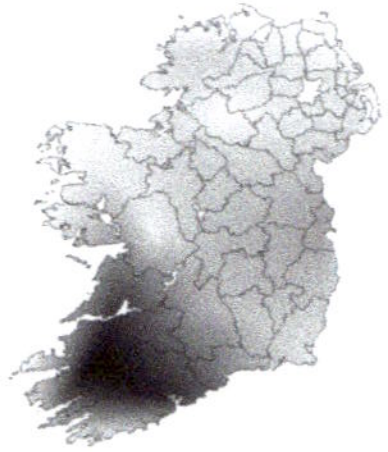

FROST/SNOW

AUGUST **30** SUNDAY

The depression centred over Ireland begins to weaken around this time with winds generally in a west to northwest flow. Mostly cloudy throughout, with some lingering showers mostly in the eastern half of the country, while clearing to mostly dry in the west and south.

Belfast: Cloudy, cooler, dry, winds easing to light, chance of fog overnight
Dublin: Overcast, brief squally showers, mild temperatures, fluctuating winds
Cork: Partly cloudy, blustery and stormy winds, mostly dry, threats of rain
Galway: Cloudy, dry, mild temperatures, strong and unsettled winds

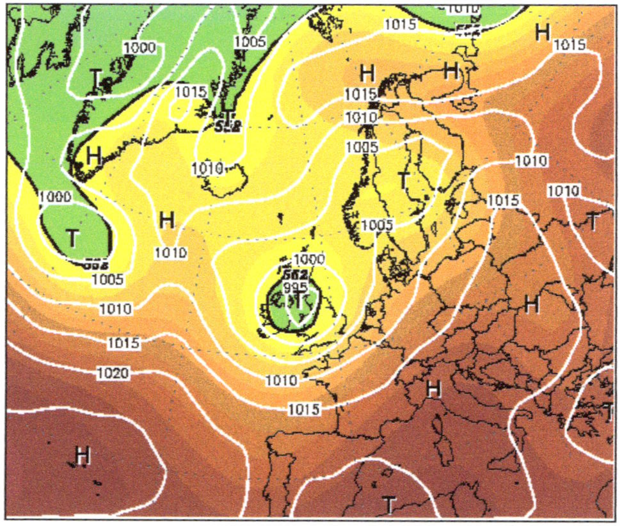

RAIN POTENTIAL

FROST/SNOW

AUGUST 31 MONDAY

Once more the depression that has brought stormy conditions over the previous week may weaken and move away to the northwest of Ireland around this time, with winds lighter in a west to northwest flow. In contrast, a high pressure ridge builds to the south of the country. Mostly cloudy throughout, and dry with possible patches of fog overnight in the northern half of the country, while showers may prevail over the southern half of Ireland.

Belfast: Dry, warmer, mostly cloudy and calm air, chance of fog
Dublin: Cloudy, mild temperatures, dry, moderate breezes
Cork: Cloudy, mild temperatures, rain, moderate breezes
Galway: Partly cloudy, winds easing, rain, cooler

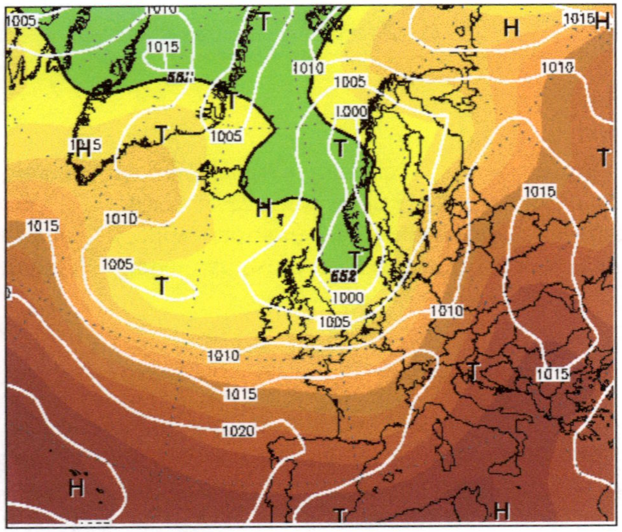

RAIN POTENTIAL

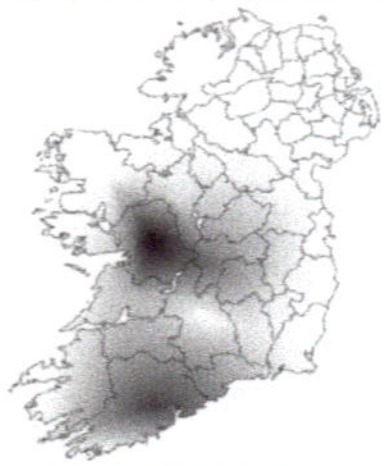

FROST/SNOW

SEPTEMBER

PHASES OF THE MOON

4th	Third Quarter
5th	Northern Declination
6th	Perigee #12
11th	New Moon
12th	Crossing Equator
18th	First Quarter
18th	Southern Declination
19th	Apogee
26th	Full Moon
26th	Crossing Equator

1	2	3	4	5	6	7	8	9	10	11	12	13	14	15	16	17	18	19	20	21	22	23	24	25	26	27	28	29	30
Tue	Wed	Thur	Fri	Sat	Sun	Mon	Tue	Wed	Thur	Fri	Sat	Sun	Mon	Tue	Wed	Thur	Fri	Sat	Sun	Mon	Tue	Wed	Thur	Fri	Sat	Sun	Mon	Tue	Wed
			3Q	^	P12					N	XhS						1Q V	A							F XhN				

MONTHLY SUMMARY

1st-11th A high pressure influence that started to build at the end of last month may continue over the next 10 days, with a more settled period prevailing. Despite this, some scattered rain activity may still occur almost daily, but any rainfall may be much lighter and more sparse than the previous couple of weeks in late August. During this period, there may be very little variation between day and night temperatures, and most places may see very little sunshine on a number of days.

3rd Chance of heavy rain in many parts of the country.

5th-7th Warmer spell. Very dull and gloomy almost everywhere.

7th A large high may cover the country.

8th Unusually warm day throughout.

9th-11th Very dull and gloomy air with widespread cloud throughout.

Over the next four weeks, the wet conditions of the previous four weeks may continue for much of this outlook period, with drier conditions finally more prevalent by mid-October. Low pressure may be the general trend for much of this outlook period, with high pressure dominating mostly from the end of September into the first couple of weeks of October. Generally, the next four weeks may be wetter than what is typical for this time of year, particularly in eastern parts, sunnier than average except in the east with temperatures nearer to average for this time of year. Overall, the next four weeks may be drier than average in some parts of the north, southeast and southwest, while average to wetter elsewhere. Generally, there may be between 12-15 wet days during this outlook period, which is near to average for this time of year. Sunshine levels may be near average in most parts of the country, except in the southwest where up to a third more hours of sunshine than average may occur. The first couple of weeks ahead may be cloudier than average, while the second half of this outlook period may be sunnier than average. Temperatures during this outlook period may be close to average throughout, with air temperatures within half a degree of normal almost everywhere. Some patches of overnight frosts may once again start to appear nearer to the end of this outlook period. During this outlook period there may be 6 days with thunderstorms, 7 days with widespread fogs, 5-7 days on which frosts are observed, and 3 days with gale gusts, but no days with snow or hail.

12th-27th Unsettled weather conditions of the previous few weeks may continue with a mix of spells of sunshine, rain or showers on most days, sometimes accompanied by strong winds.

12th-14th Very dull and gloomy spell. Chance of widespread thunderstorms likely, particularly over the eastern half of the country.

12th-18th Spells of heavy rain likely.

12th-23rd Low pressure may dominate the weather.

15th-16th Very wet spell with possible heavy falls for a time in the east.

16th-17th A very intense depression may bring very heavy rain in the east of the country, particularly in the Dublin area.

18th Unusually sunny day possible, particularly sunny in the northwest.

19th-21st Spell of widespread rain activity likely with very heavy rain likely over much of the country, possibly causing localised flooding, particularly in the east and south.

21st Chance of thunderstorms.

20th-22nd A deep depression close to the northwest coast may bring very strong and blustery south to southwest winds with gale gusts at times.

28th-9th October High pressure becomes established as the predominant weather influence bringing a respite from the recent rains, with most rainfall during this time generally confined to light coastal showers.

29th A band of rain may cross the country, with heavy falls at times mostly in the north.

30th Chance of widespread patches of fog.

IRELAND RAINFALL ESTIMATES

September 2026

d=mainly dry, n= nonrecordable, l=light shrs, s=significant shrs, r=rain, h=hvy falls

		ULSTER PROVINCE					CONNAUGHT PROVINCE						LEINSTER PROVINCE				MUNSTER PROVINCE						
		Alderg	Hillsbo	Armag	Loughl	Malin H	Sligo	Ardtari	Belmul	Drums	Galway	Roscoi	Markre	Edende	Kilkeni	Dublin	Shanni	Newpo	Tralee	Killarne	Cork	All	Moon
1st	Tues	d	d	d	n	l	d	n	d	d	d	d	d	d	s	n	n	d	n	l	l	17	
2nd	Wed	d	d	d	l	d	d	l	d	d	l	l	s	d	d	d	d	l	d	n	d	9	
3rd	Thur	l	s	d	s	s	s	s	r	s	h	s	d	s	s	s	r	r	h	s	r	153	
4th	Fri	l	d	s	l	l	l	l	l	l	s	l	s	d	d	d	d	s	l	d	l	42	3Q
5th	Sat	l	s	d	l	s	l	l	s	s	s	s	s	l	d	l	s	s	n	d	n	54	^
6th	Sun	l	l	l	l	l	l	l	l	l	l	l	l	d	d	l	d	s	l	d	n	32	P12
7th	Mon	n	d	l	l	l	l	l	l	d	l	n	l	d	d	d	l	l	l	d	d	19	
8th	Tues	d	d	d	d	n	l	d	l	n	d	l	s	d	d	d	d	d	l	d	d	10	
9th	Wed	d	d	d	d	d	l	d	l	l	l	l	l	d	d	d	l	l	d	n	n	12	
10th	Thur	d	d	d	l	l	l	l	s	s	n	s	l	d	d	d	d	s	l	s	s	51	
11th	Fri	l	s	d	l	s	s	l	r	s	r	l	l	l	s	n	l	s	r	s	r	87	N
12th	Sat	s	s	s	l	s	s	l	l	s	h	r	l	l	l	l	l	r	l	s	l	101	XhS
13th	Sun	s	r	d	r	l	l	r	s	r	r	s	s	r	l	l	s	r	s	h	s	148	
14th	Mon	s	s	s	r	s	s	r	r	s	l	l	r	l	l	s	r	r	h	h	s	167	
15th	Tues	l	d	l	d	d	l	d	l	d	d	d	d	l	r	d	s	n	l	l	s	42	
16th	Wed	s	r	d	r	l	s	r	r	s	r	r	s	h	r	h	r	s	s	s	ll	256	
17th	Thur	r	d	h	d	d	r	d	d	d	d	n	d	d	d	s	l	d	l	d	l	73	
18th	Fri	n	d	l	d	d	l	d	l	d	d	d	d	d	s	d	d	d	d	d	d	10	1Q V
19th	Sat	l	l	d	r	d	l	r	s	l	h	s	h	l	l	d	d	r	l	r	s	119	A
20th	Sun	l	s	d	r	s	s	r	r	h	l	r	s	s	l	s	h	h	r	s	h	220	
21st	Mon	s	s	s	h	s	r	h	h	r	h	r	s	s	d	l	l	r	r	h	r	226	
22nd	Tues	s	l	s	s	s	s	s	l	r	l	s	s	d	d	l	l	r	s	s	s	83	
23rd	Wed	s	s	s	d	d	s	d	l	d	s	l	d	l	l	l	l	l	l	n	l	35	
24th	Thur	l	n	s	r	d	l	r	l	s	ll	s	r	d	d	d	l	s	s	l	s	87	
25th	Fri	s	h	d	h	s	l	h	l	h	r	h	h	h	r	r	s	h	s	s	r	282	
26th	Sat	s	d	h	d	s	s	d	d	l	d	l	n	l	d	s	d	d	d	d	d	48	F XhN
27th	Sun	d	d	d	d	n	d	d	d	d	d	d	d	l	d	l	d	d	d	d	d	3	
28th	Mon	d	d	d	d	n	d	d	d	d	d	d	d	d	d	d	d	d	d	d	d	1	
29th	Tues	n	l	d	s	s	s	s	r	l	s	l	s	l	n	l	l	s	n	n	d	57	
30th	Wed	l	d	l	l	l	l	l	l	d	l	n	n	d	d	d	l	d	d	d	d	13	
Estimate:		92	93	90	149	82	110	149	129	153	172	148	148	107	78	104	92	168	119	123	152	2457	
Average:		76	81	68	99	96	104	106	102	79	100	105	104	78	70	60	76	132	128	87	95	1846	
Trend:		w tr	w tr	w tr	w tr	drr	av	w tr	w tr	w tr	w tr	w tr	w tr	w tr	av	w tr	w tr	w tr	av	w tr	w tr	w tr	

IRELAND SUNSHINE ESTIMATES

September 2026

F=fine (8-12 hours of sunshine), pc= partly cloudy (4-7 hours), c=cloudy (1-3 hours), o=overcast (0 hours)

		ULSTER PROVINCE					CONNAUGHT PROVINCE						LEINSTER PROVINCE				MUNSTER PROVINCE						
		Alderg	Hillsbo	Armag	Loughl	Malin H	Sligo	Ardtarı	Belmul	Drums	Galway	Roscoı	Markre	Edendı	Kilkenı	Dublin	Shannı	Newpo	Tralee	Killarnı	Cork	All	Moon
1st	Tues	pc	pc	pc	pc	pc	F	F	F	pc	F	F	pc	pc	pc	pc	pc	pc	pc	pc	pc	151	
2nd	Wed	c	c	c	c	c	c	c	pc	c	pc	pc	c	c	c	pc	F	pc	c	c	c	85	
3rd	Thur	o	o	o	o	o	o	o	o	o	o	o	o	o	o	o	o	o	o	o	o	8	
4th	Fri	pc	pc	pc	pc	pc	c	pc	c	pc	pc	pc	pc	pc	pc	F	pc	pc	c	c	c	112	3Q
5th	Sat	c	c	c	c	c	pc	c	pc	c	c	pc	c	c	o	o	o	o	o	o	o	53	^
6th	Sun	o	o	o	o	o	o	o	o	o	o	o	o	o	o	o	o	o	o	o	o	2	P12
7th	Mon	o	o	o	o	o	c	o	o	o	o	o	o	o	o	o	o	o	o	o	o	6	
8th	Tues	c	c	c	c	c	o	c	o	pc	c	o	c	pc	pc	pc	c	pc	c	c	F	58	
9th	Wed	o	o	o	o	o	o	o	o	o	o	o	o	o	o	o	o	o	o	o	o	6	
10th	Thur	o	o	o	o	o	c	o	c	o	c	c	o	o	o	o	o	o	o	o	o	14	
11th	Fri	pc	pc	pc	pc	pc	pc	pc	pc	pc	pc	pc	pc	pc	pc	c	pc	pc	pc	pc	pc	104	N
12th	Sat	F	F	F	F	F	pc	F	F	F	pc	F	F	F	F	F	pc	F	pc	pc	F	161	XhS
13th	Sun	pc	pc	pc	pc	pc	c	pc	F	pc	pc	F	pc	pc	pc	pc	pc	pc	pc	pc	pc	126	
14th	Mon	pc	pc	pc	pc	pc	pc	pc	pc	pc	pc	pc	pc	pc	c	pc	pc	pc	c	c	c	95	
15th	Tues	pc	pc	pc	pc	pc	F	pc	F	pc	F	F	pc	pc	pc	pc	pc	pc	F	F	pc	153	
16th	Wed	o	o	o	o	o	o	o	pc	o	c	pc	o	o	o	o	o	o	o	o	o	15	
17th	Thur	pc	pc	pc	pc	pc	F	pc	F	c	pc	pc	pc	c	c	o	pc	pc	pc	pc	pc	110	
18th	Fri	c	c	c	c	c	o	c	o	c	o	o	pc	c	pc	pc	c	c	c	c	pc	49	1Q V
19th	Sat	pc	pc	pc	pc	pc	c	pc	pc	c	c	pc	pc	c	c	pc	c	c	o	o	o	64	A
20th	Sun	c	c	c	c	c	o	c	o	c	o	o	c	c	c	c	o	c	c	c	c	32	
21st	Mon	o	o	o	o	o	o	o	o	o	o	o	o	o	o	o	o	o	o	o	o	7	
22nd	Tues	c	c	c	c	c	c	c	c	c	c	c	c	c	c	c	pc	c	pc	pc	pc	60	
23rd	Wed	c	c	c	c	c	c	c	pc	c	c	pc	c	c	c	c	c	c	c	c	c	61	
24th	Thur	pc	pc	pc	pc	pc	o	pc	o	pc	o	o	pc	pc	pc	F	o	c	o	o	o	60	
25th	Fri	o	o	o	o	o	o	o	o	o	o	o	o	o	o	o	o	o	o	o	o	1	
26th	Sat	c	c	c	c	c	pc	c	pc	c	pc	pc	c	c	c	o	pc	c	pc	pc	c	70	F XhN
27th	Sun	c	c	c	c	c	c	c	F	c	pc	pc	c	c	c	o	o	o	c	c	c	44	
28th	Mon	c	c	c	c	c	pc	c	pc	c	F	pc	c	c	c	o	F	pc	F	F	pc	112	
29th	Tues	c	c	c	c	c	o	c	o	c	o	o	c	c	pc	pc	o	c	c	c	c	50	
30th	Wed	o	o	o	o	o	o	o	o	o	o	o	o	o	o	o	o	o	c	c	o	5	
	Estimate:	91	91	91	91	91	89	91	122	90	105	114	91	90	89	92	87	88	91	91	86	1871	
	Average:	118	118	111	111	105	99	101	110	96	111	120	111	119	115	129	111	95	109	108	124	2219	
	Trend:	less	less	less	less	less	av	less	more	av	av	av	less	less	less	less	less	av	less	less	less	less	

SEPTEMBER **1** TUESDAY

Low pressure directs a light west to northwest flow over the country. Mostly dry and sunny day except for the chance of some isolated showers overnight in the northwest, midlands, and southeast.

Belfast: Mostly sunny and dry, mild temperatures, chance of fog patches
Dublin: Partly cloudy, isolated showers, mild day cool night, moderate breezes
Cork: Partly cloudy, light showers, mostly calm air, chance of fog
Galway: Sunny, dry, mild temperatures, light breezes to mostly calm air

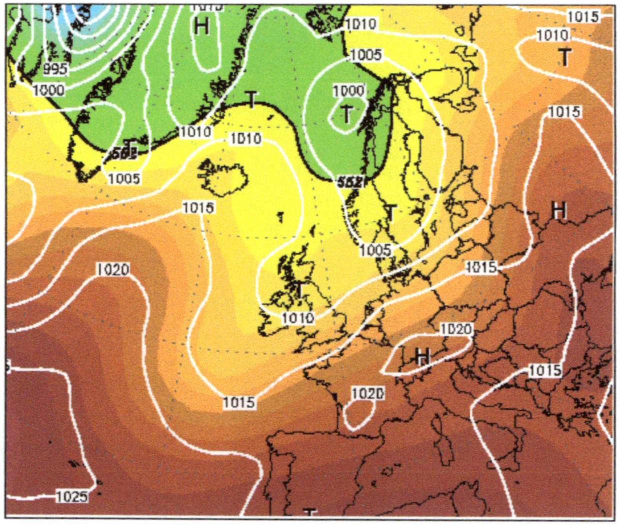

RAIN POTENTIAL

FROST/SNOW

SEPTEMBER **2** WEDNESDAY

A high pressure ridge builds to the west of Ireland while low pressure lingers to the east, with a light to moderate northwesterly flow prevailing throughout. Occasional scattered cloud, and mostly dry conditions everywhere. Fog patches possible overnight, particularly in the north and coastal southern parts.

Belfast: Mostly cloudy, dry, calm and pleasant air, chance of overnight fog
Dublin: Partly cloudy, dry, light breezes to mostly calm air
Cork: Cloudy, dry, gentle breezes to mostly calm air, chance of fog patches
Galway: Partly cloudy, spotty showers, light breezes, mild day cool night

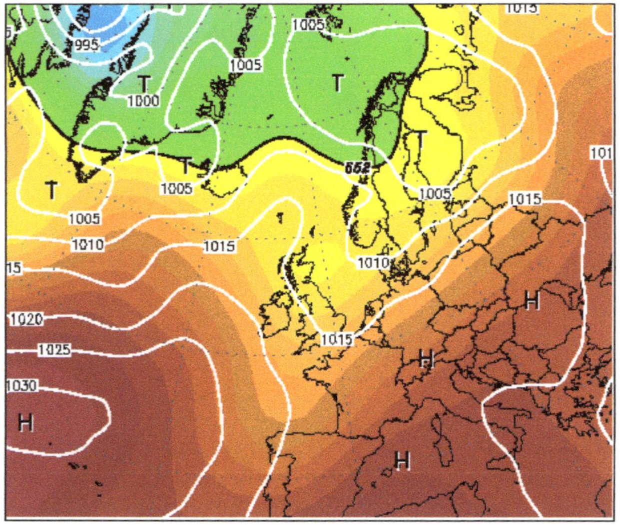

RAIN POTENTIAL

FROST/SNOW

SEPTEMBER **3** THURSDAY

A high pressure ridge passes to the south of Ireland while low pressure prevails to the north. Generally light to moderate west to southwest breezes prevail, more gusty at times along the northern coasts. Cloud likely to build to a mostly overcast and dull day throughout. Mostly dry at first before bands of rain move into the western areas of Ireland at first and spread eastwards to bring scattered outbreaks of rain everywhere. Rain may be heavier for a time in some western areas to the midlands.

Belfast: Overcast, light breezes, occasional showers
Dublin: Overcast, occasional showers, light breezes to mostly calm air
Cork: Gentle breezes, overcast, rain, cooler
Galway: Overcast, heavy rain, cooler, light breezes

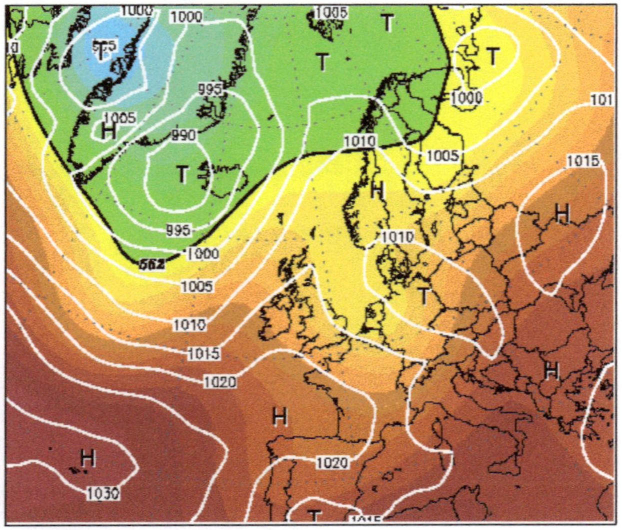

RAIN POTENTIAL

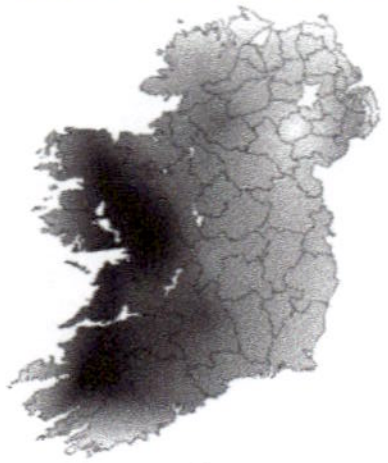

FROST/SNOW

SEPTEMBER **4** FRIDAY

An intense depression builds to the north of Ireland with a strengthening westerly flow prevailing, gusty at times along the north and western coasts. Chance of fog patches overnight mostly in the north and central parts. Mostly dry and sunny in the east, while a mixed day of sunny spells and passing showers likely elsewhere.

Belfast: Partly cloudy, cooler, strong and unsettled winds, chance of gale gusts
Dublin: Sunny, warmer, dry, strengthening winds
Cork: Cloudy, spotty showers, strong and blustery winds
Galway: Partly cloudy, scattered showers, strong and gusty winds, chance of gales

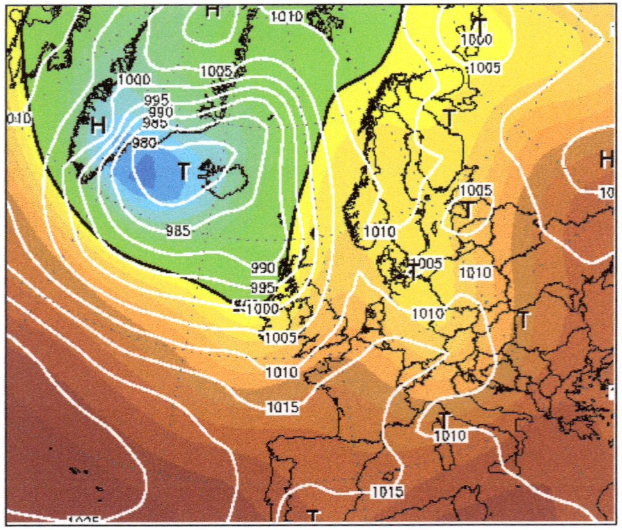

RAIN POTENTIAL

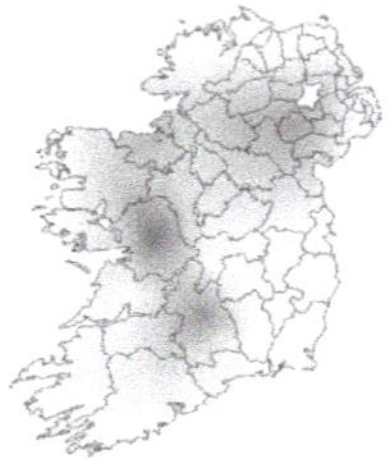

FROST/SNOW

SEPTEMBER **5** SATURDAY

An intense depression to the north of Ireland brings strong and blustery west to southwest winds across the country and a threatening stormy air. Widespread cloud prevails throughout, with rain and drizzle patches likely to spread over the western half of the country, while remaining mostly dry and cloudy with increasing threats of rain in the east.

Belfast: Stormy air, mostly cloudy, warmer, squally showers
Dublin: Overcast, mild temperatures, spotty showers, stormy winds at times
Cork: Overcast, mild temperatures, isolated showers, windy with gusts
Galway: Unsettled winds, chance of gales, mix of showers and sunny outbreaks

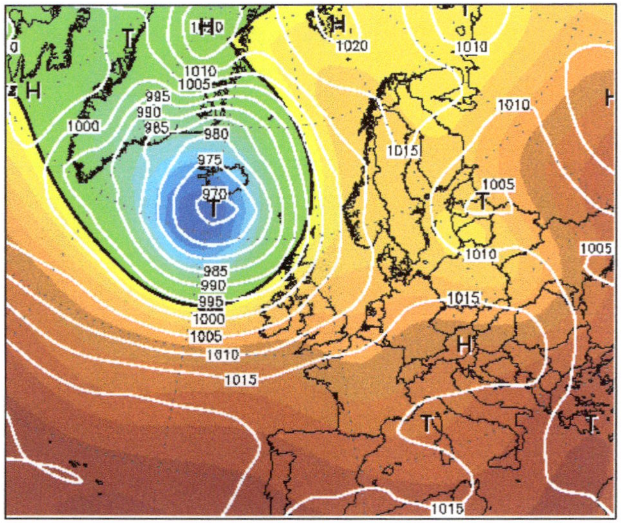

RAIN POTENTIAL

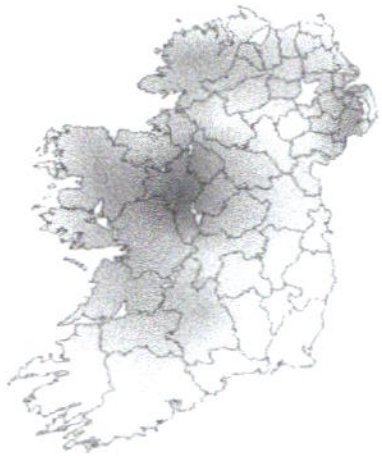

FROST/SNOW

SEPTEMBER **6** SUNDAY

The depression to the north of Ireland continues to bring a west to southwest flow across the country, but with winds decreasing throughout. Dull and gloomy with passing showers almost everywhere, but mostly dry with rain threatening along the east and southern coasts.

Belfast: Overcast, windy at times, light showers
Dublin: Overcast, brief showers, mild temperatures, fresh winds
Cork: Overcast, isolated showers, mild temperatures, windy at times
Galway: Cooler, overcast, strong winds, brief showers

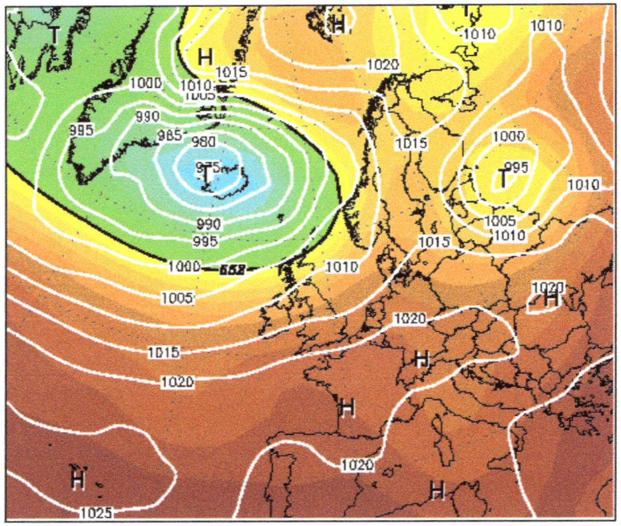

RAIN POTENTIAL

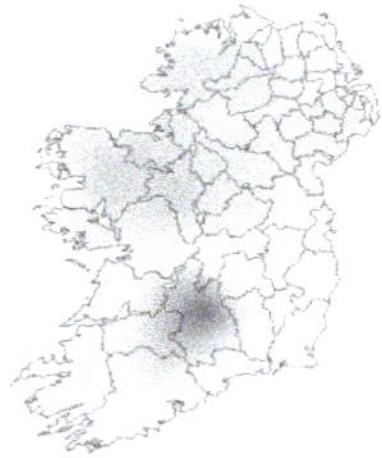

FROST/SNOW

SEPTEMBER 7 MONDAY

The depression to the north of Ireland weakens as it drifts further northwards, with a high pressure ridge building to the south of the country. Winds generally light to moderate in a southwest flow, with just the occasional spell of stronger gusts mostly around the coasts. Widespread cloud continues to prevail throughout, with some spotty showery spells and drizzle patches mostly in the western half of the country. Mostly dry in the east.

Belfast: Overcast, cooler day milder night, dry, threats of rain, windy at times
Dublin: Overcast, mostly dry and cooler, moderate breezes, odd threat of rain
Cork: Overcast, mostly dry, some threats of rain, mild temperatures, fresh breezes
Galway: Overcast, blustery at times, squally showers

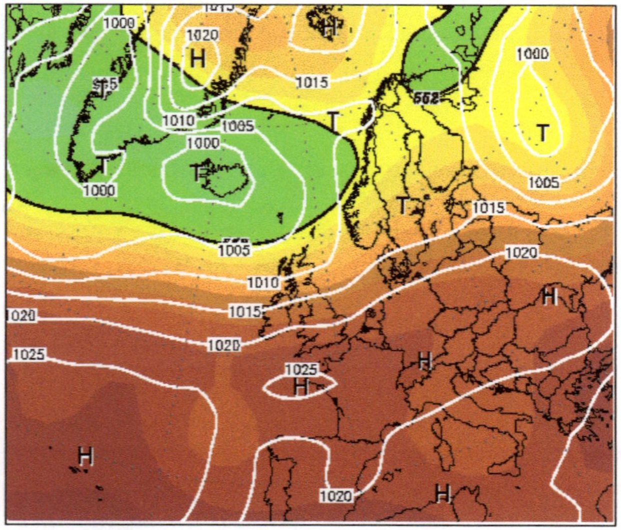

RAIN POTENTIAL

FROST/SNOW

SEPTEMBER 8 TUESDAY

The high pressure system to the southwest of Ireland ridges over the country, with a light westerly flow prevailing. Chance of fog patches overnight mostly in central parts. Occasional sunny spells and mostly dry, except for the chance of some light patchy rain or drizzle briefly in areas near the western coasts.

Belfast: Cloudy, mild temperatures, dry, moderate breezes
Dublin: Warm, dry, partly cloudy, moderate breezes
Cork: Light breezes, dry, sunny, warm
Galway: Cloudy, dry, strong winds at times

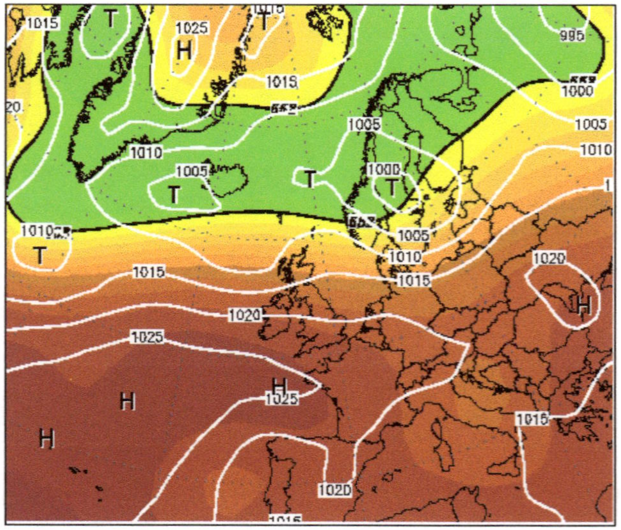

RAIN POTENTIAL

FROST/SNOW

SEPTEMBER **9** WEDNESDAY

Anticyclonic. Winds generally light in variable directions. Dull and gloomy everywhere with widespread cloud and some patches of fog overnight mostly in central and southern parts. Chance of some brief showers or drizzle patches in the central west of the country, while mostly dry elsewhere.

Belfast: Overcast, warm, dry, gentle breezes to mostly calm air
Dublin: Overcast, slightly cooler, dry with passing threats of rain, light breezes
Cork: Cooler, overcast, isolated showers, mostly calm air, fog patches possible
Galway: Overcast, slightly warmer, light showers, winds easing to light

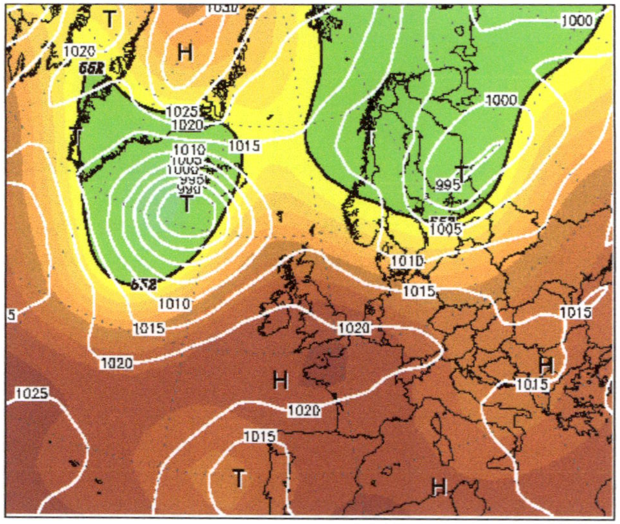

RAIN POTENTIAL

FROST/SNOW

SEPTEMBER **10** THURSDAY

A depression centred to the northwest of Ireland brings a light southwest flow over the country with widespread cloud throughout. Mostly dry in the east to the northeast, with light showers elsewhere. Chance of fog patches overnight in many places.

Belfast: Overcast, mostly dry, some threats of rain, gentle breezes, possible fog
Dublin: Overcast, dry, odd threat of rain, gentle breezes, chance of fog
Cork: Overcast, showers, cooler, light breezes, chance of drizzle or fog patches
Galway: Cloudy, isolated showers, moderate to light breezes

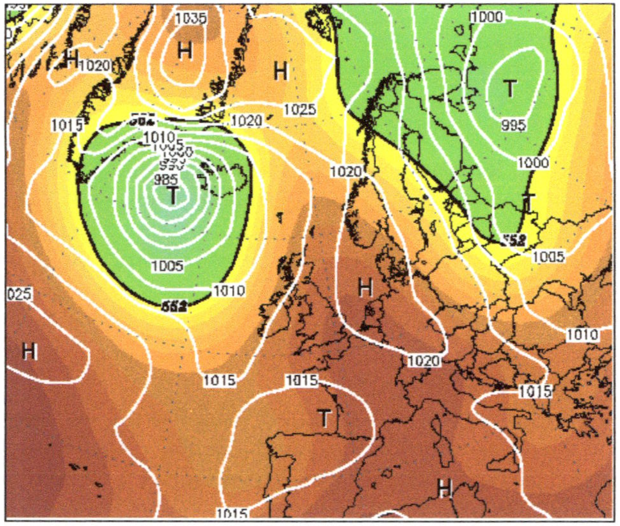

RAIN POTENTIAL

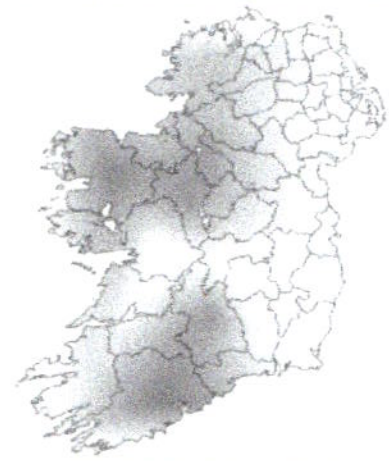

FROST/SNOW

SEPTEMBER **11** FRIDAY

Low pressure to the northwest of Ireland brings a light, mostly westerly breeze throughout, later backing south to southeasterly. A mixed day of some sunny spells and light showers and drizzle patches almost everywhere, easing to mostly dry by the afternoon. Further showers may again arrive towards evening, particularly in the west and south. Chance of fog mostly in the midlands.

Belfast: Partly cloudy, some showers, gentle shifting breezes
Dublin: Cloudy, isolated overnight showers, mostly calm air, fog patches possible
Cork: Partly cloudy, outbreaks of rain, mostly calm air
Galway: Partly cloudy, overnight rain, mostly calm air

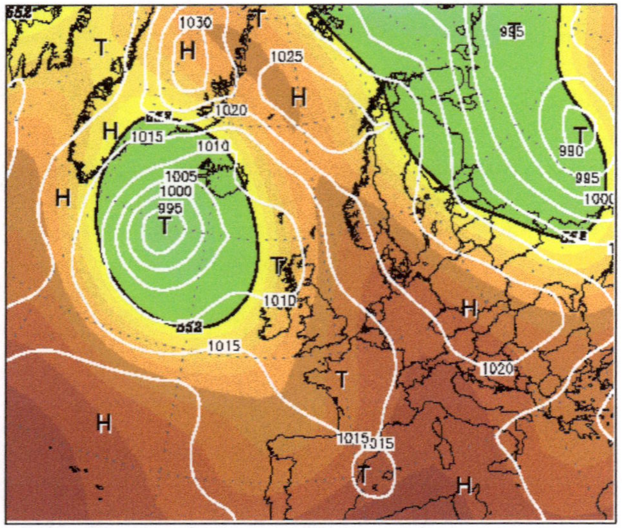

RAIN POTENTIAL

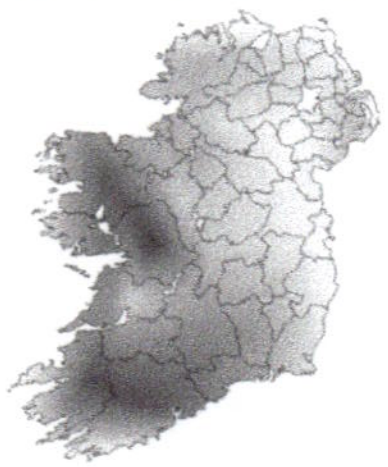

FROST/SNOW

SEPTEMBER **12** SATURDAY

The depression centred to the northwest of Ireland weakens as it moves nearer to the northwest of the country with winds generally in a light west to southwest flow throughout. Some overnight showers likely in most areas, with possible heavy falls for a time in the west, and possible thunderstorms in the north. Mostly sunny and dry day after the showers clear, before further shower activity may return later in the evening.

Belfast: Sunny, overnight showers, chance of thunderstorms, fluctuating breezes
Dublin: Sunny, gentle breezes, some overnight showers
Cork: Sunny, light overnight showers, mild day cool night, fresh breezes
Galway: Mostly sunny, heavy rain overnight, cooler, fluctuating breezes

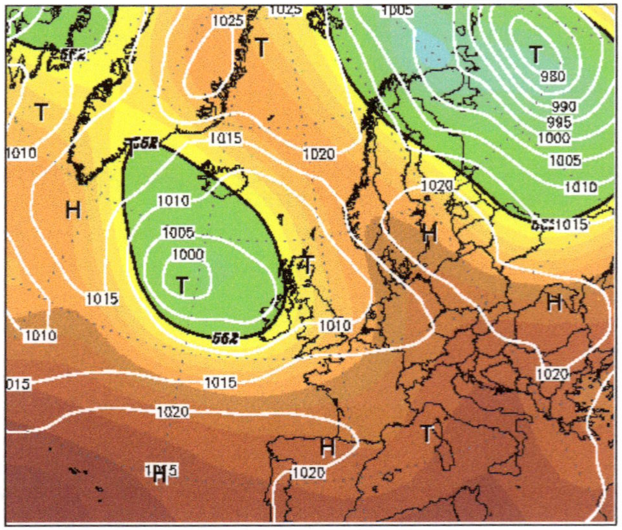

RAIN POTENTIAL

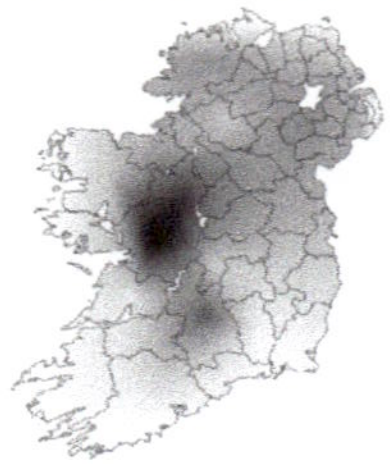

FROST/SNOW

SEPTEMBER **13** SUNDAY

A weak depression moves onto Ireland directing moderate southwesterly winds throughouth. Some showers overnight, mostly in the west, may later spread eastward into all other parts of the country. Showers may be heavy for a time. Chance of thunderstorms. Some sunny outbreaks between the bands of showers likey.

Belfast: Partly cloudy, overnight rain, cooler, moderate breezes
Dublin: Partly cloudy, cooler, light showers, moderate breezes
Cork: Partly cloudy, occasional showers, mild day cool night, fresh breezes
Galway: Mostly sunny, overnight rain, chance of thunderstorms, windy spells

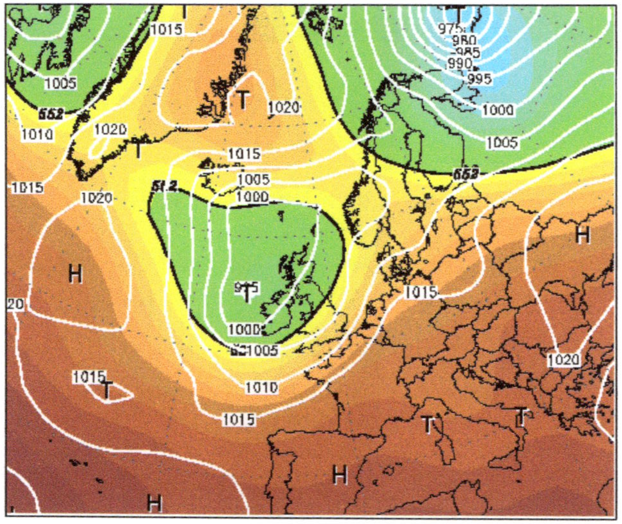

RAIN POTENTIAL

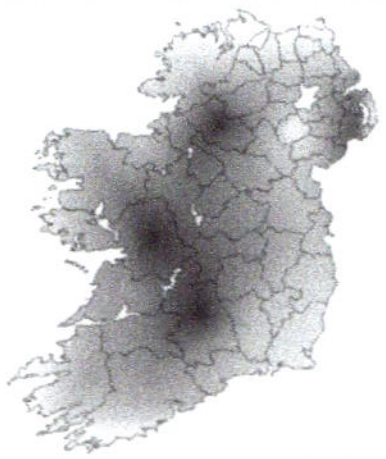

FROST/SNOW

SEPTEMBER **14** MONDAY

A depression over Ireland brings fresh to strong mostly westerly winds throughout, easing towards evening. A stormy air may prevail everywhere, with widesprad cloud and bands of rain crossing the country, sometimes accompanied by thunderstorm activity. Precipitation may be particularly heavy for a time in the southwest, while becoming more isolated in the far north.

Belfast: Stormy air, rain, some sunny spells, mild day cool night, unsettled winds
Dublin: Strong winds at times, mix of showers and some sunny spells
Cork: Unsettled windy spells, cloudy, scattered showers
Galway: Partly cloudy, brief showers, mild day colder night, unsettled winds

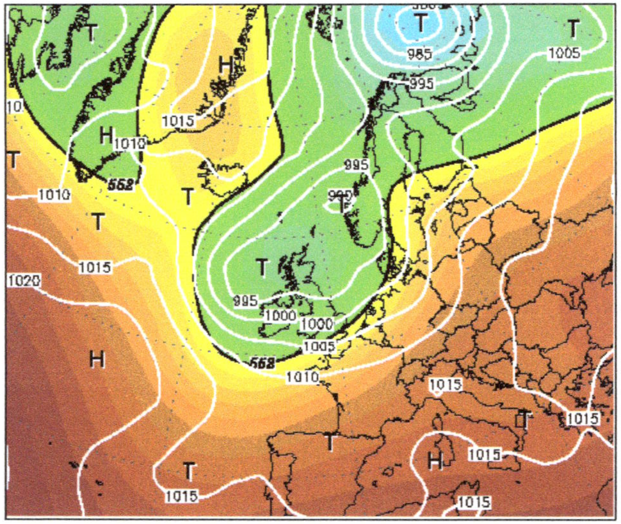

RAIN POTENTIAL

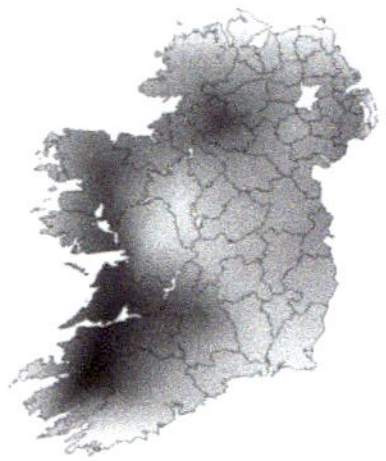

FROST/SNOW

SEPTEMBER **15** TUESDAY

A low pressure trough associated with a depression now centred to the east of Ireland brings a moderate westerly flow throughout, fresher at times around the southern coasts. A changeable day with a mix of sunny spells and passing showers, sometimes accompanying thunderstorm activity, particularly around the west and south. Mostly dry and sunny in the far northwest.

Belfast: Partly cloudy, dry, moderate and shifting breezes
Dublin: Partly cloudy, mostly dry with increasing threats of rain, moderate breezes
Cork: Partly cloudy, occasional showers, mild day cool night, moderate breezes
Galway: Sunny, dry, mild day cool night, moderate breezes

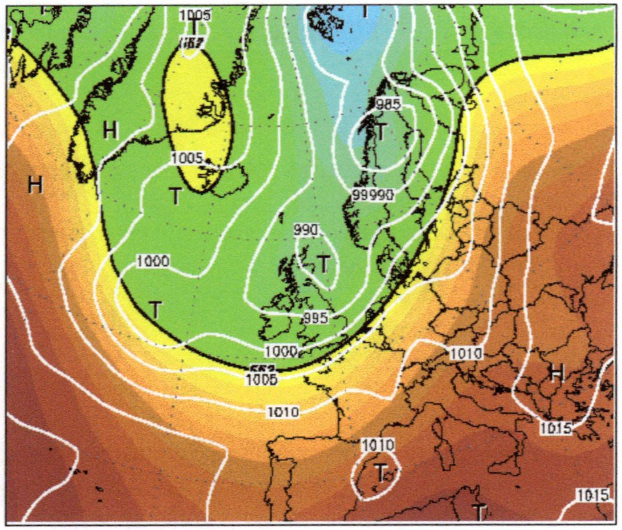

RAIN POTENTIAL

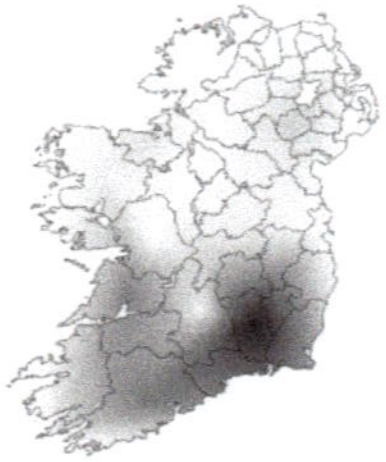

FROST/SNOW

SEPTEMBER **16** WEDNESDAY

A depression that has lingered around Ireland the past few days intensifies as it crosses the country once more, bringing widespread stormy conditions throughout, with strong winds and occasional gale gusts at times mostly in a southeasterly flow. Dull and gloomy with widespread rain everywhere. Rain may be particularly heavy at times, particularly in central parts and the east, with the chance of localised flooding in some areas.

Belfast: Overcast, rain, unsettled winds, chance of gales later in the day
Dublin: Overcast, blustery winds at times, heavy rain
Cork: Stormy air, overcast, heavy rain, chance of gales, cooler
Galway: Cloudy, rain, cooler, unsettled and fluctuating winds

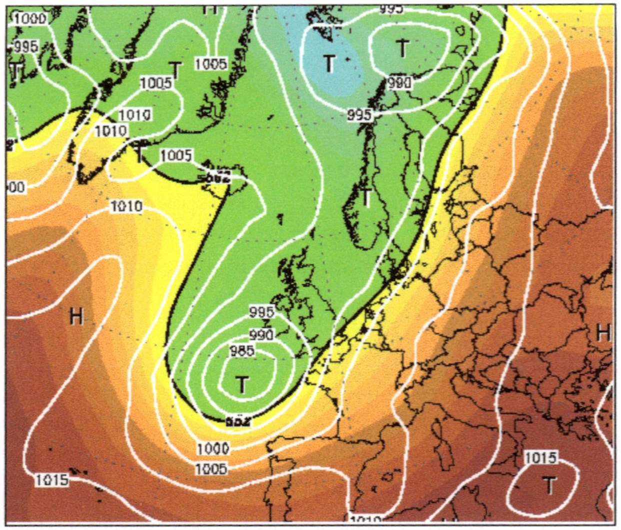

RAIN POTENTIAL

FROST/SNOW

SEPTEMBER **17** THURSDAY

A depression over Ireland around this time may continue to bring a stormy air over the country with very strong and gusty winds, southeasterly at first and then backing northeasterly with a clearing change. Overnight rain likely in many areas, with heavy falls at times in some northern areas, while only isolated showers may fall elsewhere. Later it may clear to a mostly sunny and dry day throughout, and with winds easing to light by evening.

Belfast: Partly cloudy, cooler, dry, strong and unsettled winds
Dublin: Overcast, scattered showers, stormy winds
Cork: Partly cloudy, spotty showers, slightly warmer, strong fluctuating winds
Galway: Partly cloudy, dry, warmer, strong and unsettled winds with gusts at times

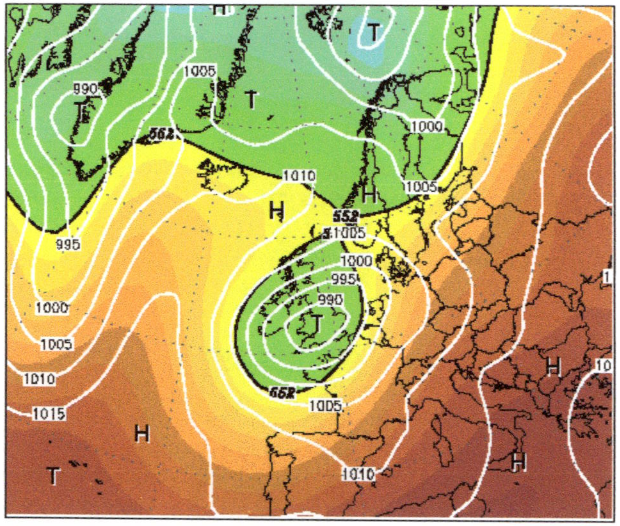

RAIN POTENTIAL

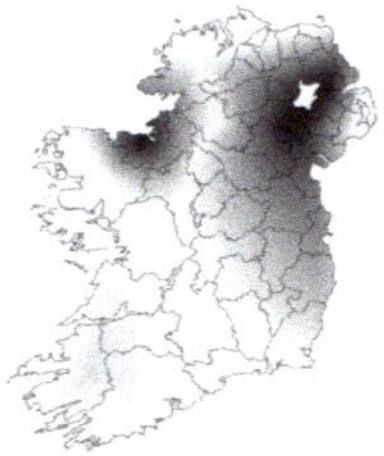

FROST/SNOW

SEPTEMBER **18** FRIDAY

A weak high pressure influence slowly builds to the west of Ireland as a depression to the east weakens as it moves away. Winds generally light in a northerly flow. Mostly dry everywhere except for some isolated showers possibly lingering in the southeast of Leinster to northeast Munster districts. Sunniest in the east and south, while cloudier in the north and west.

Belfast: Cloudy, dry, warmer day colder night, lighter winds with occasional gusts
Dublin: Partly cloudy, dry, warmer day colder night, unsettled winds easing
Cork: Partly cloudy, dry, slightly cooler, shifting windy spells
Galway: Overcast, dry with increasing threats of rain, colder, fresh breezes

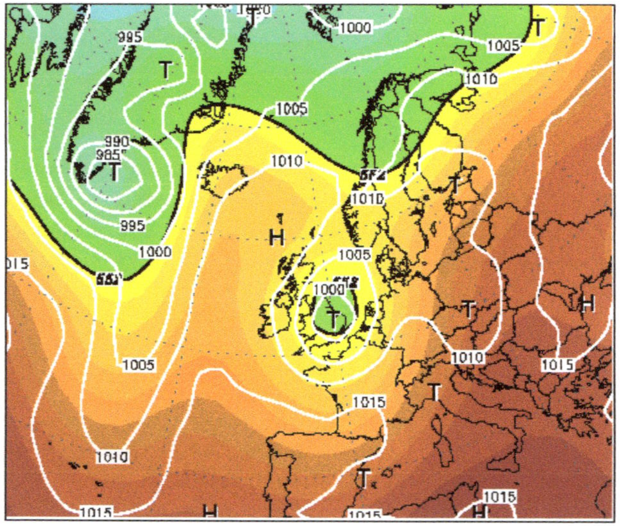

RAIN POTENTIAL

FROST/SNOW

SEPTEMBER **19** SATURDAY

The weak high pressure ridge slips away to the north of Ireland while another ridge lingers to the southeast, allowing for further low pressure troughs to return as the main influence across the country around this time. Winds generally light to moderate, in variable directions, northerly at times, backing southeasterly later. Mostly dry and sunny start to the day before cloud and showers move back onto the country later in the day.

Belfast: Partly cloudy, light showers, gentle breezes
Dublin: Partly cloudy, dry, mostly calm and pleasant air
Cork: Overcast, scattered showers, mild day cool night, gentle breezes
Galway: Mostly cloudy, heavy rain, light breezes, mild day cool night

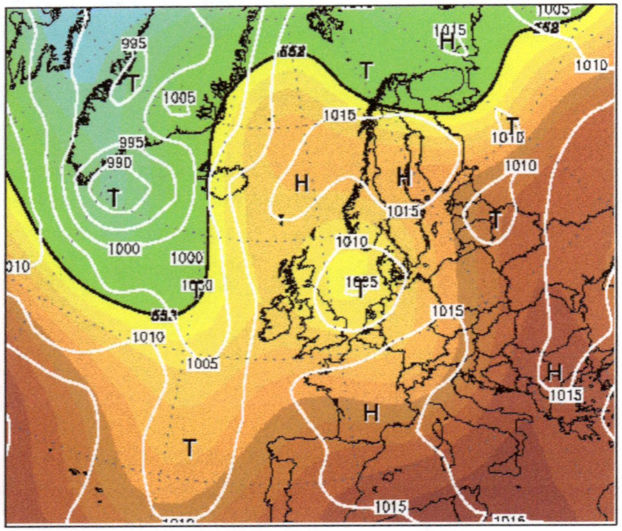

RAIN POTENTIAL

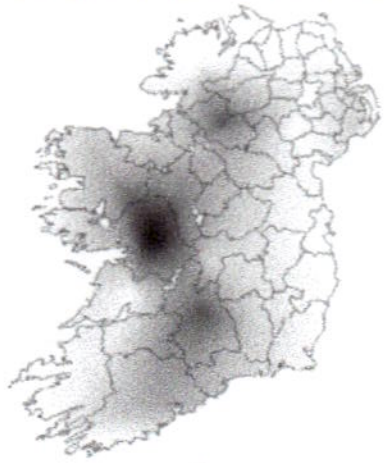

FROST/SNOW

SEPTEMBER **20** SUNDAY

Low pressure directs a moderate to strong south to southwesterly flow up through the country, with a stormy air prevailing throughout. Mostly cloudy with widespread rain in all areas, with rain particularly heavy in the south and west, possibly resulting in some pockets of localised flooding.

Belfast: Cloudy, scattered showers, mild temperatures, fresh winds
Dublin: Mostly cloudy, occasional showers, fresh breezes at times
Cork: Cloudy, heavy rain, moderate breezes
Galway: Overcast, some showers, fresh breezes

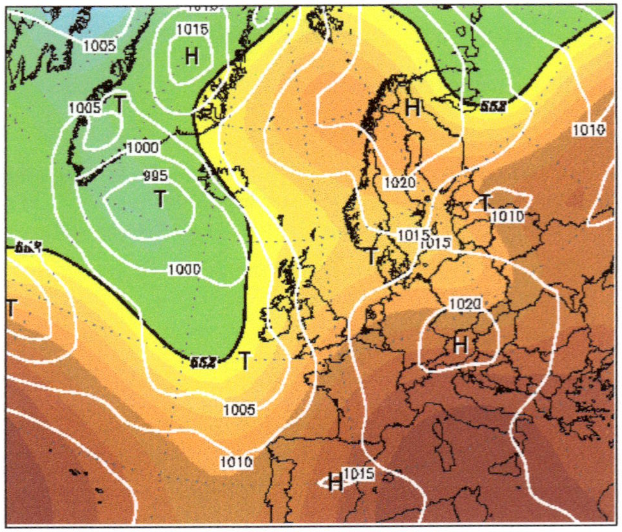

RAIN POTENTIAL

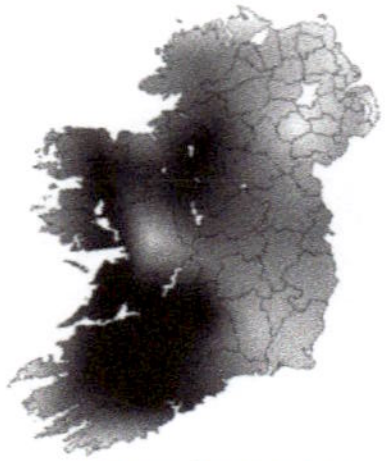

FROST/SNOW

SEPTEMBER 21 MONDAY

A depression to the west of Ireland strengthens and directs fresh to strong west to southwesterly winds throughout, with a stormy air prevailing everywhere. Unsettled conditions prevail with widespread cloud and bands of rain crossing the country. Rain may be particularly heavy for a time in the western half of the country, while lighter in the east and north.

Belfast: Unsettled and changeable, occasional gusts, overcast, scattered showers
Dublin: Fluctuating winds, overcast, brief showers
Cork: Unsettled winds, overcast, rain
Galway: Overcast, heavy rain, strong and fluctuating winds

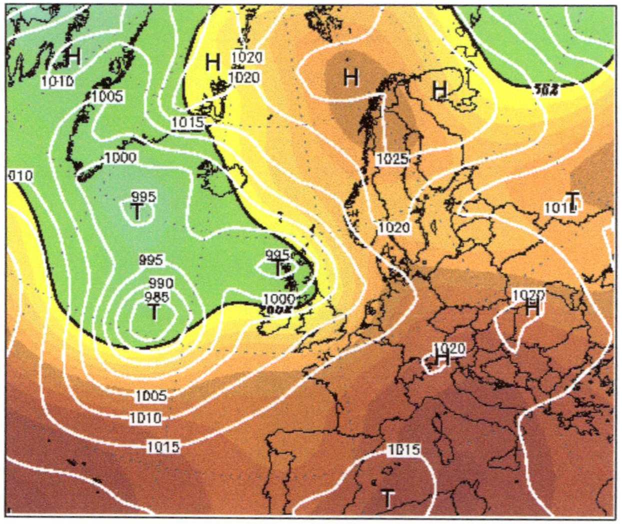

RAIN POTENTIAL

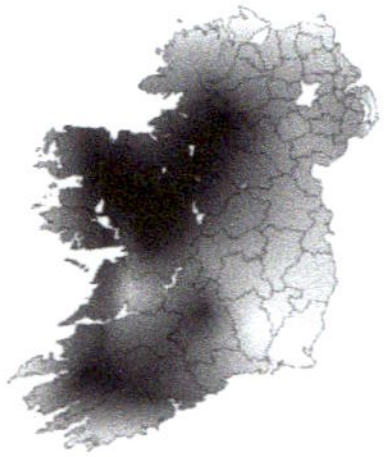

FROST/SNOW

SEPTEMBER 22 TUESDAY

The depression to the west of Ireland continues to strengthen bringing very strong and blustery southwest winds throughout, possibly up to gale force at times, later easing with a southeasterly change. Overnight showers and patches of rain in most districts, more frequent and heavier in the western half of the country. Rain activity likely to ease to light later in the day allowing for some passing sunny outbreaks.

Belfast: Cloudy, brief passing squalls, strong winds with gusts, chance of gales
Dublin: Stormy winds, cloudy, passing squalls
Cork: Partly cloudy, passing squalls, mild day cool night, stormy air, blustery winds
Galway: Partly cloudy, occasional showers, strong winds with stormy air

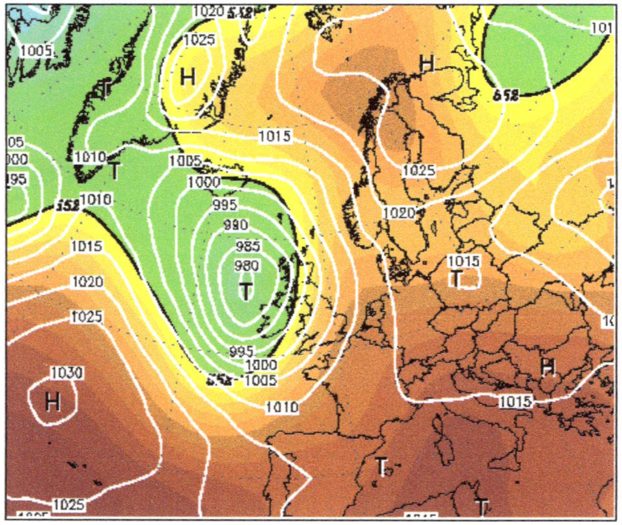

RAIN POTENTIAL

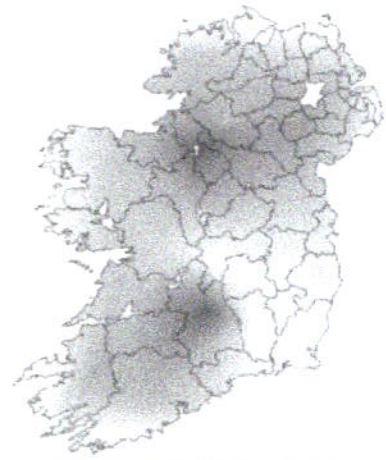

FROST/SNOW

SEPTEMBER 23 WEDNESDAY

A high pressure ridge builds to the southwest of Ireland while low pressure slips away to the north, with a light mostly northwesterly flow prevailing. A changeable day likely, with a mix of cloudy spells and isolated showers, clearing to a mostly dry afternoon, sunniest in the south and west, cloudier in the north and east.

Belfast: Cloudy, slightly warmer, scattered showers, winds easing to moderate
Dublin: Cloudy, isolated showers, winds easing to light
Cork: Winds easing, partly cloudy, light showers
Galway: Changeable, mix of showers and some sunny spells, mild day cool night

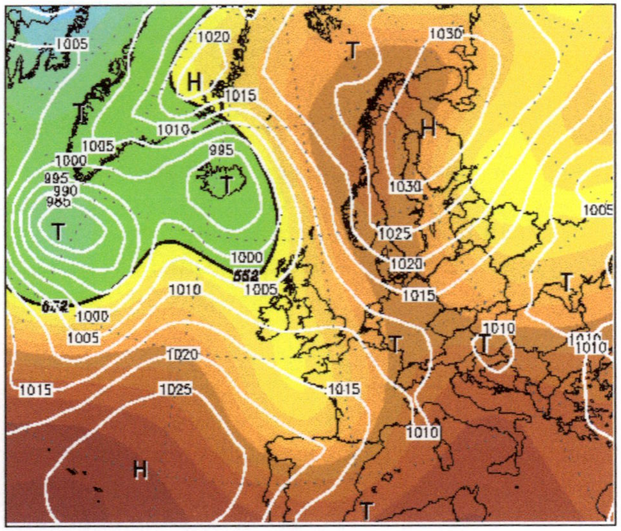

RAIN POTENTIAL

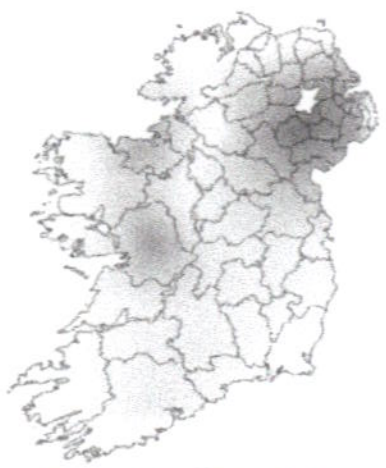

FROST/SNOW

SEPTEMBER 24 THURSDAY

The high pressure ridge nears the southern coasts of Ireland, while a low pressure trough lingers throughout. Winds generally light in a southwesterly flow, more northeasterly over the northern areas. Mostly dry and sunny in the Leinster and in the far northwest of the country, with increased cloud and occasional showers elsewhere. Chance of fog patches overnight in some southern districts.

Belfast: Partly cloudy, isolated showers, chance of overnight fog or misty patches
Dublin: Sunny, dry, light breezes to mostly calm air, threats of rain later in the day
Cork: Overcast, scattered showers, cooler, mostly calm air, chance of fog
Galway: Overcast, colder, rain, light breezes to mostly calm air

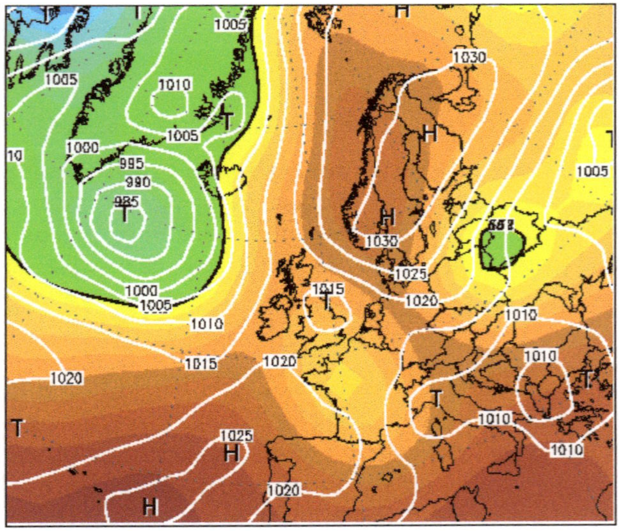

RAIN POTENTIAL

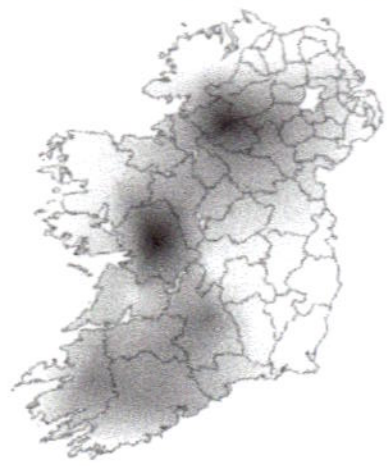

FROST/SNOW

SEPTEMBER 25 FRIDAY

A rain embedded trough over Ireland may bring light variable winds throughout. Widespread cloud with rain in all districts, starting in the west before spreading eastward. Rain may be particularly heavy at times in central to eastern areas for a time, while clearing to isolated in the west by evening.

Belfast: Overcast, heavy rain, light breezes
Dublin: Overcast, rain, cooler, mostly calm air
Cork: Overcast, rain, light breezes to mostly calm air, chance of persistent drizzles
Galway: Overcast, rain, light shifting breezes

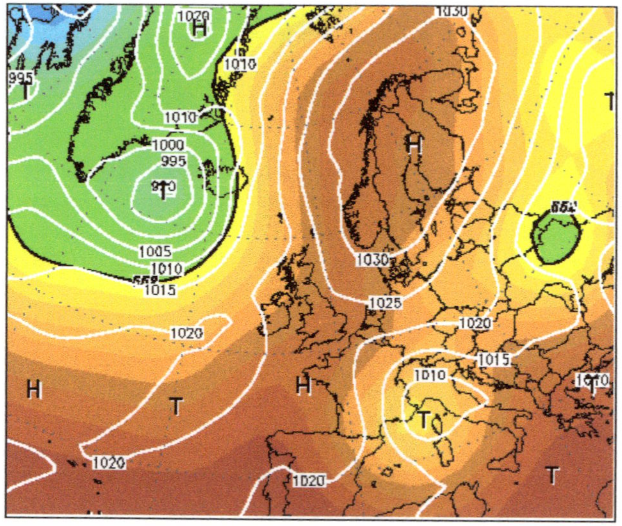

RAIN POTENTIAL

FROST/SNOW

SEPTEMBER 26 SATURDAY

Ireland may sit between two high pressure ridges around this time, with a light north to northwesterly flow prevailing. Mostly sunny in the west and south, while cloudier in the east and north. Some showers or drizzle patches over the northern half of the country and in the east, while mostly dry in the south and west. Chance of fog patches overnight in many areas.

Belfast: Cloudy, mostly dry, odd threats of rain, calm air, chance of fog patches
Dublin: Overcast, occasional showers, gentle breezes to mostly calm air
Cork: Cloudy, mostly calm air, dry, warmer
Galway: Partly cloudy, dry, mild day cool night, mostly calm air

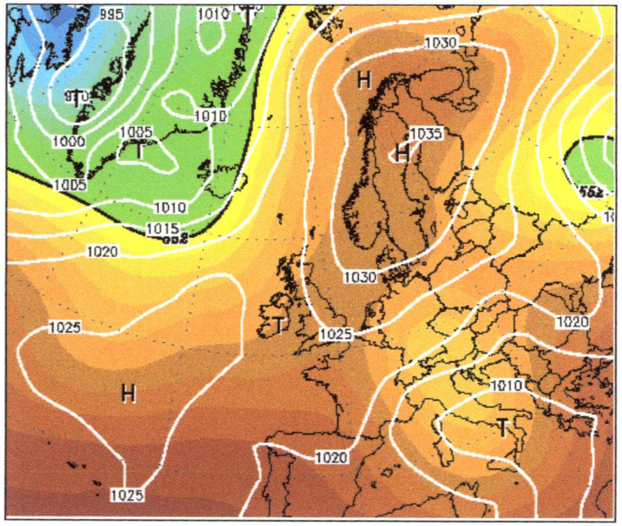

RAIN POTENTIAL

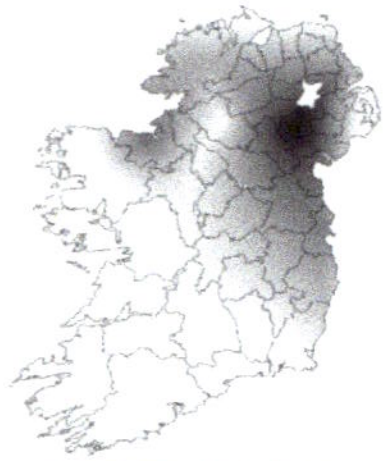

FROST/SNOW

SEPTEMBER **27** SUNDAY

A saddle between two high pressure ridges may continue to bring a light north to northwesterly flow over the country, with frequent passing cloud throughout, duller in the east while sunnier in the west. A few showers or light drizzle patches possible in the east, while mostly dry elsewhere.

Belfast: Cloudy, dry, cooler, gentle breezes to mostly calm air
Dublin: Overcast, cooler, spotty showers, light shifting breezes to mostly calm air
Cork: Light breezes to mostly calm air, cloudy, dry
Galway: Partly cloudy and dry, mild day cool night, gentle breezes

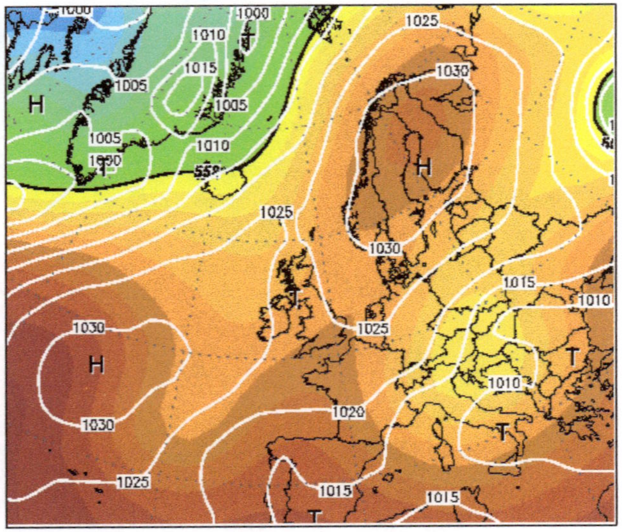

RAIN POTENTIAL

FROST/SNOW

SEPTEMBER **28** MONDAY

A large anticyclonic system covers the country with a light northerly air at times, later backing more southerly. Mostly dry and pleasant day throughout, sunny in the south and west, while cloudier in the north and east. Chance of mist or fog patches overnight particularly in northern and central areas.

Belfast: Mostly cloudy, dry, cooler, gentle shifting breezes, fog patches possible
Dublin: Overcast and cool, mostly dry, some threats of rain, gentle breezes
Cork: Mostly sunny, dry, warmer day cooler night, moderate breezes
Galway: Sunny, dry, mild day cool night, gentle shifting breezes

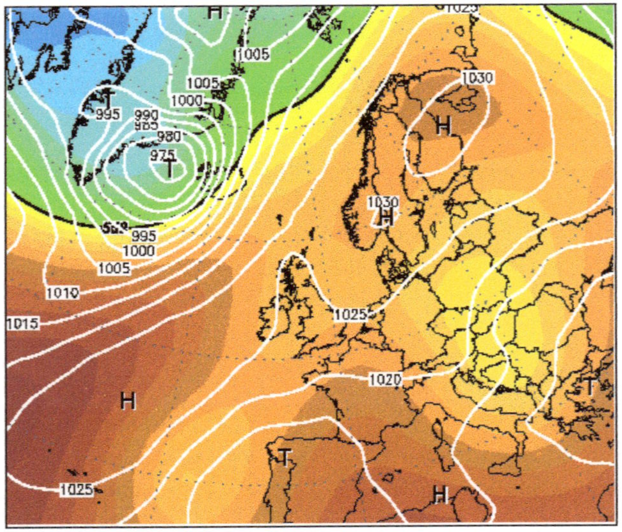

RAIN POTENTIAL

FROST/SNOW

SEPTEMBER 29 TUESDAY

An intense depression to the of Ireland builds behind the high pressure ridge situated to the west of Ireland. Light southwesterly breezes prevail, but stronger for a time in the north. Mostly cloudy throughout with chance of fog in central parts overnight, and mostly dry at first before some showers cross the western coasts and spread gradually eastwards during the day bringing some showers to most parts of the country except along the southern coasts.

Belfast: Cloudy, isolated showers, mild day cool night, mostly calm air, possible fog
Dublin: Mostly calm and serene air, partly cloudy, isolated showers
Cork: Cloudy, cooler, mostly dry, some threats of rain, calm air, fog possible
Galway: Overcast and cooler, occasional showers, mostly calm air

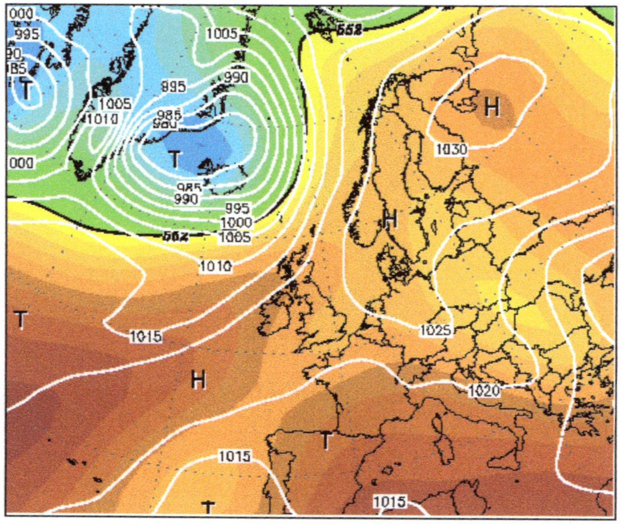

RAIN POTENTIAL

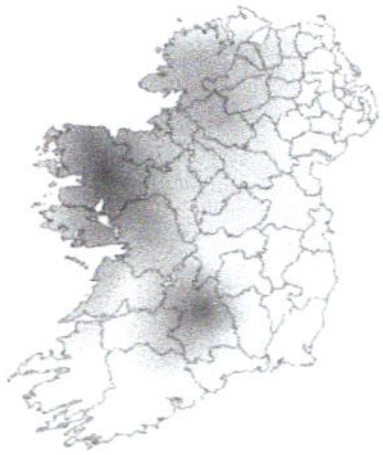

FROST/SNOW

SEPTEMBER 30 WEDNESDAY

A high pressure ridge continues to linger over the country with winds generally light in a southwesterly flow. Patchy showers likely overnight mostly around the western coasts, while patches of fog likely from the north, through central parts to the southern districts. Mostly dull and gloomy during the day, but remaining mostly dry with passing threats of rain.

Belfast: Overcast, dry, light breezes to mostly calm air, fog patches possible
Dublin: Dull and gloomy air, dry, gentle shifting breezes
Cork: Overcast, gentle breezes, dry, chance of misty patches
Galway: Overcast, light showers, light breezes

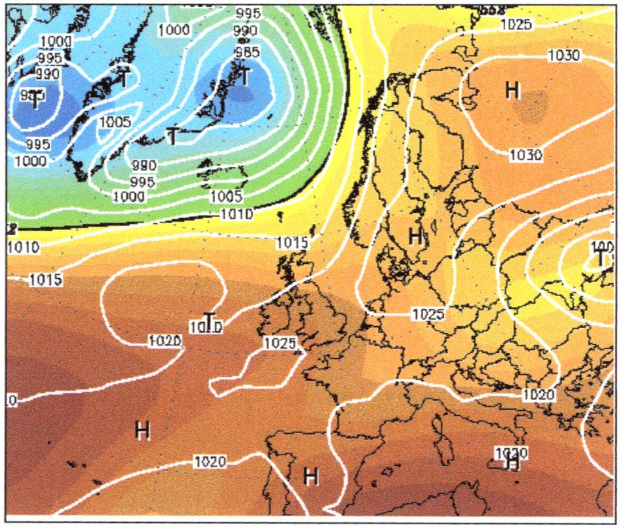

RAIN POTENTIAL

FROST/SNOW

OCTOBER

PHASES OF THE MOON

1st	Perigee #13
2nd	Northern Declination
3rd	Third Quarter
10th	New Moon
10th	Crossing Equator
16th	Apogee
16th	Southern Declination
18th	First Quarter
24th	Crossing Equator
26th	Full Moon
28th	Perigee #9
29th	Northern Declination

1	2	3	4	5	6	7	8	9	10	11	12	13	14	15	16	17	18	19	20	21	22	23	24	25	26	27	28	29	30	31
Thur	Fri	Sat	Sun	Mon	Tue	Wed	Thur	Fri	Sat	Sun	Mon	Tue	Wed	Thur	Fri	Sat	Sun	Mon	Tue	Wed	Thur	Fri	Sat	Sun	Mon	Tue	Wed	Thur	Fri	Sat
P13	^	3Q							N XhS						A V		1Q						XhN		F		P9	^		

MONTHLY SUMMARY

1st-9th October High pressure that began to establish itself as the predominant weather influence at the end of last month continues and brings a respite from the recent rains, with most rainfall during this time generally confined to light coastal showers. Sunnier spell with winds generally light and variable in direction. Clear and calm conditions during this period may see patches of widespread mist or fog on several mornings.

1st Unusually warmer than average day before cooler autumn temperatures begin to establish.

1st-2nd Chance of widespread fog patches.

3rd-5th Cooler than average overnight with the chance of patches of light ground frost in some inland areas.

4th Cold overnight with chance of some patches of frost, mostly in the west to midlands.

5th-8th Chance of widespread patches of fog.

7th A very strong high pressure system may cover the country.

10th-11th Unsettled conditions may return as high pressure slips to the south of Ireland, and Atlantic depressions return nearer to the country once again. Rainfall may be high at this time in northern parts, as well as a strong and blustery westerly flow developing. Mostly cloudy and cooler everywhere.

Over the next four weeks it may be colder, windier, and wetter, but sunnier than normal almost everywhere. Fast moving frontal systems associated with Atlantic depressions may dominant much of the upcoming weather, with some of them deepening considerably as the track to the northwest of Ireland. Overall, low pressure may prevail over most of the northern half of the country, while stronger high pressure dominates the southern half. Air pressure may be at its lowest, as an intense depression, at the end of October, while the highest pressure may be at its strongest near the end of the first week of November. Rainfall amounts during the next four weeks will be above normal throughout with some areas possibly having nearly twice as much rainfall than what is normal for this time of year, and with very few dry days. Generally, there may be up to 22-25 days with rain, particularly in the western half of the country. Despite the high number of rain days, the next four weeks may be sunnier than average everywhere, with many areas up to 30% more sunshine hours possible. Temperatures may be around one degree below average everywhere by the end of this outlook period, despite the first half of this period having near normal temperatures. This may be because in the last week of this outlook period a significant cooler change may happen, with temperatures dropping well below the norm for this time of year, and including the chance of cold precipitation of hail, sleet or snow in some places and some severe ground frosts particularly in inland parts. During this outlook period, there may be 6 days with thunderstorm activity, 14 days with hail showers, 8 days with possible gale winds, snow on 2 days, widespread fogs on 2 days, and 15-16 days with frosts mostly in inland parts, which is about twice the normal amount for this time of year.

12th-31st A series of fast moving frontal systems associated with Atlantic depressions cross the country bringing frequent spells of rain or showers almost daily, as well as the chance of some spells of heavy rain bringing localised flooding and strong and blustery winds at times.

12th-14th Cooler spell. Chance of hail showers.

12th-19th Despite the frequent spells of rain, it may be unusually sunny.

13th Chance of thunderstorms.

15th Chance of heavy rain. Temperatures may return nearer to what is typical for this time of year for the remainder of this month and into the first week of November.

16th-19th Chance of more frequent sunny spells.

20th-21st Widespread rain activity likely with heavy falls in some places that may produce some localised flooding.

16th-17th Chance of widespread ground frosts, particularly in inland areas.

17th-26th Chance of widespread fog in some areas.

24th Milder temperatures possible.

25th Heavy rain likely at times.

30th-31st A deep depression to the north of Ireland may produce very strong and unsettled winds, possibly to gale force at times. Chance of thunderstorms and hail showers.

IRELAND RAINFALL ESTIMATES
October 2026

d=mainly dry, n= nonrecordable, l=light shrs, s=significant shrs, r=rain, h=hvy falls

		ULSTER PROVINCE					CONNAUGHT PROVINCE						LEINSTER PROVINCE				MUNSTER PROVINCE						
		Alderg	Hillsbo	Armag	Loughl	Malin H	Sligo	Ardtara	Belmul	Drums	Galway	Roscoı	Markre	Edende	Kilkenn	Dublin	Shannc	Newpo	Tralee	Killarne	Cork	All	Moon
1st	Thur	d	d	d	d	d	d	d	d	d	d	d	d	d	d	d	d	d	d	d	d	0	P13
2nd	Fri	d	d	d	d	d	l	d	l	d	d	d	d	d	d	d	d	d	d	d	d	2	^
3rd	Sat	d	d	d	d	d	d	d	d	d	d	d	d	d	d	d	d	d	d	d	d	0	3Q
4th	Sun	d	d	d	d	d	d	d	d	d	d	d	d	d	d	d	d	d	d	d	d	0	
5th	Mon	d	d	d	d	d	d	d	d	d	d	d	d	d	d	d	d	d	d	d	d	0	
6th	Tues	d	d	d	d	d	d	d	d	d	d	d	d	d	d	d	d	d	d	d	d	0	
7th	Wed	d	d	d	d	d	d	d	d	d	d	d	d	d	d	d	d	d	d	d	d	0	
8th	Thur	d	n	d	l	s	l	l	l	d	l	d	l	l	n	n	d	d	d	l	d	15	
9th	Fri	n	d	l	s	l	l	s	l	n	d	d	l	d	d	l	l	l	d	d	d	21	
10th	Sat	l	l	d	h	l	s	h	r	l	s	s	h	l	l	l	l	h	l	l	d	132	N XhS
11th	Sun	s	l	r	s	h	r	s	r	r	s	r	r	l	l	r	s	h	l	l	l	177	
12th	Mon	s	s	s	s	s	s	s	r	s	s	s	s	l	l	n	s	r	l	r	l	111	
13th	Tues	l	l	l	s	l	s	s	s	r	s	r	s	l	l	n	s	s	s	s	l	81	
14th	Wed	d	d	d	n	l	l	n	l	l	l	l	l	n	l	d	d	l	l	l	d	14	
15th	Thur	s	r	n	h	r	r	h	h	h	h	h	h	h	h	r	h	h	h	h	h	369	
16th	Fri	s	n	r	l	l	s	l	d	l	s	s	l	d	n	s	l	l	l	l	l	55	A V
17th	Sat	s	s	l	s	n	l	s	s	s	h	r	r	s	s	l	s	h	r	s	s	140	
18th	Sun	s	s	s	s	r	r	s	h	s	l	s	s	l	s	r	r	s	s	l	s	137	1Q
19th	Mon	l	d	s	l	l	l	l	d	d	d	n	l	d	d	d	l	d	d	d	d	17	
20th	Tues	s	r	d	r	l	s	r	r	h	h	h	r	s	l	l	s	h	h	h	l	317	
21st	Wed	r	s	r	l	r	r	l	r	s	r	s	s	r	l	r	r	s	h	r	r	185	
22nd	Thur	l	d	s	d	l	l	d	d	d	d	n	d	d	d	d	d	d	l	d	d	14	
23rd	Fri	d	d	d	l	d	d	l	d	d	l	l	l	d	d	d	d	l	l	n	l	10	
24th	Sat	d	d	d	d	l	l	d	l	d	s	l	l	d	d	d	l	l	s	h	l	51	XhN
25th	Sun	s	s	d	d	d	l	d	l	s	s	s	l	r	h	r	h	l	r	h	h	175	
26th	Mon	l	l	s	s	s	s	s	s	s	l	s	l	l	n	l	l	s	s	s	n	70	F
27th	Tues	l	l	l	l	s	l	l	s	l	n	l	l	d	n	d	l	l	l	n	d	31	
28th	Wed	d	d	d	d	l	l	d	l	l	s	l	l	d	d	d	l	s	l	l	d	24	P9
29th	Thur	n	d	l	l	l	l	l	l	l	l	l	l	l	d	l	l	l	n	l	l	20	^
30th	Fri	l	s	n	s	l	l	s	s	l	s	r	l	l	d	d	l	l	s	h	s	86	
31st	Sat	l	l	l	r	s	s	r	r	s	s	s	s	l	l	l	r	r	s	s	l	121	
	Estimate:	76	75	75	122	106	107	122	138	124	182	139	129	70	69	77	117	222	149	181	97	2375	
	Average:	90	100	86	129	121	134	136	146	112	129	140	134	104	93	79	105	176	183	133	138	2467	
	Trend:	drr	drr	drr	av	drr	drr	drr	av	w tr	w tr	av	av	drr	drr	av	w tr	w tr	drr	w tr	drr	av	

IRELAND SUNSHINE ESTIMATES

October 2026

F=fine (8-12 hours of sunshine), pc= partly cloudy (4-7 hours), c=cloudy (1-3 hours), o=overcast (0 hours)

		ULSTER PROVINCE					CONNAUGHT PROVINCE						LEINSTER PROVINCE				MUNSTER PROVINCE						
		Alderg	Hillsbo	Armag	Loughl	Malin H	Sligo	Ardtarı	Belmul	Drums	Galway	Roscoı	Markre	Edende	Kilkenı	Dublin	Shannc	Newpo	Tralee	Killarnε	Cork	All	Moon
1st	Thur	F	F	F	F	F	pc	F	F	F	F	F	F	F	F	F	F	F	F	F	pc	184	P13
2nd	Fri	pc	pc	pc	pc	pc	c	pc	pc	pc	pc	pc	pc	pc	pc	pc	pc	pc	c	c	c	92	^
3rd	Sat	pc	pc	pc	pc	pc	F	pc	pc	pc	F	pc	pc	pc	pc	pc	F	F	F	F	F	162	3Q
4th	Sun	F	F	F	F	F	F	F	pc	F	pc	pc	F	F	pc	F	pc	pc	pc	pc	pc	158	
5th	Mon	pc	pc	pc	pc	pc	pc	pc	F	pc	F	F	pc	pc	pc	c	F	pc	F	F	F	122	
6th	Tues	pc	pc	pc	pc	pc	c	pc	o	pc	c	c	pc	pc	pc	pc	pc	pc	F	F	pc	98	
7th	Wed	pc	pc	pc	pc	pc	pc	pc	c	pc	c	c	pc	pc	pc	pc	c	pc	F	F	pc	106	
8th	Thur	c	c	c	c	c	o	c	o	pc	c	c	c	pc	pc	pc	pc	pc	c	c	pc	74	
9th	Fri	c	c	c	c	c	c	c	pc	c	pc	pc	c	c	pc	pc	pc	pc	F	F	c	93	
10th	Sat	o	o	o	o	o	o	o	o	c	o	o	o	c	c	o	c	c	c	c	pc	26	N XhS
11th	Sun	o	o	o	o	o	o	o	o	o	o	o	o	o	o	c	o	o	o	o	o	10	
12th	Mon	pc	pc	pc	pc	pc	pc	pc	pc	pc	pc	pc	pc	pc	F	F	pc	pc	pc	pc	pc	145	
13th	Tues	pc	pc	pc	pc	pc	c	pc	c	pc	pc	pc	pc	pc	pc	pc	pc	pc	pc	pc	pc	98	
14th	Wed	pc	pc	pc	pc	pc	c	pc	c	pc	pc	pc	pc	pc	pc	pc	pc	pc	pc	pc	pc	107	
15th	Thur	o	o	o	o	o	o	o	o	o	o	o	o	o	o	o	o	o	o	o	o	0	
16th	Fri	F	F	F	F	F	F	F	F	F	F	F	F	F	F	F	F	F	F	F	F	197	A V
17th	Sat	c	c	c	c	c	o	c	c	c	c	c	c	c	o	c	c	c	o	o	o	27	
18th	Sun	c	c	c	c	c	c	c	pc	c	c	pc	c	c	c	o	c	c	pc	pc	c	52	1Q
19th	Mon	F	F	F	F	F	F	F	F	F	F	F	F	F	F	F	F	F	F	F	F	176	
20th	Tues	o	o	o	o	o	o	o	o	o	o	o	o	o	o	o	o	o	o	o	o	2	
21st	Wed	o	o	o	o	o	o	o	o	o	o	o	o	o	o	o	o	o	o	o	o	1	
22nd	Thur	pc	pc	pc	pc	pc	pc	pc	pc	pc	pc	pc	pc	c	c	c	pc	c	c	c	c	84	
23rd	Fri	c	c	c	c	c	c	c	c	c	c	c	c	c	c	pc	c	c	c	c	o	52	
24th	Sat	pc	pc	pc	pc	pc	pc	pc	c	pc	c	c	pc	pc	pc	pc	o	c	o	o	c	73	XhN
25th	Sun	c	c	c	c	c	c	c	pc	o	c	pc	c	o	o	o	o	o	o	o	o	29	
26th	Mon	c	c	c	c	c	c	c	c	c	c	c	c	c	c	c	c	c	c	c	c	44	F
27th	Tues	pc	pc	pc	pc	pc	pc	pc	pc	pc	pc	pc	pc	pc	pc	pc	pc	pc	pc	pc	pc	116	
28th	Wed	o	o	o	o	o	o	o	c	c	c	c	o	c	c	o	c	c	pc	pc	c	31	P9
29th	Thur	c	c	c	c	c	c	c	c	c	c	c	c	c	c	c	pc	c	c	c	c	42	^
30th	Fri	o	o	o	o	o	o	o	c	o	o	o	o	o	o	o	o	o	o	o	o	4	
31st	Sat	c	c	c	c	c	c	c	c	c	c	c	c	c	c	c	c	c	c	c	c	60	
	Estimate:	120	120	120	120	120	108	120	114	124	123	120	120	124	128	130	130	129	137	137	125	2464	
	Average:	91	91	87	82	74	74	76	82	75	85	88	85	86	84	96	82	71	80	78	87	1653	
	Trend:	more	more	more	more	more	more	more	more	more	more	more	more	more	more	more	more	more	more	more	more	more	

OCTOBER 1 THURSDAY

A large high pressure system over Ireland directs a light west to northwest flow throughout, with chance of fog patches overnight in many areas, possibly more persistent to clear in the south. Mostly sunny and dry day throughout.

Belfast: Sunny, dry, mild temperatures, moderate breeze
Dublin: Sunny, dry, warm day cool night, light shifting breezes
Cork: Mostly sunny and dry, mild temperatures, mostly calm air, fog possible
Galway: Light breezes to mostly calm air, sunny, warm, dry

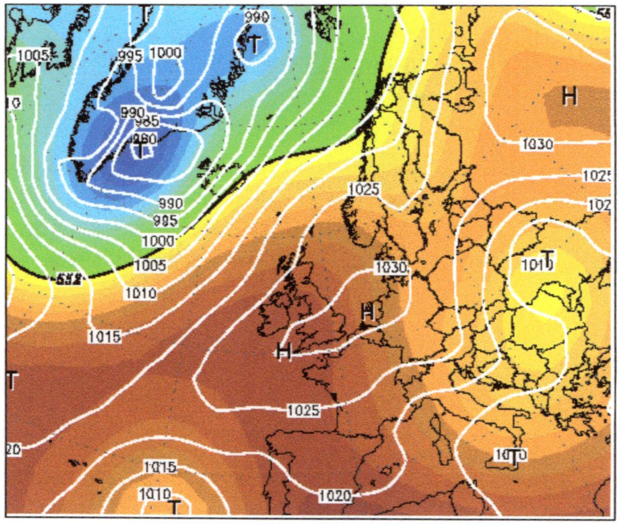

RAIN POTENTIAL

FROST/SNOW

OCTOBER 2 FRIDAY

Anticyclonic conditions continue throughout the country around this time, with a light to moderate west to northwest flow prevailing, occasional scattered cloud, and mostly dry conditions throughout. Fog patches possible overnight, particularly in the north and coastal southern parts. Mostly dry and pleasant day everywhere, cloudiest in the south, mostly sunny elsewhere. Chance of some isolated evening showers in the midlands and northwest.

Belfast: Partly cloudy, dry, warmer day cooler night, fresh breezes at times
Dublin: Mostly sunny and dry, slightly cooler, gentle shifting breezes
Cork: Mostly cloudy and dry, chance of misty patches, mild temperatures
Galway: Partly cloudy, cooler, dry, gentle breezes

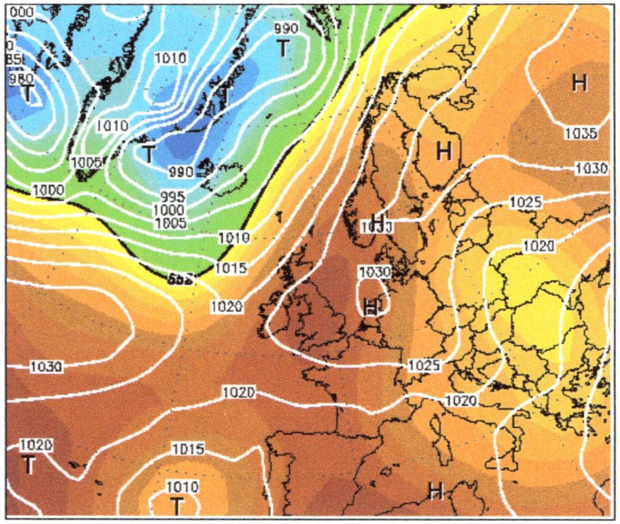

RAIN POTENTIAL

FROST/SNOW

OCTOBER **3** SATURDAY

A high pressure ridge brings a light northeasterly change across the country. Mostly sunny and dry throughout, although chance of some isolated overnight showers in the northwest. Colder with possible light ground frosts in some inland areas.

Belfast: Partly cloudy, dry, mild day cold overnight, light shifting breezes
Dublin: Light breezes, partly cloudy, dry, cooler
Cork: Sunny, slightly cooler, dry, light breezes
Galway: Sunny, dry, mild day cool night, moderate breezes

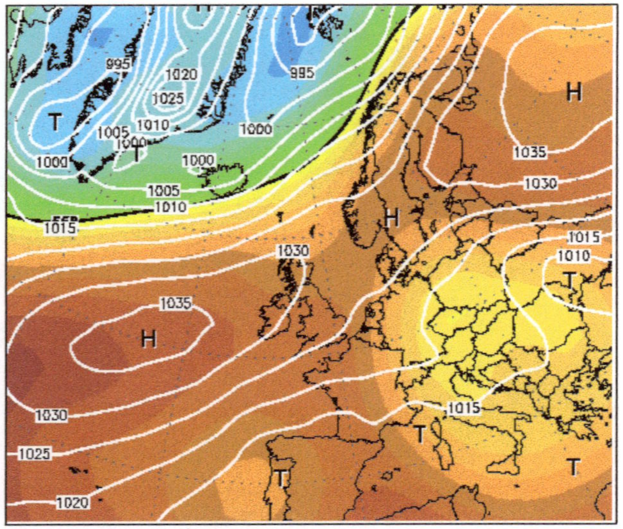

RAIN POTENTIAL

FROST/SNOW

OCTOBER **4** SUNDAY

A large high pressure system continues to linger over Ireland directing a mostly light northeasterly flow, with possible patches of fog overnight in the north and light ground frosts in inland parts. Mostly dry and sunny throughout with a pleasant air prevailing.

Belfast: Sunny, dry, mild day cool night, mostly calm air, fog patches likely
Dublin: Light breezes to mostly calm air, sunny, mild day cold night, dry
Cork: Partly cloudy and dry, moderate breezes
Galway: Partly cloudy, dry, mild day cool night, gentle breezes

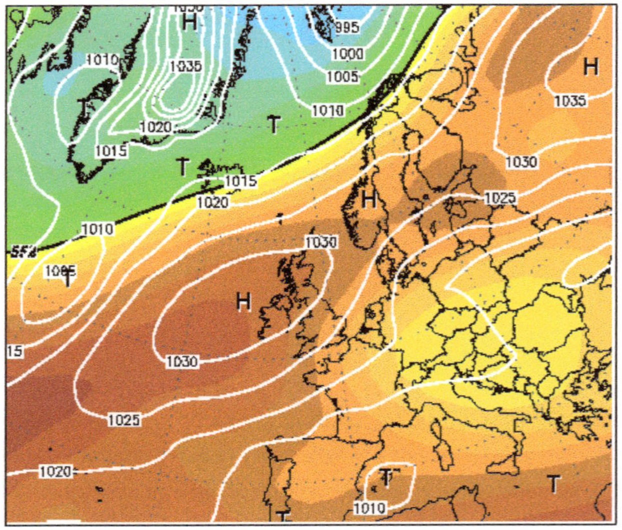

RAIN POTENTIAL

FROST/SNOW

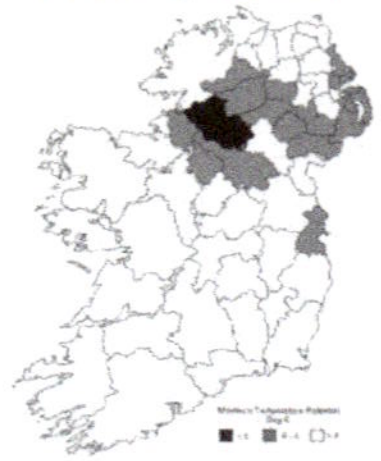

OCTOBER 5 MONDAY

High pressure continues to ridge over Ireland as it drifts slowly to be centred to the northeast of the country. Winds generally light in a southwesterly flow, with chance of fog patches overnight, mostly in northern parts to the midlands. Generally a pleasant and dry day everywhere, mostly cloudy in the northeast, while sunnier elsewhere.

Belfast: Partly cloudy, dry, gentle shifting breezes, chance of fog overnight
Dublin: Cloudy, dry, milder temperatures, mostly calm and serene air, chance of fog
Cork: Sunny, dry, calm and serene air, chance of fog overnight
Galway: Calm and pleasant air, sunny, dry, warm day cool night

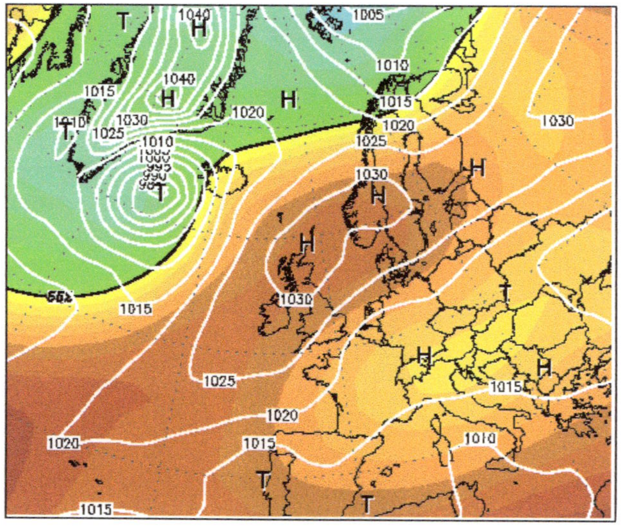

RAIN POTENTIAL

FROST/SNOW

OCTOBER 6 TUESDAY

The high pressure ridge flows to the west of Ireland with light to moderate southwesterly breezes prevailing throughout. Calm, dry, and pleasant air everywhere, with good sunny spells. Chance of widespread mist and fog overnight, some patches of fog dense and slow to clear.

Belfast: Calm and dry air, occasional sunny spells
Dublin: Partly cloudy, slightly warmer and dry, calm atmosphere, fog patches likely
Cork: Mostly sunny, dry, slightly warmer, calm and pleasant air, fog likely
Galway: Gentle breezes to mostly calm air, dry, cloudy, slightly warmer

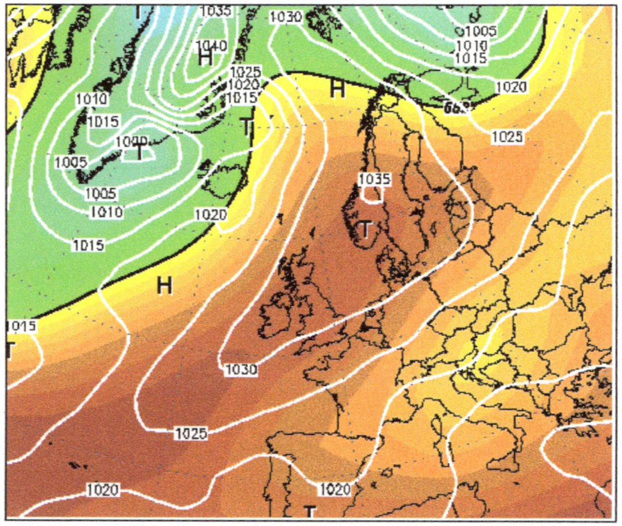

RAIN POTENTIAL

FROST/SNOW

OCTOBER **7** WEDNESDAY

High pressure continues to bring a moderate west to southwest flow across the country, fresher at times around the the coasts. Partly cloudy and dry throughout with overnight fog and mist in many districts, some patches slow to clear. Sunniest in the south and east while a little cloudier elsewhere.

Belfast: Partly cloudy, dry, mild day cool night, gentle breezes, chance of fog
Dublin: Partly cloudy, dry, mild day cool night, fog patches possible
Cork: Partly cloudy and dry, mild temperatures, calm air, fog likely
Galway: Cloudy, dry, mild day cool night, gentle breezes

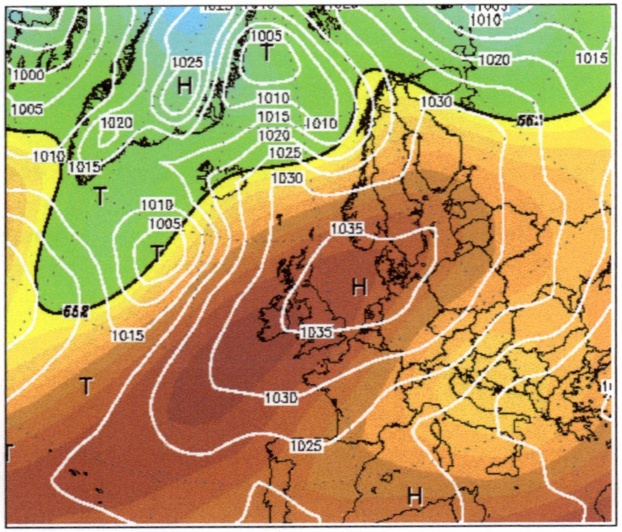

RAIN POTENTIAL

FROST/SNOW

OCTOBER **8** THURSDAY

A large anticyclonic system lingers over Ireland with a light to moderate west to southwest flow prevailing, fresher at times around the southwest coasts. Chance of fog patches overnight mostly in central parts and more scattered elsewhere. Occasional sunny spells and mostly dry, except for the chance of some light patchy rain or drizzle briefly in areas near the western coasts.

Belfast: Slightly warmer, mostly cloudy, chance of isolated showers or misty patches
Dublin: Partly cloudy, isolated showers, mostly calm air, fog likely
Cork: Partly cloudy, mild temperatures, dry, mostly calm air, fog possible
Galway: Cloudy, light showers, slightly cooler, gentle shifting breezes

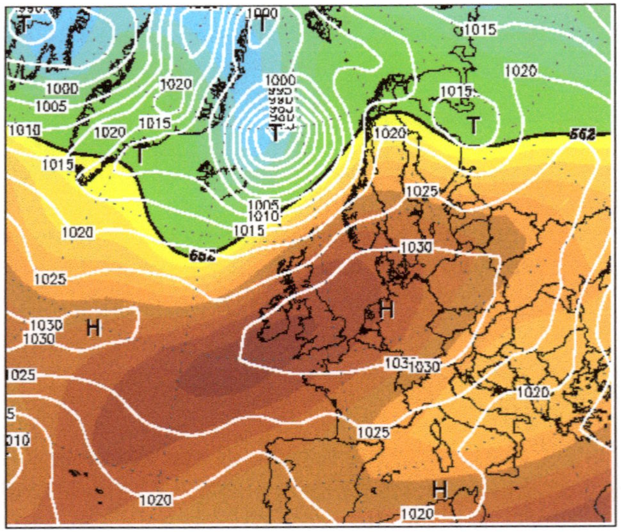

RAIN POTENTIAL

FROST/SNOW

OCTOBER 9 FRIDAY

Another high pressure system to the west of Ireland ridges over the country bringing a light south to southeast change. Passing scattered cloudy spells in most districts, with some patches of fog overnight mostly in northern, western, and central. Chance of some brief showers or drizzle patches in the west of the country, while mostly dry elsewhere.

Belfast: Gentle shifting breezes, partly cloudy, cooler, dry, chance of fog
Dublin: Isolated showers, sunny spells, cooler, gentle breezes to mostly calm air
Cork: Mostly cloudy, cooler and dry, gentle breezes
Galway: Partly cloudy and dry, gentle shifting breezes, chance of fog patches

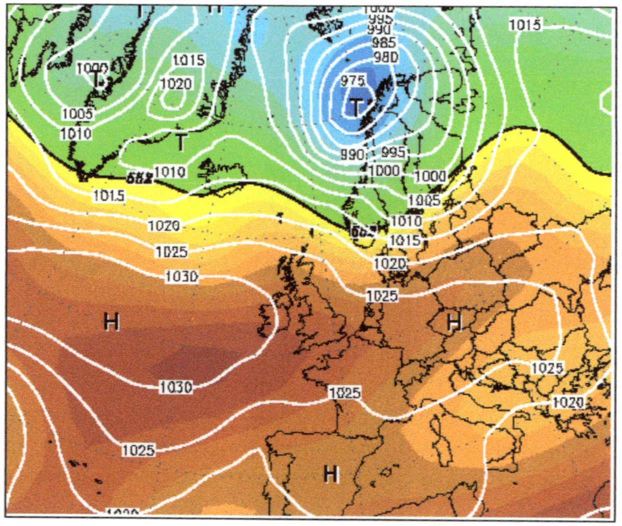

RAIN POTENTIAL

FROST/SNOW

OCTOBER 10 SATURDAY

The high pressure ridge to the west of Ireland directs a light west to northwest flow over the country, fresher at times around the northern coasts. Widespread cloud and outbreaks of drizzly showers likely throughout starting in the northwest and gradually spread southeastward, although the far southeast likely to remain mostly dry.

Belfast: Overcast, light showers, fresh windy spells
Dublin: Overcast, drizzly showers, mild temperatures, light breezes
Cork: Gentle shifting breezes, partly cloudy, dry
Galway: Light breezes, overcast, brief showers, mild day cool night

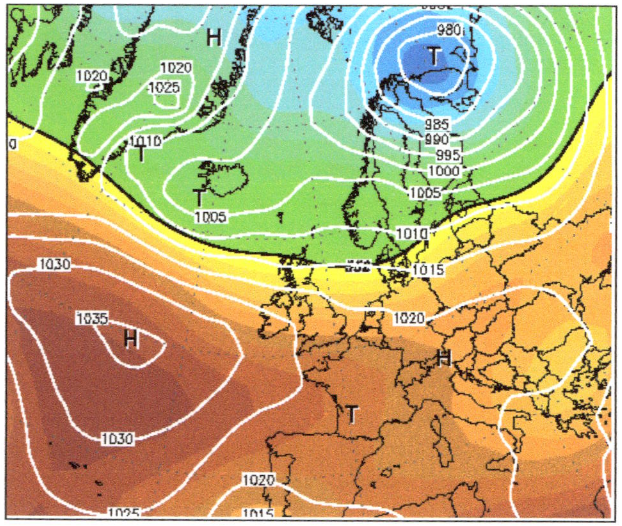

RAIN POTENTIAL

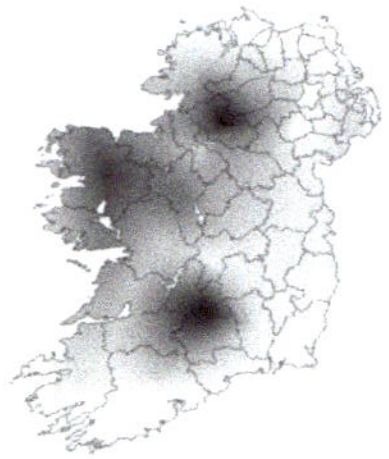

FROST/SNOW

OCTOBER 11 SUNDAY

A depression to the northeast of Ireland moves down nearer the country while the high pressure ridge slips away to the south. Winds fresh to strong in a northwesterly flow throughout, possibly up to gale force with gusts at times. Overcast with rain in most parts of the country, although possibly remaining mostly dry along the far southeast coastal fringes. Rain may be particularly heavy for a time in parts of the north, northeastern coasts, and midlands.

Belfast: Overcast and dull, some showers, cooler, strong and unsettled winds
Dublin: Cloudy, rain, mild temperatures, blustery winds at times
Cork: Overcast, brief showers overnight, mild temperatures, strengthening winds
Galway: Overcast, occasional showers, windy at times

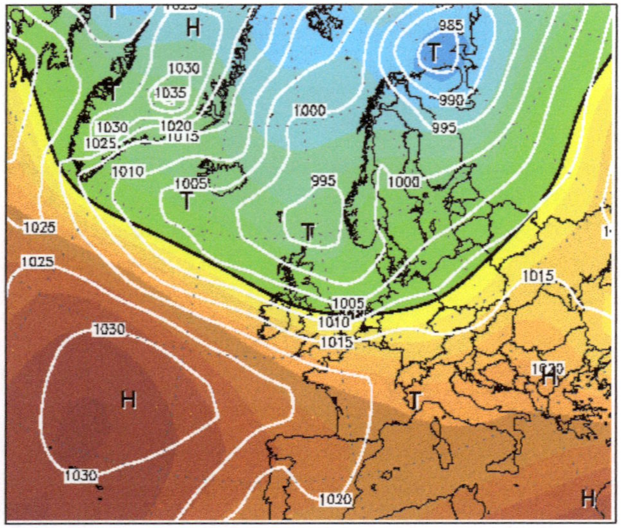

RAIN POTENTIAL

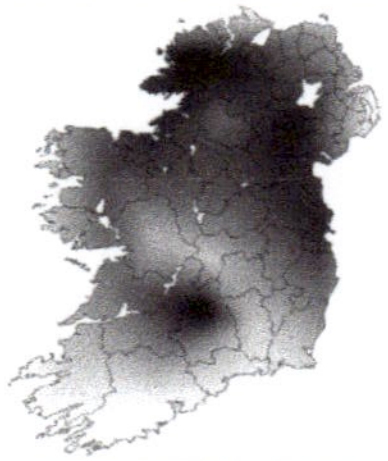

FROST/SNOW

OCTOBER 12 MONDAY

An intense depression centred to the northeast of Ireland brings a very strong to stormy northwest flow across the country. Very unsettled day, with a mix of sunny spells and bands of squally showers, starting in the north at first, and spreading throughout most of the country during the day, with only isolated showers likely in the eastern coastal areas and southern coastal fringes. Chance of brief hail showers or sleet overnight in some higher reaches in the west.

Belfast: Strong and blustery winds, mostly sunny, overnight scattered showers
Dublin: Unsettled and blustery winds, isolated showers overnight, sunny day
Cork: Unsettled and blustery winds, partly cloudy, light showers, slightly cooler
Galway: Stormy air, partly cloudy, squally showers, cooler, chance of snow overnight

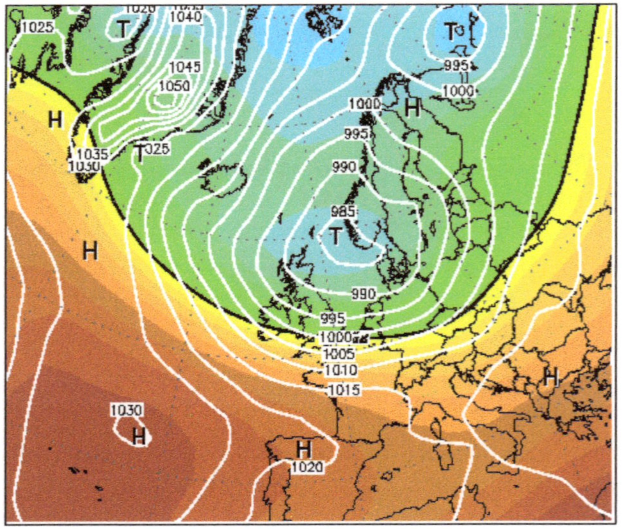

RAIN POTENTIAL

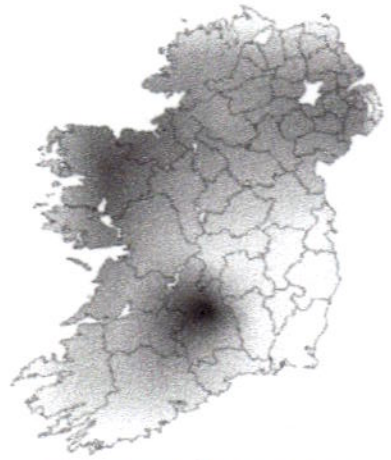

FROST/SNOW

OCTOBER **13** TUESDAY

The depression to the northeast of Ireland continues to bring very strong and blustery northwest winds across the country with further unsettled and changeable conditions prevailing. A mix of sunny spells and showers likely in all districts except along the eastern coasts, which may remain mostly dry and sunny. Chance of thunderstorms over the northern half of the country at times.

Belfast: Partly cloudy, squally showers, strong winds, gusts, chance of gales
Dublin: Partly cloudy, isolated showers, mild day cool night, strong gusty winds
Cork: Partly cloudy, light showers, mild day cool night, blustery at times
Galway: Changeable, blustery, mix of showers and some sunny spells, cooler

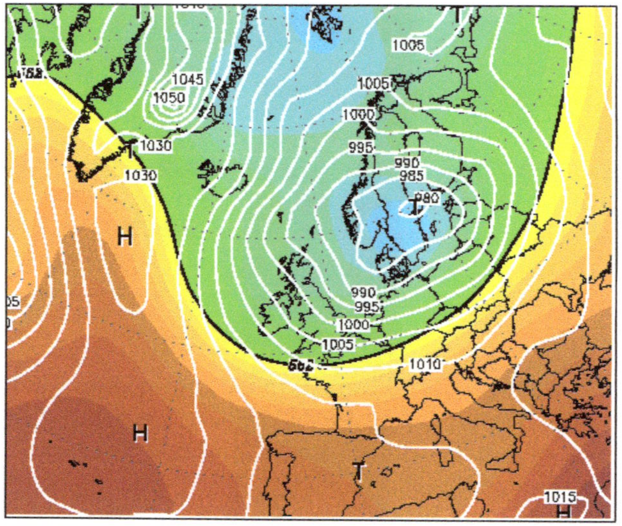

RAIN POTENTIAL

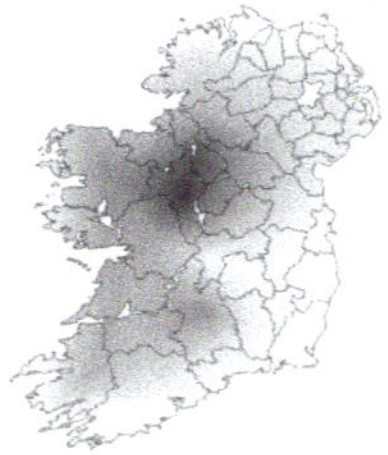

FROST/SNOW

OCTOBER **14** WEDNESDAY

A high pressure ridge builds near the south and west of Ireland while low pressure departs to the northeast. Winds generally strong and gusty in a north to northwest flow over the northern half of the country, possibly up to gale force around the northern coasts, while slightly lighter to the southeast. Some isolated showers possible mostly overnight in the west and midlands, while mostly dry with longer sunny spells elsewhere.

Belfast: Partly cloudy, dry, strong winds, chance of gales
Dublin: Unsettled and blustery winds at times, partly cloudy, dry
Cork: Unsettled winds, partly cloudy, dry, mild day cool night
Galway: Partly cloudy, some showers, warmer day colder night, strong winds at times

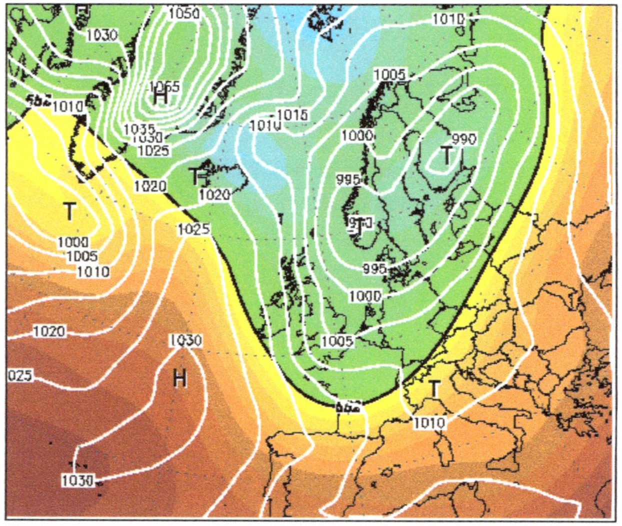

RAIN POTENTIAL

FROST/SNOW

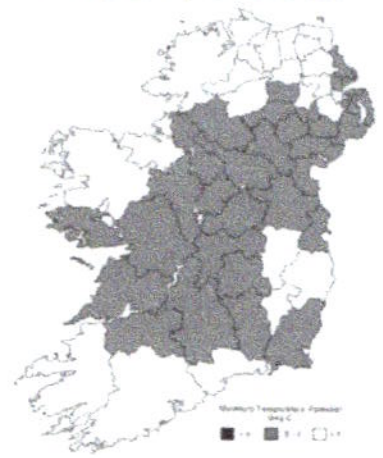

OCTOBER **15** THURSDAY

A high pressure ridge lingers to the south of Ireland, while a blustery southwest flow associated with a depression to the west of Ireland prevails and may bring a very stormy day across the country. Overcast and with heavy rain almost everywhere, with rain moving into the northwest at first and spreading southeastwardly throughout the day to reach all areas by evening. Rain may be particularly heavy in the southwest, with chance of localised flooding risks.

Belfast: Overcast, rain, unsettled winds, colder
Dublin: Overcast, rain, warmer temperatures, fluctuating and shifting breezes
Cork: Overcast, unsettled winds, warmer, heavy rain
Galway: Overcast, heavy rain, milder temperatures, unsettled winds

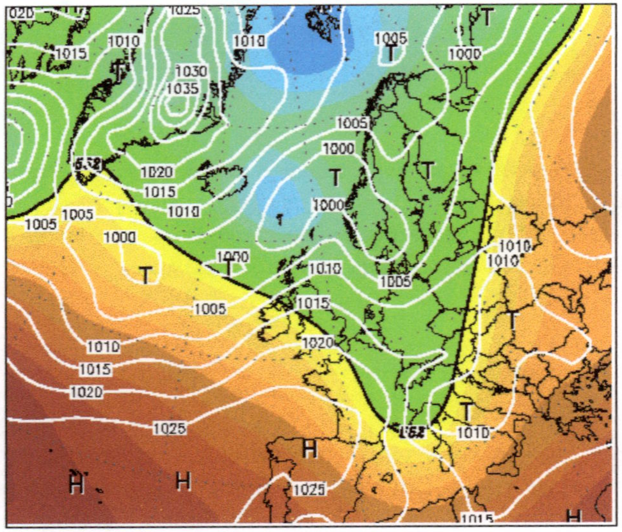

RAIN POTENTIAL

FROST/SNOW

OCTOBER **16** FRIDAY

A depression continues to linger over Ireland around this time with a moderate westerly flow at first later backing and becoming lighter with a southeasterly flow. Storm eases overnight with light showers in most districts, clearing to a mostly dry and sunny day throughout. Chance of frost patches in sheltered inland areas.

Belfast: Sunny, isolated showers, warmer day cold overnight, fluctuating winds
Dublin: Sunny, light shifting breezes, cooler, overnight showers
Cork: Sunny, dry, brief overnight showers, gentle shifting breezes
Galway: Overnight showers, sunny day, cooler, fluctuating breezes

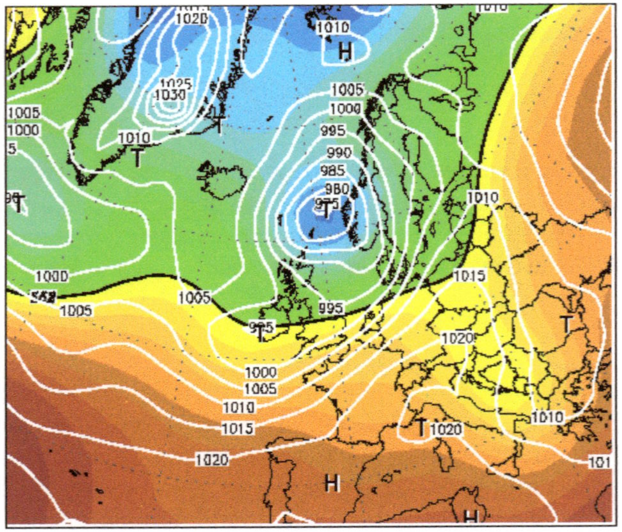

RAIN POTENTIAL

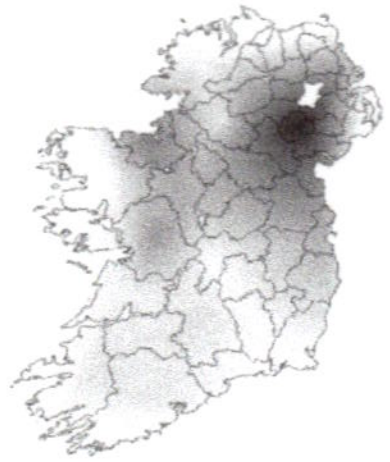

FROST/SNOW

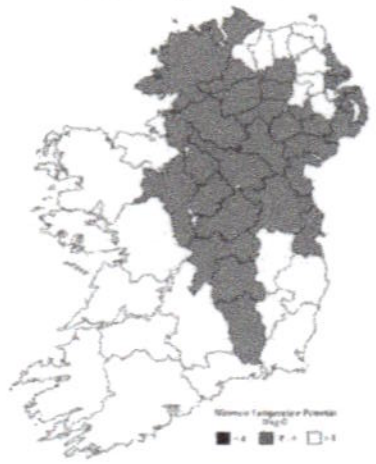

OCTOBER 17 SATURDAY

A trough between two depressions brings a light to moderate southeasterly flow throughout most of the country, but fresher and more blustery for a time around the northwest coasts. Mostly cloudy with patches of showers and drizzles everywhere, heavier in parts of the west and midlands. Chance of fog or light misty drizzle patches overnight mostly in central and southern parts.

Belfast: Cloudy, rain, fresh breezes
Dublin: Cloudy, light showers, warmer day colder night, gentle breezes, possible fog
Cork: Light breezes, fog possible, overcast and dull, some showers
Galway: Cloudy, heavy rain, warmer day, cold overnight, fresh windy spells

RAIN POTENTIAL

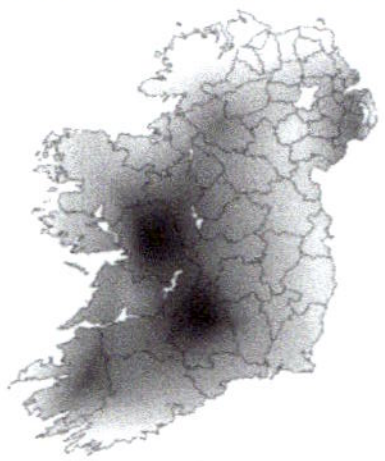

FROST/SNOW

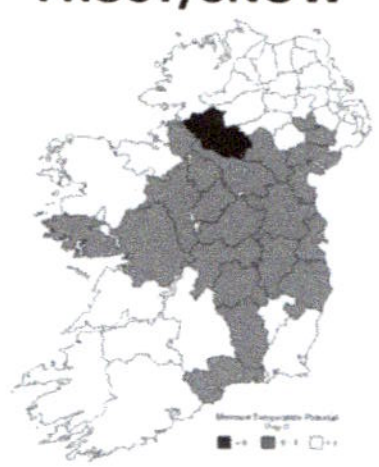

OCTOBER 18 SUNDAY

A large depression centred to the northwest of Ireland directs a moderate southwesterly flow over the country, freshening later in the day. Mostly cloudy and unsettled everywhere. Rain at first in the west, may later spread eastward to cover all areas of the country with scattered showers during the day. Showers likely to ease to isolated by evening.

Belfast: Cloudy, scattered showers, mild temperatures, strong winds at times
Dublin: Overcast, rain, strong and unsettled winds, chance of gales and gusts
Cork: Cloudy, occasional showers, mild day cold night, strong winds at times
Galway: Mostly cloudy, light showers, milder, moderate winds with occasional gusts

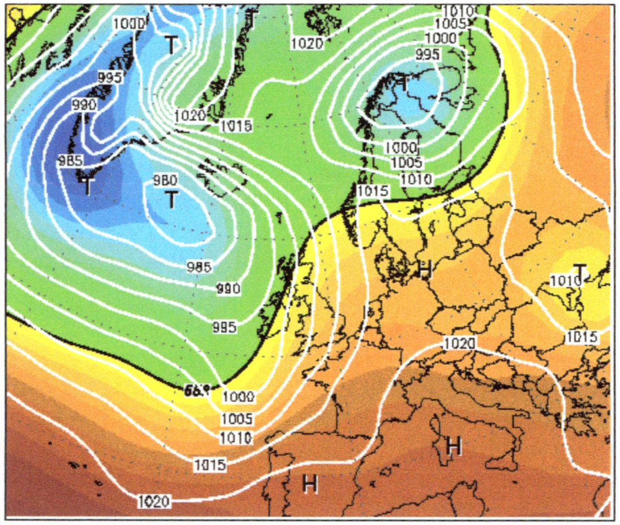

RAIN POTENTIAL

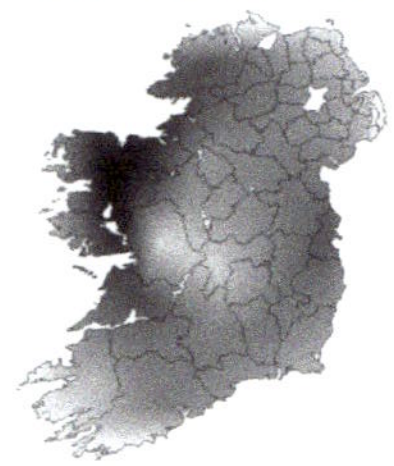

FROST/SNOW

OCTOBER **19** MONDAY

A light to moderate west to southwest flow associated with a depression to the northwest of Ireland prevails. Mostly dry and sunny day throughout except for the chance of some brief showers overnight mostly in the north.

Belfast: Sunny, dry, mild day cool night, occasional breezy spells
Dublin: Sunny, dry, mild day cool night, moderate breezes, chance of drizzle patches
Cork: Sunny, dry, light winds, misty patches possible, warmer day cold night
Galway: Moderate breezes, dry and sunny, mild day cold night

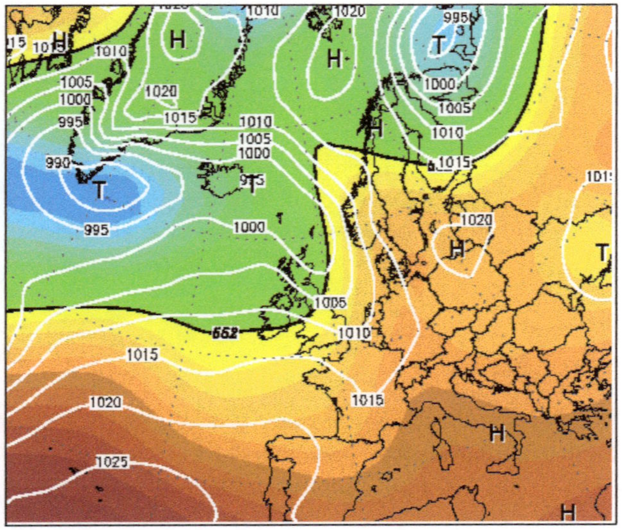

RAIN POTENTIAL

FROST/SNOW

OCTOBER **20** TUESDAY

A depression to the northwest of Ireland builds into an intense storm, while a high pressure system to the south of Ireland ridges over the country, with a southwest flow prevailing throughout, moderate at first, but becoming stronger and gustier for a time. Widespread cloud brings a dull and gloomy day everywhere, with heavy rain at times in the western half of the country, while lighter to isolated towards the eastern coasts. Heavy rain may bring some brief flooding in some western areas.

Belfast: Overcast, rain, strengthening winds later in the day, chance of gales
Dublin: Overcast, light showers, moderate breezes
Cork: Overcast, cooler, dull and showery, moderate breezes
Galway: Overcast, heavy rain, chance of localised flooding, freshening winds

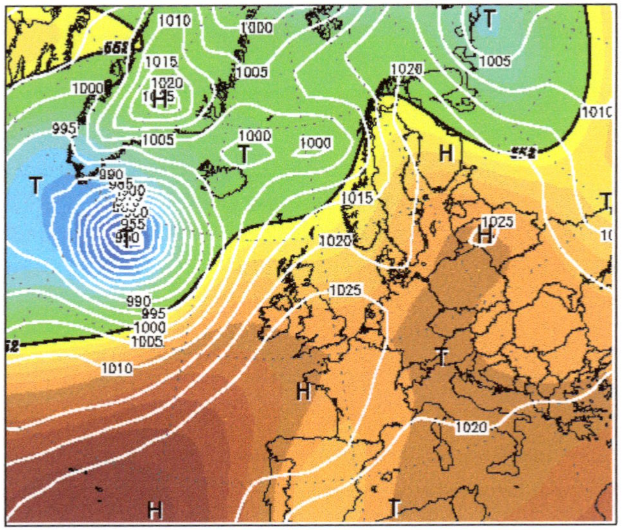

RAIN POTENTIAL

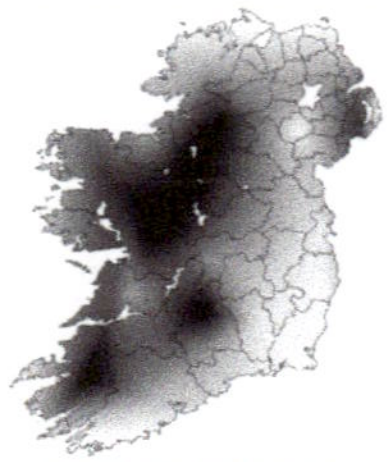

FROST/SNOW

OCTOBER 21 WEDNESDAY

The high pressure ridge may remain stable over the country while the depression to the northwest pushes up against the ridge, bringing an intense stormy day throughout, with a strong mostly southwesterly flow, very blustery at times and possibly to gale force in some areas. Widespread cloud, drizzle and rain throughout for much of the day before easing towards the evening.

Belfast: Overcast, stormy air, scattered showers, mild temperatures, chance of gales
Dublin: Stormy, dull and gloomy air, rain, gusty winds
Cork: Stormy air, strong winds with gusts at times, rain, slightly warmer
Galway: Stormy, overcast, rain, moderate temperatures, strong and blustery winds

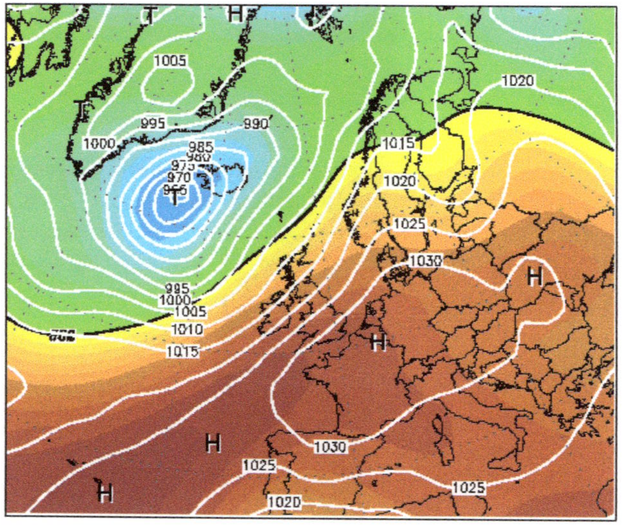

RAIN POTENTIAL

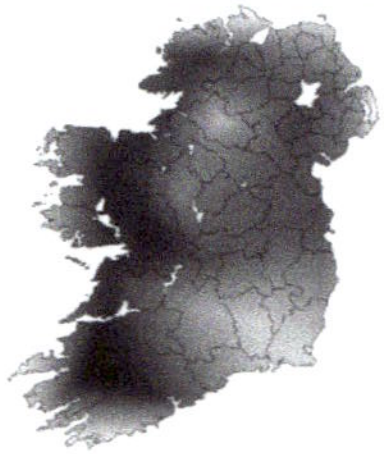

FROST/SNOW

OCTOBER 22 THURSDAY

The high pressure ridge still persists over the country, while the depression to the north moves slowly away, allowing for the stormy conditions of the last couple of days to ease throughout. Winds generally lighter in a southwest flow, with chance of some patches of fog overnight, mostly in the south, where it may be slow to clear. Some lingering showers in the far southwest and north of the country, while mostly dry with increased sunny spells elsewhere.

Belfast: Partly cloudy, dry, mild temperatures, blustery at times
Dublin: Cloudy, mostly dry with odd threat of rain, slightly cooler, moderate breezes
Cork: Moderate breezes, cloudy, dry, chance of misty patches overnight
Galway: Partly cloudy, breezy, dry, mild temperatures

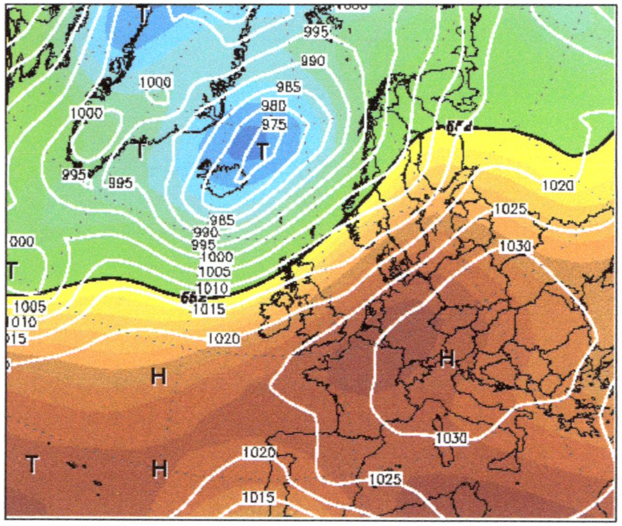

RAIN POTENTIAL

FROST/SNOW

OCTOBER 23 FRIDAY

A large anticyclonic system to the east of Ireland directs a very light southwesterly flow over the country. Mostly dry with occasional sunny spells in most parts of the country, although duller and some possible drizzle patches in the southeast. Widespread fog possible overnight, particularly in the south, which may be very slow to clear.

Belfast: Cloudy, dry, colder overnight, moderate breezes
Dublin: Partly cloudy, dry, mild day cold night, calm atmosphere
Cork: Overcast, brief showers, mild temperatures, calm atmosphere, fog likely
Galway: Cloudy, isolated showers, mild day cold night, mostly calm air

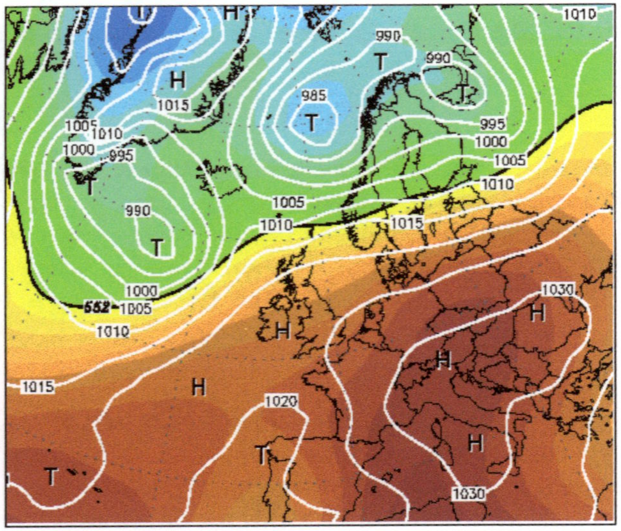

RAIN POTENTIAL

FROST/SNOW

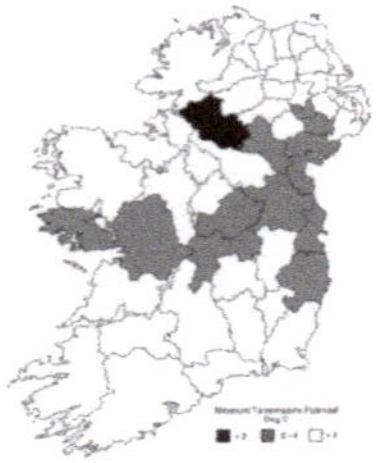

OCTOBER 24 SATURDAY

The high pressure ridge slips to the south of Ireland allowing for a depression centred to the northwest to move nearer to the country, with a moderate southwesterly flow prevailing throughout. A mixed day with some showers in the western half of the country, mostly in the morning, while mostly dry with more sunny spells in the east over most of the day. Further patches of fog likely overnight mostly in the south and possible light frost in inland parts.

Belfast: Partly cloudy, dry, mild temperatures, fresh breezes at times
Dublin: Partly cloudy, dry, slightly warmer, moderate breezes
Cork: Cloudy, milder temperatures, spotty showers, gentle breezes, chance of fog
Galway: Cloudy, mild temperatures, occasional showers, moderate breezes

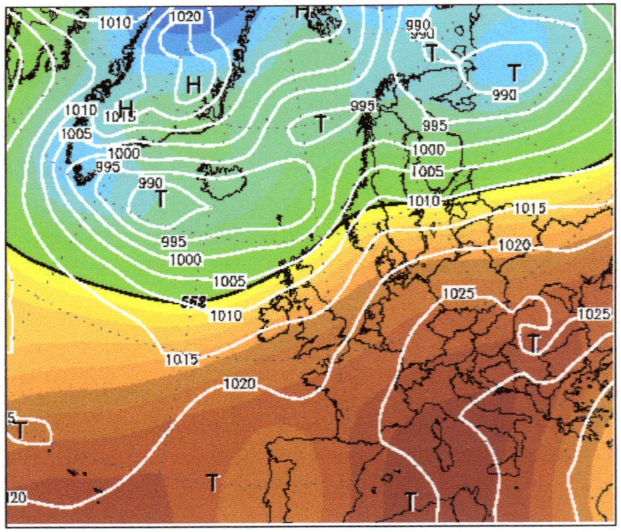

RAIN POTENTIAL

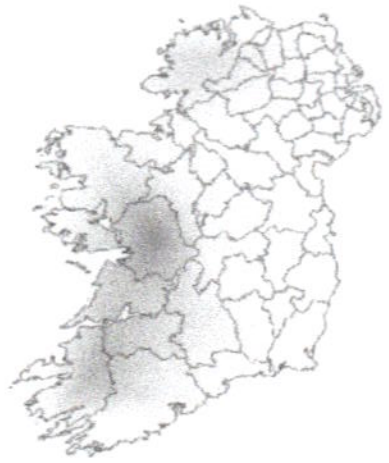

FROST/SNOW

OCTOBER 25 SUNDAY

A weak high pressure ridge sits to the south of Ireland, while a couple of depressions are to the north, with a trough between the systems bringing a light mostly west to northwest flow throughout. Mostly cloudy and dry overnight with patches of fog and light frosts here and there. Later, rain may spread up from the south and west and cross the country bringing widespread rain throughout, with very heavy falls likely in the southern half of the country.

Belfast: Cloudy, rain, gentle breezes, cold overnight with chance of fog
Dublin: Mostly calm, cooler and gloomy air, chance of fog overnight, rain
Cork: Calm atmosphere, overcast, misty patches, heavy rain
Galway: Mostly cloudy, occasional showers, cooler, mostly calm air

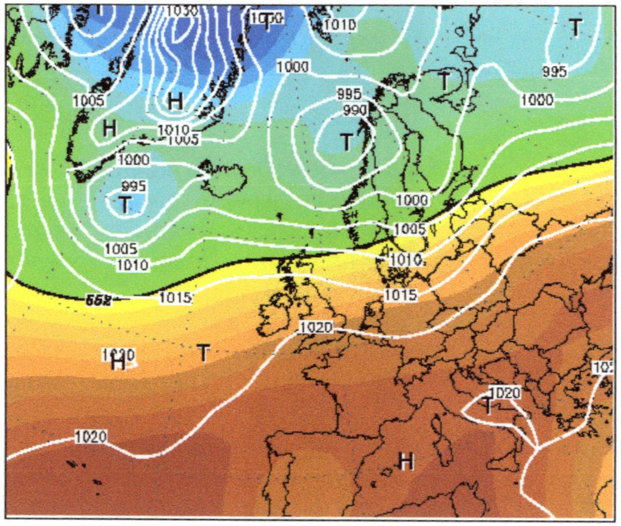

RAIN POTENTIAL

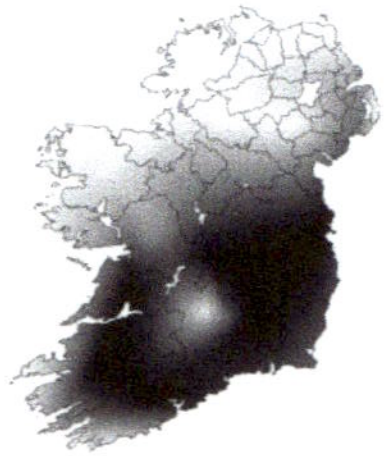

FROST/SNOW

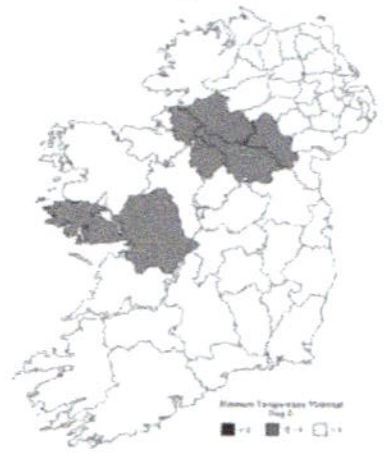

OCTOBER 26 MONDAY

A depression to the northwest of Ireland directs a gentle northwesterly air over the country, fresher at times around the coasts. Cloudy with passing shower activity almost everywhere. Chance of fog in many districts, and light frost patches overnight, particularly in central parts, which may also see some brief freezing showers of hail or sleet for a time.

Belfast: Cloudy, colder, spotty showers, gentle breezes, chance of fog patches
Dublin: Cloudy, light showers, gentle breezes, chance of fog
Cork: Cloudy, light breezes, isolated showers, misty patches possible
Galway: Cloudy, spotty showers, mild day cold night, light shifting breezes

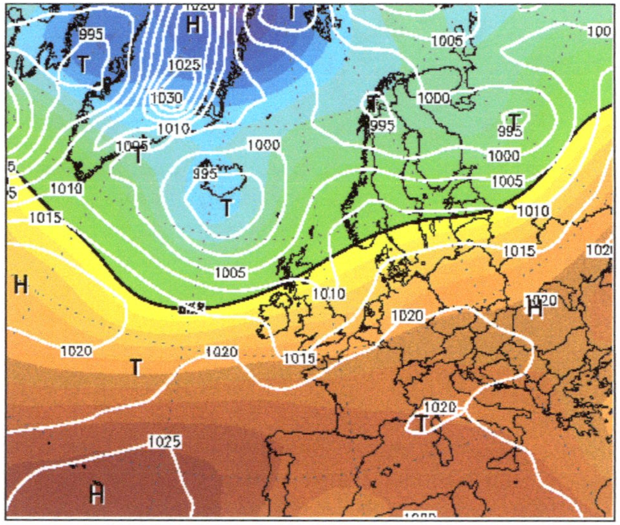

RAIN POTENTIAL

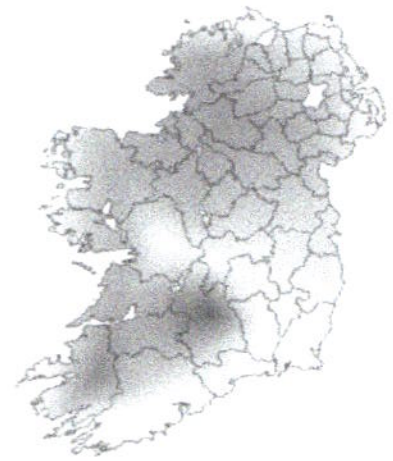

FROST/SNOW

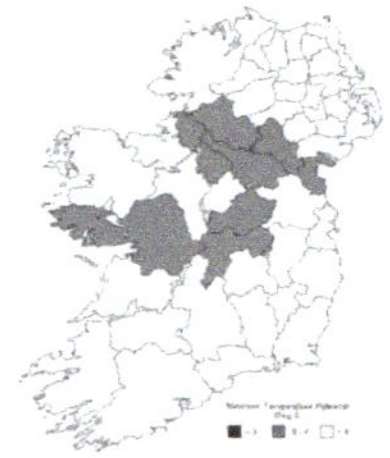

OCTOBER 27 TUESDAY

A depression to the north drifts eastward building a stronger northwesterly flow over the country, blustery at times around the coasts. Some overnight showers mostly in western and central parts and possibly of hail, clearing to a mostly dry day with scattered cloudy spells, and the occasional outbreak of isolated showers.

Belfast: Partly cloudy, isolated showers, breezy at times
Dublin: Partly cloudy, dry, moderate breezes
Cork: Dry, occasional sunny spells, mild day cool night, gentle breezes
Galway: Partly cloudy, isolated showers, mild day cold night, fresh breezy spells

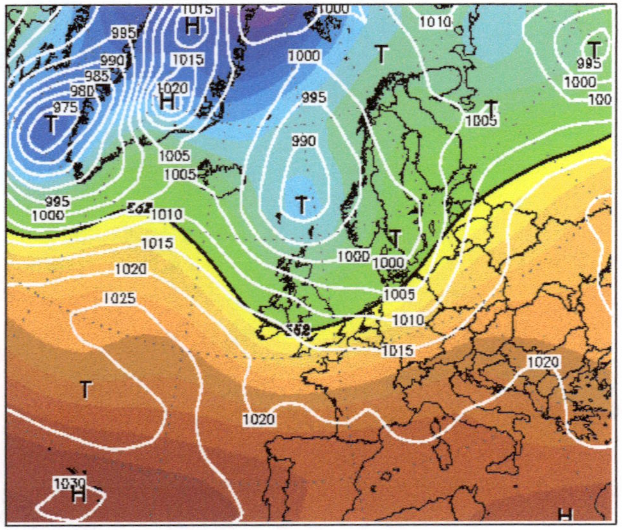

RAIN POTENTIAL

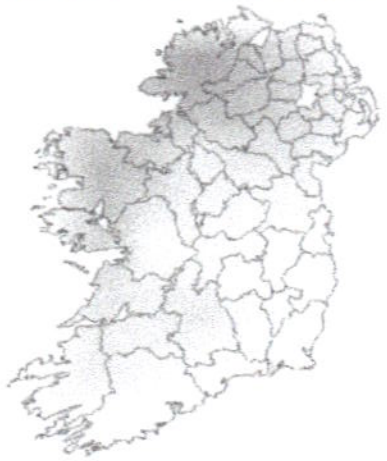

FROST/SNOW

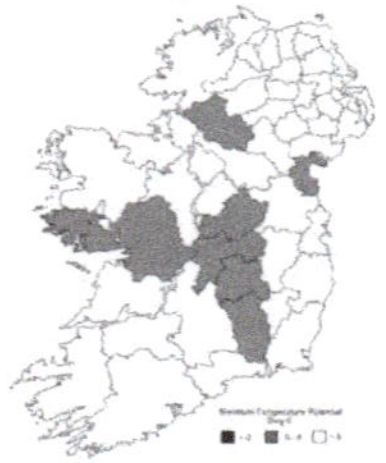

OCTOBER 28 WEDNESDAY

A weak high pressure ridge moves up over the country while low pressure prevails to the north. Winds generally very light in a gentle westerly flow with overnight patches of light frosts, mist and fog mostly in central part. Dull and gloomy throughout the day with brief showers in the west and south, while remaining mostly dry in the north and east.

Belfast: Overcast, moderate breezes, mostly dry with passing threats of rain
Dublin: Overcast, dry, light breezes
Cork: Mostly cloudy, dry, cooler, mostly calm air, chance of misty patches
Galway: Cloudy, scattered showers, mild day cold night, gentle breezes

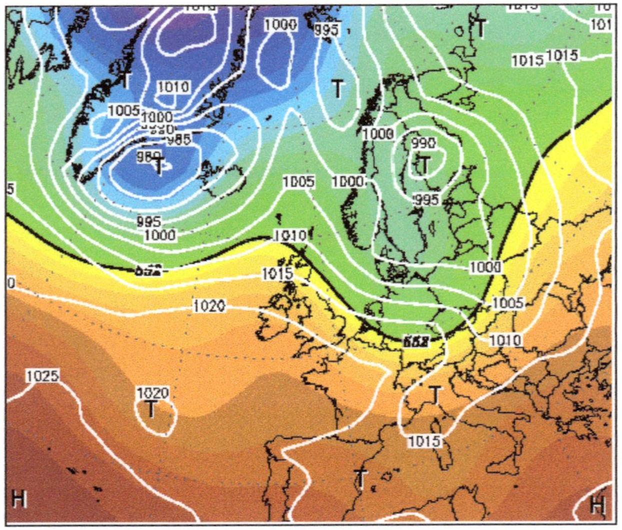

RAIN POTENTIAL

FROST/SNOW

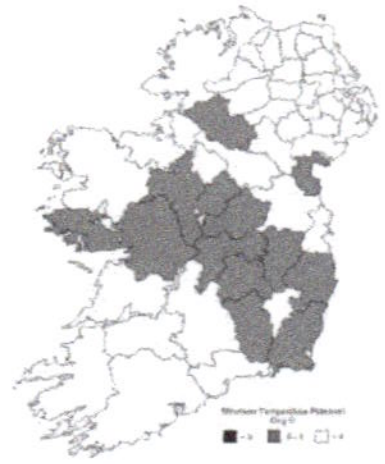

OCTOBER 29 THURSDAY

A weak high pressure ridge slips to the south of Ireland, while a depression to the north moves nearer, freshening winds to a stronger west to southwest flow, blustery on the northern coasts. Cloudy throughout with passing patches of light rain or drizzle crossing the country, in the west at first and later spreading into the east. Mostly dry in the far southwest and along the central eastern coasts.

Belfast: Cloudy, mostly dry, odd threat of rain, windy at times
Dublin: Cloudy, spotty showers, slightly warmer, light to moderate breezes
Cork: Cloudy, light showers, slightly warmer, moderate breezes
Galway: Light breezes, cloudy, warmer, increasing spotty showers

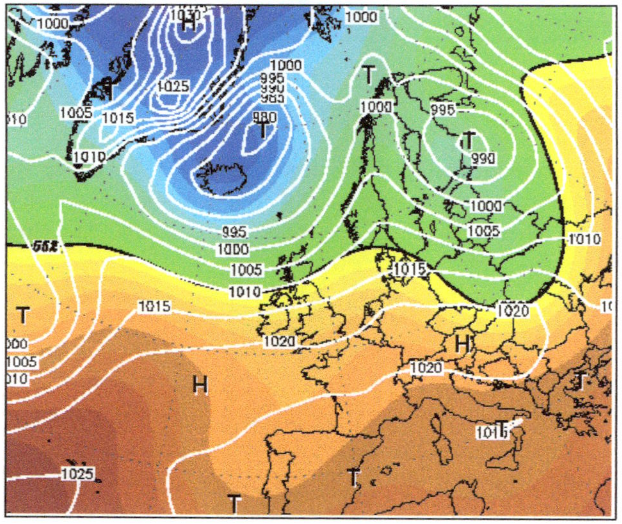

RAIN POTENTIAL

FROST/SNOW

OCTOBER 30 FRIDAY

The depression to the north slips further southwestward to be centred off the northwest of Ireland and extending strengthening southwest winds throughout. Very dull and gloomy throughout with some rain and drizzle in the northwest at first, and spreading throughout the country as the day progresses. Some showers may be briefly heavy for a time mostly in the west and southeast.

Belfast: Overcast, blustery at times, passing squalls
Dublin: Freshening winds, overcast, mostly dry with passing threats of rain
Cork: Overcast, scattered showers, windy spells
Galway: Overcast, rain, windy at times

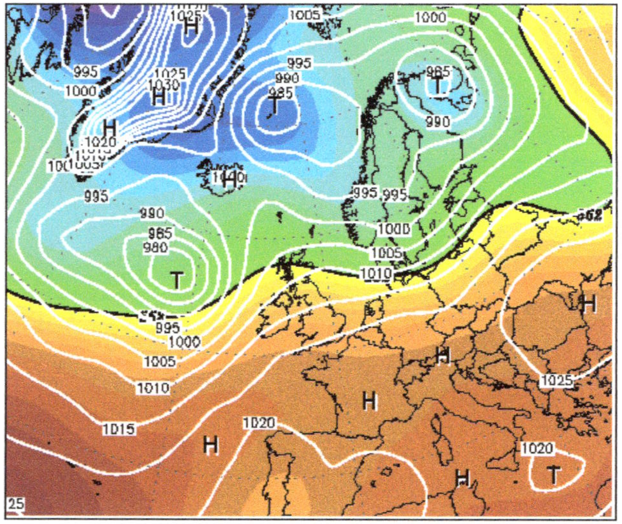

RAIN POTENTIAL

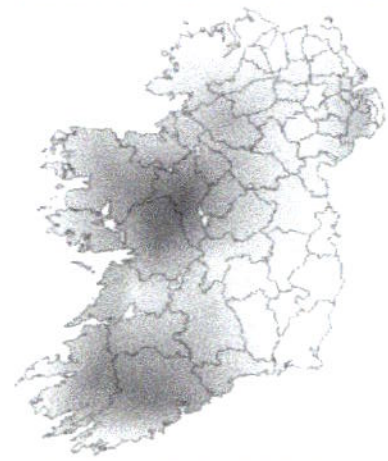

FROST/SNOW

OCTOBER 31 SATURDAY

A depression lingering to the north of Ireland intensifies and strengthens the southwest winds to very strong and gusty throughout the country, possibly reaching gale force for a time in many areas. A stormy air prevails throughout, with widespread cloud, scattered thunderstorms, squally showers, some of hail. Chance of sleet or snow occuring in some higher western and central parts.

Belfast: Stormy air, cloudy, occasional squalls, chance of gale winds and gusts
Dublin: Stormy air, cloudy, occasional squalls, mild temperatures, gales likely
Cork: Mostly cloudy, light showers, strong stormy winds, gales possible
Galway: Storm, cloudy, squalls, strong gusty winds, chance of gales and snow

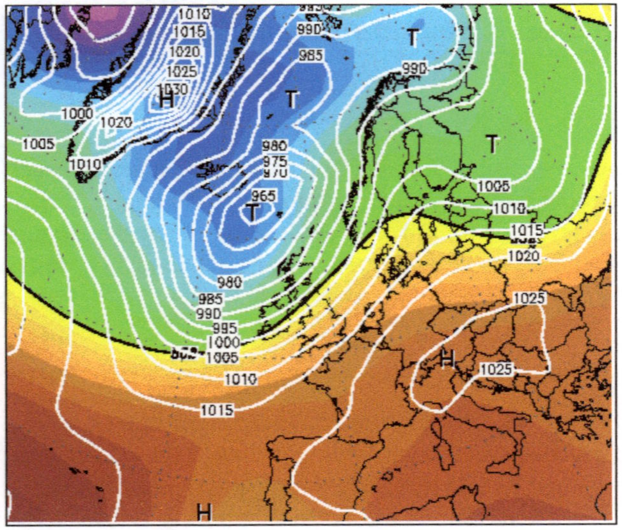

RAIN POTENTIAL

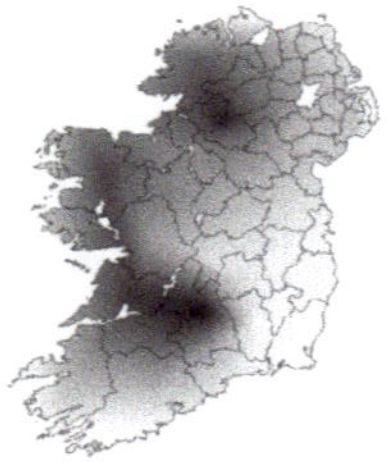

FROST/SNOW

NOVEMBER

PHASES OF THE MOON

1st	Third Quarter
6th	Crossing Equator
9th	New Moon
12th	Southern Declination
13th	Apogee
17th	First Quarter
20th	Crossing Equator
24th	Full Moon
25th	Perigee #5
27th	Northern Declination

1	2	3	4	5	6	7	8	9	10	11	12	13	14	15	16	17	18	19	20	21	22	23	24	25	26	27	28	29	30
Sun	Mon	Tue	Wed	Thur	Fri	Sat	Sun	Mon	Tue	Wed	Thur	Fri	Sat	Sun	Mon	Tue	Wed	Thur	Fri	Sat	Sun	Mon	Tue	Wed	Thur	Fri	Sat	Sun	Mon
3Q					XhS			N			V	A				1Q			XhN				F	P5	^				

MONTHLY SUMMARY

1st-6th A series of fast moving frontal systems associated with Atlantic depressions crossing the country in the past couple of weeks may continue with further frequent spells of rain or showers almost daily, as well as the chance of some spells of heavy rain bringing localised flooding and strong and blustery winds at times. Very strong and unsettled winds likely, possibly to gale force at times. Chance for hail showers at times.

1st Very sunny day likely despite the chance of thunderstorms.

3rd-5th Chance of very heavy rain at times in many parts of the country. Localised flooding possible, particularly in some areas in the southwest of the country.

4th High pressure may prevail.

4th-11th Cold spell with wintry showers of hail, sleet or snow returning at times.

5th Very strong winds in the northwest likely.

6th-7th Chance of thunderstorms and hail showers.

7th-11th A cold front crosses the country in a strong southeasterly flow at first, later backing to a north to northwest flow and then a northeasterly change later, bringing a spell of wintry showers, some heavy, and some of sleet or snow. Overnight temperatures may be significantly below average, with widespread frosts likely, some severe in places.

8th-10th Chance of frosts in inland and eastern areas, some severe. Wintry showers, some of hail, snow or sleet possible, with chance of snow accumulating for a time in some northern and higher reaches.

10th Thunderstorms possible.

11th Sunny day likely. Chance of severe frosts in some areas.

Most of the next four weeks may see little change in the air pressure levels, slightly lower in the north and higher in the south, with a strong high pressure system prevailing at the end of this month and depressions in the first couple of weeks and at the end of this outlook period. Overall, the next four weeks may be drier and milder than average throughout and particularly sunnier in southern areas. These next four weeks may see a change in rainfall with many areas, particularly in the east, likely to receive only average to below average amounts, despite a fairly wet and windy start, with more settled conditions dominating the period and more days on which rainfall may occur, amounts may be less than normal. Overall, there may be between 10-13 wet days in the east and south, mostly around the coasts, and up to 15-20 wet days elsewhere. Sunshine amounts may also be variable, with more than average sunshine in the east and south, with the rest of the country around 20% cloudier than what is typical for this time of year. Temperatures may be colder than average with widespread frosts likely in the first and last weeks of this outlook period, with the middle two weeks warmer than average, but only by around half a degree above the norm. During this outlook period, there may be 7 days with thunderstorms, 9 days with hail, 6 days with widespread fogs, around 12 days with ground frosts and 4 days with severe frosts, 5 days with sleet or snow, and 4 days with gales possible.

12th-16th A ridge of high pressure over Ireland brings a spell of mostly dry and sunny weather with light winds, generally northeasterly or variable in direction, and only the brief isolated showery spell. Daytime temperatures may be higher than average but fall swiftly after dark, producing widespread frost. It may be particularly fair in the south and east.

13th-16th Chance of widespread fog, clearing to a mostly sunny day.

17th-23rd Wet fronts accompanying slow-moving depressions near the western coast may bring stormy conditions throughout, with bands of rain, sometimes heavy and accompanied by thunderstorms, and strong winds at times, in most areas.

18th-22nd Very strong and unsettled winds.

19th A cold turn develops as a very active cold may cross the country at this time bringing squally showers and strong winds that may cause some damage to local property, particularly in the southeast. Showers may turn more wintry around this time. Chance of heavy rain and hail showers within a strong westerly flow in particular around southern parts of the country.

19th-22nd Chance of thunderstorms and hail showers.

20th Chance of snow in some places.

21st An intense depression may pass to the north of the country bringing further strong winds, including up to gale force at times.

24th-30th Anticyclonic. Winds ease to a light to moderate west to southwest flow over the country during this period, with very little sunshine and drizzly shower conditions prevailing, especially in the east and south. Warmer than average temperatures may prevail with few frosts.

IRELAND RAINFALL ESTIMATES
November 2026

d=mainly dry, n= nonrecordable, l=light shrs, s=significant shrs, r=rain, h=hvy falls

		ULSTER PROVINCE					CONNAUGHT PROVINCE						LEINSTER PROVINCE				MUNSTER PROVINCE						
		Alderg	Hillsbo	Armag	Loughl	Malin I	Sligo	Ardtarı	Belmul	Drums	Galway	Roscoı	Markre	Edende	Kilkenı	Dublin	Shann	Newpo	Tralee	Killarne	Cork	All	Moon
1st	Sun	s	l	s	r	s	s	r	l	l	l	l	s	d	l	d	s	s	s	s	n	85	3Q
2nd	Mon	l	l	l	l	l	l	l	l	s	h	s	s	l	l	d	l	s	l	h	d	75	
3rd	Tues	l	s	l	l	s	s	l	r	s	r	s	l	s	l	s	r	s	h	h	h	196	
4th	Wed	l	l	l	s	s	l	s	l	s	h	s	s	s	r	d	d	r	l	s	d	94	
5th	Thur	l	s	l	s	h	r	s	h	r	r	r	r	r	l	r	h	r	h	h	r	231	
6th	Fri	l	l	s	s	r	s	s	l	s	s	l	s	l	s	l	l	s	s	s	l	91	XhS
7th	Sat	l	l	l	s	s	l	s	s	r	l	s	s	d	l	d	s	s	s	r	l	77	
8th	Sun	l	l	l	s	s	l	s	s	l	s	l	s	l	n	n	l	r	l	l	l	53	
9th	Mon	s	r	l	h	h	s	h	r	r	s	r	h	r	s	s	s	r	r	h	s	219	N
10th	Tues	l	d	s	d	d	l	d	l	d	d	d	l	d	l	s	n	l	l	d	d	26	
11th	Wed	d	d	d	d	l	l	d	l	d	d	d	d	d	d	d	d	l	d	d	d	3	
12th	Thur	d	d	d	n	d	d	n	n	d	d	d	d	d	d	s	d	d	d	d	d	6	V
13th	Fri	d	d	d	d	n	l	d	l	d	d	d	d	d	d	d	d	d	d	l	d	4	A
14th	Sat	d	d	d	d	d	l	d	s	d	d	d	d	d	d	d	d	d	n	d	d	7	
15th	Sun	d	n	d	l	d	d	l	d	n	d	d	l	d	d	d	d	d	d	d	l	6	
16th	Mon	d	d	n	d	l	d	d	d	d	d	d	d	d	d	d	d	d	d	d	n	2	
17th	Tues	l	s	d	s	d	s	s	r	s	h	r	s	s	s	n	h	r	r	h	h	193	1Q
18th	Wed	s	s	s	r	s	s	r	r	s	h	s	s	s	s	l	r	r	h	r	r	195	
19th	Thur	s	s	l	r	s	s	r	r	r	r	h	r	r	r	r	r	h	s	h	r	229	
20th	Fri	s	s	s	l	r	s	l	r	s	s	s	s	l	r	l	r	s	s	r	r	137	XhN
21st	Sat	l	s	n	s	r	s	s	r	h	l	r	s	l	l	l	s	h	s	s	s	141	
22nd	Sun	l	d	l	l	l	l	l	s	s	l	l	l	l	n	l	s	l	l	s	l	36	
23rd	Mon	l	s	d	s	l	n	s	n	s	s	s	r	s	l	d	d	h	l	l	d	78	
24th	Tues	l	d	s	l	s	s	l	s	l	l	l	l	l	d	s	s	s	l	d	l	56	F
25th	Wed	l	l	n	l	s	l	l	l	l	s	l	l	l	l	d	d	l	l	s	n	37	P5
26th	Thur	l	l	n	s	d	l	s	l	s	h	s	s	s	l	l	r	s	s	l	l	74	^
27th	Fri	l	l	l	l	d	s	l	s	l	l	l	l	l	l	l	l	s	l	l	n	49	
28th	Sat	l	l	l	l	s	l	l	s	s	s	s	s	l	l	l	l	r	s	s	l	74	
29th	Sun	n	d	l	d	l	n	d	d	d	d	n	l	d	d	d	d	d	l	d	d	5	
30th	Mon	d	d	d	l	l	d	l	d	l	d	l	l	l	d	d	d	l	d	l	d	15	
	Estimate:	60	70	49	125	146	104	125	159	140	176	119	144	77	77	67	132	210	177	224	113	2492	
	Average:	80	88	75	119	109	128	130	134	102	120	129	128	88	86	73	94	170	168	121	120	2263	
	Trend:	drr	drr	drr	av	w tr	drr	av	w tr	w tr	w tr	av	w tr	drr	av	av	w tr	w tr	av	w tr	av	w tr	

IRELAND SUNSHINE ESTIMATES

November 2026

F=fine (8-12 hours of sunshine), pc= partly cloudy (4-7 hours), c=cloudy (1-3 hours), o=overcast (0 hours)

		ULSTER PROVINCE					CONNAUGHT PROVINCE						LEINSTER PROVINCE				MUNSTER PROVINCE						
		Alderg	Hillsbo	Armag	Loughl	Malin H	Sligo	Ardtarı	Belmul	Drums	Galway	Roscoı	Markre	Edende	Kilkenı	Dublin	Shannc	Newpo	Tralee	Killarne	Cork	All	Moon
1st	Sun	pc	pc	pc	pc	pc	pc	pc	pc	F	pc	pc	pc	F	F	F	pc	pc	pc	pc	F	151	3Q
2nd	Mon	pc	pc	pc	pc	pc	c	pc	o	pc	c	o	pc	pc	pc	pc	c	pc	c	c	pc	73	
3rd	Tues	o	o	o	o	o	o	o	o	o	o	o	o	o	o	o	o	o	o	o	o	6	
4th	Wed	pc	pc	pc	pc	pc	pc	pc	pc	pc	pc	pc	pc	pc	F	F	pc	pc	pc	pc	F	137	
5th	Thur	o	o	o	o	o	o	o	o	o	o	o	o	o	o	o	o	o	o	o	o	0	
6th	Fri	pc	pc	pc	pc	pc	pc	pc	pc	pc	pc	pc	pc	pc	pc	pc	pc	pc	pc	pc	pc	117	XhS
7th	Sat	pc	pc	pc	pc	pc	c	pc	pc	pc	pc	pc	pc	pc	pc	pc	pc	pc	c	c	pc	101	
8th	Sun	pc	pc	pc	pc	pc	c	pc	pc	pc	pc	pc	pc	pc	pc	pc	pc	pc	pc	pc	pc	99	
9th	Mon	o	o	o	o	o	o	o	o	o	o	o	o	o	o	o	o	o	o	o	c	10	N
10th	Tues	pc	pc	pc	pc	pc	F	pc	pc	c	pc	pc	pc	pc	c	c	c	c	c	c	o	84	
11th	Wed	pc	pc	pc	pc	pc	F	pc	pc	pc	pc	pc	pc	pc	pc	pc	F	F	F	F	F	154	
12th	Thur	c	c	c	c	c	c	c	o	pc	c	o	c	pc	pc	pc	c	pc	o	o	pc	70	V
13th	Fri	pc	pc	pc	pc	pc	c	pc	c	pc	c	c	pc	pc	pc	pc	c	c	c	c	c	76	A
14th	Sat	pc	pc	pc	pc	pc	pc	pc	o	pc	c	c	pc	pc	pc	pc	pc	pc	F	F	F	118	
15th	Sun	o	o	o	o	o	o	o	pc	o	c	c	o	o	o	o	o	o	c	c	o	18	
16th	Mon	o	o	o	o	o	o	o	o	o	o	o	o	o	o	o	o	o	o	o	o	0	
17th	Tues	c	c	c	c	c	o	c	c	c	o	c	c	c	o	c	o	o	o	o	o	18	1Q
18th	Wed	o	o	o	o	o	o	o	o	c	o	o	o	c	c	o	o	c	c	c	c	17	
19th	Thur	c	c	c	c	c	o	c	o	c	o	o	c	c	c	c	o	c	o	o	o	19	
20th	Fri	c	c	c	c	c	o	c	c	c	c	c	c	c	c	c	c	c	o	o	o	27	XhN
21st	Sat	pc	pc	pc	pc	pc	c	pc	o	pc	c	c	pc	pc	pc	pc	pc	pc	c	c	pc	96	
22nd	Sun	pc	pc	pc	pc	pc	pc	pc	c	pc	c	c	pc	pc	pc	pc	pc	pc	c	c	pc	87	
23rd	Mon	c	c	c	c	c	c	c	c	c	c	c	c	c	pc	c	pc	pc	c	c	pc	60	
24th	Tues	o	o	o	o	o	o	o	o	o	o	o	o	o	o	o	o	o	o	o	o	0	F
25th	Wed	c	c	c	c	c	o	c	o	c	o	o	c	c	c	c	o	o	o	o	o	21	P5
26th	Thur	o	o	o	o	o	o	o	o	o	o	o	o	o	o	c	o	o	o	o	o	9	^
27th	Fri	c	o	o	o	o	o	o	o	o	o	o	o	o	o	o	o	o	o	o	o	6	
28th	Sat	o	o	o	o	o	o	o	o	o	o	o	o	o	o	o	o	o	o	o	o	1	
29th	Sun	c	c	c	c	c	o	c	c	c	c	c	c	c	c	c	c	c	c	c	c	34	
30th	Mon	o	o	o	o	o	o	o	o	c	o	o	o	c	c	o	o	c	o	o	pc	16	
	Estimate:	83	83	83	83	83	66	83	58	92	68	63	83	92	102	98	77	89	69	69	105	1625	
	Average:	59	59	57	52	45	51	49	52	55	60	72	65	70	67	72	58	50	53	52	65	1161	
	Trend:	more	more	more	more	more	more	more	av	more	av	av	more	more	more	more	more	more	more	more	more	more	

NOVEMBER 1 SUNDAY

A depression centred to the far north of Ireland directs an unsettled and blustery northwesterly flow over the country, light in southern parts. Unsettled overnight with bands of showers, some of hail and accompanying thunderstorms possible, particularly in western and northern parts. Mostly dry in the eastern coastal areas. Good sunny spells between the showers likely.

Belfast: Mostly sunny, strong gusty winds, isolated showers, mild day cold night
Dublin: Sunny, colder, dry, stormy winds easing later in the day
Cork: Sunny, cooler, isolated showers, windy with gusts
Galway: Partly cloudy, showers, colder, blustery winds, chance of thunderstorms

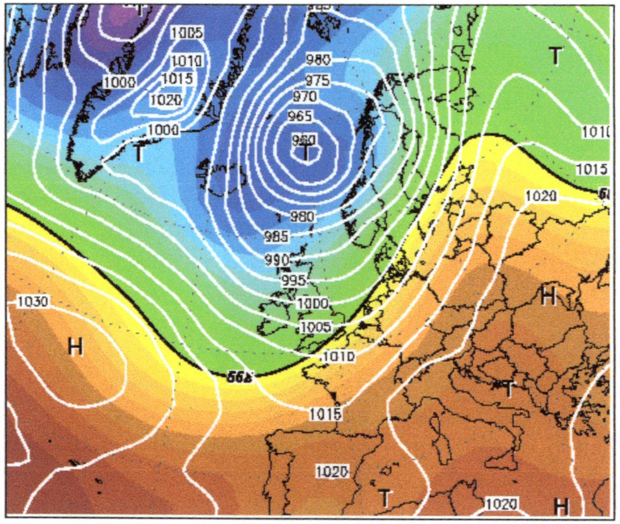

RAIN POTENTIAL

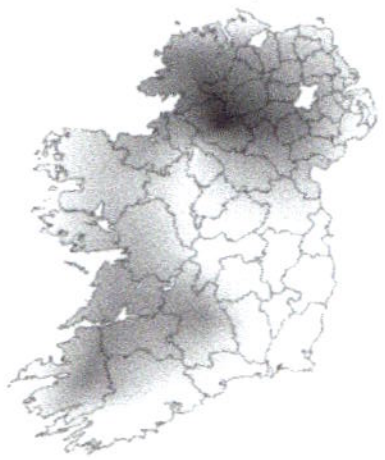

FROST/SNOW

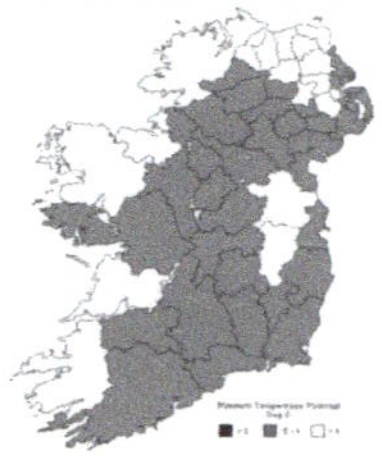

NOVEMBER 2 MONDAY

High pressure ridges to the south of Ireland while low pressure prevails over the northern half, with winds moderate in a westerly flow, fresher and more unsettled at times along the northern coasts. A mixed day with sunny spells and occasional showers almost everywhere, although mostly dry along the eastern coastal fringes. Chance of frost patches overnight in some central areas.

Belfast: Partly cloudy, brief showers, colder, blustery at times
Dublin: Partly cloudy, dry, slightly warmer, fresh winds with occasional gusts
Cork: Partly cloudy, mostly dry, odd threat of rain, windy spells, mild day cold night
Galway: Cloudy, heavy rain, slightly warmer, blustery at times

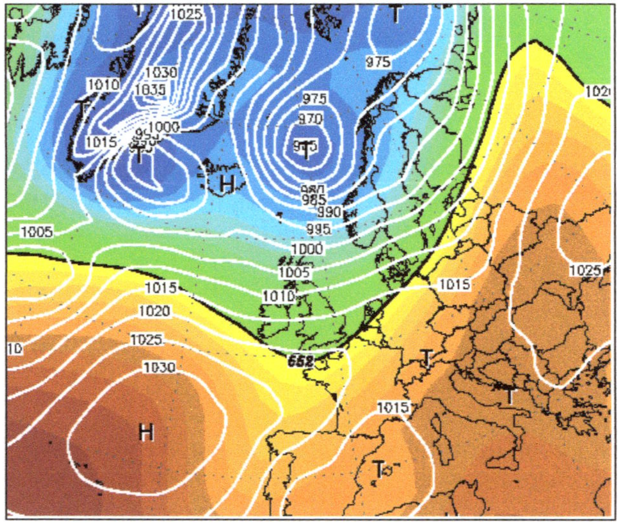

RAIN POTENTIAL

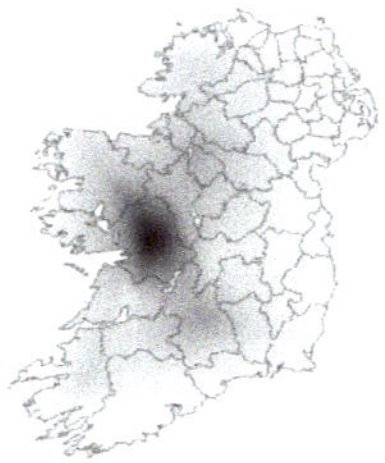

FROST/SNOW

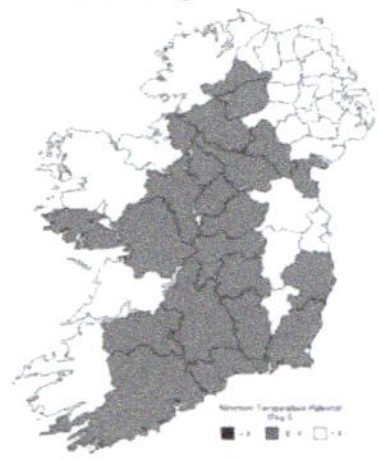

NOVEMBER 3 TUESDAY

A very large and intense depression centred to the north of Ireland brings a stormy air throughout the country, with very strong and gusty southwest winds prevailing, up to gale force at times in many parts, particularly around the coasts. Overcast with rain everywhere, with rain persistent and heavy at times, particularly in western and southern parts where localised flooding may occur.

Belfast: Overcast, occasional showers, slightly warmer, strong winds, possible gales
Dublin: Overcast, scattered showers, stormy air with gusts, gales likely
Cork: Overcast, heavy rain, stormy air with strong winds and gusts
Galway: Overcast, rain, mild temperatures, strong and blustery winds, gales likely

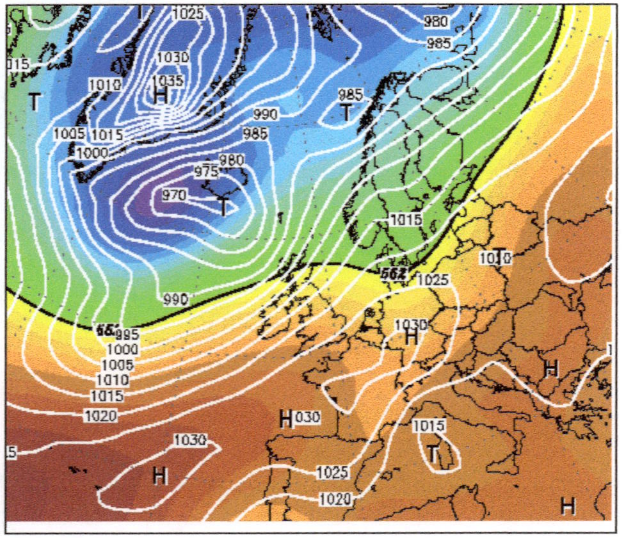

RAIN POTENTIAL

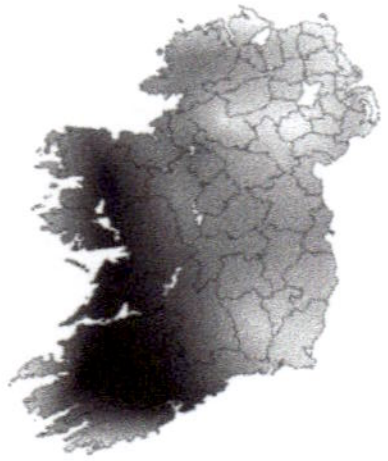

FROST/SNOW

NOVEMBER 4 WEDNESDAY

An intense storm system to the north of Ireland continues to bring very strong winds across the country in a gusty westerly flow. Overnight rain likely to persist in many areas, before clearing to a mostly dry and sunny day, particularly in the east and south, although shower activity may persist longer in some parts of the midlands.

Belfast: Partly cloudy, blustery, possible gales, some squalls, milder day cooler night
Dublin: Sunny, dry, cooler, strong and blustery winds
Cork: Sunny, dry, mild day cold night, strong winds with occasional gusts
Galway: Partly cloudy, heavy rain, slightly cooler, strong and unsettled winds

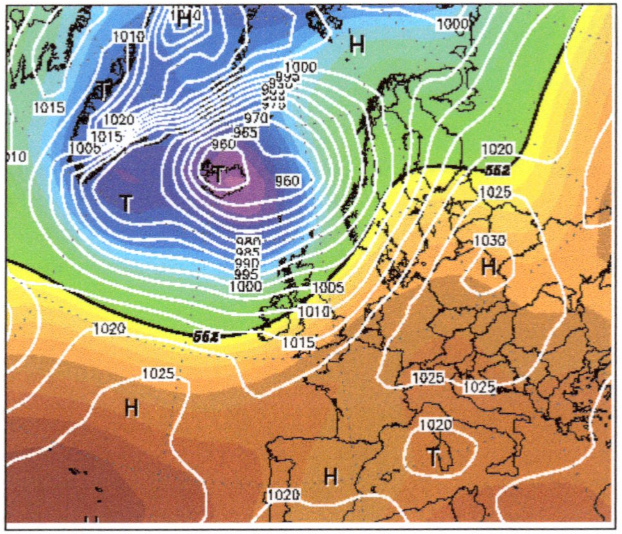

RAIN POTENTIAL

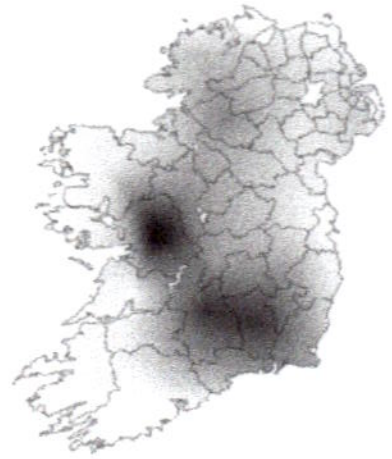

FROST/SNOW

NOVEMBER **5** THURSDAY

Storm conditions associated with an intense depression north of Ireland returns throughout the country once more with winds returning to a southwest flow, very strong and blustery and up to gale to storm force levels at times, particularly over the northern half of the country. Overcast everywhere with widespread rain, heaviest in the west and south, which may once again cause localised flooding, and lighter but persistent in the east and north.

Belfast: Overcast, colder, scattered showers, strong gusty winds, stormy air
Dublin: Unsettled and windy, overcast, rain
Cork: Wet and windy, overcast, milder temperatures
Galway: Overcast, rain, slightly warmer, strong winds, gales likely

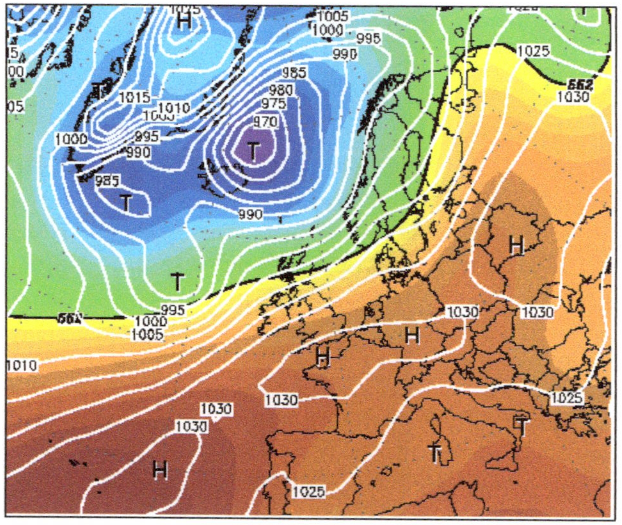

RAIN POTENTIAL

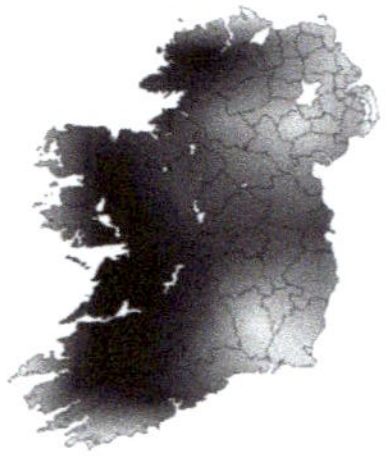

FROST/SNOW

NOVEMBER **6** FRIDAY

The large storm system to the north of Ireland may still continue to bring stormy winds across the country but easing with a northwest change. A mixed day as rain eases to scattered showers throughout with some sunny spells now and then. Shower activity may turn wintry in the north overnight, possibly turning to sleet, while temperatures may drop to light frost levels for a time in some inland parts.

Belfast: Partly cloudy, light showers, slightly warmer, blustery at times
Dublin: Partly cloudy, light showers, strong winds with occasional gusts
Cork: Partly cloudy, occasional showers, mild day colder night, windy at times
Galway: Partly cloudy, colder, showers, blustery, chance of gales and light snow

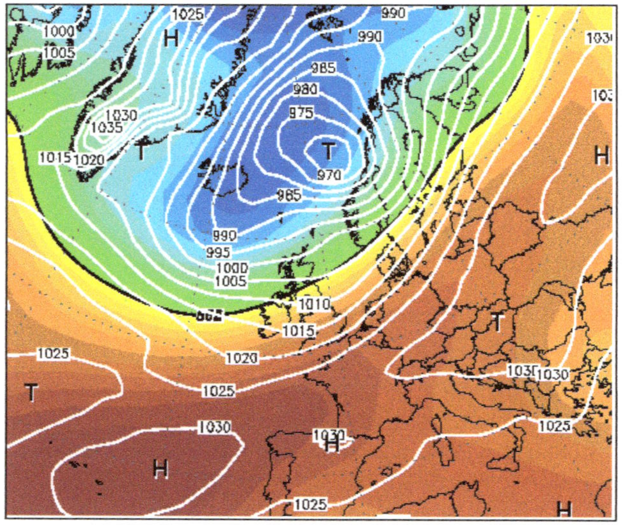

RAIN POTENTIAL

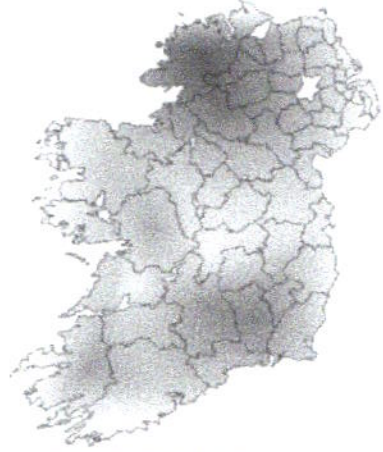

FROST/SNOW

NOVEMBER 7 SATURDAY

The storm system to the north of Ireland begins to weaken allowing for winds to ease to moderate while still in a northwesterly flow, although some blustery spells still likely around the northern coasts. A changeable day with a mix of light showers and sunny spells almost everywhere except for mostly dry along the eastern coastal fringes. Showers may become confined mostly to the western coasts later in the day. Chance of frosts overnight in inland parts, and possible snow flurries in the west.

Belfast: Partly cloudy, isolated showers, colder, strong unsettled winds
Dublin: Partly cloudy, dry, colder, unsettled winds with occasional gusts
Cork: Partly cloudy, isolated showers, colder, windy
Galway: Partly cloudy, light showers, chance of frost and snow flurries, windy spells

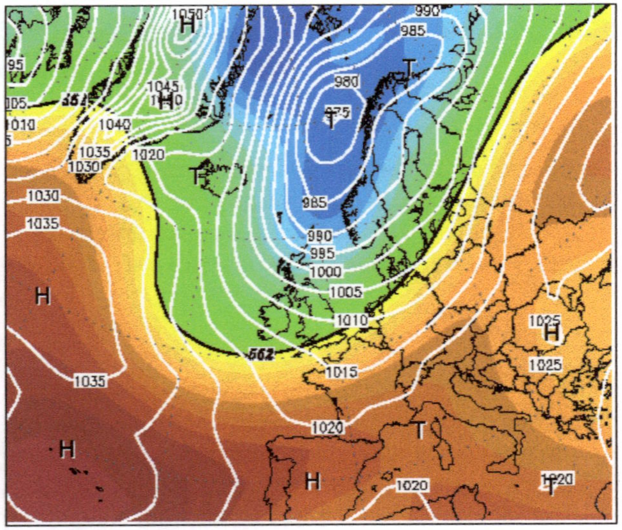

RAIN POTENTIAL

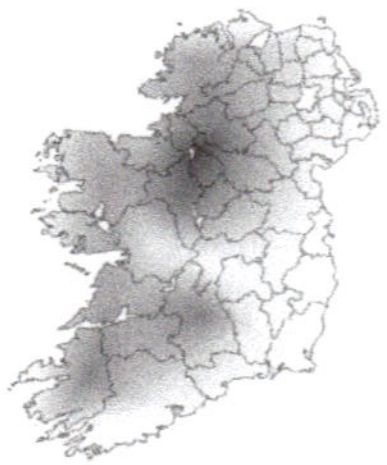

FROST/SNOW

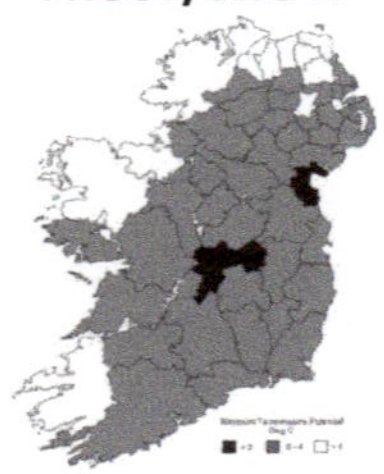

NOVEMBER 8 SUNDAY

A high pressure ridge builds to the west of Ireland while low pressure prevails to the east, with moderate to fresh north to northwesterly winds prevailing, gusty for a time before easing later in the day. A changeable day with a mix of sunny spells and brief showers in most parts of the country, although mostly dry and sunny around the east and southeast coastal fringes. Cold overnight with frosts in many districts, and chance of showers being very wintry and possibly light snow.

Belfast: Partly cloudy, light showers, colder, strong winds, chance of light snow
Dublin: Cold with frosts, sunny spells, isolated showers, chance of snow flurries
Cork: Moderate winds, partly cloudy, colder, light showers
Galway: Mix of showers and sunny spells, cold, chance of frost and overnight snow

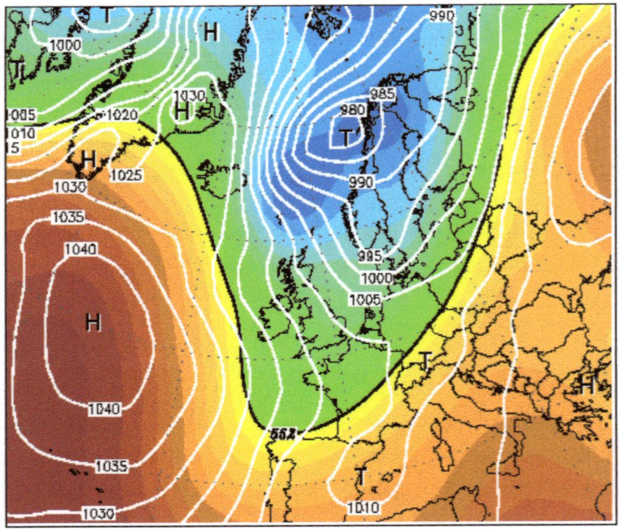

RAIN POTENTIAL

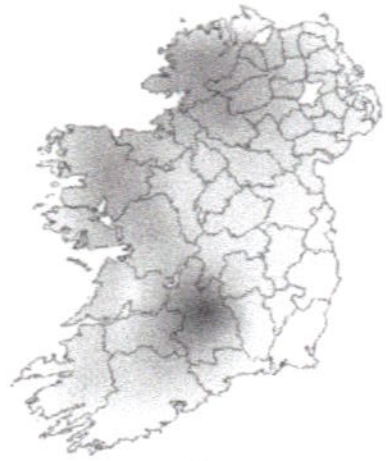

FROST/SNOW

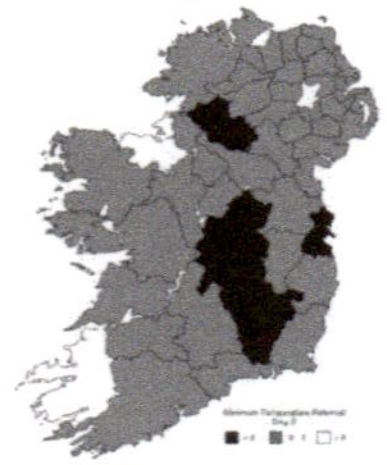

NOVEMBER **9** MONDAY

The high pressure ridge slips away to the southeast of the country while a weak depression in the northeast drifts nearer. Dull and gloomy everywhere with a mix of wintry showers throughout, some possibly of hail, sleet, or light snow developing during the day. Overnight temperatures may see severe frosts in numerous inland areas. Precipitation may be very heavy for a time, particularly in northern parts.

Belfast: Overcast, rain, cold, chance of hail, sleet or snow overnight
Dublin: Overcast, cold, frosts, rain, overnight snow likely, moderate shifting winds
Cork: Cloudy, fresh winds, occasional showers, chance of frost pockets
Galway: Overcast, rain, warmer day overnight frosts, fresh breezes, snow likely

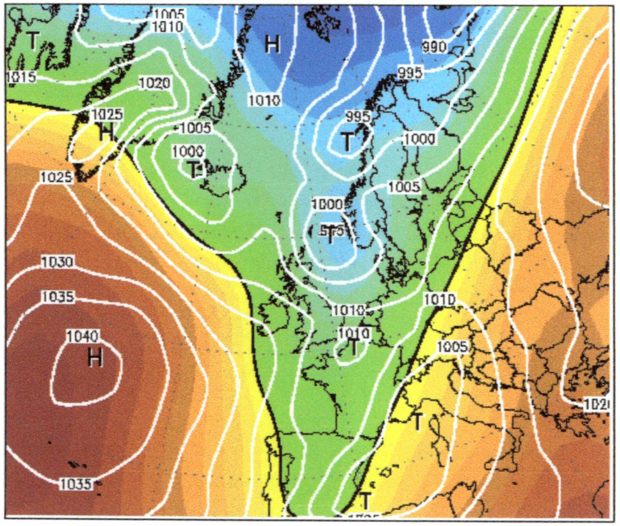

RAIN POTENTIAL

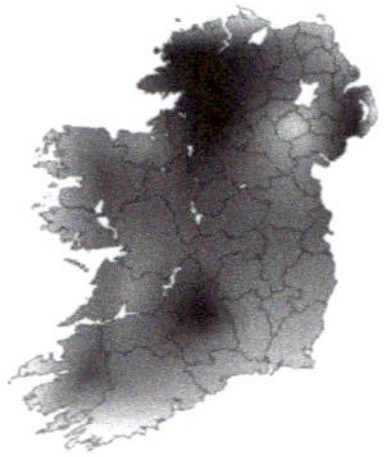

FROST/SNOW

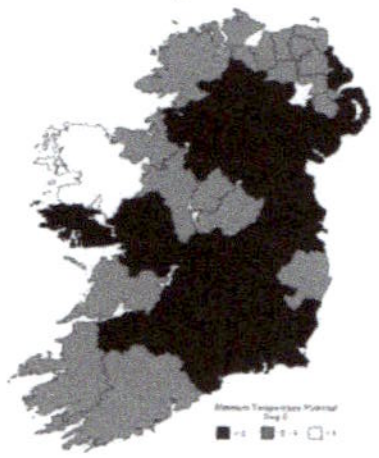

NOVEMBER **10** TUESDAY

A depression crosses the country bringing unsettled east to northeasterly winds throughout, generally moderate but occasionally gusty. Mostly cloudy to overcast with some wintry showers in most districts, and including the chance of some light snow or hail, particularly in the east and inland parts.

Belfast: Cold, partly cloudy, dry, strong and unsettled winds
Dublin: Cloudy, cold, showers, fluctuating winds, thunderstorms and hail possible
Cork: Overcast, dry, colder, odd threat of rain, unsettled and fluctuating winds
Galway: Fluctuating winds, partly cloudy, dry, colder, chance of snow flurries

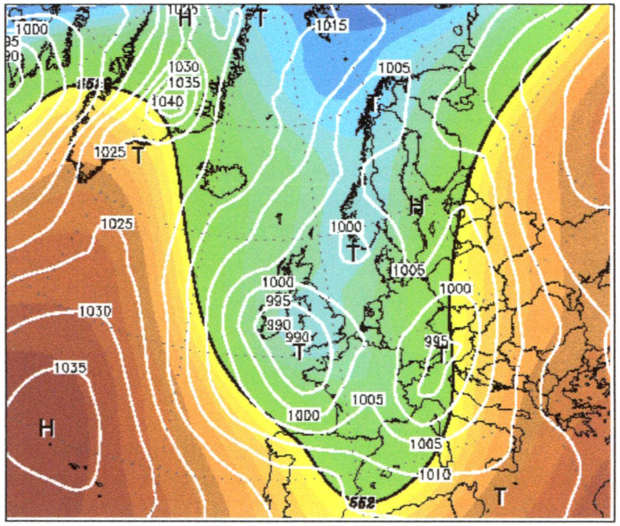

RAIN POTENTIAL

FROST/SNOW

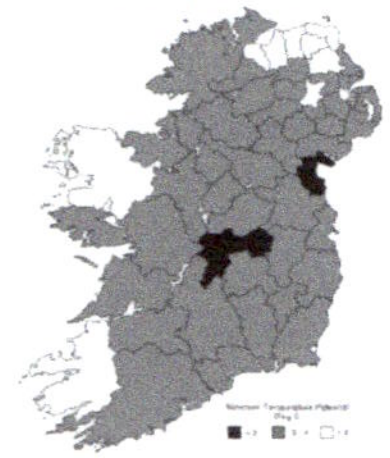

NOVEMBER **11** WEDNESDAY

The depression that brought unsettled conditions the past couple of days may now have moved to be centred to the south of Ireland while a high pressure ridge builds once more to the west of the country. Mostly dry and sunny throughout with widespread frosts overnight. Winds generally moderate to fresh in a northeasterly flow.

Belfast: Mostly sunny, dry, cold, frosts, strong winds
Dublin: Partly cloudy, cold, mostly dry with threats of rain, blustery at times
Cork: Sunny, dry, cold, chance of frost in sheltered pockets, fresh windy spells
Galway: Partly cloudy, dry, cold, fresh winds, chance of gales

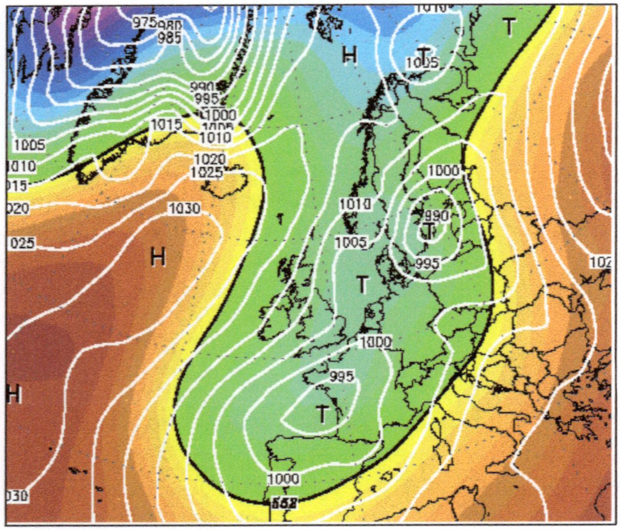

RAIN POTENTIAL

FROST/SNOW

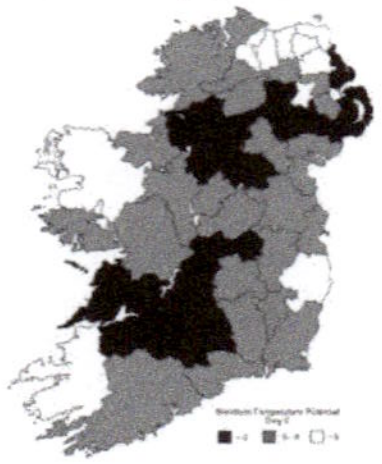

NOVEMBER **12** THURSDAY

High pressure ridges to the west, while low pressure prevails to the southeast, with a light northeasterly flow prevailing. Mostly sunny in the east with some isolated showers possible, mostly sunny and dry in the south and cloudier and dry in the north and west. Widespread frosts likely overnight.

Belfast: Mostly cloudy, dry, cold, winds easing to mostly light
Dublin: Partly cloudy, scattered showers, cold, winds easing to light for a time
Cork: Mostly sunny and cold, dry, moderate breezes
Galway: Cloudy, dry, cold, light to moderate breezes, chance of frost pockets

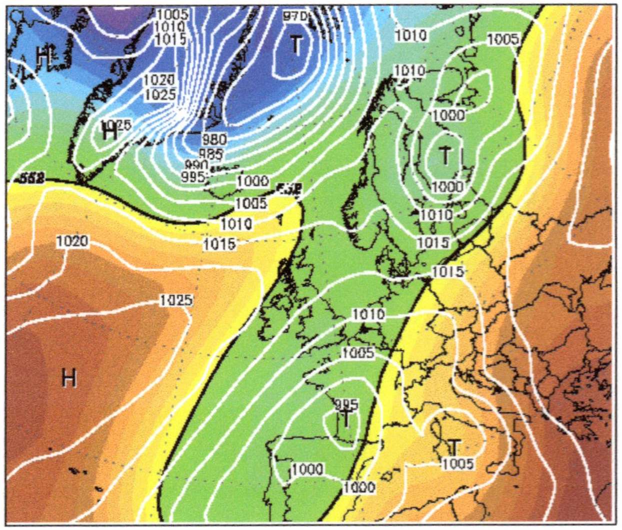

RAIN POTENTIAL

FROST/SNOW

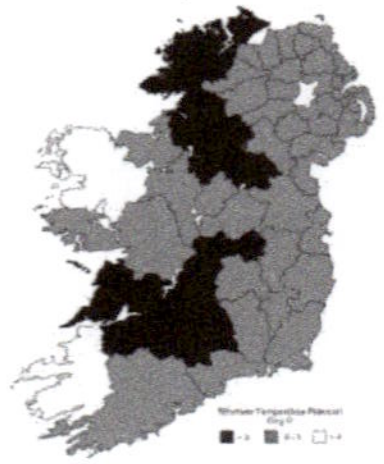

NOVEMBER **13** FRIDAY

A large anticyclone to the southwest of Ireland extends a ridge over the country, with a very light, mostly northerly flow prevailing. Partly cloudy and mostly dry with possible patches of fog and frosts in the north and east. Mostly cloudy and milder in the south and west, dry in the south and possible isolated overnight showers in the west.

Belfast: Partly cloudy, cold and dry, light shifting breezes, chance of fog patches
Dublin: Partly cloudy, warmer day colder night, dry, moderate breezes
Cork: Mostly cloudy and dry, slightly warmer, moderate shifting breezes
Galway: Cloudy, slightly warmer, dry, moderate breezes, chance of fog overnight

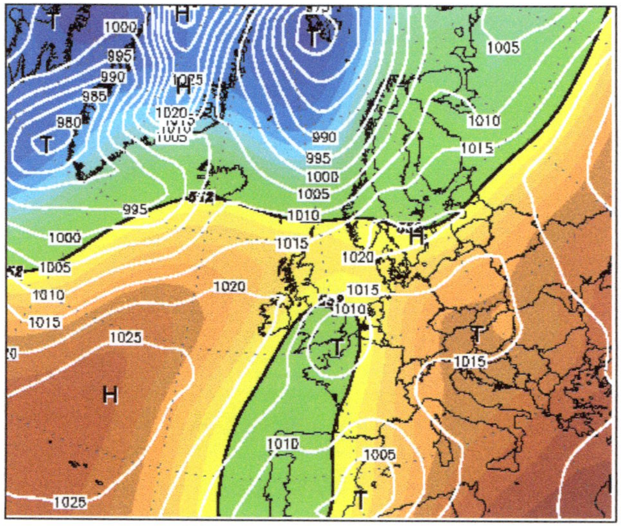

RAIN POTENTIAL

FROST/SNOW

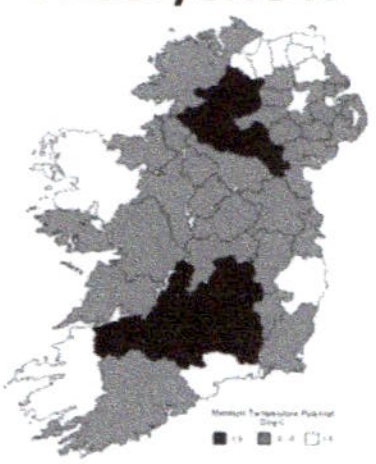

NOVEMBER **14** SATURDAY

A saddle between two high pressure ridges crosses the country while a weak depression lingers to the south of Ireland. Winds generally very light to mostly calm throughout, with early morning mist and fog patches along the coasts in the morning, while returning and being more widespread in the evening. Mostly sunny and dry day throughout, except for in the west, which may be mostly cloudy with some light shower activity for a time. Frosts likely in inland parts.

Belfast: Partly cloudy, milder, dry, gentle breezes, chance of fog or freezing mists
Dublin: Partly cloudy, mostly dry, threats of rain, gentle breezes, chance of mists
Cork: Sunny, dry, slightly warmer, gentle breezes
Galway: Cloudy, dry, slightly warmer, light breezes to mostly calm air, fog likely

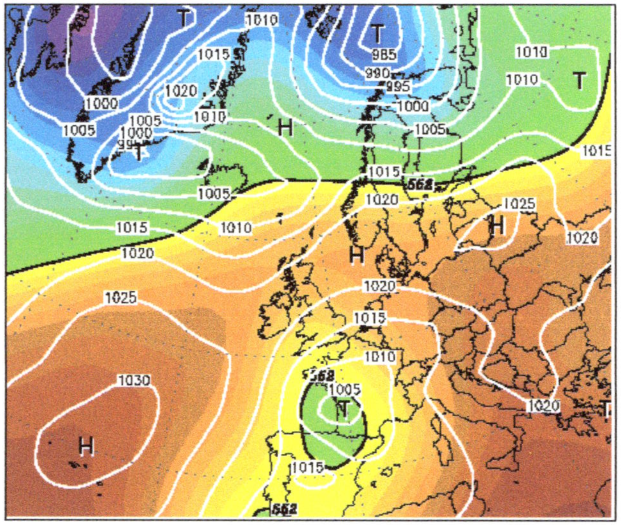

RAIN POTENTIAL

FROST/SNOW

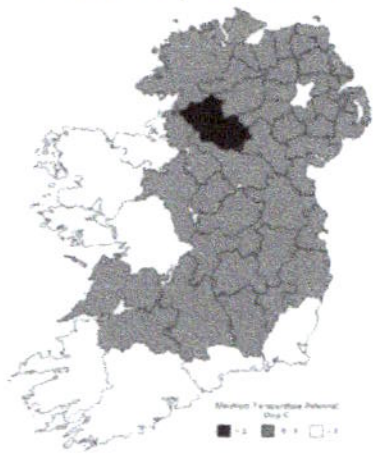

NOVEMBER **15** SUNDAY

A very light to mostly calm anticyclonic air lingers over the country bringing patches of mist, fog, and light drizzle overnight in many areas and scattered patches of frost, mostly in inland parts. Mostly cloudy to overcast throughout the day, with chance of some isolated showers, mostly in northern parts; dry elsewhere, slightly sunnier in the west. Chance of fog or light drizzle patches likely to return after dark.

Belfast: Overcast, cold, frosts, isolated showers, chance of snow flurries and fog
Dublin: Overcast and dull, dry, threats of rain, gentle breezes
Cork: Isolated showers, dull and gloomy air, cooler, gentle breezes
Galway: Mostly calm air, cloudy, dry, fog likely

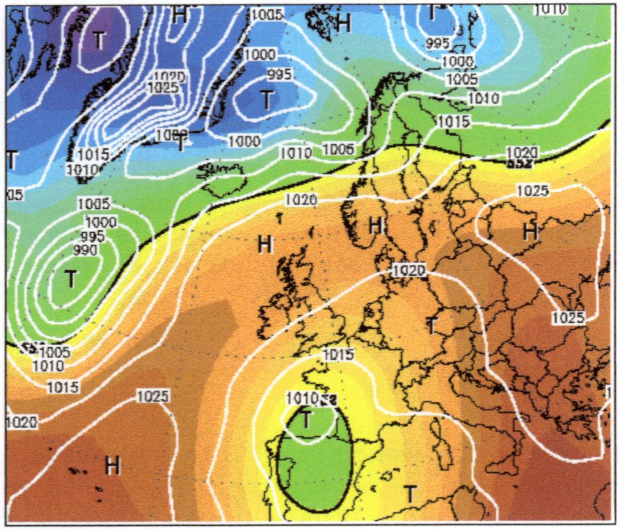

RAIN POTENTIAL

FROST/SNOW

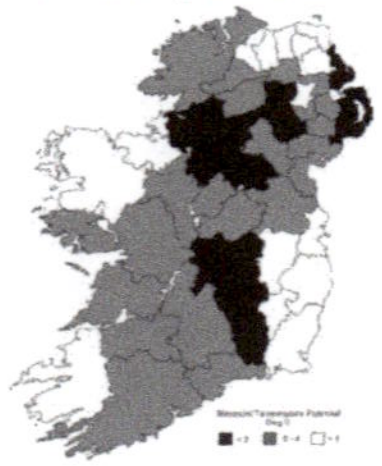

NOVEMBER **16** MONDAY

A gentle anticyclonic atmosphere may continue to prevail throughout the country with overnight patches of mist, fog, or light drizzle in most areas at both ends of the day, slow to clear in the morning. Mostly dry with widespread cloud throughout. Milder temperatures likely with only a small chance of frost. Winds generally light in variable directions.

Belfast: Calm, cold, and gloomy air, dry with passing threats of rain, chance of fog
Dublin: Overcast, dry, threats of rain, mild temperatures, mostly calm air
Cork: Overcast, spotty showers, warmer, mostly calm air, fog patches likely
Galway: Overcast, dry with increasing threats of rain, mostly calm air

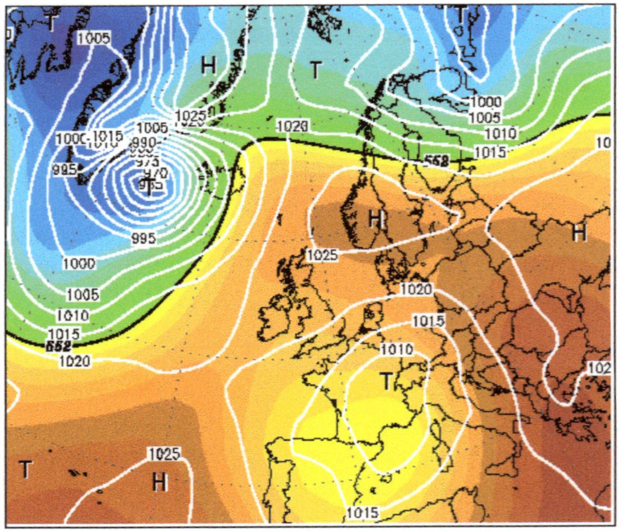

RAIN POTENTIAL

FROST/SNOW

NOVEMBER **17** TUESDAY

Low pressure trough moves over Ireland, with light, mostly southwesterly breezes prevailing. Cloudy to overcast everywhere. Patchy rain moves into the southern half of the country at first, spreading showers northwards during the day, with falls heavy in the south and west at times, while remaining mostly dry to isolated showers along the far northern coasts.

Belfast: Cloudy, calm air, occasional showers, milder temperatures, fog possible
Dublin: Cloudy, calm atmosphere, isolated showers
Cork: Mild and calm air, overcast, heavy rain
Galway: Overcast, heavy rain, milder, gentle breezes freshening later in the day

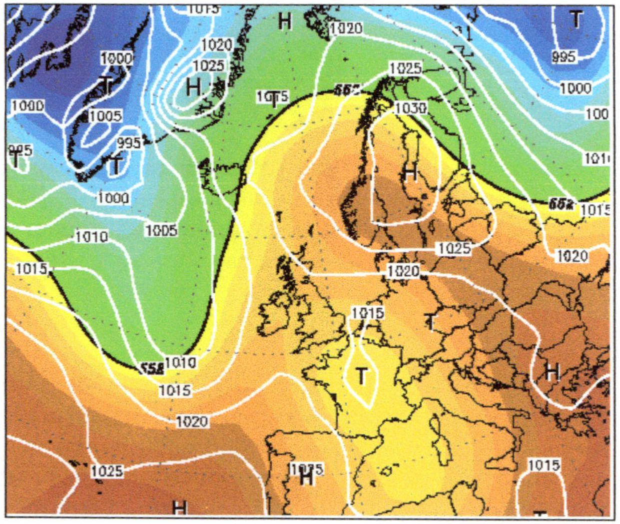

RAIN POTENTIAL

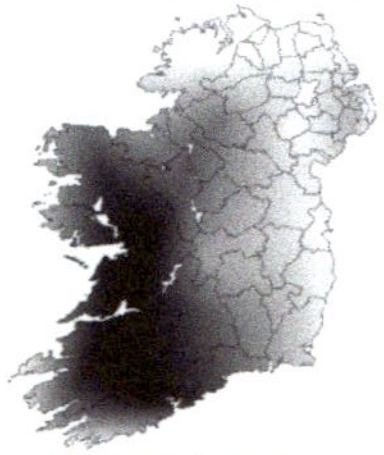

FROST/SNOW

NOVEMBER **18** WEDNESDAY

A depression forms in the low pressure trough over Ireland, with dense fog overnight at first before moderate to strong west to southwest winds develop, more gusty and possibly up to gale force in some areas. A stormy air prevails throughout, with widespread cloud to overcast in most districts and outbreaks of rain everywhere, heavier in the south and west.

Belfast: Stormy air, overcast, mild temperatures, scattered showers
Dublin: Overcast, brief showers, mild temperatures, fresh winds
Cork: Freshening breezes, stormy air, partly cloudy, rain
Galway: Overcast, heavy rain, mild day cold night, fluctuating winds

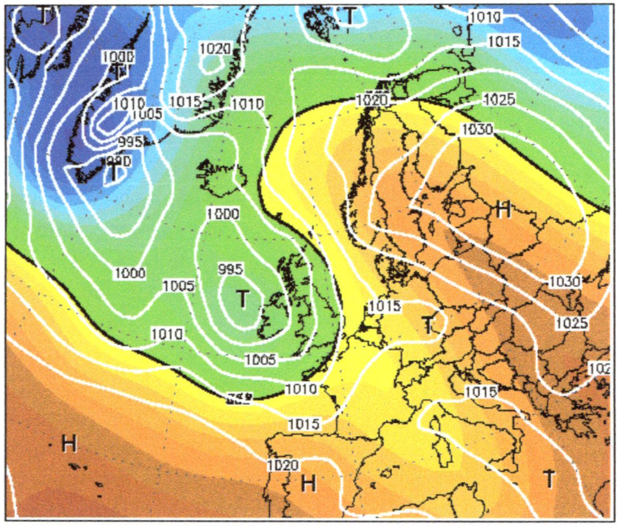

RAIN POTENTIAL

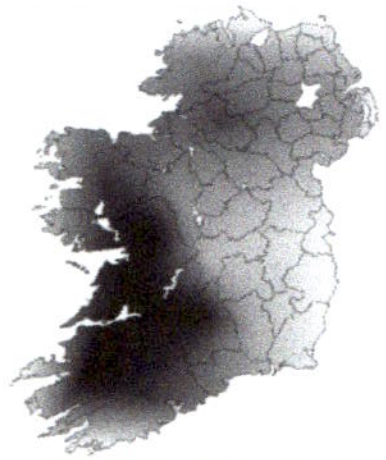

FROST/SNOW

NOVEMBER 19 THURSDAY

A large depression centred to the northwest of Ireland directs a strong and blustery southwest flow over the country, possibly up to gale force at times. Another stormy day likely everywhere, with widespread cloud and rain developing by the afternoon and spreading throughout with chance of heavy falls in the southern half of the country. Afternoon thunderstorms and hail likely, particularly in western parts, and possible sleet or snow may be observed in the north.

Belfast: Cloudy, rain, cooler, strong winds with gusts
Dublin: Warmer day cooler night, cloudy, rain, strong winds with occasional gusts
Cork: Unsettled and blustery winds, overcast, rain, mild day cold night
Galway: Overcast, rain, windy, chance of snow and thunderstorms overnight

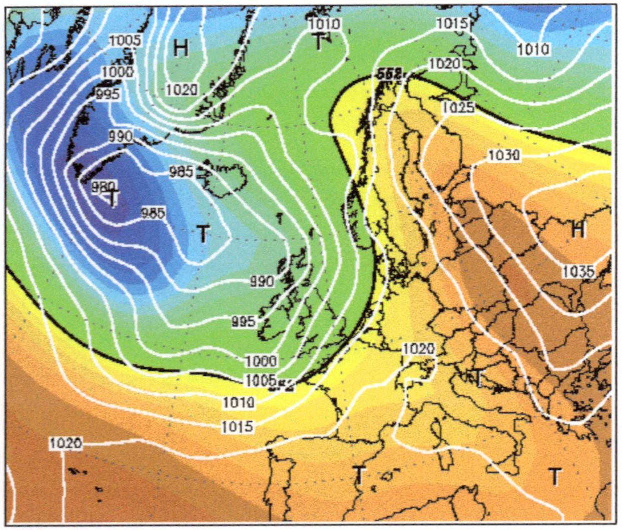

RAIN POTENTIAL

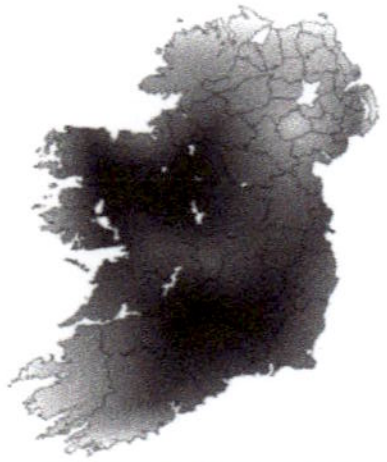

FROST/SNOW

NOVEMBER 20 FRIDAY

The depression intensifies as It nears the northern coast of Ireland bringing further widespread stormy conditions throughout and very strong west to southwest winds, possibly with gale gusts in northern parts. Cloudy and dull everywhere, with heavy showers overnight, and bands of showers during the day, some of hail, and more frequent in the west. Chance of thunderstorms, hail, and possible snow, mostly along the northern and southern coasts.

Belfast: Cloudy, scattered showers, mild day cold night, stormy air with gales likely
Dublin: Cooler, cloudy, light showers, strong stormy winds with gales likely
Cork: Overcast, rain, colder, stormy air
Galway: Cloudy, showers, cooler, blustery, thunderstorms, hail and snow possible

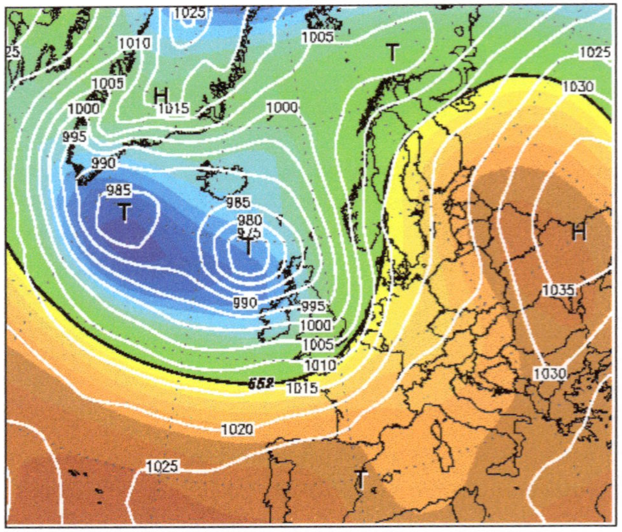

RAIN POTENTIAL

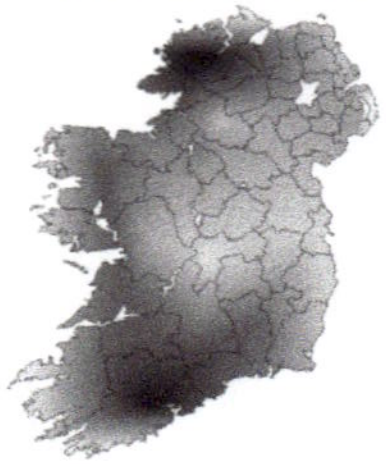

FROST/SNOW

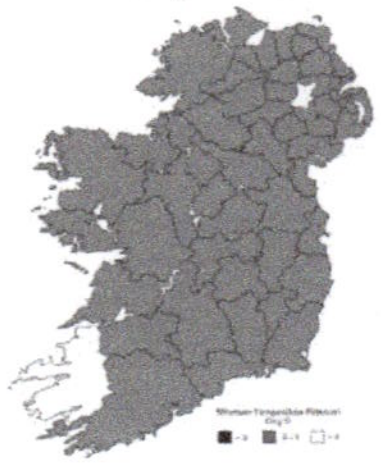

NOVEMBER **21** SATURDAY

An intense, almost stationary depression, sits to the immediate north of Ireland around this time, maintaining stormy conditions throughout. Winds very strong and blustery in a west to southwest flow, possibly up to gale force at times. An unsettled and mixed day with brief sunny outbreaks between spells of squally rain. Showers likely to be heavy at times in some western parts. Chance of further thunderstorms and hail showers. Possible snow or sleet in the west for a time.

Belfast: Colder, stormy, mix of showers and some sunny spells, gales possible
Dublin: Partly cloudy, cold, passing squalls, strong and stormy winds
Cork: Mix of showers and sunny spells, strong winds, gales likely, stormy air
Galway: Cloudy, some showers, stormy air, gales, thunderstorms, and snow possible

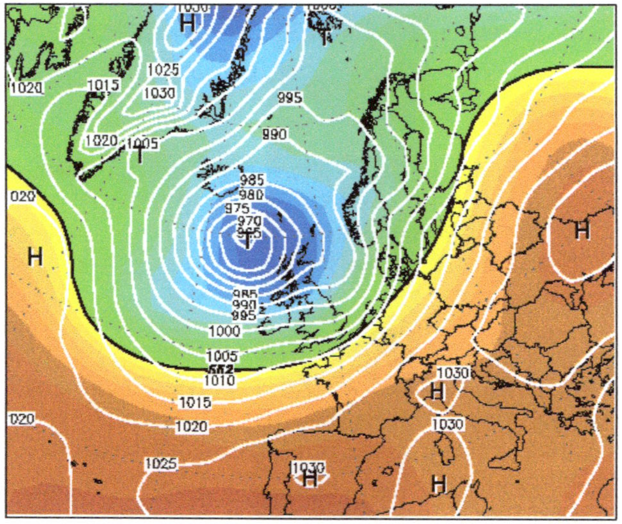

RAIN POTENTIAL

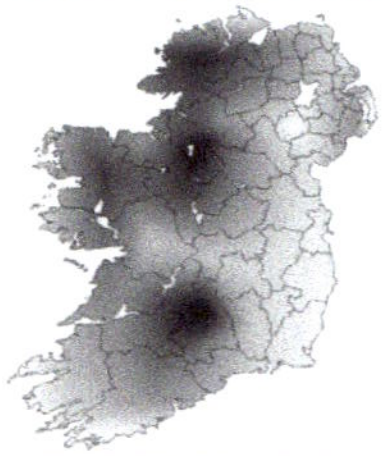

FROST/SNOW

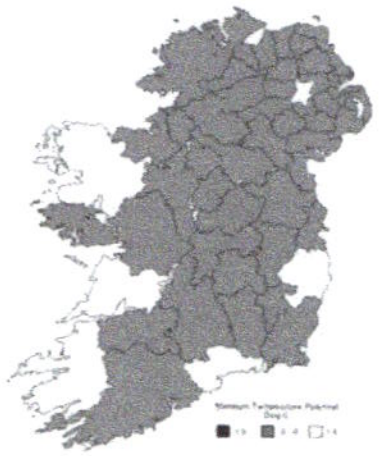

NOVEMBER **22** SUNDAY

An intense depression continues to almost stall centred to the immediate northeast of Ireland, maintaining very strong and blustery northwest winds throughout. While a stormy air may continue to prevail throughout, rain eases to lighter showers between increasing and longer sunny spells almost everywhere. Clearing to mostly dry along the northern and eastern coasts, with only the chance of some isolated showers around the far southern coasts.

Belfast: Partly cloudy, mostly dry, unsettled winds, milder temperatures
Dublin: Stormy and unsettled air, partly cloudy, brief squalls, gale gusts possible
Cork: Partly cloudy, light showers, stormy, gusty winds, slightly warmer
Galway: Mostly cloudy, brief showers, milder temperatures, strong gusty winds

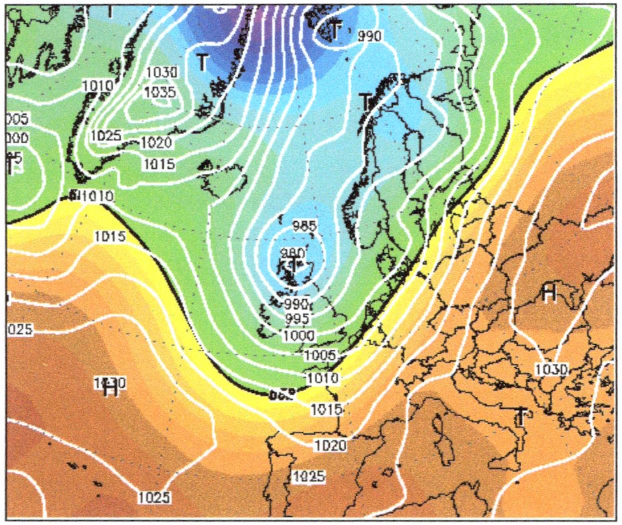

RAIN POTENTIAL

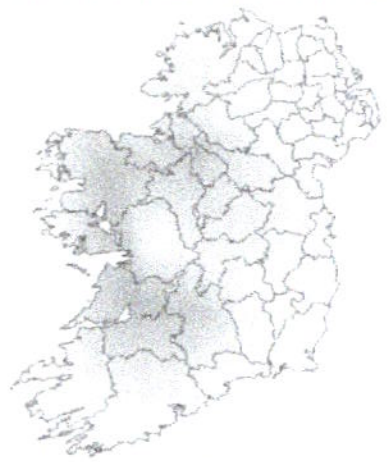

FROST/SNOW

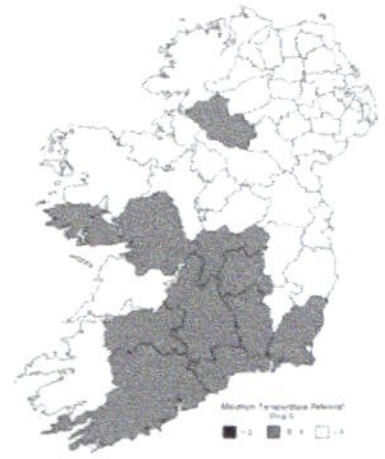

NOVEMBER 23 MONDAY

The depression that had brought unstable conditions over the previous few days may weaken around now as it slowly moves to the northeast of Ireland, while a high pressure system builds a ridge to the southwest of the country. Winds lighter in a northwesterly flow. Scattered cloudy spells and some showers likely during the day, although mostly dry around the coasts. Colder with chance of frosts and fog patches overnight.

Belfast: Cloudy, occasional showers, breezy spells
Dublin: Cloudy, dry, cool, winds easing to light
Cork: Partly cloudy, mostly dry, odd threat of rain, moderate breezes
Galway: Scattered showers, mostly cloudy, moderate breezes, chance of frost

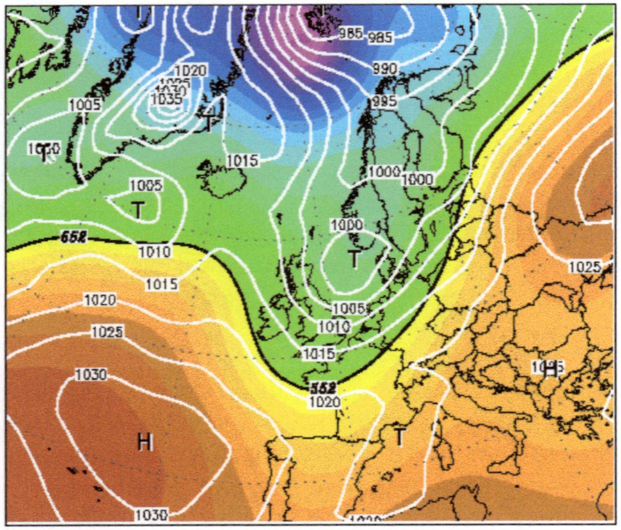

RAIN POTENTIAL

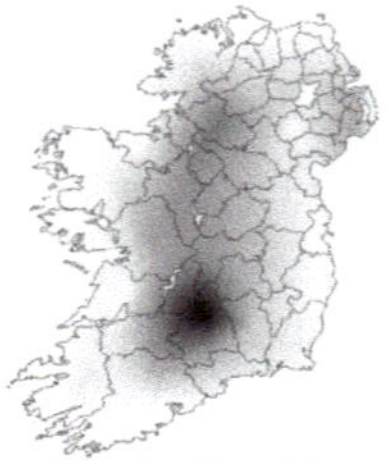

FROST/SNOW

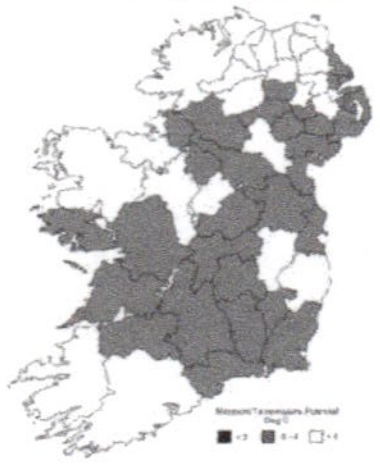

NOVEMBER 24 TUESDAY

A high pressure ridge covers the southern half of the country while low pressure prevails over the northern half. Widespread cloud prevails throughout with patches of drizzle and light rain almost everywhere. Winds generally moderate in a westerly flow. Chance of fog patches here and there.

Belfast: Overcast, mostly dry with chance of misty patches, windy at times
Dublin: Overcast, scattered showers, warmer temperatures, moderate breezes
Cork: Gentle breezes, chance of fog, overcast, milder temperatures, brief showers
Galway: Overcast, brief showers, windy at times

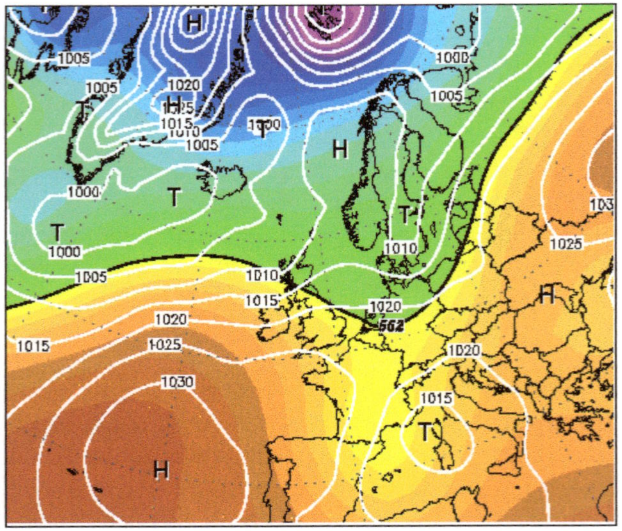

RAIN POTENTIAL

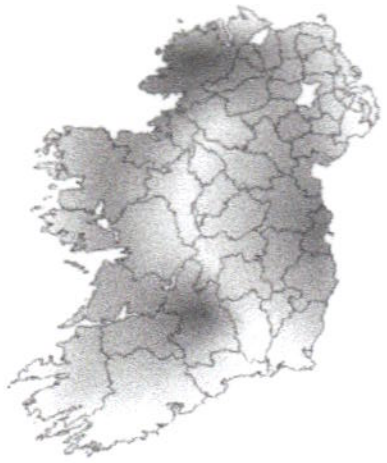

FROST/SNOW

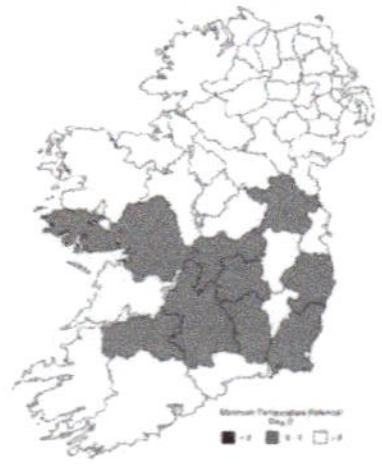

NOVEMBER **25** WEDNESDAY

A high pressure system to the south of Ireland builds a high pressure ridge over the country and brings fresh westerly breezes throughout, strong and gusty at times, particularly in the west and north. Mostly cloudy and dull throughout, with some isolated showers here and there. Mostly dry around the eastern and southern coasts.

Belfast: Cloudy, blustery, milder temperatures, spotty showers
Dublin: Cloudy, mostly dry with passing threats of rain, mild temperatures, blustery
Cork: Overcast, mild temperatures, isolated showers, strong gusty winds at times
Galway: Fluctuating winds, warmer temperatures, overcast, scattered showers

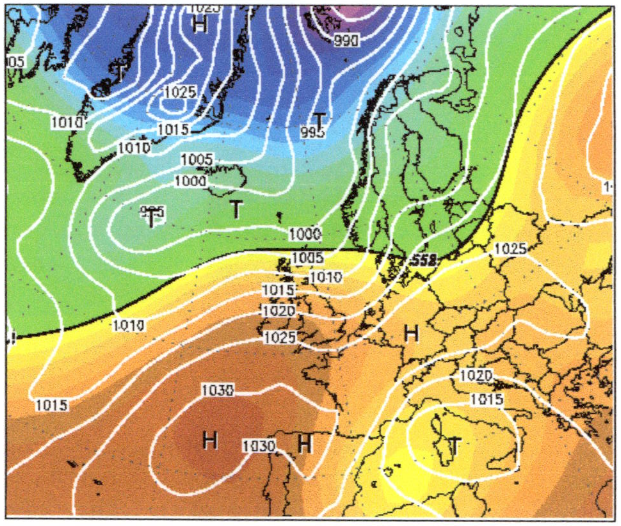

RAIN POTENTIAL

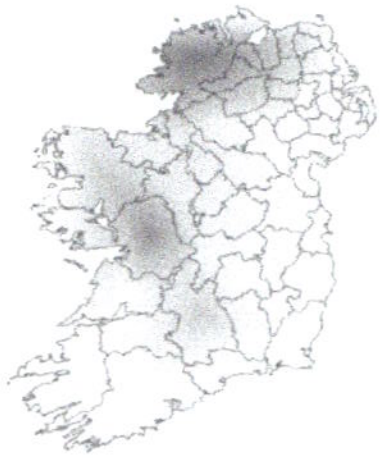

FROST/SNOW

NOVEMBER **26** THURSDAY

Low pressure prevails to the north of Ireland while the high pressure ridge lingers over the southern half of the country, with a moderate to fresh west to southwest flow prevailing, more gusty at times particularly in the north. Widespread cloud throughout with drizzle patches and showers in most areas, although mostly dry in the far northwest and northern coastal fringes. Chance of fog patches overnight mostly in north and central parts.

Belfast: Overcast, light showers, mild temperatures, strong and gusty winds
Dublin: Mild temperatures, cloudy, light showers, blustery winds
Cork: Overcast, mild temperatures, brief showers, blustery at times
Galway: Overcast, rain, mild temperatures, blustery

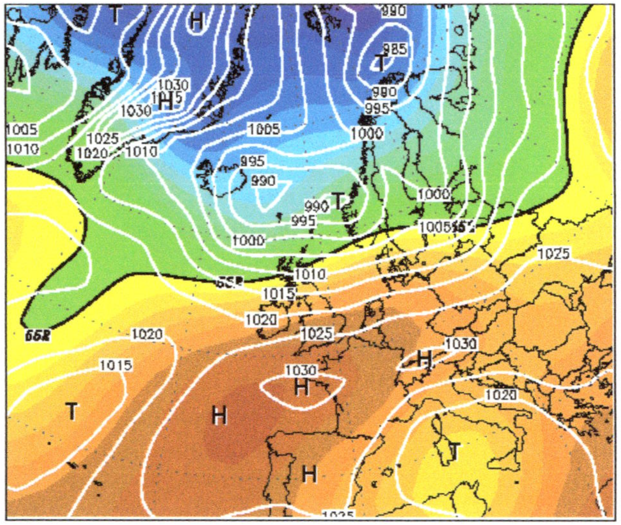

RAIN POTENTIAL

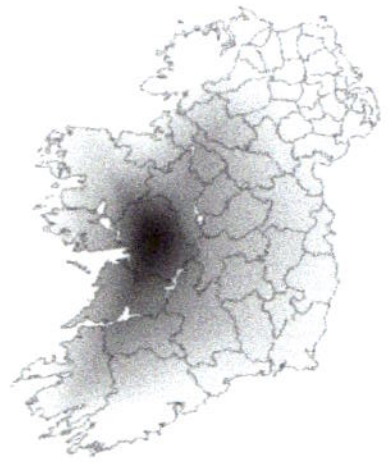

FROST/SNOW

NOVEMBER 27 FRIDAY

A high pressure ridge covers the country bringing a light mostly west to northwest flow throughout with patches of fog patches likely overnight in many places. Widespread cloud to overcast with some light showers almost everywhere, more persistent and drizzly in central parts, but mostly dry along the northern coastal fringes.

Belfast: Winds easing, gloomy air, some showers, mild temperatures, fog possible
Dublin: Overcast, cooler, some showers, light winds to mostly calm air
Cork: Light shifting breezes, overcast, isolated showers, fog patches possible
Galway: Winds easing to light, overcast, brief showers, mild temperatures

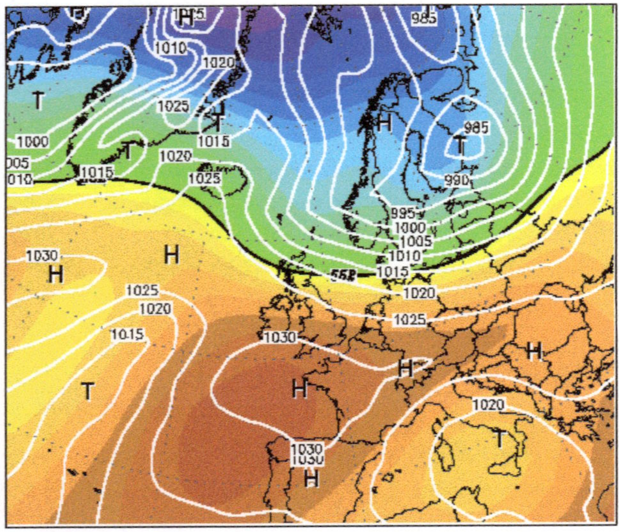

RAIN POTENTIAL

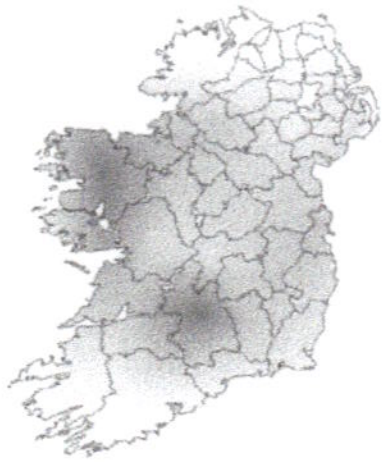

FROST/SNOW

NOVEMBER 28 SATURDAY

An anticyclonic system centred to the immediate south of Ireland maintains a very light west to southwest flow throughout, and with widespread cloud and overnight fog almost everywhere. Chance a band of rain may cross the country during the day bringing some shower activity in all districts, particularly in the north and west, more isolated in the east and south.

Belfast: Shifting breezy spells, overcast, light showers, mild temperatures
Dublin: Overcast, light showers, slightly warmer, mostly calm air
Cork: Overcast, light showers, gentle breezes to mostly calm air, chance of fog
Galway: Overcast, some showers, light shifting breezes, chance of fog patches

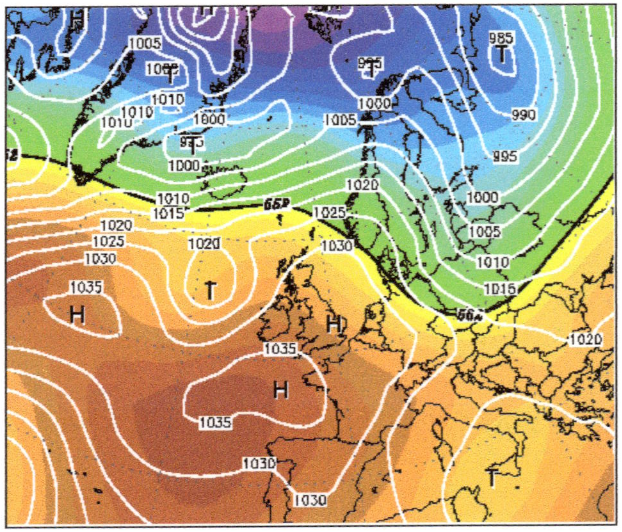

RAIN POTENTIAL

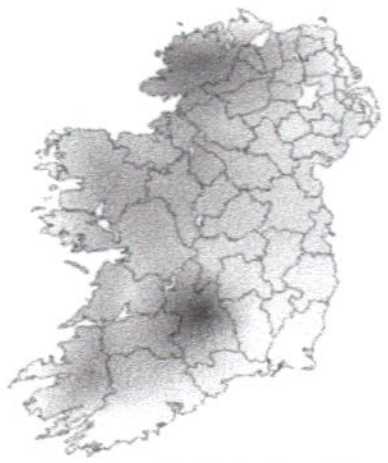

FROST/SNOW

NOVEMBER **29** SUNDAY

A large anticyclonic system strengthens to the west of Ireland while low pressure flows through the eastern half of the country. Winds generally light to moderate in a northwest flow. Cloudy but mostly dry everywhere after overnight shower activity clears and patches of fog mostly in the north and central parts lift.

Belfast: Unsettled and blustery winds, cloudy, dry
Dublin: Cloudy, dry, blustery winds at times
Cork: Cloudy, cooler, dry, windy
Galway: Cloudy, dry, blustery at times, mild temperatures

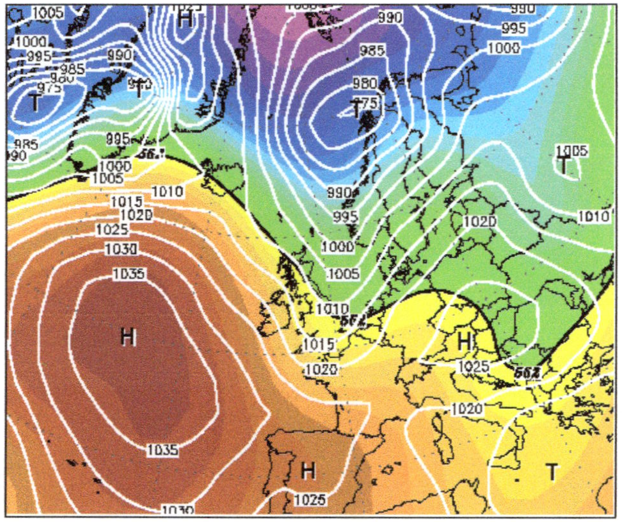

RAIN POTENTIAL

FROST/SNOW

NOVEMBER **30** MONDAY

A large anticyclonic system continues to linger to the west of Ireland bringing a moderate northwesterly flow throughout, fresher at times in the north. Mostly sunny and dry in the south of Ireland. Cloudier to overcast and dull elsewhere but remaining mostly dry in the west, east, and northeast. Some light showers and drizzle patches elsewhere through inland and central parts to the northwest. Chance of further overnight patches of fog and light frosts mostly in the north and central areas

Belfast: Overcast, dry, threats of rain, windy with occasional gusts likely
Dublin: Overcast, dry, some threats of rain, fresh breezes
Cork: Mostly sunny and dry, fresh to strong winds
Galway: Overcast, dry with some threats of rain, windy at times

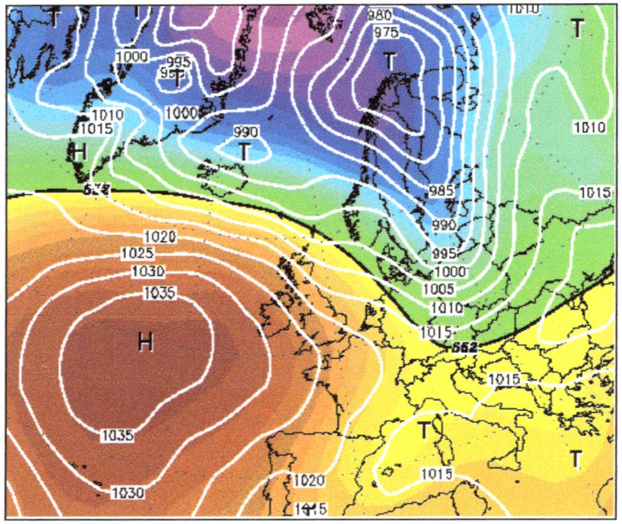

RAIN POTENTIAL

FROST/SNOW

DECEMBER

PHASES OF THE MOON

1st	Third Quarter
3rd	Crossing Equator
9th	New Moon
9th	Southern Declination
11th	Apogee
17th	First Quarter
17th	Crossing Equator
23rd	Northern Declination
24th	Full Moon
24th	Perigee #1
30th	Third Quarter
30th	Crossing Equator

1	2	3	4	5	6	7	8	9	10	11	12	13	14	15	16	17	18	19	20	21	22	23	24	25	26	27	28	29	30	31
Tue	Wed	Thur	Fri	Sat	Sun	Mon	Tue	Wed	Thur	Fri	Sat	Sun	Mon	Tue	Wed	Thur	Fri	Sat	Sun	Mon	Tue	Wed	Thur	Fri	Sat	Sun	Mon	Tue	Wed	Thur
3Q		XhS						N V		A						1Q XhN						^	F P1						3Q XhS	

MONTHLY SUMMARY

1st-7th The light to moderate west to southwest flow that became established at the end of November may continue during this week, with very little sunshine and drizzly shower conditions continuing, especially in the east and south. Warmer than average temperatures may persist with few frosts.

4th Cooler change with overnight frosts returning in some areas, mostly in central parts. Chance of thunderstorms and hail showers.

4th-5th Very strong and unsettled winds throughout with the chance of gale gusts in most areas.

8th-11th A cold front may cross the country in a light westerly flow, later turning northerly. Mostly dry, cold, and sunny with widespread frost and fog overnight, lingering long into the day in some places. Chance of some showers mostly around the coastal fringes, including hail showers and snow some places.

8th Chance of thunderstorms.

9th-11th Widespread frosts likely, particularly in the east and some severe in inland parts.

10th Very cold day across the country, particularly in the east, where some areas maximum temperatures may not rise far above zero.

10th-11th Chance of widespread fog and frost.

11th Chance of thunderstorms. Frosts may be very severe in some parts of the southeast.

Over the next four weeks it may be drier and sunnier than what is typical for this time of year, as well as colder than average almost everywhere. Depressions near the country over the first couple of weeks ahead may bring widespread unsettled conditions before a high pressure influence becomes established over the second half of the outlook period. Many days ahead may see some rain, sleet or snow in places, but despite this, overall rainfall amounts may be below normal throughout. Rainfall may be heavy in the first week ahead, before being lighter for the remaining three weeks, with most parts of the country possibly only seeing around 60-80% of their average rainfall amounts for this time of year. Overall, there may only be around 12 days with rain in the east and south, and 14-18 days elsewhere. Subsequently, accompanying the below average rainfall will be above average sunshine during this outlook period. Temperatures may be a little above normal in the west, while up to a degree below average in the east. Ground frosts may be widespread and severe in the first and last weeks of this outlook period with up to 20-25 days with frosts in inland parts overall, which is well above what is typical for this time of year. During this outlook period, there may be 1 day with thunderstorms, 12 days with hail, snow or sleet on 14 days, widespread fogs on 4 days, frosts on 18-20 days, and gale winds on 4 days.

12th-31st A series of deep depressions near to the country bring unsettled conditions throughout. Winds generally northwest to northerly in direction, and occasionally very strong and blustery, with temperatures close to average or slightly below average for this time of year. Chance of widespread frosts on numerous days.

12th-14th Chance of snow or sleet.

12th-16th Occasional showers of hail likely.

12th-26th Frontal systems crossing the country bring spells of wintry showers or rain on most days.

13th-15th Chance of heavy rain and thunderstorms in many parts of the country.

17th-19th Chance of widespread fog and severe frosts.

21st-22nd Chance of further patches of widespread fog overnight.

22nd-24th Chance of widespread wintry showers.

23rd A deep depression near Iceland around this time may extend an active frontal system over Ireland bringing very unsettled conditions throughout, with widespread rain likely, heaviest in the west and south and very strong southwest winds.

24th-25th Chance of wintry showers including some of sleet and snow possible.

24th-30th Chance of hail showers.

27th-1st January 2027 Milder conditions may develop as a south to southwest flow prevails. Cloudy and damp throughout, with spells of rain or drizzle but no frost likely.

30th A deep depression near the north of Ireland may extend colder temperatures and very strong to gale force winds over much of the country.

IRELAND RAINFALL ESTIMATES
December 2026

d=mainly dry, n= nonrecordable, l=light shrs, s=significant shrs, r=rain, h=hvy falls

		ULSTER PROVINCE					CONNAUGHT PROVINCE						LEINSTER PROVINCE				MUNSTER PROVINCE						
		Alderg	Hillsbo	Armag	Loughl	Malin H	Sligo	Ardtarı	Belmul	Drums	Galway	Roscoı	Markre	Edende	Kilkenı	Dublin	Shannı	Newpo	Tralee	Killarnı	Cork	All	Moon
1st	Tues	l	l	d	s	d	l	s	l	n	l	l	s	l	n	d	l	l	l	l	d	27	3Q
2nd	Wed	d	d	d	l	l	n	l	l	l	l	l	l	n	d	l	d	l	l	l	l	24	
3rd	Thur	l	l	n	s	l	l	s	s	s	l	r	s	l	s	l	d	r	l	s	d	92	XhS
4th	Fri	s	s	l	s	s	s	s	s	r	s	h	s	s	s	l	s	s	s	r	s	139	
5th	Sat	s	d	s	d	s	s	d	l	l	l	n	s	d	l	s	l	l	l	l	n	46	
6th	Sun	d	d	d	s	l	n	s	n	l	d	l	l	l	d	d	d	l	l	d	d	22	
7th	Mon	n	l	d	s	l	n	s	l	s	s	s	s	l	d	d	d	s	l	s	d	51	
8th	Tues	l	d	l	l	r	s	l	r	l	s	l	d	d	d	l	s	r	s	l	l	68	
9th	Wed	d	d	d	s	s	l	s	l	d	d	n	l	d	d	d	l	l	s	l	d	34	N V
10th	Thur	d	d	d	l	l	l	l	s	d	l	n	l	d	d	d	d	s	s	d	d	27	
11th	Fri	d	d	d	l	s	l	l	s	l	l	l	l	d	d	d	d	l	s	l	d	29	A
12th	Sat	l	l	d	l	l	l	l	s	l	l	l	s	l	d	d	n	s	l	l	d	34	
13th	Sun	l	d	l	d	l	l	d	l	d	l	d	s	d	l	l	s	d	r	s	r	54	
14th	Mon	s	r	d	r	s	s	r	r	r	h	r	r	h	r	l	s	r	r	h	s	208	
15th	Tues	s	d	s	l	s	s	l	s	s	s	s	s	d	s	s	r	h	r	r	r	136	
16th	Wed	d	d	d	l	l	l	l	s	l	l	l	l	l	d	d	l	l	s	s	l	30	
17th	Thur	d	d	d	d	d	d	d	d	d	d	d	d	d	l	d	d	d	d	d	d	2	1Q XhN
18th	Fri	l	l	d	s	l	l	s	s	s	r	s	s	l	l	d	d	s	d	s	d	65	
19th	Sat	l	d	l	l	s	l	l	s	s	l	s	s	d	l	l	s	l	l	s	s	66	
20th	Sun	d	d	d	l	l	l	l	l	l	n	l	s	d	l	d	d	l	l	l	d	20	
21st	Mon	d	d	d	l	d	l	l	l	l	l	l	l	s	n	d	d	l	n	l	l	26	
22nd	Tues	l	s	d	s	l	l	s	s	s	s	s	l	s	s	l	l	s	l	l	l	73	
23rd	Wed	r	h	s	r	s	r	r	r	r	h	h	r	h	h	r	h	h	h	h	r	311	^
24th	Thur	s	d	r	r	h	r	r	r	s	n	l	r	d	d	d	l	l	s	s	d	126	F P1
25th	Fri	d	d	d	l	s	d	l	d	l	l	l	l	d	d	d	n	l	l	d	d	23	
26th	Sat	n	l	d	s	l	l	s	l	l	s	s	l	d	d	d	d	s	l	l	l	45	
27th	Sun	l	l	l	s	s	s	s	s	l	l	l	l	l	l	l	l	l	r	s	s	71	
28th	Mon	l	n	l	l	l	l	l	l	l	l	l	r	l	d	d	l	r	n	n	d	44	
29th	Tues	l	l	l	s	l	l	s	s	l	s	l	s	l	l	l	s	s	l	l	l	63	
30th	Wed	s	l	s	s	r	s	s	s	s	s	r	r	l	d	l	l	r	d	l	l	101	3Q XhS
31st	Thur	s	l	s	s	s	s	s	s	r	s	s	l	n	d	s	l	r	l	l	l	88	
Estimate:		66	62	67	137	129	102	137	137	128	117	120	151	68	64	48	81	179	132	147	70	2141	
Average:		79	90	77	124	116	126	128	137	110	123	138	126	95	84	73	104	180	155	112	133	2309	
Trend:		drr	drr	av	w tr	w tr	drr	av	av	w tr	av	drr	w tr	drr	drr	drr	drr	av	drr	w tr	drr	drr	

IRELAND SUNSHINE ESTIMATES

December 2026

F=fine (8-12 hours of sunshine), pc= partly cloudy (4-7 hours), c=cloudy (1-3 hours), o=overcast (0 hours)

		ULSTER PROVINCE					CONNAUGHT PROVINCE						LEINSTER PROVINCE				MUNSTER PROVINCE						
		Alderg	Hillsbo	Armag	Loughl	Malin H	Sligo	Ardtarı	Belmul	Drums	Galway	Roscoı	Markre	Edende	Kilkenı	Dublin	Shannı	Newpo	Tralee	Killarnı	Cork	All	Moon
1st	Tues	c	c	c	c	c	o	c	o	c	o	o	c	c	c	c	o	c	o	o	c	31	3Q
2nd	Wed	c	c	c	c	c	c	c	o	c	o	o	c	c	c	c	o	c	o	o	o	30	
3rd	Thur	o	o	o	o	o	o	o	o	c	o	o	o	c	c	o	o	c	o	o	c	14	XhS
4th	Fri	c	c	c	c	c	c	c	c	c	c	c	c	c	c	c	c	c	c	c	c	43	
5th	Sat	c	c	c	c	c	pc	c	pc	c	c	pc	c	c	c	o	c	c	c	c	c	64	
6th	Sun	c	c	c	c	c	o	c	o	c	o	o	c	c	pc	pc	c	c	c	c	c	49	
7th	Mon	o	o	o	o	o	o	o	o	c	o	o	o	c	c	o	o	c	c	c	c	20	
8th	Tues	c	c	c	c	c	c	c	c	pc	pc	pc	c	pc	pc	pc	pc	pc	pc	pc	pc	84	
9th	Wed	pc	pc	pc	pc	pc	pc	pc	c	pc	pc	pc	pc	pc	pc	pc	pc	pc	c	c	pc	101	N V
10th	Thur	c	c	c	c	c	c	c	c	c	c	c	c	c	c	o	pc	c	c	c	pc	41	
11th	Fri	pc	pc	pc	pc	pc	pc	pc	o	pc	c	c	pc	pc	pc	pc	pc	pc	c	c	pc	102	A
12th	Sat	pc	pc	pc	pc	pc	pc	pc	c	pc	pc	pc	pc	pc	pc	pc	pc	pc	pc	pc	pc	118	
13th	Sun	o	o	o	o	o	o	o	o	o	o	o	o	o	o	o	o	o	o	o	o	4	
14th	Mon	c	c	c	c	c	c	c	o	c	o	o	c	c	c	pc	c	c	o	o	c	45	
15th	Tues	c	c	c	c	c	o	c	o	c	o	o	c	c	c	pc	o	c	c	c	c	46	
16th	Wed	c	c	c	c	c	c	c	c	c	c	c	c	c	pc	c	c	pc	c	c	pc	65	
17th	Thur	pc	pc	pc	pc	pc	pc	pc	pc	pc	pc	pc	pc	pc	pc	pc	pc	pc	pc	pc	pc	119	1Q XhN
18th	Fri	c	c	c	c	c	o	c	o	c	c	o	c	c	c	4.c	c	c	o	o	o	35	
19th	Sat	o	o	o	o	o	o	o	c	o	c	c	o	o	o	o	c	c	o	o	o	14	
20th	Sun	c	c	c	c	c	o	c	o	c	o	o	c	c	c	pc	c	c	o	o	o	39	
21st	Mon	o	o	o	o	o	o	o	o	o	o	o	o	o	o	c	o	o	o	o	o	7	
22nd	Tues	o	o	o	o	o	o	o	o	o	o	o	o	o	o	o	o	o	o	o	o	0	
23rd	Wed	o	o	o	o	o	o	o	o	o	o	o	o	o	o	o	o	o	o	o	o	0	^
24th	Thur	c	c	c	c	c	c	c	o	c	o	o	c	c	c	c	c	c	o	o	pc	46	F P1
25th	Fri	c	c	c	c	c	c	c	c	c	pc	c	c	c	c	o	pc	pc	pc	pc	pc	53	
26th	Sat	o	o	o	o	o	o	o	o	o	o	o	o	o	o	o	o	o	o	o	o	1	
27th	Sun	o	o	o	o	o	o	o	c	o	o	c	o	o	o	o	o	o	o	o	o	9	
28th	Mon	o	o	o	o	o	o	o	o	o	o	o	o	o	c	o	o	o	c	c	c	12	
29th	Tues	o	o	o	o	o	o	o	o	o	o	o	o	o	o	o	o	o	o	o	o	2	
30th	Wed	o	o	o	o	o	o	o	o	o	o	o	o	o	o	o	o	o	o	o	o	0	3Q XhS
31st	Thur	o	o	o	o	o	o	o	o	o	o	o	o	o	o	o	c	o	o	o	o	4	
	Estimate:	63	63	63	63	63	50	63	33	71	47	41	63	71	78	75	60	69	41	41	80	1198	
	Average:	39	39	34	34	28	34	34	33	33	36	42	47	45	46	53	44	30	36	36	48	774	
	Trend:	more	more	more	more	more	more	more	av	more	av	av	more	more	more	more	more	more	av	av	more	more	

DECEMBER 1 TUESDAY

Ireland may sit between a large high pressure system centred to the southwest, and an intense depression centred well to the northeast, with a high pressure ridge directing a fresh to blustery northwesterly flow throughout. Mostly cloudy with light showers and passing drizzle patches mostly in the west and central parts, while mostly cloudy and dry around the north, east, and southern coasts.

Belfast: Cloudy, brief squalls, strong winds, gales possible
Dublin: Mostly cloudy and dry, strong winds, chance of brief drizzle patches
Cork: Dry, cloudy, mild temperatures, strong and unsettled winds at times
Galway: Overcast, isolated showers, blustery winds, chance of gales at times

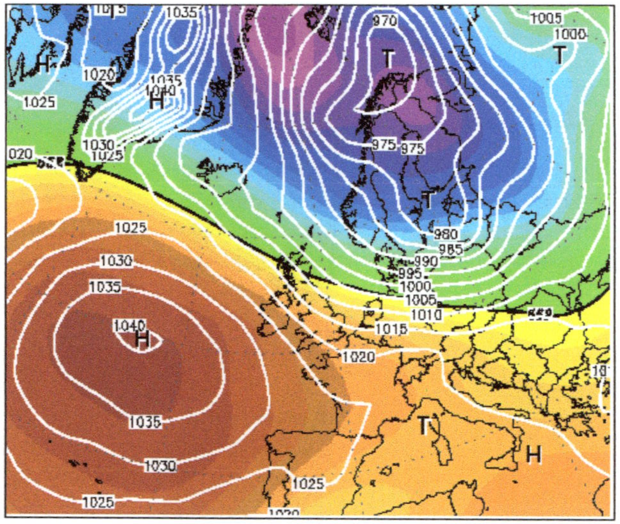

RAIN POTENTIAL

FROST/SNOW

DECEMBER 2 WEDNESDAY

A fresh to moderate northwesterly flow continues through Ireland directed along a high pressure ridge associated with a large high to the west of the country, while a large depression to the northeast Ireland brings widespread cloud throughout. Some light showers or passing drizzle spells in western and central parts, with possible thunderstorms and spotty showers in the south. Mostly dry along the northern and eastern coastal fringes.

Belfast: Cloudy, dry, strong unsettled winds, gales possible
Dublin: Cloudy, isolated showers, strong winds with occasional gusts
Cork: Overcast, blustery, squally showers, chance of thunderstorms
Galway: Overcast, brief squally showers, mild temperatures, near gale winds at times

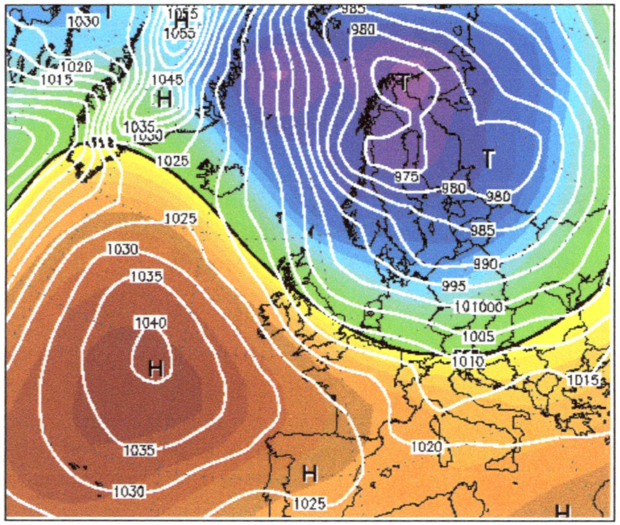

RAIN POTENTIAL

FROST/SNOW

DECEMBER 3 THURSDAY

There may be little change from the past couple of days with the high pressure ridge over the country bringing a moderate to strong northwesterly flow throughout, and the depression to the northeast of Ireland maintaining widespread cloud and bands of patchy rain and drizzle into almost all districts, although it may be mostly dry in the far northern and southeastern coasts.

Belfast: Overcast, slightly cooler, scattered showers, strong winds
Dublin: Overcast, spotty showers, colder, fresh winds
Cork: Mostly cloudy, dry, milder, strong winds, chance of thunderstorms
Galway: Mild temperatures, light showers, dull and gloomy air, unsettled winds

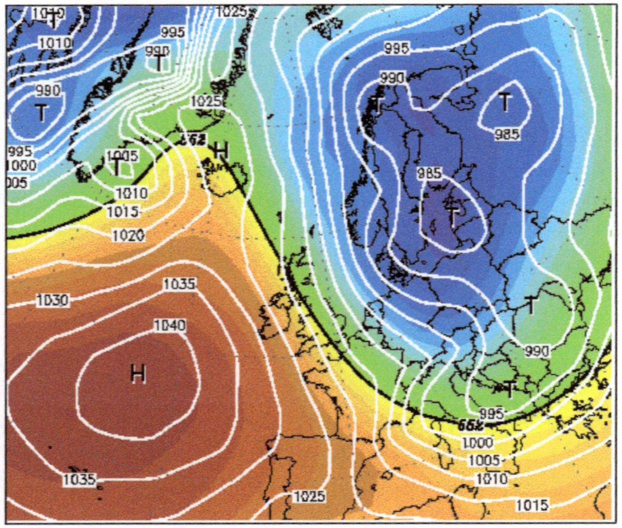

RAIN POTENTIAL

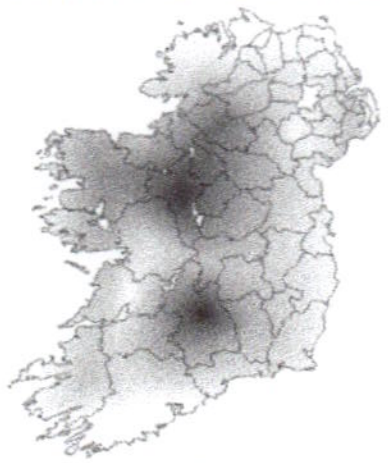

FROST/SNOW

DECEMBER 4 FRIDAY

A small depression to the north of Ireland may push the high pressure ridge away to the southwest of the country, with prevailing winds still mostly fresh to gusty in a northwesterly flow, more blustery along the northern coasts. Mostly cloudy with showers everywhere, some heavy for a time in inland western areas. Chance of brief sunny breakthroughs between the showers. Showers may be particularly wintry in the west for a time, possibly falling as light snow.

Belfast: Cloudy, fluctuating winds with strong gusts, squally outbreaks of rain
Dublin: Unsettled winds, cloudy, passing showers, warmer day colder night
Cork: Mostly cloudy, scattered showers, blustery at times, cooler
Galway: Fluctuating winds, colder, cloudy, showers, snow possible overnight

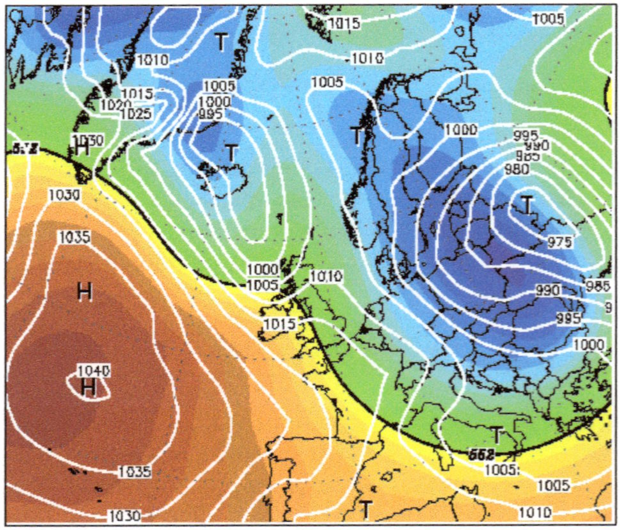

RAIN POTENTIAL

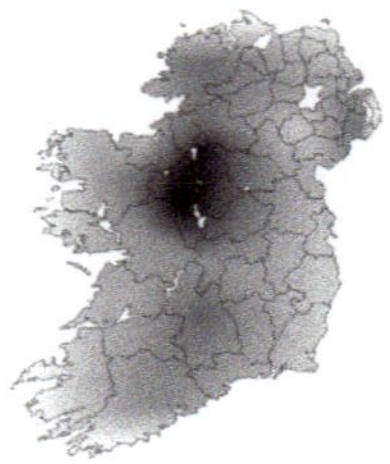

FROST/SNOW

DECEMBER **5** SATURDAY

A depression crosses the country in a strong and gusty northerly flow, with winds possibly to gale force at times overnight. Widespread cloud and outbreaks of showers throughout, some wintry and accompanied by thunderstorm activity. Chance of light snow overnight in western parts. Mostly dull and gloomy day everywhere, but with some sunny spells mostly in the afternoon. Overnight temperatures may be very cold with severe ground frosts in many inland areas.

Belfast: Cloudy, mostly dry with passing threats of rain, colder, stormy air
Dublin: Overcast, colder, showers, strong winds, chance of frost in sheltered parts
Cork: Cloudy, cooler, isolated showers, strong and forceful winds
Galway: Strong winds, cold, isolated showers, chance of snow flurries overnight

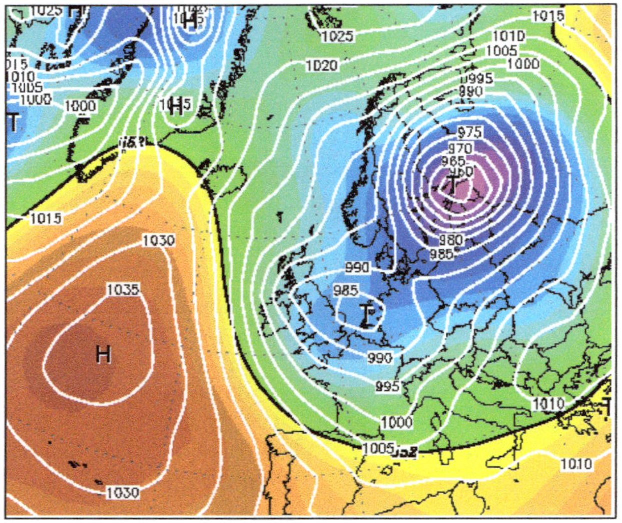

RAIN POTENTIAL

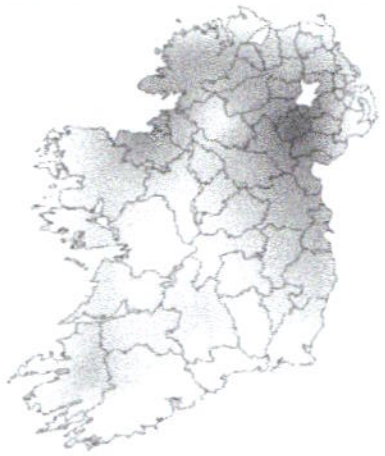

FROST/SNOW

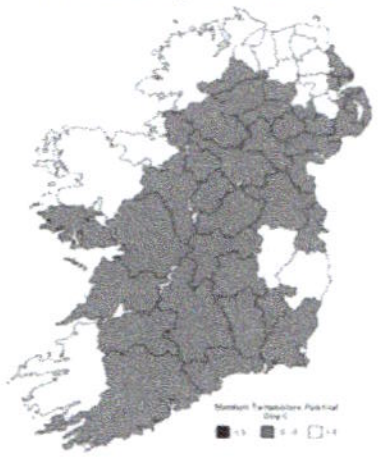

DECEMBER **6** SUNDAY

The large Atlantic high pressure system to the west of Ireland ridges back over the country bringing a cold north to northwesterly flow throughouth, with winds mostly moderate with occasional gusts at times. Mostly cloudy and dry throughout except for some brief mostly overnight showers in the northwest of the country.

Belfast: Cloudy, mostly dry with chance of misty drizzles, cold, strong winds
Dublin: Unsettled winds, partly cloudy, mostly dry, cold, chance of frost overnight
Cork: Mostly cloudy, cold and dry, windy at times
Galway: Windy at times, overcast, dry with passing threats of rain, mild day cold night

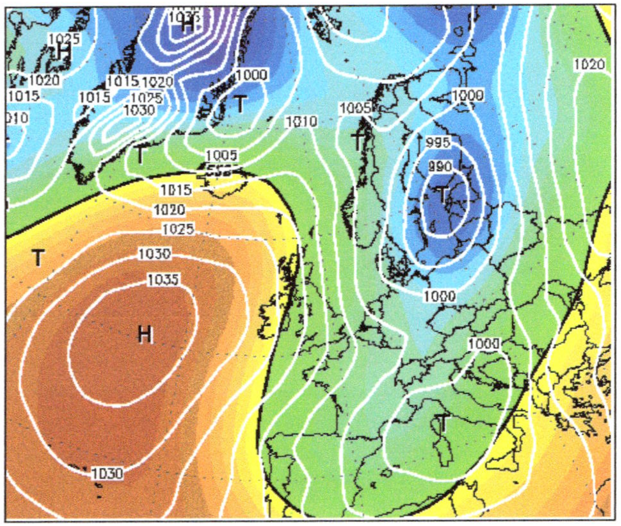

RAIN POTENTIAL

FROST/SNOW

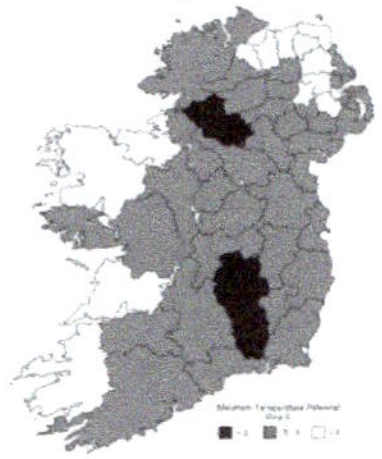

DECEMBER 7 MONDAY

A large high to the southwest of Ireland extends a light to moderate northwesterly flow over the country with some patches of fog likely mostly in the north and central parts before winds strengthen with a southwest change. Partly cloudy and dry in the south of the country, while cloudier with outbreaks of light rain or drizzle elsewhere, particularly in inland parts.

Belfast: Windy, overcast and dull, spotty showers, mild day cold night
Dublin: Dull and gloomy air, milder, mostly dry with threats of rain, blustery at times
Cork: Partly cloudy and dry, warmer, light breezes, chance of misty patches
Galway: Overcast, scattered showers, slightly warmer, fresh breezes

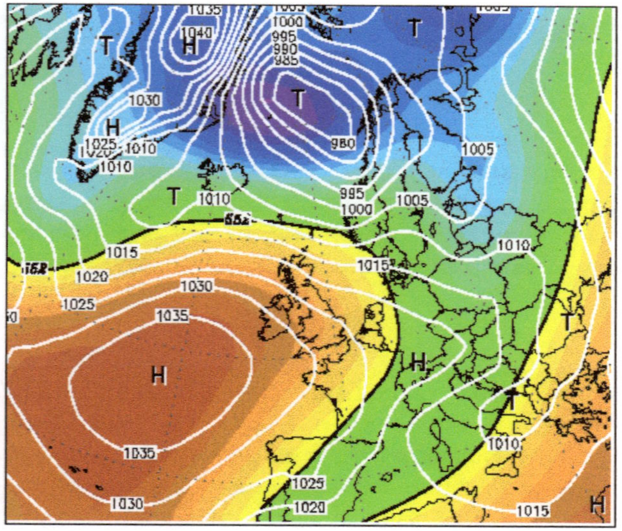

RAIN POTENTIAL

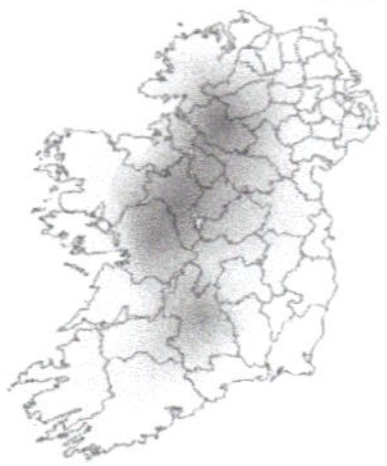

FROST/SNOW

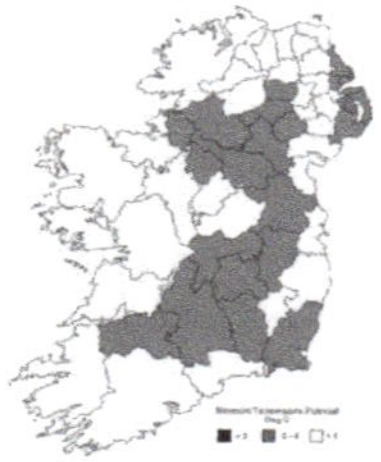

DECEMBER 8 TUESDAY

An intense depression to the north of Ireland directs fresh to strong west to southwesterly winds across the country, possibly up to gale force for a time before easing. A mixed day of sunny spells between bands of rain, with some accompanying thunderstorm activity and hail, particularly in the west, and heavier rain in the north. Frosty temperatures may bring the chance of brief snow overnight in the west. Mostly dry along the eastern coastal fringes.

Belfast: Mostly cloudy, dry, milder, strong and stormy winds, gales possible
Dublin: Partly cloudy, brief showers, cooler, strong winds with gusts
Cork: Partly cloudy, colder, light showers, chance of frost, strong winds with gusts
Galway: Colder, sunny, showers, gusty winds, chance of snow and thunderstorms

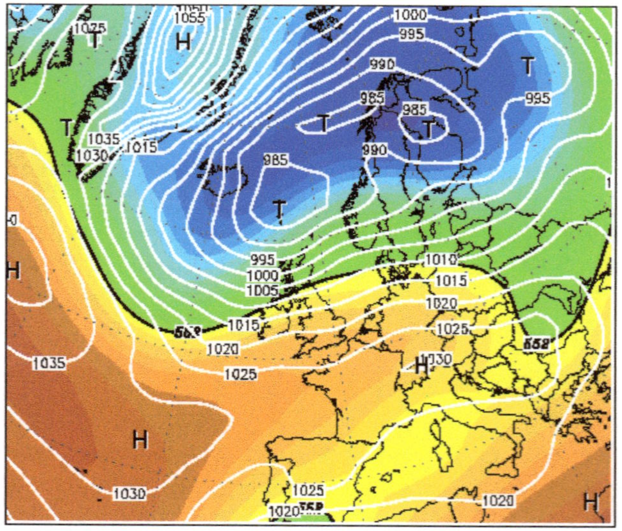

RAIN POTENTIAL

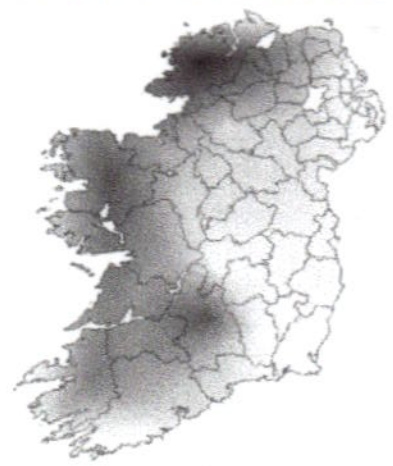

FROST/SNOW

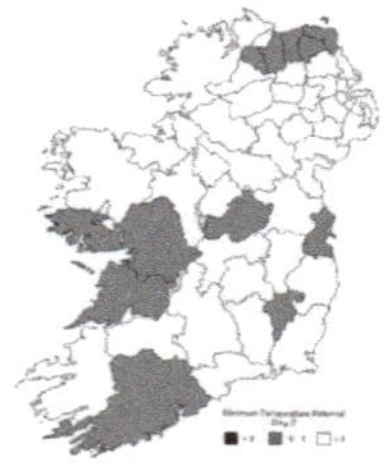

DECEMBER **9** WEDNESDAY

A depression to the north of Ireland continues to bring fresh to strong southwesterly breezes, easing to lighter as the day progresses. Another mixed day in the western half of the country with sunny spells between wintry showers, some of which may be of hail. Mostly dry with passing sunny spells in the eastern half of the country. Chance of frosts setting in after dark. Possible snow flurries for a time in central parts.

Belfast: Partly cloudy, colder, shifting and blustery winds, dry
Dublin: Partly cloudy, dry, cold with frosts, moderate breezes
Cork: Dry, partly cloudy, very cold with frosts, winds easing to light
Galway: Windy, partly cloudy, dry, cold, frosts

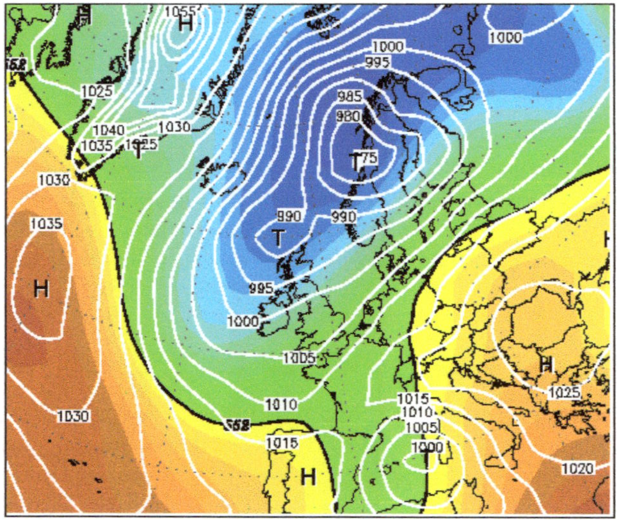

RAIN POTENTIAL

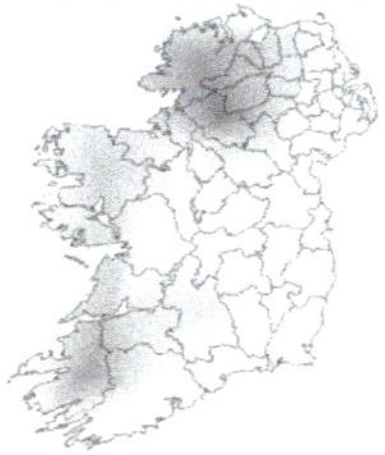

FROST/SNOW

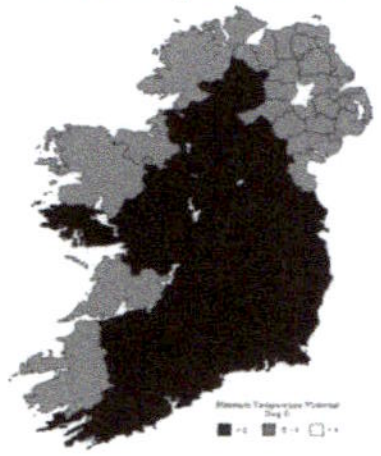

DECEMBER **10** THURSDAY

Low pressure centred over the country brings a cold northerly to variable air throughout. Mostly dry and cloudy in the east, while mostly cloudy with wintry showers in the west and south. Very cold throughout, with widespread frosts and scattered freezing fogs likely, with fog being slow to clear and possibly lingering well into the day in the midlands and eastern parts. Chance of light snow in the north, central, and western areas.

Belfast: Very cold, frosts, cloudy, mostly dry, chance of snow flurries, light breezes
Dublin: Mostly calm air, overcast, very cold, frosts, chance of freezing fog
Cork: Partly cloudy, dry, cold and frosty, gentle breezes
Galway: Cloudy, isolated showers, cold, severe frosts, light snow flurries possible

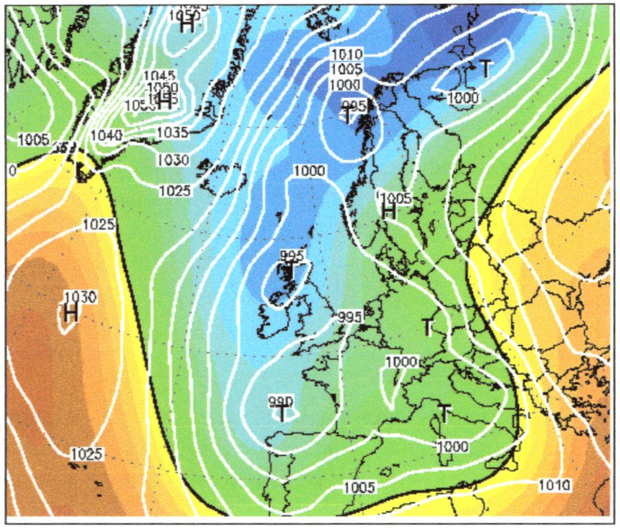

RAIN POTENTIAL

FROST/SNOW

DECEMBER **11** FRIDAY

Low pressure continues to bring a light and cold northerly flow over the country. Mostly dry with good sunny spells in central eastern parts, while cloudier with light wintry showers elsewhere. Widespread severe frosts and freezing fogs likely overnight, with the fog very slow to lift, lingering around almost all day in some places and frosts and fog returning again after dark. Chance of further light snow in the west and north.

Belfast: Partly cloudy, dry, very cold northerly air, frosts, chance of snow flurries
Dublin: Partly cloudy, cold, frosts, mostly dry, chance of patches of freezing fog
Cork: Cold, frosts, some sunny spells, dry, fresh breezes at times
Galway: Mostly cloudy, light showers, cold, frosts, windy, chance of snow overnight

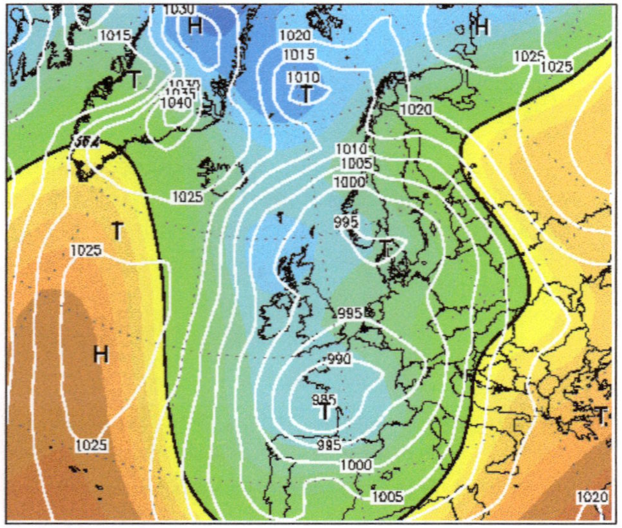

RAIN POTENTIAL

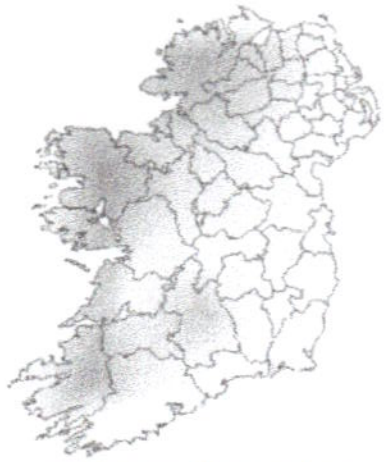

FROST/SNOW

DECEMBER **12** SATURDAY

Low pressure weakens to the east of Ireland while a high pressure ridge nears the west. Winds generally light in a cold northerly flow throughout. Further mostly isolated wintry showers likely in most areas except for mostly dry around the south and east coasts. Some long sunny spells likely in many areas. Frosts may be severe and widespread overnight. Chance of further light snow overnight in the west and north.

Belfast: Partly cloudy, cold, frosts, brief showers, chance of light snow overnight
Dublin: Partly cloudy, mostly dry, cold, frosts, light breezes
Cork: Partly cloudy, dry, cold, moderate breezes, chance of frosts
Galway: Partly cloudy, light showers, moderate breezes, frosts, chance of snow

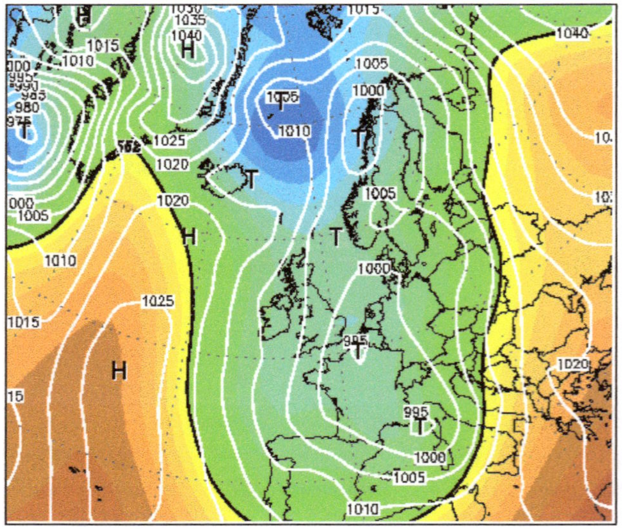

RAIN POTENTIAL

FROST/SNOW

DECEMBER **13** SUNDAY

A low pressure trough associated with a weak depression to the immediate north of Ireland directs a fresh to strong northwest flow throughout, possibly up to gale force at times in the north. Very dull and gloomy with wintry showers throughout, some of sleet and light snow. Precipitation may clear for a time before becoming more persistent while spreading from the southwest into the east. Very cold with frosts overnight.

Belfast: Overcast, dry, threats of rain, cold, frosts, chance of snow overnight
Dublin: Overcast, brief showers, cold, frosts, fresh breezes, chance of snow flurries
Cork: Overcast, rain, slightly warmer, windy spells
Galway: Overcast, cold with frosts, some showers, windy at times, snow possible

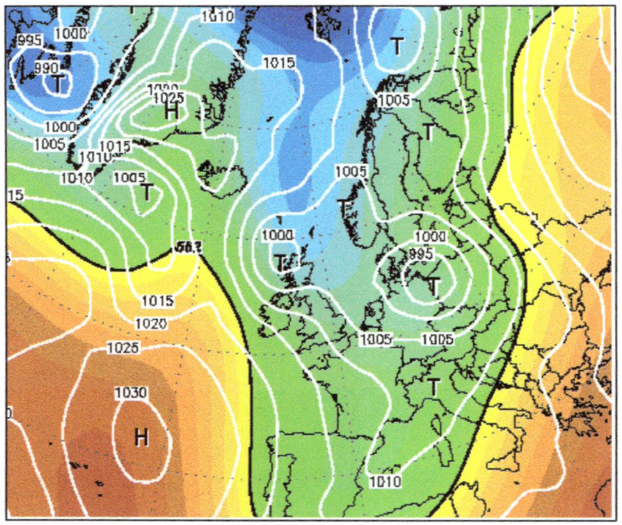

RAIN POTENTIAL

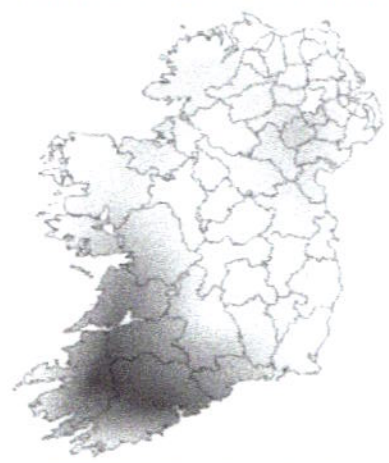

FROST/SNOW

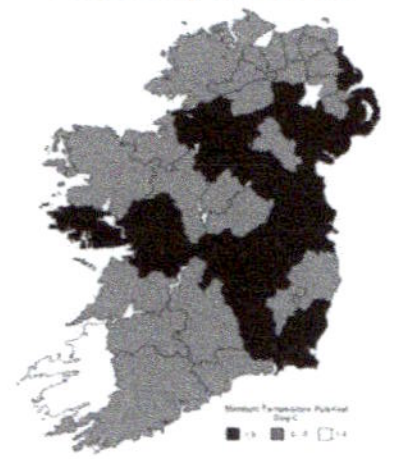

DECEMBER **14** MONDAY

A large depression lingers over Ireland bringing light to variable breezes, fresher at times with a northeast change. Widespread cloud throughout, with severe frosts and possible freezing fog patches overnight. Bands of wintry rain likely to spread across the country during the day, first in the east and central parts, becoming heavy for a time, as well as spreading elsewhere and lighter. Some sleet or snow likely in the north of Ireland in the higher reaches.

Belfast: Cloudy, cold with severe frosts, snow, lighter winds, chance of freezing fog
Dublin: Mix of showers and sunny spells, severe frosts, light breezes, snow possible
Cork: Cloudy, showers, slightly warmer day overnight frosts, gentle breezes
Galway: Overcast, heavy rain, severe frosts overnight, snow likely, light breezes

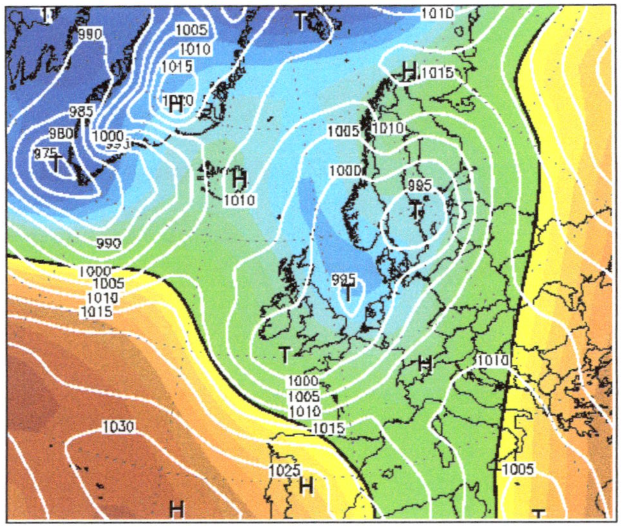

RAIN POTENTIAL

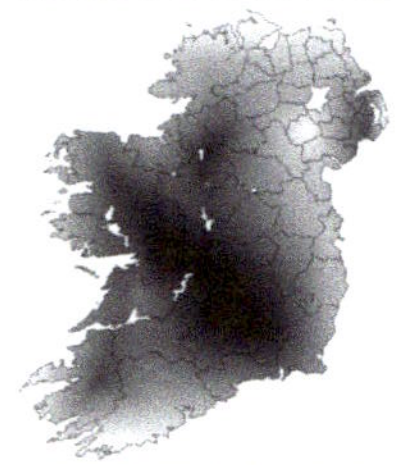

FROST/SNOW

DECEMBER **15** TUESDAY

Low pressure over Ireland intensifies into a strong depression centred to the north of the country, with strong and gusty southwest winds bringing a stormy air throughout. Scattered frosts likely overnight particularly in inland parts. Cloudy to overcast with wintry showers everywhere, some accompanied by thunderstorm activity particularly in the west.

Belfast: Cloudy, dry, cold with frosts, strong and stormy winds
Dublin: Storm, mix of showers and sunny spells, cold, chance of snow overnight
Cork: Stormy air, cloudy, rain, milder temperatures, gales likely
Galway: Overcast, showers, frosts, gusty winds, thunderstorms and gales possible

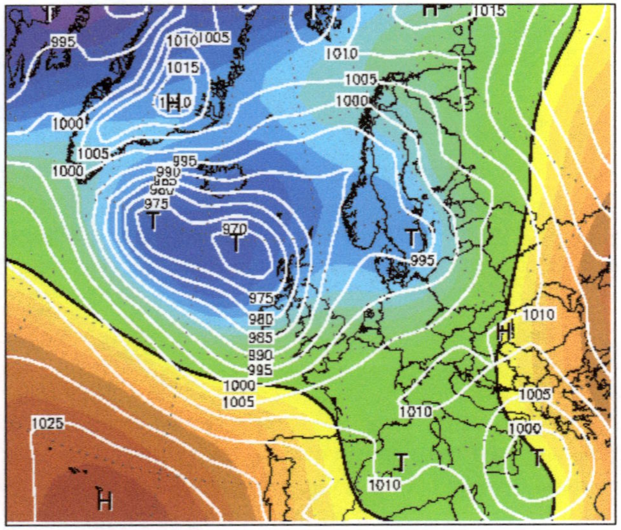

RAIN POTENTIAL

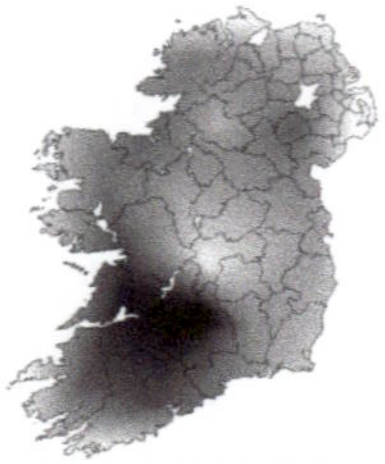

FROST/SNOW

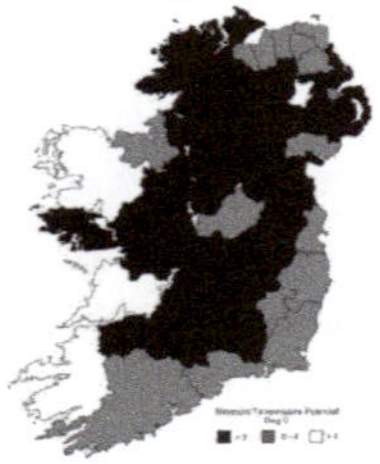

DECEMBER **16** WEDNESDAY

A depression sits over Ireland directing moderate to fresh north to northwest winds, lighter in the east, stronger in the west. Very cold overnight with patches of fog mostly in the east and some overnight wintry showers or hail, mostly in the western half of Ireland, with some accompanying thunderstorm activity. Mostly dry during the day, with frosts returning after dark. Chance of some snow flurries in the north for a time.

Belfast: Cloudy, dry, cold, light breezes, chance of snow flurries and fog patches
Dublin: Mostly cloudy, dry, cold, mostly calm air, chance of misty patches
Cork: Strong gusty winds at times, partly cloudy, brief showers
Galway: Cloudy, light showers, slightly milder, strong and unsettled winds

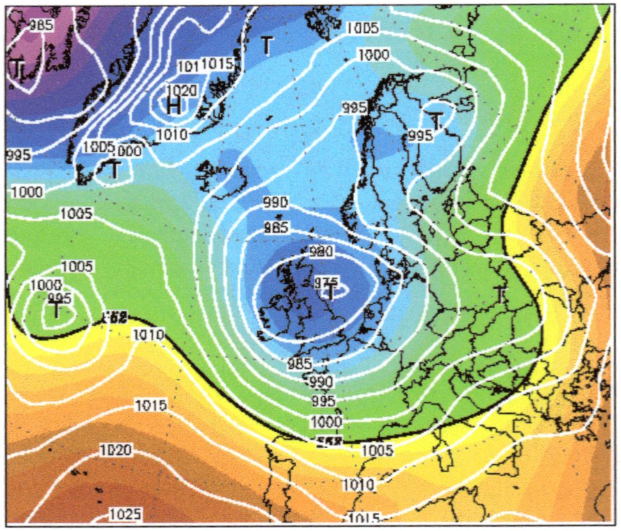

RAIN POTENTIAL

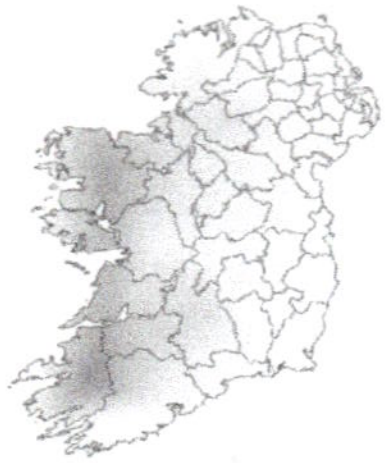

FROST/SNOW

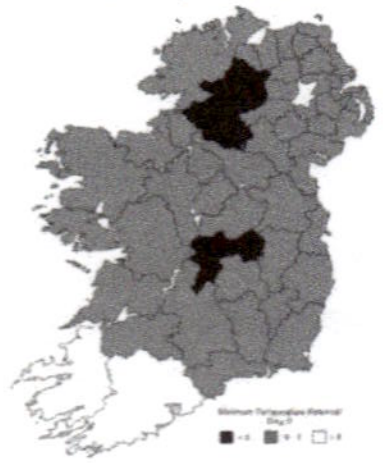

DECEMBER **17** THURSDAY

A low pressure trough over Ireland brings a gentle flow throughout, variable in direction. Very cold with patches of severe frosts and freezing fog in inland parts. Mostly sunny and dry day almost everywhere. Some patches of fog may persist well into the day in some districts before clearing and then returning after dark. Chance of isolated snow flurries briefly in some central parts.

Belfast: Partly cloudy, dry, cold with frosts and fog patches
Dublin: Cold with frosts, occasional sunny spells, mostly dry, calm air, fog likely
Cork: Partly cloudy, cold, dry, light breezes, chance of frost and fog patches
Galway: Partly cloudy, dry, cold, moderate breezes

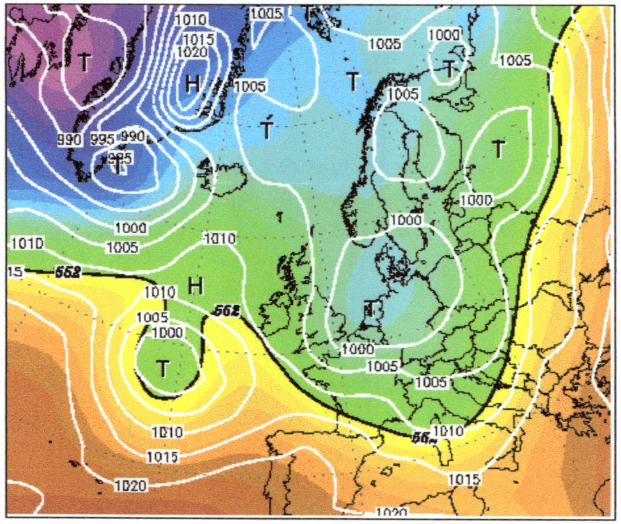

RAIN POTENTIAL

FROST/SNOW

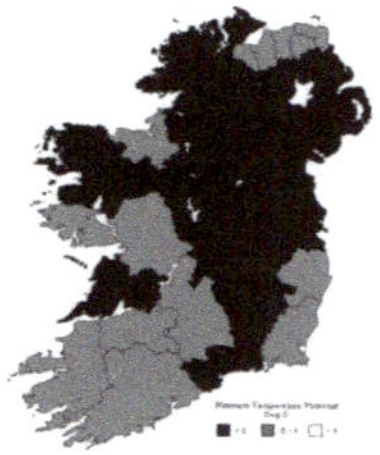

DECEMBER **18** FRIDAY

A weak high pressure ridge moves over the country bringing a light westerly flow, more variable for a time. Very cold with severe frosts overnight and further fog and freezing misty patches. Mostly dry day with occasional sunny spells before another band of rain may move into the northwest at first and spread light wintry showers into most districts, although the coastal areas in the east and southern areas may remain mostly dry.

Belfast: Cloudy, cold, frosts, some freezing showers, mostly calm air, chance of fog
Dublin: Partly cloudy, dry, cold with severe frosts, calm atmosphere, fog likely
Cork: Mostly calm, dull and gloomy, dry, threats of rain, chance of misty patches
Galway: Cloudy, rain, mild day, severe frosts overnight, mostly calm air

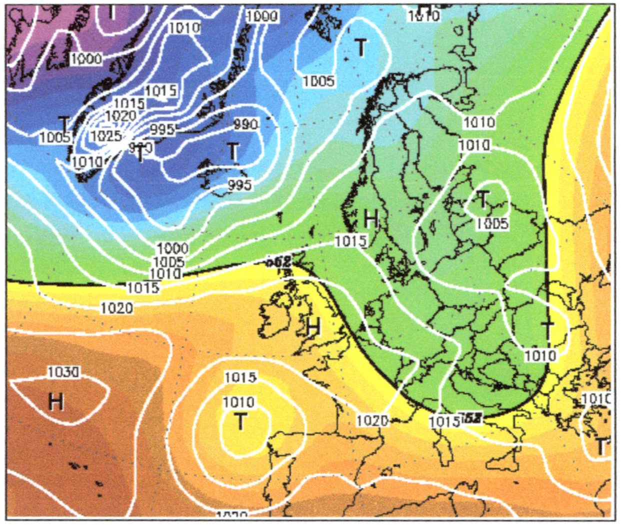

RAIN POTENTIAL

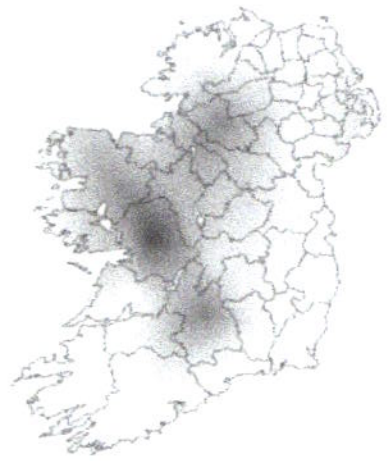

FROST/SNOW

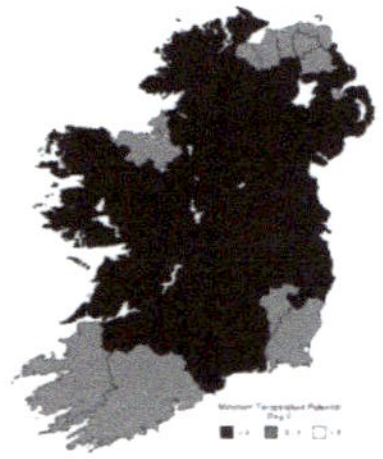

DECEMBER **19** SATURDAY

An intense depression to the north of Ireland begins to push against the high pressure ridge that may still be lingering over the country, bringing strengthening westerly winds, particularly in the north. A cold front associated with the depression moves into the northwest and spreads wintry shower activity down through the country in a southeasterly tract reaching the south of the country by the end of the day. Some wintry showers may be of snow, particularly in the east.

Belfast: Dry with passing threats of rain, dull and gloomy air, frosts, unsettled winds
Dublin: Overcast, showers, windy at times, chance of overnight brief snow flurries
Cork: Overcast and dull, cold, some showers, windy spells
Galway: Cloudy, brief showers, unsettled winds, milder temperatures overnight

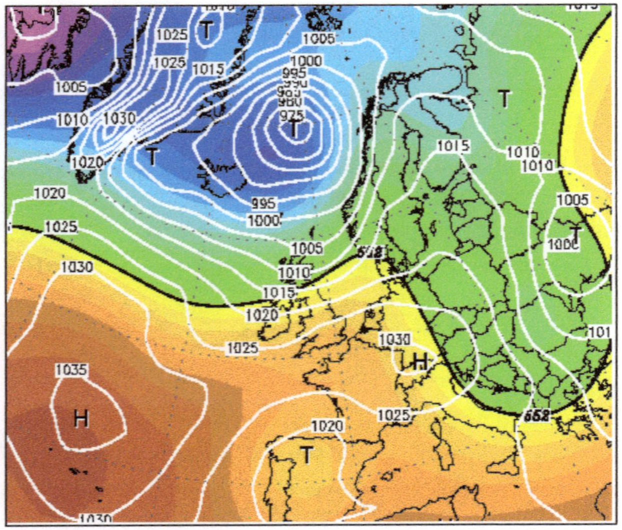

RAIN POTENTIAL

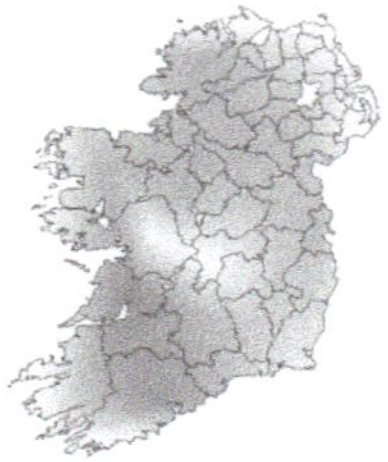

FROST/SNOW

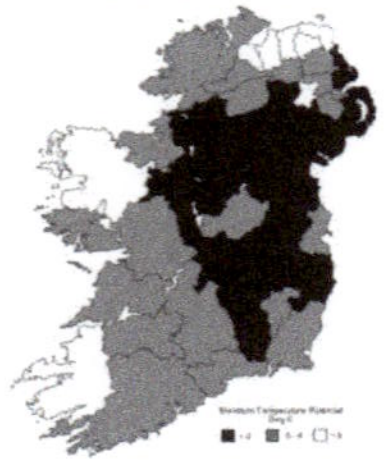

DECEMBER **20** SUNDAY

A high pressure ridge moves onto the country bringing a light to moderate northwesterly flow throughout, fresher at times along the western coasts. Partly cloudy and dry in eastern parts, overcast and mostly dry in the south, while cloudier with isolated showers in the north and west. Slight frosts overnight in many inland parts. Chance of brief overnight snow flurries in the west.

Belfast: Cloudy, mostly dry, odd threat of rain, blustery at times
Dublin: Partly cloudy, mostly dry and colder, some threats of rain, windy at times
Cork: Overcast, cold, mostly dry with odd threat of rain, fresh breezes
Galway: Overcast, isolated showers, strong winds, frosts, chance of snow flurries

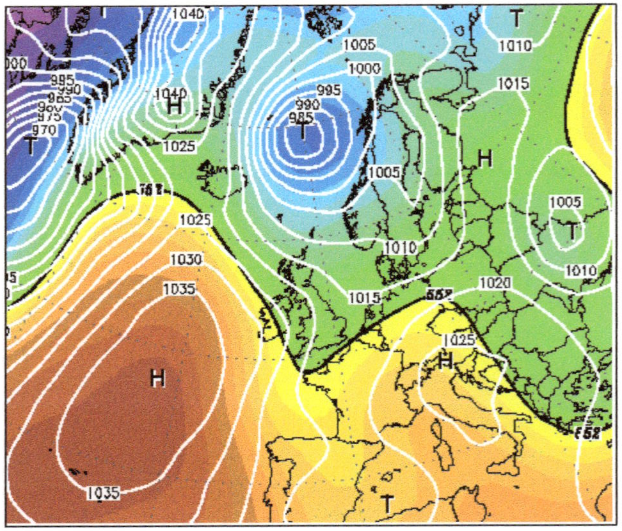

RAIN POTENTIAL

FROST/SNOW

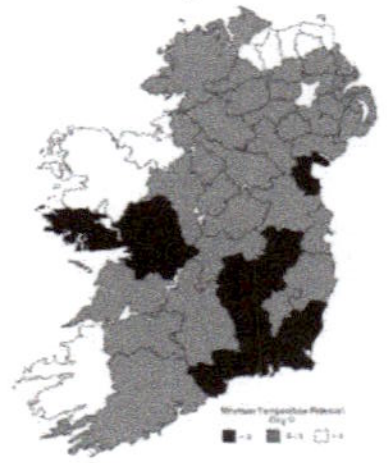

DECEMBER **21** MONDAY

High pressure continues to bring a light north to northwesterly flow throughout with very cold temperatures, and patches of frost and freezing fog overnight mostly in the north, midlands, and east. Widespread cloud everywhere, with brief showers and passing spells of drizzle mostly in the west and south, and possible isolated showers elsewhere. Chance of light snow in central parts.

Belfast: Overcast, dry, cold with frosts, light breezes, chance of fog overnight
Dublin: Very cold, frosts, cloudy, mostly dry, gentle breezes, chance of freezing fog
Cork: Slightly warmer, overcast, light showers, moderate breezes
Galway: Overcast, moderate breezes, isolated showers, chance of snow flurries

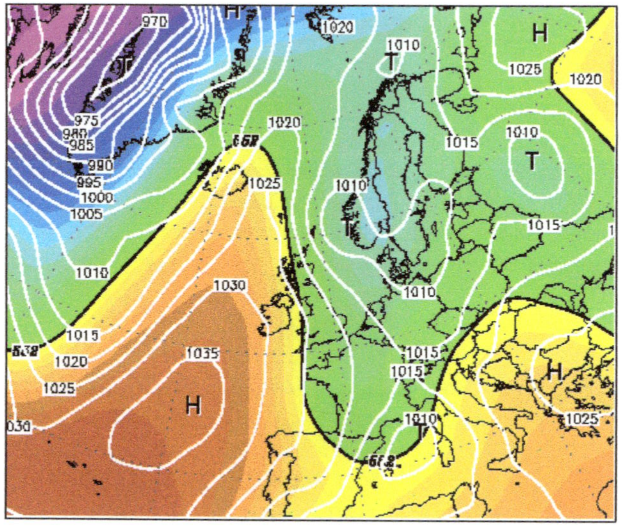

RAIN POTENTIAL

FROST/SNOW

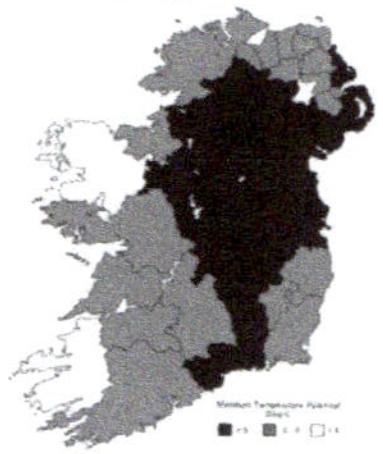

DECEMBER **22** TUESDAY

High pressure slips to the south of Ireland while low pressure moves onto the country, with light variable breezes prevailing. Dull and gloomy with spells of rain and drizzle throughout. Mist, fog, and frosts likely in many parts of the country overnight. Chance of light snow in the north and central areas. Some drier spells in the afternoon.

Belfast: Overcast, scattered showers, cold, severe frosts, chance of fog and snow
Dublin: Overcast, cold, severe frosts, some showers, chance of snow flurries
Cork: Overcast, cold, brief showers, gentle shifting breezes, chance of fog patches
Galway: Overcast, scattered showers, fluctuating breezes, cold

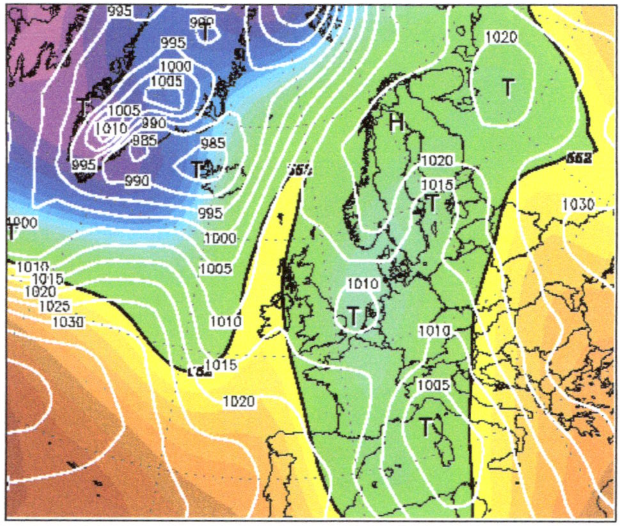

RAIN POTENTIAL

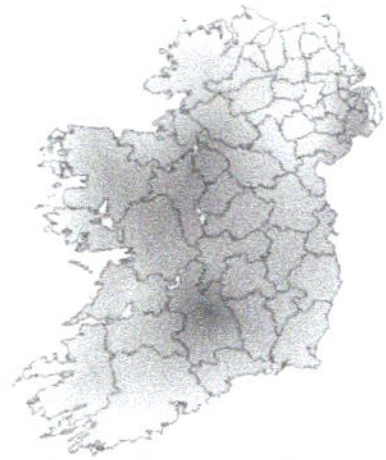

FROST/SNOW

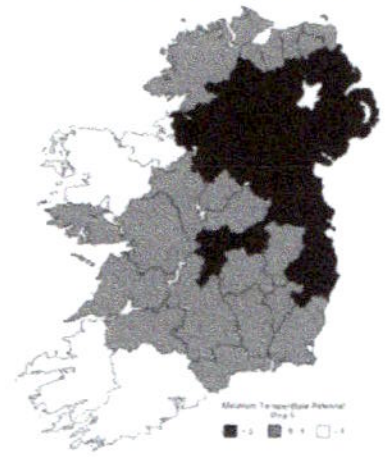

DECEMBER **23** WEDNESDAY

An intense depression to the northwest of Ireland bears down on the country around this time bringing fresher west to southwest breezes, more blustery along the northern coasts. Very dull and gloomy throughout. An active cold front may spread across the country during the day bringing widespread rain and unsettled winds everywhere, and possible snow in the north and west. Wintry precipitation may be particularly heavy in the southern half of Ireland and in the northeast.

Belfast: Overcast and dull, cold, frosts, heavy precipitation, fluctuating winds, snow
Dublin: Warmer, rain, overcast, gentle shifting breezes
Cork: Overcast, warmer day cold night, rain
Galway: Overcast, heavy precipitation, warmer day, overnight frosts, chance of snow

RAIN POTENTIAL

FROST/SNOW

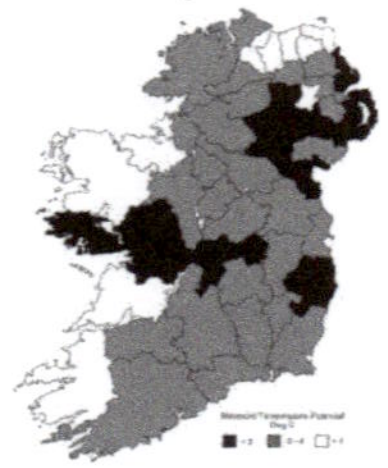

DECEMBER **24** THURSDAY

A large depression moves onto Ireland bringing widespread stormy conditions throughout with strong and unsettled northwest winds prevailing. Heavy wintry precipitation overnight in the west and north, with lighter, wintry showers elsewhere. Some clearer spells at times during the day. Chance of severe ground frosts overnight. Chance of snow in the north and west.

Belfast: Cloudy, mostly dry, threats of rain, strong gusty winds, chance of snow
Dublin: Cloudy, colder, dry, strong blustery winds at times
Cork: Cold, frosts, some sunny spells, dry, strong and unsettled winds
Galway: Overcast, cold with frosts, isolated showers, chance of snow flurries and gale winds

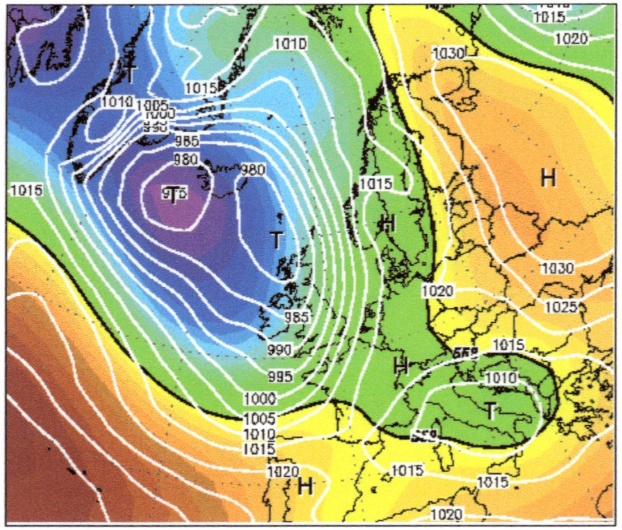

RAIN POTENTIAL

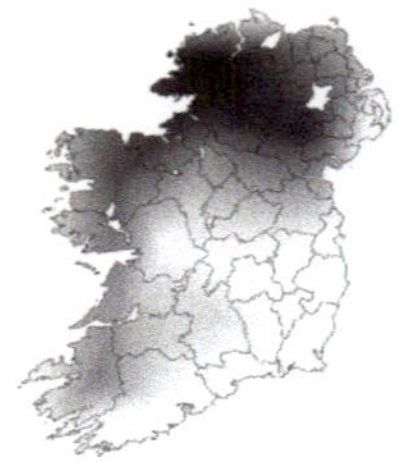

FROST/SNOW

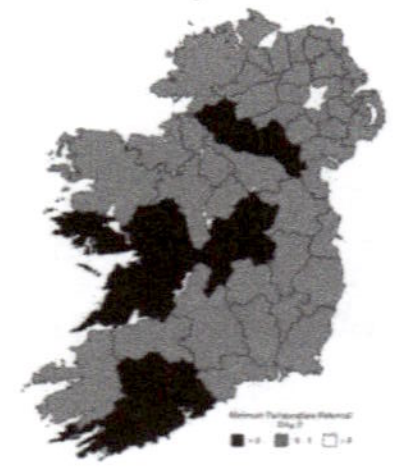

DECEMBER **25** FRIDAY

A large depression weakens as it crosses Ireland in a lighter north to northwest flow throughout. Cold and mostly cloudy everywhere with severe frosts in many areas, and freezing fog here and there, slow to clear. Some wintry showers may still persist in the north and west, with further possible snow in central and western areas. Some brief sunny spells may develop later in the day in the east and western coastal areas.

Belfast: Cold and cloudy, dry, light breezes to mostly calm air, chance of freezing fog
Dublin: Overcast, very cold and calm air, dry with odd threats of rain
Cork: Partly cloudy, dry, cold, frosts, possible fog, gentle breezes
Galway: Partly cloudy, brief showers, cold, frosts, chance of overnight snow flurries

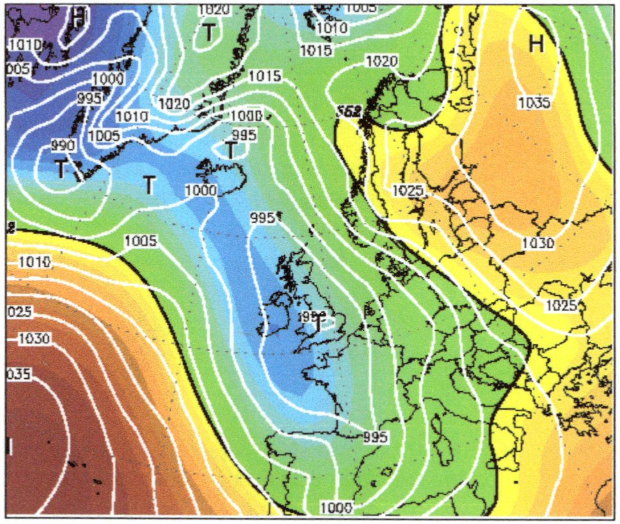

RAIN POTENTIAL

FROST/SNOW

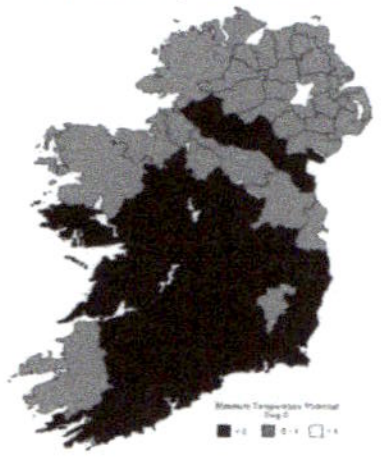

DECEMBER **26** SATURDAY

A low pressure trough moves across Ireland in a light to moderate southerly flow. Widespread cloud throughout, with isolated wintry showers in most areas, except in the east which may be mostly dry and frosty. Some severe ground frosts overnight accompanied by freezing fog patches in many parts. Chance of light snow in southern parts.

Belfast: Overcast, isolated showers, milder day overnight frosts, possible fog
Dublin: Light breezes, overcast, dry with increasing threats of rain, slightly warmer
Cork: Overcast, brief showers, slightly warmer, chance of snow flurries
Galway: Overcast, rain, mild day overnight frosts, light to moderate breezes

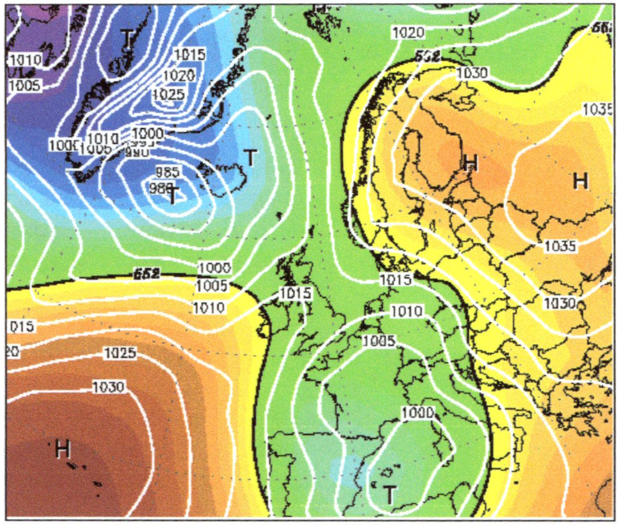

RAIN POTENTIAL

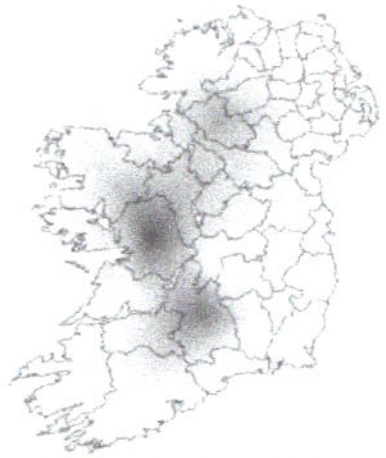

FROST/SNOW

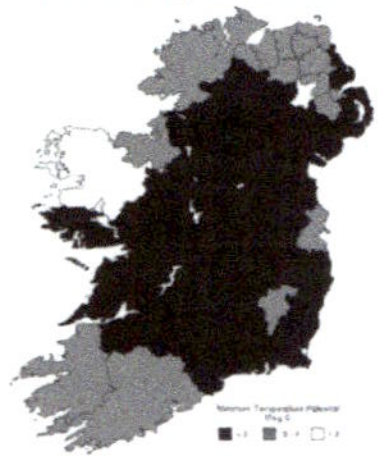

DECEMBER **27** SUNDAY

A large depression to the far north of Ireland directs an unsettled and blustery southwest flow over the country, later backing westerly and easing. A band of rain crosses the country bringing widespread cloud and scattered showers throughout, heaviest for a time in the southwest and later in the north. Milder temperatures likely.

Belfast: Overcast and dull, light showers, milder temperatures, blustery at times
Dublin: Overcast, freshening winds, warmer temperatures, light showers
Cork: Fresh breezes, milder temperatures, overcast, scattered showers
Galway: Overcast, light showers, milder, blustery at times

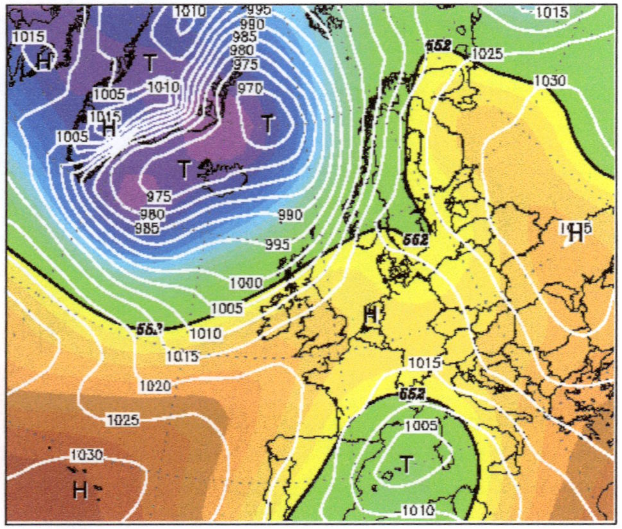

RAIN POTENTIAL

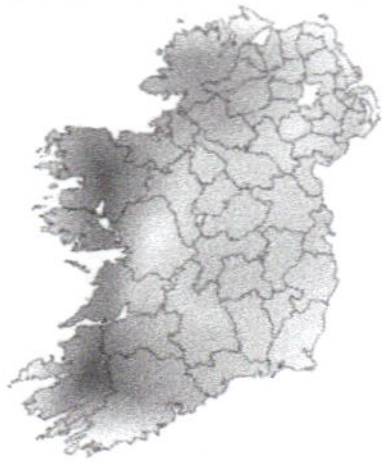

FROST/SNOW

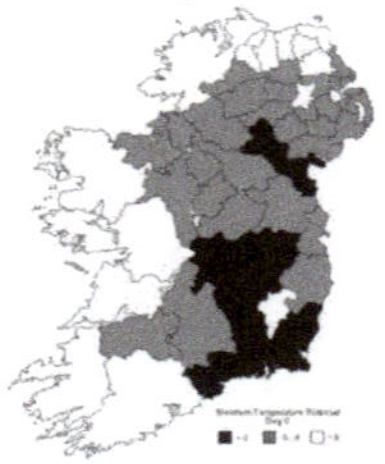

DECEMBER **28** MONDAY

An intense depression centred over Greenland brings a very strong and unsettled westerly flow over the country, stronger in the north with gale gusts at times, more moderate to the south. Dull and gloomy everywhere with isolated showers at first in the west and north, becoming more persistent for a time in the northern half of the country. Mostly dry in the far south and eastern coastal areas.

Belfast: Overcast, isolated showers, mild day cold night, strong and unsettled winds
Dublin: Overcast, mostly dry with passing threats of rain, mild day cold night, windy
Cork: Cloudy, mostly dry, threats of rain, mild day cold night, blustery at times
Galway: Mild day, overnight frosts, dull and gloomy air, light showers, blustery

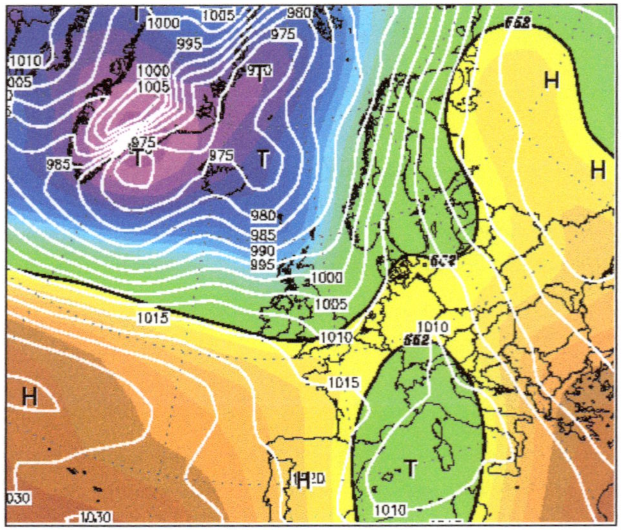

RAIN POTENTIAL

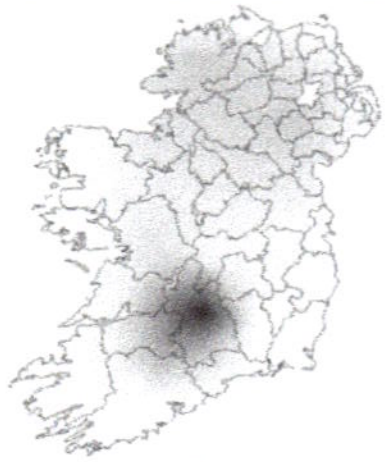

FROST/SNOW

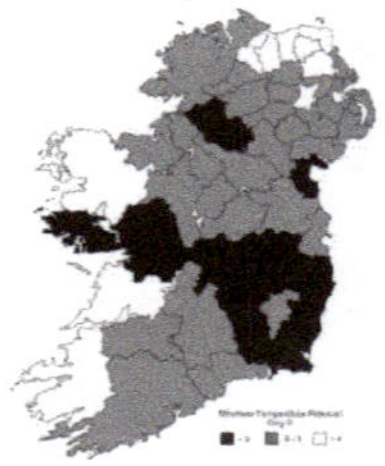

DECEMBER 29 TUESDAY

An intense depression to the north of Ireland continues to bring very strong and unsettled westerly winds over the country, very strong and blustery at times particularly in the west and north. A stormy air with squally showers and drizzle likely in most parts of the country, as well as the chance of hail showers and light snow overnight in the west.

Belfast: Stormy air with gales likely, overcast, scattered showers, mild temperatures
Dublin: Overcast, light showers, strong and unsettled winds, chance of gales
Cork: Overcast, light showers, mild day cold night, strong winds, occasional gusts
Galway: Strong winds, overcast, squalls, milder, chance of snow overnight

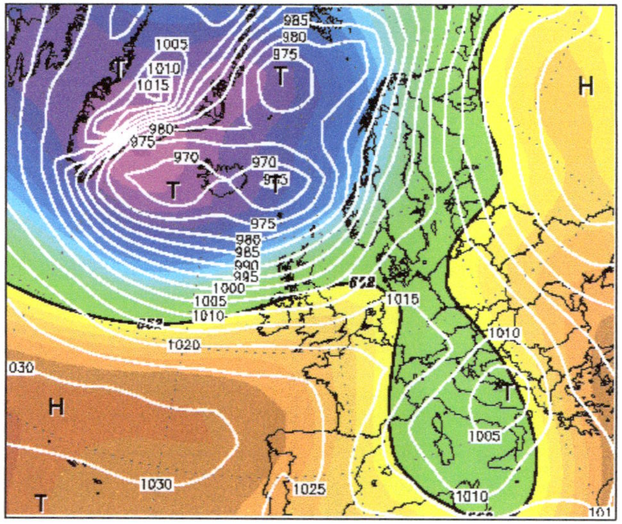

RAIN POTENTIAL

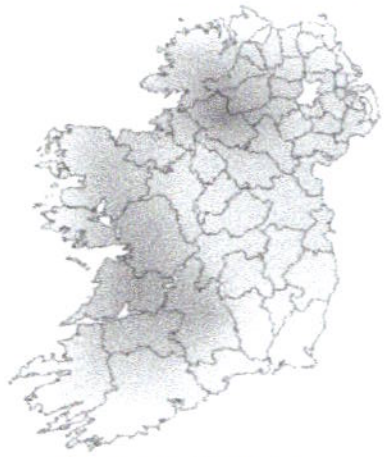

FROST/SNOW

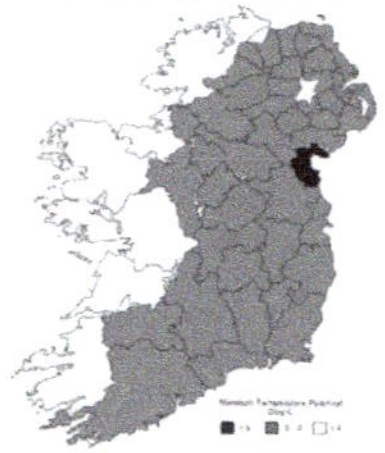

DECEMBER 30 WEDNESDAY

Low pressure and a stormy air may continue to persist over the northern half of the country while a high pressure ridge moves over the southern half easing storm conditions. West to southwest winds prevail throughout, moderate in the south while strong and stormy in the north. Rain activity may spread through the west and northwest at first, and reaching the eastern coasts later in the day, while only patchy drizzle may reach the south. Chance of snow overnight in the west.

Belfast: Stormy air, overcast, slightly cooler, squally showers
Dublin: Unsettled fluctuating winds, dull and gloomy air, brief showers
Cork: Overcast, brief showers, cold, chance of frost overnight, fluctuating winds
Galway: Overcast, scattered showers, frosts, blustery, chance of snow overnight

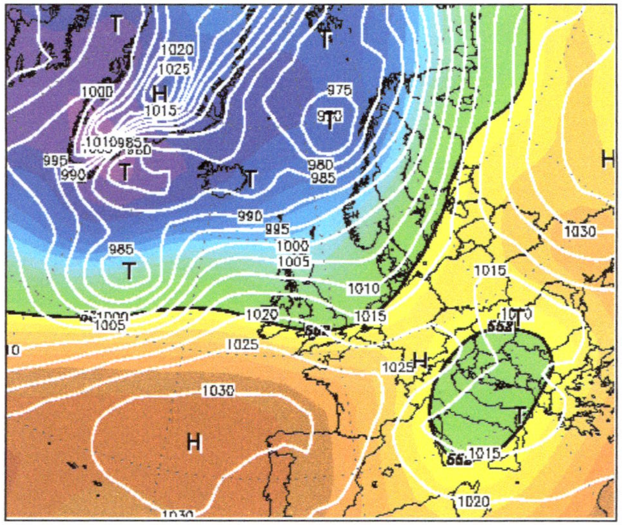

RAIN POTENTIAL

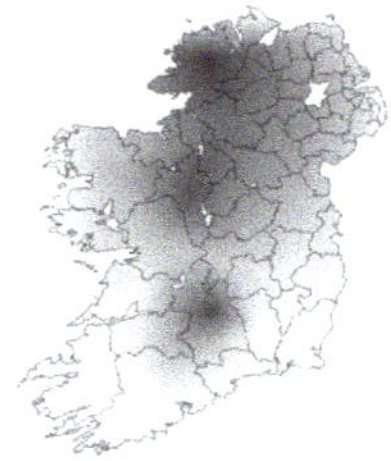

FROST/SNOW

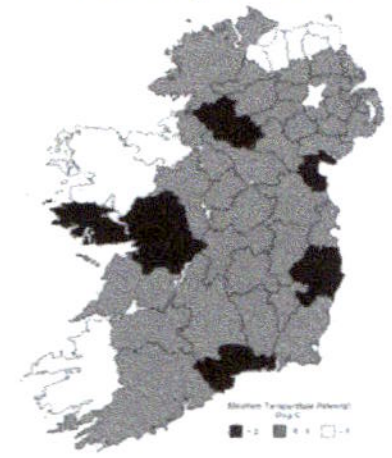

DECEMBER **31** THURSDAY

High pressure to the south of Ireland extends a ridge over the southern half of the country while a depression centred to the northeast brings low pressure over the northern half. Winds generally moderate to fresh in the south, while blustery and strong in the north, with a stormy air persisting. Widespread cloud and outbreaks of showers almost everywhere, with some spells of rain more persistent and lengthy.

Belfast: Milder temperatures, overcast and dull, brief showers, strong gusty winds
Dublin: Stormy, overcast, mild temperatures, squally showers, gale winds possible
Cork: Stormy, overcast, light showers, unsettled winds, milder temperatures
Galway: Overcast, showers, milder, strong and unsettled winds, chance of gale gusts

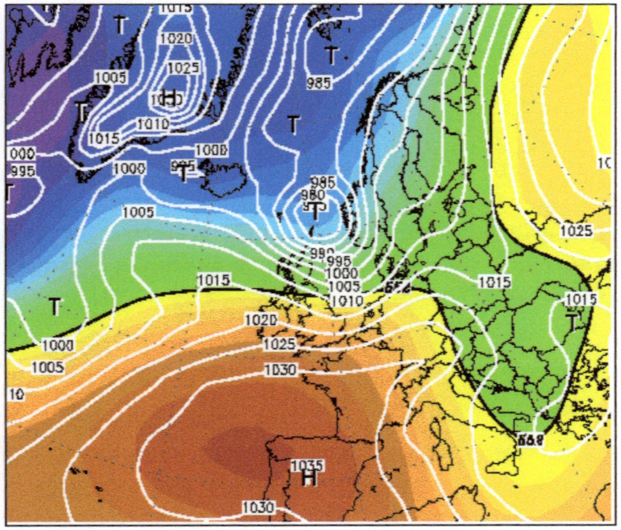

RAIN POTENTIAL

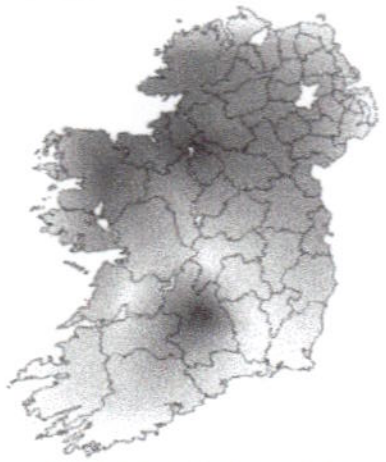

FROST/SNOW

APPENDICES

RAINFALL EXPECTATIONS

WETTER OR DRIER BY 10MM THAN LONG-TERM AVERAGES

	Ulster Province					Connaught Province						Leinster Province				MUNSTER PROVINCE					
Date	Aldergrove	Hillsborough	Armagh	Lough Navar	Malin Head	Sligo	Ardtarmon	Belmullet	Drumsna	Galway	Roscommon	Markree Castle	Edenderry	Kilkenny	Dublin	Shannon	Newport	Tralee	Killarney	Cork	All
JAN	w tr	w tr	w tr	av	w tr	w tr	av	w tr	w tr	w tr	w tr	w tr	w tr	w tr	w tr	w tr	w tr	w tr	w tr	w tr	w tr
FEB	drr	drr	av	av	w tr	w tr	av	w tr	w tr	w tr	drr	w tr	drr	drr	drr	w tr	av	av	w tr	drr	av
MAR	av	av	w tr	w tr	w tr	w tr	av	w tr	w tr	w tr	w tr	w tr	drr	drr	av	av	w tr	w tr	w tr	drr	w tr
APR	drr	drr	drr	av	av	w tr	drr	w tr	av	av	av	drr	av	av	w tr	av	av	drr	av	drr	av
MAY	drr	drr	drr	drr	drr	drr	drr	drr	drr	drr	drr	drr	drr	drr	drr	drr	drr	drr	w tr	av	drr
JUNE	drr	drr	drr	av	drr	drr	drr	drr	av	w tr	drr	drr	av	av	av	av	drr	av	av	av	drr
JULY	w tr	w tr	w tr	w tr	w tr	w tr	w tr	w tr	w tr	w tr	w tr	w tr	w tr	w tr	w tr	w tr	w tr	w tr	w tr	w tr	w tr
AUG	w tr	w tr	w tr	av	w tr	w tr	av	w tr	w tr	w tr	w tr	w tr	w tr	w tr	w tr	w tr	w tr	w tr	w tr	w tr	w tr
SEP	w tr	w tr	w tr	w tr	drr	av	w tr	w tr	w tr	w tr	w tr	w tr	w tr	av	w tr	w tr	w tr	av	w tr	w tr	w tr
OCT	drr	drr	drr	av	drr	drr	drr	av	w tr	w tr	av	av	drr	drr	av	w tr	w tr	drr	w tr	drr	av
NOV	drr	drr	drr	av	w tr	drr	av	w tr	w tr	w tr	av	w tr	drr	av	av	w tr	w tr	av	w tr	av	w tr
DEC	drr	drr	av	w tr	w tr	drr	av	av	w tr	av	drr	w tr	drr	drr	drr	drr	av	drr	w tr	drr	drr
Overall	av	av	av	w tr	w tr	av	av	w tr	w tr	w tr	av	w tr	w tr	w tr	w tr	w tr	w tr	w tr	w tr	w tr	w tr
year	861	944	812	1179	1107	1260	1284	1245	1047	1193	1289	1260	971	840	758	978	1607	1583	1131	1228	22578
estimates	**936**	**966**	**890**	**1286**	**1284**	**1338**	**1286**	**1489**	**1456**	**1751**	**1330**	**1424**	**1172**	**989**	**963**	**1325**	**1890**	**1744**	**2022**	**1391**	**26933**

	Ulster Province					Connaught Province						Leinster Province				MUNSTER PROVINCE					
Date	Aldergrove	Hillsborough	Armagh	Lough Navar	Malin Head	Sligo	Ardtarmon	Belmullet	Drumsna	Galway	Roscommon	Markree Castle	Edenderry	Kilkenny	Dublin	Shannon	Newport	Tralee	Killarney	Cork	All
WINTER	w tr	w tr	w tr	w tr	w tr	w tr	av	w tr	w tr	w tr	w tr	w tr	av	w tr	av	w tr	w tr	w tr	w tr	av	w tr
SPRING	av	drr	drr	av	w tr	w tr	drr	w tr	av	av	av	drr	drr	drr	av	av	w tr	av	w tr	drr	av
SUMMER	w tr	w tr	w tr	w tr	w tr	av	av	w tr	w tr	w tr	av	w tr	w tr	w tr	w tr	w tr	w tr	w tr	w tr	w tr	w tr
AUTUMN	av	av	av	w tr	av	drr	av	w tr	w tr	w tr	w tr	w tr	av	av	w tr	w tr	w tr	drr	w tr	av	w tr
Overall	av	av	av	w tr	w tr	av	av	w tr	w tr	w tr	av	w tr	w tr	w tr	w tr	w tr	w tr	w tr	w tr	w tr	w tr
year	861	944	812	1179	1107	1260	1284	1245	1047	1193	1289	1260	971	840	758	978	1607	1583	1131	1228	22578
estimates	**936**	**966**	**890**	**1286**	**1284**	**1338**	**1286**	**1489**	**1456**	**1751**	**1330**	**1424**	**1172**	**989**	**963**	**1325**	**1890**	**1744**	**2022**	**1391**	**26933**

RAINFALL EXPECTATIONS

WETTER OR DRIER BY 25% THAN LONG-TERM AVERAGES

	Ulster Province					Connaught Province						Leinster Province				MUNSTER PROVINCE					
Date	Aldergrove	Hillsborough	Armagh	Lough Navar	Malin Head	Sligo	Ardtarmon	Belmullet	Drumsna	Galway	Roscommon	Markree Castle	Edenderry	Kilkenny	Dublin	Shannon	Newport	Tralee	Killarney	Cork	All
JAN	w tr	w tr	w tr	av	w tr	w tr	av	av	w tr	w tr	w tr	av	w tr	w tr	w tr	w tr	w tr	w tr	w tr	w tr	w tr
FEB	av	drr	av	av	av	w tr	av	w tr	w tr	w tr	av	av	drr	drr	drr	av	av	av	w tr	drr	av
MAR	av	av	av	av	w tr	w tr	av	w tr	w tr	w tr	av	av	drr	av	av	av	w tr	av	w tr	av	w tr
APR	av	av	av	av	av	av	av	w tr	av	av	av	av	av	av	av	av	av	av	av	drr	av
MAY	drr	drr	drr	drr	drr	drr	drr	drr	av	drr	drr	drr	drr	drr	drr	av	drr	av	w tr	av	drr
JUNE	drr	drr	drr	av	drr	drr	av	drr	av	av	av	drr	av	av	av	av	drr	av	av	av	av
JULY	av	av	w tr	w tr	w tr	av	av	w tr	w tr	w tr	av	w tr	w tr	w tr	w tr	w tr	av	w tr	w tr	w tr	w tr
AUG	w tr	w tr	w tr	av	w tr	w tr	av	w tr	w tr	w tr	av	w tr	w tr	w tr	w tr	w tr	w tr	w tr	w tr	w tr	w tr
SEP	av	av	w tr	w tr	av	av	w tr	w tr	w tr	w tr	w tr	w tr	w tr	av	w tr	av	w tr	av	w tr	w tr	w tr
OCT	av	av	av	av	av	av	av	av	av	w tr	av	av	drr	drr	av	av	w tr	av	w tr	drr	av
NOV	av	av	drr	av	w tr	av	av	av	w tr	w tr	av	av	av	av	av	w tr	av	av	w tr	av	av
DEC	av	drr	av	av	av	av	av	av	av	av	av	av	drr	av	drr	av	av	av	w tr	drr	av
Overall	av	av	av	av	av	av	av	av	w tr	w tr	av	av	av	av	w tr	w tr	av	av	w tr	av	av
year	861	944	812	1179	1107	1260	1284	1245	1047	1193	1289	1260	971	840	758	978	1607	1583	1131	1228	22578
estimates	936	966	890	1286	1284	1338	1286	1489	1456	1751	1330	1424	1172	989	963	1325	1890	1744	2022	1391	26933

	Ulster Province					Connaught Province						Leinster Province				MUNSTER PROVINCE					
Date	Aldergrove	Hillsborough	Armagh	Lough Navar	Malin Head	Sligo	Ardtarmon	Belmullet	Drumsna	Galway	Roscommon	Markree Castle	Edenderry	Kilkenny	Dublin	Shannon	Newport	Tralee	Killarney	Cork	All
WINTER	av	av	av	av	av	w tr	av	av	w tr	w tr	av	av	av	av	av	w tr	av	av	w tr	av	w tr
SPRING	av	av	av	av	av	av	av	w tr	av	av	av	av	drr	av	av	av	av	av	w tr	av	av
SUMMER	w tr	w tr	w tr	av	av	av	av	w tr	w tr	w tr	av	w tr	w tr	w tr	w tr	w tr	av	w tr	w tr	w tr	w tr
AUTUMN	av	av	av	av	av	av	av	av	w tr	w tr	av	av	av	av	av	av	av	av	w tr	av	av
Overall	av	av	av	av	av	av	av	av	w tr	w tr	av	av	av	av	w tr	w tr	av	av	w tr	av	av
year	861	944	812	1179	1107	1260	1284	1245	1047	1193	1289	1260	971	840	758	978	1607	1583	1131	1228	22578
estimates	936	966	890	1286	1284	1338	1286	1489	1456	1751	1330	1424	1172	989	963	1325	1890	1744	2022	1391	26933

RAINFALL ESTIMATION SUMMARIES

% WETTER OR DRIER OF LONG-TERM AVERAGES

Date	Ulster Province					Connaught Province						Leinster Province				MUNSTER PROVINCE					All
	Aldergrove	Hillsborough	Armagh	Lough Navar	Malin Head	Sligo	Ardtarmon	Belmullet	Drumsna	Galway	Roscommon	Markree Castle	Edenderry	Kilkenny	Dublin	Shannon	Newport	Tralee	Killarney	Cork	
JANUARY	87%	78%	83%	5%	41%	28%	-1%	24%	63%	107%	44%	20%	44%	103%	69%	132%	30%	71%	206%	71%	62%
	w tr	w tr	w tr	w tr	w tr	w tr	drr	w tr	w tr	w tr	w tr	w tr	w tr	w tr	w tr	w tr	w tr	w tr	w tr	w tr	w tr
FEBRUARY	-22%	-39%	-12%	11%	19%	46%	9%	30%	34%	34%	-16%	21%	-39%	-48%	-56%	20%	6%	-5%	50%	-41%	4%
	drr	drr	drr	w tr	w tr	w tr	w tr	w tr	w tr	w tr	drr	w tr	drr	drr	drr	w tr	w tr	drr	w tr	drr	w tr
MARCH	9%	-11%	16%	25%	71%	51%	9%	52%	38%	60%	18%	14%	-31%	-20%	17%	0%	73%	9%	70%	-12%	26%
	w tr	drr	w tr	w tr	w tr	w tr	w tr	w tr	w tr	w tr	w tr	w tr	drr	drr	w tr	av	w tr	w tr	w tr	drr	w tr
APRIL	-21%	-24%	-23%	-7%	-2%	19%	-18%	44%	7%	-7%	-1%	-17%	-8%	-11%	24%	0%	1%	-19%	7%	-27%	-4%
	drr	drr	drr	drr	drr	w tr	drr	w tr	w tr	drr	drr	drr	drr	drr	w tr	av	w tr	drr	w tr	drr	drr
MAY	-40%	-31%	-54%	-38%	-43%	-34%	-50%	-28%	-20%	-45%	-54%	-47%	-55%	-43%	-65%	-28%	-63%	-16%	102%	-11%	-34%
	drr	drr	drr	drr	drr	drr	drr	drr	drr	drr	drr	drr	drr	drr	drr	drr	drr	drr	w tr	drr	drr
JUNE	-62%	-56%	-72%	-11%	-51%	-65%	-24%	-42%	0%	19%	-22%	-30%	-11%	1%	-10%	6%	-42%	2%	8%	5%	-22%
	drr	drr	drr	drr	drr	drr	drr	drr	av	w tr	drr	drr	drr	w tr	drr	w tr	drr	w tr	w tr	w tr	drr
JULY	25%	9%	51%	38%	80%	12%	24%	44%	46%	90%	13%	30%	123%	154%	109%	59%	24%	37%	127%	121%	57%
	w tr	w tr	w tr	w tr	w tr	w tr	w tr	w tr	w tr	w tr	w tr	w tr	w tr	w tr	w tr	w tr	w tr	w tr	w tr	w tr	w tr
AUGUST	116%	122%	119%	9%	26%	68%	6%	69%	134%	109%	16%	72%	252%	126%	192%	156%	56%	82%	170%	110%	94%
	w tr	w tr	w tr	w tr	w tr	w tr	w tr	w tr	w tr	w tr	w tr	w tr	w tr	w tr	w tr	w tr	w tr	w tr	w tr	w tr	w tr
SEPTEMBER	22%	15%	33%	49%	-15%	5%	40%	27%	93%	72%	41%	42%	37%	12%	75%	22%	27%	-7%	41%	60%	33%
	w tr	w tr	w tr	w tr	drr	w tr	w tr	w tr	w tr	w tr	w tr	w tr	w tr	w tr	w tr	w tr	w tr	drr	w tr	w tr	w tr
OCTOBER	-15%	-25%	-13%	-5%	-12%	-21%	-11%	-5%	11%	41%	-1%	-4%	-33%	-26%	-3%	11%	26%	-19%	36%	-30%	-4%
	drr	drr	drr	drr	drr	drr	drr	drr	w tr	w tr	drr	drr	drr	drr	drr	w tr	w tr	drr	w tr	drr	drr
NOVEMBER	-25%	-20%	-35%	5%	34%	-19%	-4%	19%	37%	46%	-8%	12%	-12%	-10%	-9%	40%	23%	5%	85%	-6%	10%
	drr	drr	drr	w tr	w tr	drr	drr	w tr	w tr	w tr	drr	w tr	drr	drr	drr	w tr	w tr	w tr	w tr	drr	w tr
DECEMBER	-17%	-31%	-13%	11%	11%	-19%	8%	-1%	16%	-5%	-13%	20%	-29%	-23%	-34%	-22%	-1%	-15%	31%	-48%	-7%
	drr	drr	drr	w tr	w tr	drr	w tr	drr	w tr	drr	drr	w tr	drr	drr	drr	drr	drr	drr	w tr	drr	drr
WINTER	25%	19%	19%	9%	22%	36%	5%	17%	41%	62%	13%	23%	4%	16%	-3%	61%	20%	25%	128%	8%	28%
	w tr	w tr	w tr	w tr	w tr	w tr	w tr	w tr	w tr	w tr	w tr	w tr	w tr	w tr	drr	w tr	w tr	w tr	w tr	w tr	w tr
SPRING	-16%	-21%	-19%	-3%	17%	16%	-17%	27%	10%	7%	-9%	-13%	-31%	-25%	-10%	-9%	13%	-8%	56%	-16%	-2%
	drr	drr	drr	drr	w tr	w tr	drr	w tr	w tr	w tr	drr	drr	drr	drr	drr	drr	w tr	drr	w tr	drr	drr
SUMMER	35%	33%	41%	13%	22%	10%	3%	30%	64%	77%	4%	27%	128%	95%	100%	79%	19%	40%	103%	80%	47%
	w tr	w tr	w tr	w tr	w tr	w tr	w tr	w tr	w tr	w tr	w tr	w tr	w tr	w tr	w tr	w tr	w tr	w tr	w tr	w tr	w tr
AUTUMN	-7%	-11%	-6%	14%	2%	-13%	6%	12%	42%	52%	9%	15%	-6%	-10%	17%	24%	25%	-7%	55%	2%	11%
	drr	drr	drr	w tr	w tr	drr	w tr	w tr	w tr	w tr	w tr	w tr	drr	drr	w tr	w tr	w tr	drr	w tr	w tr	w tr
OVERALL	9%	2%	10%	9%	16%	6%	0%	20%	39%	47%	3%	13%	21%	18%	27%	36%	18%	10%	79%	13%	19%
	w tr	w tr	w tr	w tr	w tr	w tr	av	w tr	w tr	w tr	w tr	w tr	w tr	w tr	w tr	w tr	w tr	w tr	w tr	w tr	w tr

ESTIMATED SUNSHINE HOURS TREND

TRENDS BY MONTHS FOR SELECTED TOWNS

Date	Ulster Province						Connaught Province							Leinster Province					MUNSTER PROVINCE						
	Aldergrove	Hillsborough	Armagh	Lough Navar	Malin Head	Ulster Overall	Sligo	Ardtarmon	Belmullet	Drumsna	Galway	Roscommon	Connaught Overall	Markree Castle	Edenderry	Kilkenny	Dublin	Leinster Overall	Shannon	Newport	Tralee	Killarney	Cork	Munster Overall	Ireland Overall
Dec 2025	av	av	av	av	snnr	snnr	av	av	av	snnr	av	av	snnr	av	av	av	av	av	av	snnr	av	av	snnr	snnr	snnr
Jan	av	av	av	av	av	av	av	av	av	av	av	av	av	av	av	av	av	av	av	av	av	av	av	av	average
Feb	av	av	av	av	av	snnr	av	av	av	av	av	snnr	snnr	av	av	av	snnr	snnr	av	av	snnr	snnr	av	snnr	snnr
Winter	av	av	snnr	snnr	snnr	snnr	av	av	av	snnr	av	av	snnr	av	av	snnr	av	av	av	snnr	av	av	av	snnr	snnr
Mar	av	av	av	av	av	av	av	av	av	av	cldr	cldr	cldr	av	av	av	av	av	cldr	av	cldr	av	av	cldr	cldr
Apr	snnr	snnr	snnr	snnr	snnr	snnr	snnr	snnr	av	snnr	snnr	snnr	snnr	snnr	snnr	snnr	snnr	snnr	snnr	snnr	av	av	snnr	snnr	snnr
May	snnr	snnr	snnr	snnr	snnr	snnr	av	snnr	snnr	snnr	snnr	snnr	snnr	snnr	snnr	snnr	snnr	snnr	snnr	snnr	snnr	snnr	cldr	snnr	snnr
Spring	snnr	snnr	snnr	snnr	snnr	snnr	snnr	snnr	snnr	snnr	snnr	snnr	snnr	snnr	snnr	snnr	snnr	snnr	snnr	snnr	av	av	av	snnr	snnr
June	snnr	snnr	snnr	av	av	snnr	av	snnr	snnr	snnr	snnr	snnr	snnr	snnr	snnr	snnr	av	snnr	snnr	snnr	snnr	snnr	snnr	snnr	snnr
July	cldr	cldr	av	av	av	cldr	av	av	av	av	av	av	av	av	av	av	cldr	av	av	av	cldr	cldr	cldr	cldr	cldr
Aug	cldr	cldr	cldr	cldr	cldr	cldr	cldr	cldr	cldr	av	cldr	cldr	cldr	cldr	cldr	av	cldr	cldr	cldr	av	cldr	cldr	cldr	cldr	cldr
Summer	cldr	cldr	av	cldr	av	cldr	av	av	snnr	snnr	av	av	snnr	av	av	av	cldr	av	av	snnr	cldr	cldr	cldr	cldr	cldr
Sep	cldr	cldr	cldr	cldr	av	cldr	av	av	av	av	av	av	av	cldr	cldr	cldr	cldr	cldr	cldr	av	av	av	cldr	cldr	cldr
Oct	snnr	snnr	snnr	snnr	snnr	snnr	snnr	snnr	snnr	snnr	snnr	snnr	snnr	snnr	snnr	snnr	snnr	snnr	snnr	snnr	snnr	snnr	snnr	snnr	snnr
Nov	snnr	snnr	snnr	snnr	snnr	snnr	av	snnr	av	snnr	av	av	snnr	av	snnr	snnr	snnr	snnr	av	snnr	av	av	snnr	snnr	snnr
Autumn	av	av	snnr	snnr	snnr	snnr	snnr	snnr	snnr	snnr	snnr	av	snnr	snnr	snnr	snnr	av	snnr	snnr	snnr	snnr	snnr	snnr	snnr	snnr
Dec	snnr	snnr	snnr	snnr	snnr	snnr	av	snnr	av	snnr	av	av	snnr	av	snnr	snnr	snnr	snnr	av	snnr	av	av	snnr	snnr	snnr
Year	av	av	snnr	snnr	snnr	snnr	av	snnr	snnr	snnr	snnr	av	snnr	snnr	snnr	snnr	av	snnr	av	snnr	av	av	av	snnr	snnr

ESTIMATED SUNSHINE HOURS

Date	Ulster Province						Connaught Province							Leinster Province					MUNSTER PROVINCE						
	Aldergrove	Hillsborough	Armagh	Lough Navar	Malin Head	Ulster Overall	Sligo	Ardtarmon	Belmullet	Drumsna	Galway	Roscommon	Connaught Overall	Markree Castle	Edenderry	Kilkenny	Dublin	Leinster Overall	Shannon	Newport	Tralee	Killarney	Cork	Munster Overall	Ireland Overall
Jan	av	av	av	av	av	av	av	av	av	av	av	av	av	av	av	av	av	av	av	av	av	av	av	av	average
Feb	av	av	av	av	av	snnr	av	av	av	av	av	snnr	snnr	av	av	av	snnr	snnr	av	av	snnr	snnr	av	snnr	snnr
Mar	av	av	av	av	av	av	av	av	av	av	cldr	cldr	cldr	av	av	av	av	av	cldr	av	cldr	av	av	cldr	cldr
Apr	snnr	snnr	snnr	snnr	snnr	snnr	snnr	snnr	av	snnr	snnr	snnr	snnr	snnr	snnr	snnr	snnr	snnr	snnr	snnr	av	av	snnr	snnr	snnr
May	snnr	snnr	snnr	snnr	snnr	snnr	av	snnr	snnr	snnr	snnr	snnr	snnr	snnr	snnr	snnr	snnr	snnr	snnr	snnr	snnr	snnr	cldr	snnr	snnr
June	snnr	snnr	snnr	av	av	snnr	av	snnr	snnr	snnr	snnr	snnr	snnr	snnr	snnr	snnr	av	snnr	snnr	snnr	snnr	snnr	snnr	snnr	snnr
July	cldr	cldr	av	av	av	cldr	av	av	av	av	av	av	av	av	av	av	cldr	av	av	av	cldr	cldr	cldr	cldr	cldr
Aug	cldr	cldr	cldr	cldr	cldr	cldr	cldr	cldr	cldr	av	cldr	cldr	cldr	cldr	cldr	av	cldr	cldr	cldr	av	cldr	cldr	cldr	cldr	cldr
Sep	cldr	cldr	cldr	cldr	av	cldr	av	av	av	av	av	av	av	cldr	cldr	cldr	cldr	cldr	cldr	av	av	av	cldr	cldr	cldr
Oct	snnr	snnr	snnr	snnr	snnr	snnr	snnr	snnr	snnr	snnr	snnr	snnr	snnr	snnr	snnr	snnr	snnr	snnr	snnr	snnr	snnr	snnr	snnr	snnr	snnr
Nov	snnr	snnr	snnr	snnr	snnr	snnr	av	snnr	av	snnr	av	av	snnr	av	snnr	snnr	snnr	snnr	av	snnr	av	av	snnr	snnr	snnr
Dec	snnr	snnr	snnr	snnr	snnr	snnr	av	snnr	av	snnr	av	av	snnr	av	snnr	snnr	snnr	snnr	av	snnr	av	av	snnr	snnr	snnr
Year	av	av	snnr	snnr	snnr	snnr	av	snnr	snnr	snnr	snnr	av	snnr	snnr	snnr	snnr	av	snnr	av	snnr	av	av	av	snnr	snnr

ESTIMATED AVERAGE MONTHLY MAXIMUM TEMPERATURES

	Ulster Province					Connaught Province						Leinster Province				Munster Province					
	Aldergrove	Hillsborough	Armagh	Lough Navar	Malin Head	Sligo	Ardtarmon	Belmullet	Drumsna	Galway	Roscommon	Markree Castle	Edenderry	Kilkenny	Dublin	Shannon	Newport	Tralee	Killarney	Cork	All
Jan	8.6	8.6	8.6	8.4	8.5	8.7	8.6	9.4	8.9	9.3	8.9	8.7	9.1	9.6	9.5	10.0	9.3	9.1	10.7	9.4	**9.1**
Feb	10.2	10.2	10.2	10.0	9.6	9.6	9.5	10.3	9.3	9.3	9.3	9.6	9.3	10.3	9.7	10.4	9.9	9.3	11.0	8.9	**9.8**
Mar	10.1	10.1	10.1	9.7	9.1	9.7	10.5	10.4	10.0	9.7	10.0	9.7	9.8	10.6	9.9	10.4	10.1	9.8	11.3	9.9	**10.0**
Apr	11.4	11.4	11.4	10.9	9.1	10.6	12.1	11.0	11.1	10.4	11.1	10.6	11.0	11.0	10.7	11.3	11.1	11.0	11.1	10.7	**10.9**
May	16.7	16.7	16.7	16.8	14.2	16.1	18.2	16.6	16.8	16.2	16.8	16.1	16.3	14.5	15.0	16.9	17.0	16.3	16.6	14.1	**16.2**
Jun	18.2	18.2	18.2	18.0	14.8	16.1	19.0	16.8	17.6	16.9	17.6	16.1	17.4	16.8	16.7	18.0	18.1	17.4	17.8	16.9	**17.3**
Jul	18.0	18.0	18.0	17.2	15.7	16.8	18.2	16.4	17.4	16.8	17.4	16.8	17.8	17.4	17.7	18.1	17.1	17.8	17.9	17.2	**17.4**
Aug	19.9	19.9	19.9	19.4	17.7	18.7	20.5	18.6	19.3	18.1	19.3	18.7	19.7	18.5	19.1	19.3	18.9	19.7	18.8	18.0	**19.1**
Sep	17.2	17.2	17.2	17.2	16.0	16.5	17.4	16.6	17.1	16.3	17.1	16.5	17.2	16.8	17.3	17.4	17.2	17.2	17.5	16.3	**17.0**
Oct	14.7	14.7	14.7	14.8	14.1	14.6	15.2	14.7	14.7	14.6	14.7	14.6	14.7	14.6	14.7	15.5	15.2	14.7	15.5	14.5	**14.8**
Nov	10.5	10.5	10.5	10.6	10.6	10.9	10.6	11.2	10.6	11.0	10.6	10.9	10.3	10.5	10.4	11.2	11.4	10.3	11.8	10.6	**10.8**
Dec	7.5	7.5	7.5	8.0	8.9	9.0	7.4	9.7	8.0	8.8	8.0	9.0	7.5	7.9	7.6	9.1	8.9	7.5	9.8	8.3	**8.3**
Year	13.6	13.6	13.6	13.4	12.4	13.1	13.9	13.5	13.4	13.1	13.4	13.1	13.3	13.2	13.2	14.0	13.7	13.3	14.1	12.9	13.4

ESTIMATED AVERAGE MONTHLY MINIMUM TEMPERATURES

	Ulster Province					Connaught Province						Leinster Province				Munster Province					
	Aldergrove	Hillsborough	Armagh	Lough Navar	Malin Head	Sligo	Ardtarmon	Belmullet	Drumsna	Galway	Roscommon	Markree Castle	Edenderry	Kilkenny	Dublin	Shannon	Newport	Tralee	Killarney	Cork	All
Jan	3.3	2.6	2.6	0.5	3.9	3.8	3.8	4.3	2.0	1.8	2.4	3.8	1.1	2.3	3.1	4.2	3.7	5.8	5.8	3.3	**3.2**
Feb	3.3	2.1	2.1	1.0	4.4	4.1	4.1	4.1	2.0	1.7	2.7	4.1	0.5	1.3	2.2	3.8	4.5	5.3	5.3	3.8	**3.1**
Mar	3.6	2.6	2.6	0.7	4.5	4.3	4.3	4.9	2.2	2.1	2.8	4.3	0.9	1.8	2.5	3.9	4.5	5.6	5.6	3.0	**3.3**
Apr	3.5	2.5	2.5	0.5	4.5	4.0	4.0	4.1	2.3	1.2	2.8	4.0	0.8	2.0	2.2	3.7	4.0	5.0	5.0	3.1	**3.1**
May	7.7	7.1	7.1	4.0	8.3	8.1	8.1	8.6	6.6	5.8	7.2	8.1	5.4	5.9	5.8	8.1	8.8	9.3	9.3	7.7	**7.3**
Jun	9.3	9.0	9.0	6.1	9.7	9.5	9.5	9.8	8.8	7.5	8.8	9.5	7.4	7.6	8.0	9.9	10.1	10.4	10.4	9.2	**9.0**
Jul	11.1	11.0	11.0	8.7	11.3	11.2	11.2	11.3	10.8	8.8	10.7	11.2	9.3	9.5	10.1	11.8	11.4	12.1	12.1	10.6	**10.8**
Aug	12.8	12.4	12.4	10.1	13.2	12.8	12.8	12.5	12.1	10.8	12.2	12.8	11.1	11.1	11.9	13.0	13.3	13.4	13.4	11.5	**12.3**
Sep	11.3	10.9	10.9	8.2	11.7	11.3	11.3	11.3	10.2	9.3	10.7	11.3	9.5	9.5	10.4	11.2	11.9	11.8	11.8	10.1	**10.7**
Oct	8.1	7.1	7.1	4.3	9.2	8.3	8.3	8.5	5.9	5.5	6.6	8.3	5.2	5.7	6.2	8.0	8.8	9.1	9.1	7.8	**7.4**
Nov	5.5	4.3	4.3	2.5	6.6	6.2	6.2	6.9	3.9	3.8	4.8	6.2	2.8	3.2	4.8	5.7	6.5	7.5	7.5	4.7	**5.2**
Dec	2.5	1.0	1.0	-0.6	4.0	3.6	3.6	4.7	0.2	1.3	1.4	3.6	-0.4	0.6	1.2	3.2	4.2	5.4	5.4	2.8	**2.4**
Year	6.8	6.0	6.0	3.8	7.6	7.3	7.3	7.6	5.6	5.0	6.1	7.3	4.5	5.0	5.7	7.2	7.6	8.4	8.4	6.5	6.5

ESTIMATED MAXIMUM AVERAGE TEMPERATURES – QUICK REFERENCE

MONTHLY

	Ulster Province						Connaught Province							Leinster Province					Munster Province						
	Aldergrove	Hillsborough	Armagh	Lough Navar	Malin Head	Ulster Overall	Sligo	Ardtarmon	Belmullet	Drumsna	Galway	Roscommon	Connaught Overall	Markree Castle	Edenderry	Kilkenny	Dublin	Leinster Overall	Shannon	Newport	Tralee	Killarney	Cork	Munster Overall	Overall
Jan	wmr	wmr	a-w	av	av	a-w	wmr	a-w	av	wmr	wmr	wmr	wmr	wmr	wmr	wmr	wmr	wmr	wmr	wmr	a-c	wmr	wmr	wmr	wmr
Feb	wmr	wmr	wmr	wmr	wmr	wmr	wmr	wmr	wmr	wmr	a-w	wmr	wmr	wmr	wmr	wmr	wmr	wmr	wmr	wmr	a-c	wmr	a-w	wmr	wmr
Mar	av	wmr	av	a-c	av	av	wmr	a-w	a-c	av	av	a-w	av	av	av	av	av	av	a-c	av	clr	wmr	av	av	av
Apr	a-c	av	clr	clr	clr	clr	a-c	a-w	clr	a-c	clr	a-c	a-c	a-c	a-c	clr	clr	clr	clr	clr	clr	a-c	clr	clr	clr
May	wmr	wmr	a-w	wmr	wmr	wmr	wmr	wmr	wmr	wmr	wmr	wmr	wmr	wmr	wmr	clr	av	a-w	a-w	wmr	av	wmr	av	wmr	wmr
Jun	a-w	wmr	av	a-w	av	a-w	av	wmr	a-w	wmr	av	av	a-w	av	av	clr	a-c	a-c	av	av	av	wmr	av	av	av
Jul	clr	av	clr	clr	clr	a-c	av	wmr	clr	av	clr	clr	a-c	a-c	clr	clr	clr	clr	clr	clr	clr	av	clr	clr	clr
Aug	wmr	wmr	av	a-w	a-w	wmr	wmr	wmr	a-w	wmr	av	a-w	wmr	wmr	a-w	clr	av	av	av	av	a-w	wmr	av	av	a-w
Sep	a-w	wmr	av	av	av	a-w	wmr	wmr	av	wmr	av	a-w	a-w	a-w	a-w	a-c	av	av	av	av	av	wmr	av	av	av
Oct	wmr	wmr	a-w	wmr	wmr	wmr	wmr	wmr	a-w	wmr	wmr	wmr	wmr	wmr	wmr	av	wmr	wmr	wmr	wmr	av	wmr	wmr	wmr	wmr
Nov	a-w	wmr	av	av	av	a-w	wmr	a-w	av	wmr	a-w	a-w	wmr	wmr	wmr	av	av	a-w	av	wmr	clr	wmr	av	av	a-w
Dec	av	av	av	av	av	av	wmr	a-c	av	wmr	a-w	av	a-w	wmr	av	a-c	a-c	av	av	a-w	clr	wmr	av	av	av
Year	a-w	wmr	av	av	av	av	wmr	wmr	av	wmr	av	a-w	a-w	a-w	a-w	a-c	av	av	av	av	a-c	wmr	av	av	av

wmr = 1.0> degrees above average | a-w = 0.6-1.0 degrees above average | av = 0.0-0.5 degrees from average | a-c = 0.6-1.0 degrees below average | clr = 1.0 > degrees below average

SEASONAL

	Ulster Province						Connaught Province							Leinster Province					Munster Province						
	Aldergrove	Hillsborough	Armagh	Lough Navar	Malin Head	Ulster Overall	Sligo	Ardtarmon	Belmullet	Drumsna	Galway	Roscommon	Connaught Overall	Markree Castle	Edenderry	Kilkenny	Dublin	Leinster Overall	Shannon	Newport	Tralee	Killarney	Cork	Munster Overall	Overall
Winter	wmr	wmr	wmr	a-w	wmr	wmr	wmr	wmr	wmr	wmr	wmr	wmr	wmr	wmr	wmr	wmr	wmr	wmr	wmr	wmr	a-c	wmr	wmr	wmr	wmr
Spring	av	wmr	av	av	av	av	a-w	wmr	av	a-w	av	a-w	a-w	av	av	a-c	av	av	a-c	av	clr	wmr	av	av	av
Summer	av	wmr	av	av	av	av	av	wmr	av	wmr	av	av	a-w	av	av	clr	a-c	a-c	a-c	a-c	av	a-w	a-c	av	av
Autumn	a-w	wmr	av	av	a-w	a-w	wmr	wmr	av	wmr	a-w	wmr	wmr	wmr	wmr	av	av	a-w	av	wmr	a-c	wmr	av	a-w	a-w
December	av	av	av	av	av	av	wmr	a-c	av	wmr	a-w	av	a-w	wmr	av	a-c	a-c	av	av	a-w	clr	wmr	av	av	av
Year	a-w	wmr	av	av	av	av	wmr	wmr	av	wmr	av	a-w	a-w	a-w	a-w	a-c	av	av	av	av	a-c	wmr	av	av	av

wmr = 1.0> degrees above average | a-w = 0.6-1.0 degrees above average | av = 0.0-0.5 degrees from average | a-c = 0.6-1.0 degrees below average | clr = 1.0 > degrees below average

ESTIMATED MINIMUM AVERAGE TEMPERATURES – QUICK REFERENCE

MONTHLY

	Ulster Province						Connaught Province							Leinster Province					Munster Province						
	Aldergrove	Hillsborough	Armagh	Lough Navar	Malin Head	Ulster Overall	Sligo	Ardtarmon	Belmullet	Drumsna	Galway	Roscommon	Connaught Overall	Markree Castle	Edenderry	Kilkenny	Dublin	Leinster Overall	Shannon	Newport	Tralee	Killarney	Cork	Munster Overall	Overall
Jan	wmr	wmr	wmr	av	av	wmr	wmr	wmr	a-w	av	av	a-w	a-w	wmr	av	a-w	a-w	a-w	a-w	wmr	wmr	wmr	av	wmr	wmr
Feb	wmr	wmr	a-w	av	a-w	wmr	wmr	wmr	av	a-w	av	wmr	wmr	wmr	a-c	a-c	av	av	a-w	wmr	wmr	wmr	a-w	wmr	wmr
Mar	wmr	a-w	av	clr	av	av	wmr	wmr	av	av	clr	av	av	wmr	clr	clr	a-c	av	a-c	wmr	wmr	wmr	a-c	a-w	av
Apr	av	a-c	clr	clr	clr	clr	av	av	clr	clr	clr	a-c	clr	av	clr	clr	clr	clr	clr	av	wmr	av	clr	a-c	clr
May	wmr	wmr	a-w	clr	av	av	wmr	wmr	a-w	av	clr	wmr	a-w	wmr	av	a-c	clr	av	av	wmr	wmr	wmr	av	wmr	a-w
Jun	av	av	av	clr	a-c	a-c	a-w	a-w	a-c	av	clr	av	av	a-w	clr	clr	clr	a-c	a-c	a-w	wmr	a-w	a-c	av	av
Jul	av	av	av	clr	a-c	a-c	av	av	a-c	av	clr	av	av	a-w	clr	clr	clr	a-c	clr	av	wmr	av	clr	av	av
Aug	wmr	wmr	wmr	av	a-w	wmr	wmr	wmr	av	wmr	a-c	wmr	wmr	wmr	a-w	av	av	a-w	av	wmr	wmr	wmr	av	wmr	wmr
Sep	wmr	wmr	wmr	clr	a-w	wmr	wmr	wmr	a-w	wmr	a-c	wmr	wmr	wmr	a-w	av	a-w	wmr	av	wmr	wmr	wmr	av	wmr	wmr
Oct	wmr	av	av	clr	a-w	av	wmr	wmr	av	a-c	clr	av	av	wmr	clr	a-c	clr	av	av	wmr	wmr	a-w	av	a-w	av
Nov	wmr	wmr	wmr	a-c	av	a-w	wmr	wmr	a-w	a-w	a-c	wmr	wmr	wmr	av	av	av	av	av	wmr	wmr	wmr	av	wmr	wmr
Dec	av	a-c	clr	clr	av	a-c	a-w	a-w	av	clr	clr	clr	a-c	a-w	clr	clr	clr	clr	av	a-w	wmr	wmr	a-c	a-w	av
Year	wmr	a-w	av	clr	av	av	wmr	wmr	av	av	clr	a-w	av	wmr	a-c	a-c	a-c	av	av	wmr	wmr	wmr	av	a-w	av

wmr = 1.0> degrees above average | a-w = 0.6-1.0 degrees above average | av = 0.0-0.5 degrees from average | a-c = 0.6-1.0 degrees below average | clr = 1.0 > degrees below average

SEASONAL

	Ulster Province						Connaught Province							Leinster Province					Munster Province						
	Aldergrove	Hillsborough	Armagh	Lough Navar	Malin Head	Ulster Overall	Sligo	Ardtarmon	Belmullet	Drumsna	Galway	Roscommon	Connaught Overall	Markree Castle	Edenderry	Kilkenny	Dublin	Leinster Overall	Shannon	Newport	Tralee	Killarney	Cork	Munster Overall	Overall
Winter	wmr	wmr	wmr	a-w	wmr	wmr	wmr	wmr	wmr	a-w	av	wmr	wmr	wmr	av	a-w	a-w	a-w	wmr	wmr	wmr	wmr	a-w	wmr	wmr
Spring	a-w	av	av	clr	av	av	a-w	a-w	av	av	clr	av	av	wmr	clr	clr	clr	a-c	a-c	a-w	wmr	wmr	a-c	a-w	av
Summer	a-w	a-w	a-w	clr	av	av	wmr	wmr	av	a-w	clr	a-w	av	wmr	a-c	clr	a-c	av	a-c	a-w	wmr	a-w	a-c	av	av
Autumn	wmr	wmr	wmr	clr	a-w	a-w	wmr	wmr	av	av	clr	wmr	a-w	wmr	av	av	av	av	av	wmr	wmr	wmr	av	wmr	a-w
December	av	a-c	clr	clr	av	a-c	a-w	a-w	av	clr	clr	clr	a-c	a-w	clr	clr	clr	clr	av	a-w	wmr	wmr	a-c	a-w	av
Year	wmr	a-w	av	clr	av	av	wmr	wmr	av	av	clr	a-w	av	wmr	a-c	a-c	a-c	av	av	wmr	wmr	wmr	av	a-w	av

wmr = 1.0> degrees above average | a-w = 0.6-1.0 degrees above average | av = 0.0-0.5 degrees from average | a-c = 0.6-1.0 degrees below average | clr = 1.0 > degrees below average

GARDENING GUIDE 2026

REST PERIOD:
The Rest Period in the garden is as important as the activity times. It's a time when the Moon is Void of Course, that is, traveling between one sign and another and during this time the Moon is deemed to have "no energizing power", also Rest Periods on the day of the Full Moon, New Moon and when the moon is at Southern and Northern Declinations. Hence it is a good time to sit and enjoy your garden or to attend to non-plant activities such as tidy your shed, repair or maintain outdoor structures or even to sit and plan your next "to do" list when the Moon is once more energized. Eclipse days and the day after the eclipse are also deemed days to rest in the garden.

ALL ABOVE-GROUND ACTIVITY SHOULD TAKE PLACE when the Moon is Waxing – i.e. building from New Moon through the 1st Quarter and up to the Full Moon. This is a time of increasing light and with the Moon pulling the sap upwards into the leaves and flowers.

ALL BELOW-GROUND ACTIVITY SHOULD TAKE PLACE when the Moon is Waning – i.e. from the Full Moon through the 3rd Quarter and up to the New Moon. This is a phase of decreasing light and the general trend for sap to be pulled downwards and into the root and tuber systems.

THE NEW MOON PHASE:
Activities can begin in the garden once the crescent of the New Moon begins to be visible.

At this time:
DO start sowing, transplanting, grafting or strike cuttings of all ABOVE-GROUND producers. Flowering annuals, brassicas (cabbage, broccoli etc) cover crops (mustard, lupins etc) and annual fruiting crops (cucumber, pumpkin, tomatoes etc) and house-plants all benefit from attention during this phase. Also, mow lawns should you wish to encourage growth or sow any large areas such as lawns.
DON'T plant root crops (as they will only produce plenty of leaves and little roots, or go to seed too early), nor prune at this time – your plants, and trees in particular, will be prone to excessive sap loss and suffer from die-back.

THE 1ST QUARTER PHASE:
Continue on with the New Moon phase activities and especially sow any seeds two days before the Full Moon to yield optimum results. During drought times, sow during this phase when the Moon is in Capricorn or Taurus for the strongest of survival qualities.

At this time:
DO sow and transplant all legume, grain, flowering plants (especially roses) and pot plants and start harvesting any above-ground produce, with 2-3 days before the Full Moon being the most beneficial time (especially for herbs and grape harvesting).
DO give plants a food boost if required, but do water it down and apply as close to the Full Moon as possible, particularly for those plants that require phosphorus.
DON'T harvest with the Moon in a water sign (Cancer, Scorpio or Pisces) unless for immediate use, nor prune or plant out root crops.

THE FULL MOON PHASE:
The Moon will have reached its full potency by the time of the Full Moon, and from this point to the New Moon, there is a gradual decrease in its vitality and the sap flow is now downwards into the roots – it is a time when the plants build up their root systems and store their goodness. However, you can continue to harvest above-ground producers for a few more days after the Full Moon. Leeks, onions, garlic etc are at their best potency when harvested with the Full Moon in Leo.

At this time:
DO sow or plant out any root crops (potatoes, kumaras, carrots etc) especially during the two days in the middle of the Full Moon phase. Also sow or plant out fruting plants (e.g. strawberries, passion fruit etc) as well as perennial flowering plants, bulbs and tubers (e.g. daffodils, dahlias, crocus etc) and this is also a good time to spray, prune and trim hedges.
DO perform drying activities of herbs, flowers and fruits.
DO dig herb roots or harvest leaves and bark intended for medicinal teas.
DO spray fruit trees, mulch your garden and kill weeds.

THE 3RD QUARTER PHASE:

This is considered to be the resting period of the Lunar cycle, the time when the sap in the plants runs at its slowest.

DO harvest all root crops that you wish to keep for long term storage (e.g. potatoes, garlic, onions and root herbs etc), especially if the Moon is in Taurus or Capricorn. Also lift your bulbs and tubers for their seasonal storage. You can also use this time to build up your compost, especially when the Moon is in Scorpio and attend to maintenance chores such as spreading compost, mulching, staking taller plants etc, as well as attend to all pruning and trimming of trees, vines and bushes, as there is less chance of die back and excessive sap bleeding. Cut your lawns during this phase to discourage growth.

DO plant trees and saplings during this phase for good strong foundations of root growth and bark development as well as plant strawberries or their runners so their growth is well developed before fruiting.

DO fertilize your crops for potassium if needed, as potassium absorption is at its peak at this phase.

When the Moon is in Cancer, Scorpio or Pisces:

This is a beneficial time for attending to vegetables such as lettuce, cabbage, herbs, broccoli, leafy pot plants, etc. Avoid any harvesting except for immediate use during these Moon signs as anything harvested for long term storage will deteriorate faster and be prone to mildew, except of course, unless you are attempting to build compost – choose when the Moon is in Scorpio for best results, as vegetation decomposes fastest under this influence.

When the Moon is in Taurus, Virgo or Capricorn:

This is a beneficial time for attending to vegetables such as potatoes, carrots, leeks etc. as well as to non-flowering shrubs and trees and ground covers such as lawns all respond best attended to under these signs. A Virgo Moon (particularly in the 3rd Quarter phase) is best for harvesting any medicinal leafy herbs due to the usual desire for long term storage.

When the Moon is in Gemini, Libra, Aquarius:

This is a beneficial time to attend to vegetables such as globe artichokes, flowers for medicinal purposes, and flowers in general. (Brocolli and cauliflower etc are considered "leafy" as if planted during a Flower phase they are more likely to bolt straight to seed). The Moon's focus is more directed towards developing perfume, fragrance, flower vibrancy and beauty as well as medicinal properties with the Full Moon in one of these signs.

When the Moon is in Aries, Leo or Sagittarius:

This is a beneficial time to attend to any fruit, berry, seed or grain crops such as tomatoes, peas, beans, peppers, wheat or corn, pumpkin, apples, strawberries, raspberries etc. Any seeds you wish to store for next year's planting, or for long term storage are best attended to during these Moon signs, as well as harvesting crops such as onions and garlic for their most potent flavour.

Above-ground - Leafy, Fruits and Seeds examples:

Vegetables	beans, broccoli, brussels sprouts, cabbages, cauliflower, chicory, celery, cress, cucumber, endive, lettuces, marrow, mustard, parsley, rhubarb, sage, silverbeet, corn, etc.
Flowers	cornflowers, forget-me-nots, marigolds, pansies, poppies, roses etc.
Fruits and seeds	apples, blue-berries, grapes, passion-fruit, pears, peas, peppers, raspberries, strawberries, tomatoes, wheat

Below-ground - Roots, Bulbs and Non-Flowering shrubs examples:

Vegetables	beetroot, carrots, garlic, leeks, onions, parsnips, radish, salsify, shallots, swedes, turnips etc.
Flowers	agapanthus, anemone, crocus, daffodils, dahlias, freesias, gladioli, hyacinth, ranunculus, red-hot pokers, tuberose, tulips
Fruits and Seeds	garlic, onions, peanuts

BASIC MOON IN SIGN GUIDE

Moon in Aries – Harvest root and fruit for storage. Good for destroying weeds and pests.

Moon in Taurus – Excellent time for planting and transplanting, particularly root crops. Also good for leaf crops.

Moon in Gemini – Good for harvesting root and fruit for storage. Good for destroying weeds and pests. Melon seeds respond well.

Moon in Cancer – Best time for all planting and transplanting. Good for grafting and irrigation.

Moon in Leo – Good for harvesting root and fruit for storage. Excellent for destroying weeds and pest.

Moon in Virgo – Good for destroying weeds and pests. Vines and some flowers respond well.

Moon in Libra – Best time for planting for fragrance and beauty of flowers, vines and herbs.

Moon in Scorpio – Best time for planting sturdy plants and vines. Especially favoured by tomatoes, corn and squash. Good for irrigation and transplanting.

Moon in Sagittarius – Good for harvesting roots. Especially favoured by onions and fruit trees. Good time to cultivate the soil.

Moon in Capricorn – Plant potatoes and root crops and encouraging hardy growth. Good for grafting and pruning to promote healing and applying organic fertilizers.

Moon in Aquarius – Harvest root and fruit for storage. Good for destroying weeds and pest. Good for planting onions.

Moon in Pisces – Good for planting and transplanting, especially for root growth. Good for irrigation.

JANUARY		2026	Hour:	1am – Midnight
P6	1st	Thur	Abo	Gemini
^	2nd	Fri	Abo	REST DAY - Northern Declination
F	3rd	Sat	Below	REST DAY - Full Moon
	4th	Sun	Below	Cancer / REST / Leo
	5th	Mon	Below	Leo
	6th	Tue	Below	Leo / REST / Virgo
	7th	Wed	Below	Virgo
	8th	Thur	Below	Virgo / REST
XhS	9th	Fri	Below	Libra
3Q	10th	Sat	Below	Libra / REST
	11th	Sun	Below	REST / Scorpio
	12th	Mon	Below	Scorpio
A	13th	Tue	Below	Scorpio / REST / Sag
	14th	Wed	Below	Sagittarius
	15th	Thur	Below	Sagittarius
V	16th	Fri	Below	REST DAY - Southern Declination
	17th	Sat	Below	Capricorn
N	18th	Sun	Above	REST DAY - New Moon
	19th	Mon	Above	Aquarius
	20th	Tue	Above	Aquarius
	21st	Wed	Above	Aquarius / REST / Pisces
	22nd	Thur	Above	Pisces
	23rd	Fri	Above	Pisces / RES / Aries
XhN	24th	Sat	Above	Aries / REST
	25th	Sun	Above	REST / Taurus
1Q	26th	Mon	Above	Taurus
	27th	Tue	Above	Taurus / REST / Gemini
	28th	Wed	Above	Gemini
P10 ^	29th	Thur	Above	REST DAY - Northern Declination
	30th	Fri	Above	Cancer
	31st	Sat	Above	Cancer / REST

FEBRUARY		2026	Hour:	1am – Midnight
F	1st	Sun	Below	REST DAY - Full Moon
	2nd	Mon	Below	Leo / REST
	3rd	Tue	Below	REST / Virgo
	4th	Wed	Below	Virgo
	5th	Thur	Below	Virgo / REST / Libra
XhS	6th	Fri	Below	Libra
	7th	Sat	Below	Libra / REST / Scorpio
	8th	Sun	Below	Scorpio
3Q	9th	Mon	Below	Scorpio
A	10th	Tue	Below	Scorpio / RES / Sagittarius
	11th	Wed	Below	Sagittarius
V	12th	Thur	Below	REST DAY - Southern Declination
	13th	Fri	Below	Capricorn
	14th	Sat	Below	Capricorn
	15th	Sun	Below	Capricorn / REST / Aquarius
	16th	Mon	Below	Aquarius

FEBRUARY	2026		Hour:	1am 2am 3am 4am 5am 6am 7am 8am 9am 10am 11am Noon 1pm 2pm 3pm 4pm 5pm 6pm 7pm 8pm 9pm 10pm 11pm Midnight
N	17th	Tue	Above	REST DAY - New Moon - Annual Solar Eclipse
	18th	Wed		REST DAY - Day after Eclipse Day
	19th	Thur		Pisces — REST — Aries
XhN	20th	Fri		Aries
	21st	Sat		Aries — RES — Taurus
	22nd	Sun		Taurus
	23rd	Mon		Taurus — REST
1Q P14	24th	Tue		REST — Gemini
^	25th	Wed		REST DAY - Northern Declination
	26th	Thur		REST — Cancer
	27th	Fri		Cancer
	28th	Sat		Cancer — REST — Leo

MARCH	2026		Hour:	1am 2am 3am 4am 5am 6am 7am 8am 9am 10am 11am Noon 1pm 2pm 3pm 4pm 5pm 6pm 7pm 8pm 9pm 10pm 11pm Midnight
	1st	Sun	Abo	Leo
	2nd	Mon		Leo — RES — Virgo
F	3rd	Tue	Below	REST DAY - Full Moon - Lunar Total Eclipse
	4th	Wed		REST DAY - Day after Eclipse Day
XhS	5th	Thur		Libra — REST
	6th	Fri		REST DAY - Moon Void of Course all day
	7th	Sat		REST — Scorpio
	8th	Sun		Scorpio
	9th	Mon		Scorpio — REST — Sagittarius
A	10th	Tue		Sagittarius
3Q V	11th	Wed		REST DAY - Southern Declination
	12th	Thur		REST — Capricorn
	13th	Fri		Capricorn
	14th	Sat		Capricorn — REST — Aquarius
	15th	Sun		Aquarius
	16th	Mon		Aquarius — REST — Pisces
	17th	Tue		Pisces
	18th	Wed		Pisces
N XhN	19th	Thur	Above	REST DAY - New Moon
	20th	Fri		Aries — REST
	21st	Sat		REST — Taurus
P11	22nd	Sun		Taurus
	23rd	Mon		Taurus — REST — Gemini
	24th	Tue		Gemini — REST
1Q ^	25th	Wed		REST DAY - Northern Declination
	26th	Thur		Cancer
	27th	Fri		Cancer — REST — Leo
	28th	Sat		Leo
	29th	Sun		Leo — REST — Virgo
	30th	Mon		Virgo
	31st	Tue		Virgo

APRIL		2026	Hour:	1am – Midnight
	1st	Wed	A	Virgo · REST · Libra
F XhS	2nd	Thur	Below	REST DAY - Full Moon
	3rd	Fri		REST · Scorpio
	4th	Sat		Scorpio
	5th	Sun		Scorpio · REST
	6th	Mon		RES · Sagittarius
A	7th	Tue		Sagittarius
V	8th	Wed		REST DAY - Southern Declination
	9th	Thur		Capricorn
3Q	10th	Fri		Capricorn · RES
	11th	Sat		REST · Aquarius
	12th	Sun		Aquarius
	13th	Mon		Aquarius · REST · Pisces
	14th	Tue		Pisces
	15th	Wed		Pisces · REST · Aries
XhN	16th	Thur		Aries
N	17th	Fri	Above	REST DAY - New Moon
	18th	Sat		Taurus
P7	19th	Sun		Taurus · RES · Gemini
	20th	Mon		Gemini · REST
^	21st	Tue		REST DAY - Northern Declination
	22nd	Wed		Cancer
	23rd	Thur		Cancer · RES · Leo
1Q	24th	Fri		Leo · REST
	25th	Sat		REST DAY - Moon Void of Course all day
	26th	Sun		REST · Virgo
	27th	Mon		Virgo · REST
	28th	Tue		REST · Libra
XhS	29th	Wed		Libra
	30th	Thur		Libra · REST · Scorpio

MAY		2026	Hour:	1am – Midnight
F	1st	Fri	Below	REST DAY - Full Moon
	2nd	Sat		Scorpio · REST
	3rd	Sun		REST · Sagittarius
A	4th	Mon		Sagittarius · REST
V	5th	Tue		REST DAY - Southern Declination
	6th	Wed		Capricorn
	7th	Thur		Capricorn · REST
	8th	Fri		REST · Aquarius
3Q	9th	Sat		Aquarius
	10th	Sun		Aquarius · REST · Pisces
	11th	Mon		Pisces
	12th	Tue		Pisces · REST
XhN	13th	Wed		REST · Aries
	14th	Thur		Aries · REST
	15th	Fri		REST · Taurus
N	16th	Sat	Above	REST DAY - New Moon
P3	17th	Sun		Taurus · REST · Gemini

MAY	2026		Hour:	1am – Midnight
^	18th	Mon	Above	REST DAY - Northern Declination
	19th	Tue		REST · Cancer
	20th	Wed		Cancer · REST
	21st	Thur		REST · Leo · REST
	22nd	Fri		REST DAY - Moon Void of Course all day
1Q	23rd	Sat		REST · Virgo
	24th	Sun		Virgo
	25th	Mon		Virgo · REST · Libra
XhS	26th	Tue		Libra
	27th	Wed		Libra · REST
	28th	Thur		REST · Scorpio
	29th	Fri		Scorpio
	30th	Sat		Scorpio · REST · Sagittarius
F	31st	Sun	B	REST DAY - Full Moon

JUNE	2026		Hour:	1am – Midnight
A V	1st	Mon	Below	REST DAY - Southern Declination
	2nd	Tue		REST · Capricorn
	3rd	Wed		Capricorn
	4th	Thur		Capricorn · REST · Aquarius
	5th	Fri		Aquarius · REST
	6th	Sat		REST DAY - Moon Void of Course all day
	7th	Sun		REST · Pisces
3Q	8th	Mon		Pisces
XhN	9th	Tue		Pisces · REST · Aries
	10th	Wed		Aries
	11th	Thur		Aries · REST · Taurus
	12th	Fri		Taurus
	13th	Sat		Taurus · REST · Gemini
P2	14th	Sun		Gemini
N ^	15th	Mon	Above	REST DAY - New Moon / Northern Declination
	16th	Tue		Cancer
	17th	Wed		Cancer · REST · Leo
	18th	Thur		Leo
	19th	Fri		Leo · REST · Virgo
	20th	Sat		Virgo
1Q	21st	Sun		Virgo · REST · Libra
XhS	22nd	Mon		Libra
	23rd	Tue		Libra
	24th	Wed		Libra · REST · Scorpio
	25th	Thur		Scorpio
	26th	Fri		Scorpio · REST · Sagittarius
	27th	Sat		Sagittarius
A V	28th	Sun		REST DAY - Southern Declination
F	29th	Mon	Belo	REST DAY - Full Moon
	30th	Tue		Capricorn

JULY		2026	Hour:	1am – Midnight
	1st	Wed	Below	Capricorn / REST / Aquarius
	2nd	Thur	Below	Aquarius
	3rd	Fri	Below	Aquarius / REST
	4th	Sat	Below	REST / Pisces
	5th	Sun	Below	Pisces
	6th	Mon	Below	Pisces / REST / Aries
3Q XhN	7th	Tue	Below	Aries
	8th	Wed	Below	Aries / REST / Taurus
	9th	Thur	Below	Taurus
	10th	Fri	Below	Taurus / REST / Ge
	11th	Sat	Below	Gemini / RES
^	12th	Sun	Below	REST DAY - Northern Declination
P4	13th	Mon	Below	Cancer
N	14th	Tue	Above	REST DAY - New Moon
	15th	Wed	Above	Leo / RES
	16th	Thur	Above	REST DAY - Moon Void of Course all day
	17th	Fri	Above	REST / Taurus
	18th	Sat	Above	Taurus / RES
	19th	Sun	Above	REST / Libra
XhS	20th	Mon	Above	Libra
1Q	21st	Tue	Above	Libra / REST / Scorpio
	22nd	Wed	Above	Scorpio / REST
	23rd	Thur	Above	REST DAY - Moon Void of Course all day
	24th	Fri	Above	REST / Sagittarius
A	25th	Sat	Above	Sagittarius / REST
V	26th	Sun	Above	REST DAY - Southern Declination
	27th	Mon	Above	Capricorn
	28th	Tue	Above	Capricorn / REST
F	29th	Wed	Below	REST DAY - Full Moon
	30th	Thur	Below	Aquarius / REST
	31st	Fri	Below	REST / Pisces

AUGUST		2026	Hour:	1am – Midnight
	1st	Sat	Below	Pisces
	2nd	Sun	Below	Pisces / REST / Aries
XhN	3rd	Mon	Below	Aries
	4th	Tue	Below	Aries / REST
	5th	Wed	Below	REST / Taurus
3Q	6th	Thur	Below	Taurus
	7th	Fri	Below	Tau / REST / Gemini
^	8th	Sat	Below	REST DAY - Northern Declination
	9th	Sun	Below	Gemini / REST / Cancer
P8	10th	Mon	Below	Cancer / REST
	11th	Tue	Below	REST / Leo
N	12th	Wed	Above	REST DAY - New Moon - Total Solar Eclipse
	13th	Thur	Above	REST DAY - Day after Eclipse Day
	14th	Fri	Above	REST DAY - Moon Void of Course all day
	15th	Sat	Above	REST / Libra
XhS	16th	Sun	Above	Libra
	17th	Mon	Above	Libra / REST / Scorpi

AUGUST	2026		Hour: 1am – Midnight
	18th	Tue	Above	Scorpio
	19th	Wed	Above	Scorpio
1Q	20th	Thur	Above	Scorpio / REST / Sagittarius
	21st	Fri	Above	Sagittarius
A V	22nd	Sat	Above	REST DAY - Southern Declination
	23rd	Sun	Above	Capricorn
	24th	Mon	Above	Capricorn / REST
	25th	Tue	Above	REST / Aquarius
	26th	Wed	Above	Aquarius / REST
	27th	Thur	Above	REST / Pisces
F	28th	Fri	Below	REST DAY - Full Moon - Lunar Partial Eclipse
	29th	Sat	Below	REST DAY - Day after Eclipse Day - Moon Void of Course all day
XhN	30th	Sun	Below	REST / Aries
	31st	Mon	Below	Aries / REST

SEPTEMBER	2026		Hour: 1am – Midnight
	1st	Tue	Below	REST / Taurus
	2nd	Wed	Below	Taurus / REST
	3rd	Thur	Below	REST / Gemini
3Q	4th	Fri	Below	Gemini
^	5th	Sat	Below	REST DAY - Northern Declination
P12	6th	Sun	Below	Cancer
	7th	Mon	Below	Cancer / REST / Leo
	8th	Tue	Below	Leo
	9th	Wed	Below	Leo / REST / Virgo
	10th	Thur	Below	Virgo
N	11th	Fri	Above	REST DAY - New Moon
XhS	12th	Sat	Above	REST / Libra
	13th	Sun	Above	Libra / REST
	14th	Mon	Above	REST / Scorpio
	15th	Tue	Above	Scorpio
	16th	Wed	Above	Scorpio / REST / Sagittarius
	17th	Thur	Above	Sagittarius
1Q V	18th	Fri	Above	REST DAY - Southern Declination
A	19th	Sat	Above	REST / Capricorn
	20th	Sun	Above	Capricorn
	21st	Mon	Above	Capricorn / REST / Aquarius
	22nd	Tue	Above	Aquarius
	23rd	Wed	Above	Aquarius / REST
	24th	Thur	Above	REST / Pisces
	25th	Fri	Above	Pisces
F XhN	26th	Sat	Below	REST DAY - Full Moon
	27th	Sun	Below	Aries
	28th	Mon	Below	Aries / REST / Taurus
	29th	Tue	Below	Taurus
	30th	Wed	Below	Tau / REST / Gemini

OCTOBER		2026	Hour:	1am – Midnight
P13	1st	Thur	Below	Gemini
^	2nd	Fri	Below	REST DAY - Northern Declination
3Q	3rd	Sat	Below	Cancer / REST
	4th	Sun	Below	REST DAY - Moon Void of Course all day
	5th	Mon	Below	Leo
	6th	Tue	Below	Leo / REST
	7th	Wed	Below	REST / Virgo / REST
	8th	Thur	Below	REST DAY - Moon Void of Course all day
	9th	Fri	Below	REST / Libra
N XhS	10th	Sat	Above	REST DAY - New Moon
	11th	Sun	Above	REST / Scorpio
	12th	Mon	Above	Scorpio
	13th	Tue	Above	Scorpio / REST
	14th	Wed	Above	REST / Sagittarius
	15th	Thur	Above	Sagittarius / REST
A V	16th	Fri	Above	REST DAY - Southern Declination
	17th	Sat	Above	Capricorn
1Q	18th	Sun	Above	Capricorn / REST
	19th	Mon	Above	REST / Aquarius
	20th	Tue	Above	Aquarius
	21st	Wed	Above	Aquarius / REST / Pisces
	22nd	Thur	Above	Pisces
	23rd	Fri	Above	Pisces / REST / Aries
XhN	24th	Sat	Above	Aries
	25th	Sun	Above	Aries / REST / Tau
F	26th	Mon	Below	REST DAY - Full Moon
	27th	Tue	Below	Taurus / REST
P9	28th	Wed	Below	REST / Gemini
^	29th	Thur	Below	REST DAY - Northern Declination
	30th	Fri	Below	REST / Cancer
	31st	Sat	Below	Cancer / REST

NOVEMBER		2026	Hour:	1am – Midnight
3Q	1st	Sun	Below	REST / Leo
	2nd	Mon	Below	Leo
	3rd	Tue	Below	REST / Virgo
	4th	Wed	Below	Virgo / REST
	5th	Thur	Below	REST / Libra
XhS	6th	Fri	Below	Libra
	7th	Sat	Below	Libra / REST / Scorpio
	8th	Sun	Below	Scorpio
N	9th	Mon	Above	REST DAY - New Moon
	10th	Tue	Above	REST / Sagittarius
	11th	Wed	Above	Sagittarius
V	12th	Thur	Above	REST DAY - Southern Declination
A	13th	Fri	Above	Capricorn
	14th	Sat	Above	Capricorn / REST
	15th	Sun	Above	REST / Aquarius
	16th	Mon	Above	Aquarius
1Q	17th	Tue	Above	Aquarius / REST / Pisces

NOVEMBER	2026		Hour:	1am – Midnight
	18th	Wed	Above	Pisces
	19th	Thur		Pisces
XhN	20th	Fri		Pisces / REST / Aries
	21st	Sat		Aries
	22nd	Sun		Aries / REST / Taurus
	23rd	Mon		Taurus
F	24th	Tue	Below	REST DAY - Full Moon
P5	25th	Wed		Gemini
^	26th	Thur		REST DAY - Northern Declination
	27th	Fri		Cancer
	28th	Sat		Cancer / REST / Leo
	29th	Sun		Leo
	30th	Mon		Leo / REST / Virgo

DECEMBER	2026		Hour:	1am – Midnight
3Q	1st	Tue	Below	Virgo
	2nd	Wed		Virgo / REST / Libra
XhS	3rd	Thur		Libra
	4th	Fri		Libra / REST
	5th	Sat		REST / Scorpio
	6th	Sun		Scorpio
	7th	Mon		Scorpio / REST / Sagittarius
	8th	Tue		Sagittarius
N V	9th	Wed	Above	REST DAY - New Moon / Southern Declination
	10th	Thur		REST / Capricorn / REST
A	11th	Fri		REST DAY - Moon Void of Course all day
	12th	Sat		REST / Aquarius
	13th	Sun		Aquarius
	14th	Mon		Aquarius / REST
	15th	Tue		REST / Pisces
	16th	Wed		Pisces
1Q XhN	17th	Thur		Pisces / REST / Aries
	18th	Fri		Aries
	19th	Sat		Aries / REST / Taurus
	20th	Sun		Taurus
	21st	Mon		Taurus / REST / Gemini
	22nd	Tue		Gemini
^	23rd	Wed		REST DAY - Northern Declination
F P1	24th	Thur	Below	REST DAY - Full Moon
	25th	Fri		REST / Leo
	26th	Sat		Leo
	27th	Sun		Leo / REST / Virgo
	28th	Mon		Virgo
	29th	Tue		Virgo / REST
3Q XhS	30th	Wed		REST / Libra
	31st	Thur		Libra

FISHING CALENDAR 2026

AV	VG	EX
1 fish	2 fish	3 fish

JANUARY	1st	2nd	3rd	4th	5th	6th	7th	8th	9th	10th	11th	12th	13th	14th	15th	16th	17th	18th	19th	20th	21st	22nd	23rd	24th	25th	26th	27th	28th	29th	30th	31st
MOON	P6		Full Moon							Third Q			Ap					New Moon								First Q			P10		
Best Bite Time am/pm	10.00	11.00	12.00	1.00	2.00	3.00	3.30	4.00	5.00	5.30	6.00	7.00	7.30	8.00	9.00	10.00	10.30	11.00	12.00	1.00	2.00	3.00	3.30	4.00	5.00	6.00	7.00	8.00	8.30	9.30	10.00
Next Best Bite Time am/pm	4.00	5.00	6.00	7.00	8.00	9.00	9.30	10.00	11.00	11.30	12.00	1.00	1.30	2.00	3.00	4.00	4.30	5.00	6.00	7.00	8.00	9.00	9.30	10.00	11.00	12.00	1.00	2.00	2.30	3.30	4.00
RATING	AV	EX	EX	EX	AV	AV	AV	AV	VG	VG	VG	AV	AV	AV	AV	AV	EX	EX	EX	AV	AV	AV	AV	AV	VG	VG	VG	AV	AV	AV	EX

FEBRUARY	1st	2nd	3rd	4th	5th	6th	7th	8th	9th	10th	11th	12th	13th	14th	15th	16th	17th	18th	19th	20th	21st	22nd	23rd	24th	25th	26th	27th	28th
MOON	Full Moon								Third Q	Ap							New Moon							First Q P14				
Best Bite Time am/pm	11.00	12.00	1.00	2.00	3.00	3.30	4.00	5.00	5.30	6.00	7.00	8.00	9.00	9.30	10.00	10.30	11.00	12.00	1.00	2.00	2.30	3.00	4.00	5.00	6.00	6.30	7.00	8.00
Next Best Bite Time am/pm	5.00	6.00	7.00	8.00	9.00	9.30	10.00	11.00	11.30	12.00	1.00	2.00	3.00	3.30	4.00	4.30	5.00	6.00	7.00	8.00	8.30	9.00	10.00	11.00	12.00	12.30	1.00	2.00
RATING	EX	EX	AV	AV	AV	AV	AV	VG	VG	VG	AV	AV	AV	AV	AV	EX	EX	EX	AV	AV	AV	AV	VG	VG	VG	AV	AV	AV

MARCH	1st	2nd	3rd	4th	5th	6th	7th	8th	9th	10th	11th	12th	13th	14th	15th	16th	17th	18th	19th	20th	21st	22nd	23rd	24th	25th	26th	27th	28th	29th	30th	31st
MOON			Full Moon							Ap	Third Q								New Moon			P11			First Q						
Best Bite Time am/pm	9.00	10.00	11.00	12.00	1.00	2.00	2.30	3.00	4.00	5.00	6.00	6.30	7.00	8.00	9.00	10.00	11.00	12.00	12.30	1.00	1.30	2.00	3.00	4.00	5.00	6.00	7.00	8.00	9.00	9.30	10.00
Next Best Bite Time am/pm	3.00	4.00	5.00	6.00	7.00	8.00	8.30	9.00	10.00	11.00	12.00	12.30	1.00	2.00	3.00	4.00	5.00	6.00	6.30	7.00	7.30	8.00	9.00	10.00	11.00	12.00	1.00	2.00	3.00	3.30	4.00
RATING	AV	EX	EX	EX	AV	AV	AV	AV	AV	VG	VG	VG	AV	AV	AV	AV	AV	EX	EX	EX	AV	AV	AV	VG	VG	VG	AV	AV	AV	AV	AV

APRIL	1st	2nd	3rd	4th	5th	6th	7th	8th	9th	10th	11th	12th	13th	14th	15th	16th	17th	18th	19th	20th	21st	22nd	23rd	24th	25th	26th	27th	28th	29th	30th
MOON		Full Moon					Ap			Third Q							New Moon		P7					First Q						
Best Bite Time am/pm	11.30	12.00	12.30	1.00	1.30	2.00	3.00	4.00	5.00	6.00	7.00	8.00	9.00	9.30	10.00	11.30	12.00	1.00	2.00	2.30	3.00	4.00	5.00	6.00	6.30	7.00	8.00	9.00	10.00	11.00
Next Best Bite Time am/pm	5.30	6.00	6.30	7.00	7.30	8.00	9.00	10.00	11.00	12.00	1.00	2.00	3.00	3.30	4.00	5.30	6.00	7.00	8.00	8.30	9.00	10.00	11.00	12.00	12.30	1.00	2.00	3.00	4.00	5.00
RATING	EX	EX	EX	AV	AV	AV	AV	AV	VG	VG	VG	AV	AV	AV	AV	EX	EX	EX	AV	AV	AV	AV	VG	VG	VG	AV	AV	AV	AV	EX

MAY	1st	2nd	3rd	4th	5th	6th	7th	8th	9th	10th	11th	12th	13th	14th	15th	16th	17th	18th	19th	20th	21st	22nd	23rd	24th	25th	26th	27th	28th	29th	30th	31st
MOON	Full Moon			Ap					Third Q							New Moon	P3						First Q								Full Moon
Best Bite Time am/pm	11.30	12.00	12.30	1.00	2.00	3.00	4.00	5.00	5.30	6.00	7.00	8.00	9.00	10.00	11.00	12.00	1.00	2.00	3.00	3.30	4.30	5.00	6.00	6.30	7.00	8.00	9.00	10.00	11.00	11.30	12.00
Next Best Bite Time am/pm	5.30	6.00	6.30	7.00	8.00	9.00	10.00	11.00	11.30	12.00	1.00	2.00	3.00	4.00	5.00	6.00	7.00	8.00	9.00	9.30	10.30	11.00	12.00	12.30	1.00	2.00	3.00	4.00	5.00	5.30	6.00
RATING	EX	EX	AV	AV	AV	AV	AV	VG	VG	VG	AV	AV	AV	AV	EX	EX	EX	AV	AV	AV	AV	VG	VG	VG	AV	AV	AV	AV	AV	EX	EX

JUNE	1st	2nd	3rd	4th	5th	6th	7th	8th	9th	10th	11th	12th	13th	14th	15th	16th	17th	18th	19th	20th	21st	22nd	23rd	24th	25th	26th	27th	28th	29th	30th
MOON	Ap							Third Q						P2	New Moon						First Q							Ap	Full Moon	
Best Bite Time am/pm	12.30	1.00	1.30	2.00	3.00	4.00	5.00	6.00	7.00	7.30	8.00	9.00	10.00	11.00	12.00	1.00	2.00	3.00	4.00	5.00	5.30	6.00	6.30	7.00	8.00	9.00	10.00	11.00	11.30	12.00
Next Best Bite Time am/pm	6.30	7.00	7.30	8.00	9.00	10.00	11.00	12.00	1.00	1.30	2.00	3.00	4.00	5.00	6.00	7.00	8.00	9.00	10.00	11.00	11.30	12.00	12.30	1.00	2.00	3.00	4.00	5.00	5.30	6.00
RATING	EX	AV	AV	AV	AV	AV	VG	VG	VG	AV	AV	AV	AV	EX	EX	EX	AV	AV	AV	VG	VG	VG	AV	AV	AV	AV	AV	EX	EX	EX

AV	VG	EX
1 fish	2 fish	3 fish

JULY	1st	2nd	3rd	4th	5th	6th	7th	8th	9th	10th	11th	12th	13th	14th	15th	16th	17th	18th	19th	20th	21st	22nd	23rd	24th	25th	26th	27th	28th	29th	30th	31st
MOON							Third Q						P4	New Moon							First Q				Ap				Full Moon		
Best Bite Time am/pm	12.30	1.00	1.30	2.00	3.00	4.00	5.00	6.00	7.00	8.00	9.00	10.00	11.00	12.00	1.00	2.00	3.00	4.00	5.00	5.30	6.00	6.30	7.00	8.00	9.00	10.00	11.00	11.30	12.00	1.00	2.00
Next Best Bite Time am/pm	6.30	7.00	7.30	8.00	9.00	10.00	11.00	12.00	1.00	2.00	3.00	4.00	5.00	6.00	7.00	8.00	9.00	10.00	11.00	11.30	12.00	12.30	1.00	2.00	3.00	4.00	5.00	5.30	6.00	7.00	8.00
RATING	AV	AV	AV	AV	AV	VG	VG	VG	AV	AV	AV	AV	EX	EX	EX	AV	AV	AV	AV	VG	VG	VG	AV	AV	AV	AV	AV	EX	EX	EX	AV

AUGUST	1st	2nd	3rd	4th	5th	6th	7th	8th	9th	10th	11th	12th	13th	14th	15th	16th	17th	18th	19th	20th	21st	22nd	23rd	24th	25th	26th	27th	28th	29th	30th	31st
MOON						Third Q				P8		New Moon								First Q		Ap						Full Moon			
Best Bite Time am/pm	2.30	3.00	4.00	5.00	5.30	6.00	7.00	8.00	9.00	9.30	10.00	11.00	12.00	1.00	2.00	3.00	3.30	4.30	5.00	6.00	7.00	8.00	9.00	10.00	10.30	11.00	11.30	12.00	1.00	2.00	3.00
Next Best Bite Time am/pm	8.30	9.00	10.00	11.00	11.30	12.00	1.00	2.00	3.00	3.30	4.00	5.00	6.00	7.00	8.00	9.00	9.30	10.30	11.00	12.00	1.00	2.00	3.00	4.00	4.30	5.00	5.30	6.00	7.00	8.00	9.00
RATING	AV	AV	AV	AV	VG	VG	VG	AV	AV	AV	EX	EX	EX	AV	AV	AV	AV	AV	VG	VG	VG	AV	AV	AV	AV	AV	EX	EX	EX	AV	AV

SEPTEMBER	1st	2nd	3rd	4th	5th	6th	7th	8th	9th	10th	11th	12th	13th	14th	15th	16th	17th	18th	19th	20th	21st	22nd	23rd	24th	25th	26th	27th	28th	29th	30th
MOON				Third Q		P12					New Moon							First Q	Ap							Full Moon				
Best Bite Time am/pm	4.00	5.00	5.30	6.00	7.00	7.30	8.00	9.00	10.00	11.00	12.00	1.00	2.00	3.00	3.30	4.30	5.00	5.30	6.00	6.30	7.00	8.00	9.00	10.00	11.00	11.30	12.00	1.00	2.00	3.00
Next Best Bite Time am/pm	10.00	11.00	11.30	12.00	1.00	1.30	2.00	3.00	4.00	5.00	6.00	7.00	8.00	9.00	9.30	10.30	11.00	11.30	12.00	12.30	1.00	2.00	3.00	4.00	5.00	5.30	6.00	7.00	8.00	9.00
RATING	AV	AV	VG	VG	VG	AV	AV	AV	AV	EX	EX	EX	AV	AV	AV	AV	VG	VG	VG	AV	AV	AV	AV	AV	EX	EX	EX	AV	AV	AV

OCTOBER	1st	2nd	3rd	4th	5th	6th	7th	8th	9th	10th	11th	12th	13th	14th	15th	16th	17th	18th	19th	20th	21st	22nd	23rd	24th	25th	26th	27th	28th	29th	30th	31st
MOON	P13		Third Q							New Moon						Ap		First Q								Full Moon		P9			
Best Bite Time am/pm	4.00	5.00	5.30	6.00	7.00	7.30	8.00	9.00	10.00	11.00	12.00	1.00	2.00	3.00	3.30	4.30	5.00	5.30	6.00	7.00	7.30	8.00	9.00	10.00	11.00	12.00	1.00	2.00	3.00	4.00	4.30
Next Best Bite Time am/pm	10.00	11.00	11.30	12.00	1.00	1.30	2.00	3.00	4.00	5.00	6.00	7.00	8.00	9.00	9.30	10.30	11.00	11.30	12.00	1.00	1.30	2.00	3.00	4.00	5.00	6.00	7.00	8.00	9.00	10.00	10.30
RATING	AV	VG	VG	VG	AV	AV	AV	AV	EX	EX	EX	AV	AV	AV	AV	AV	VG	VG	VG	AV	AV	AV	AV	AV	EX	EX	EX	AV	AV	AV	VG

NOVEMBER	1st	2nd	3rd	4th	5th	6th	7th	8th	9th	10th	11th	12th	13th	14th	15th	16th	17th	18th	19th	20th	21st	22nd	23rd	24th	25th	26th	27th	28th	29th	30th
MOON	Third Q								New Moon				Ap				First Q							Full Moon	P5					
Best Bite Time am/pm	5.00	6.00	7.00	7.30	8.00	9.00	10.00	11.00	11.30	12.00	1.00	2.00	2.30	3.00	4.00	5.00	5.30	6.00	6.30	7.00	8.00	9.00	10.00	11.00	12.00	1.00	2.00	3.00	4.00	5.00
Next Best Bite Time am/pm	11.00	12.00	1.00	1.30	2.00	3.00	4.00	5.00	5.30	6.00	7.00	8.00	8.30	9.00	10.00	11.00	11.30	12.00	12.30	1.00	2.00	3.00	4.00	5.00	6.00	7.00	8.00	9.00	10.00	11.00
RATING	VG	VG	AV	AV	AV	AV	AV	EX	EX	EX	AV	AV	AV	AV	AV	VG	VG	VG	AV	AV	AV	AV	EX	EX	EX	AV	AV	AV	AV	VG

DECEMBER	1st	2nd	3rd	4th	5th	6th	7th	8th	9th	10th	11th	12th	13th	14th	15th	16th	17th	18th	19th	20th	21st	22nd	23rd	24th	25th	26th	27th	28th	29th	30th	31st
MOON	Third Q								New Moon		Ap						First Q							Full Moon P1						Third Q	
Best Bite Time am/pm	6.00	6.30	7.00	7.30	8.00	9.00	10.00	11.00	12.00	1.00	2.00	3.00	4.00	4.30	5.00	5.30	6.00	7.00	7.30	8.00	9.00	10.00	11.00	12.00	1.00	2.00	2.30	3.00	4.00	5.00	6.00
Next Best Bite Time am/pm	12.00	12.30	1.00	1.30	2.00	3.00	4.00	5.00	6.00	7.00	8.00	9.00	10.00	10.30	11.00	11.30	12.00	1.00	1.30	2.00	3.00	4.00	5.00	6.00	7.00	8.00	8.30	9.00	10.00	11.00	12.00
RATING	VG	VG	AV	AV	AV	AV	AV	EX	EX	EX	AV	AV	AV	AV	AV	VG	VG	VG	AV	AV	AV	AV	EX	EX	EX	AV	AV	AV	VG	VG	VG

www.ingramcontent.com/pod-product-compliance
Lightning Source LLC
LaVergne TN
LVHW070850210926
848482LV00016B/905